Clinical Electrophysiology

Electrotherapy and Electrophysiologic Testing

Second Edition

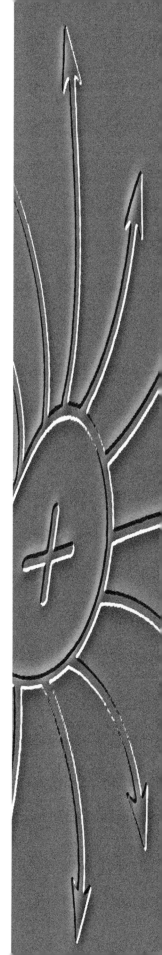

Clinical
Electrophysiology

Electrotherapy and
Electrophysiologic Testing
Second Edition

Andrew J. Robinson

Associate Professor
Department of Physical Therapy
School of Health Sciences and
Human Performance
Ithaca College
Ithaca, New York

Lynn Snyder-Mackler

Assistant Professor
Department of Physical Therapy
University of Delaware
Newark, Delaware

WILLIAMS & WILKINS

BALTIMORE·HONG KONG·LONDON·MUNICH
PHILADELPHIA·SYDNEY·TOKYO

Editor: John Butler
Managing Editor: Linda Napora
Production Coordinator: Anne Stewart Seitz

Copyright©1995
Williams & Wilkins
428 East Preston Street
Baltimore, Maryland 21202 USA

Accurate indications, adverse reactions, dosage schedules for drugs, and electrotherapeutic and electrophysiologic testing procedures are provided in this book, but it is possible that they may change. The reader is urged to review the package information data of the manufacturers of the medications and electrotherapeutic devices mentioned.

Printed in the United States of America

First edition 1989

Library of Congress Cataloging-in-Publication Data

Robinson, Andrew J.
Clinical electrophysiology : electrotherapy and electrophysiologic
testing / Andrew J. Robinson, Lynn Snyder-Mackler. —2nd ed.
    p.   cm.
Includes bibliographical references and index.
ISBN 0-683-07817-8
1. Electrotherapeutics. 2. Electrodiagnosis. 3. Electrophysiology. I. Robinson, An-
    drew J. II. Title [DNLM: 1. Electric Stimulation Therapy. 2. Electric Stimulation.
    3. Electrodiagnosis. WB 495 S675c 1994]
RM872.S68 1994
615.8'45–dc20
DNLM/DLC
for Library of Congress   94-32314
                                                                           CIP
                                                                         97 98
                                                            3 4 5 6 7 8 9 10

ISBN 0-683-07817-8

90000

9 780683 078176

to sandy, tracy, and kelly
a. j. r.

to scott, alexander, and noah and to la
l. s.-m.

Foreword

When I attended a physical therapy program in the early 1970s, electrotherapy was a seven-week course stressing evaluation and treatment. Peripheral nerve integrity was evaluated using such tests as reaction of degeneration and strength duration, and the instructor demonstrated electroneuromyography on a fellow student. I recall learning electrical stimulation treatment protocols to prevent atrophy in denervated muscle and to relax muscle spasm. The practical examination was passed by demonstrating the ability to operate two machines: a "low volt unit," which provided direct or alternating current with a pulse duration of at least 1 msec; and a chronaximeter. A period of 20 years of progress, however, has made a significant difference in the content of a course that is now often entitled Electrophysiological Measurement and Treatment. A semester does not seem adequate to cover the myriad of treatment and evaluation techniques, the categories of instrumentation, and the growing body of knowledge related to the efficacy of electrical stimulation as a clinical modality.

Traditional to physical therapy education has been the strong belief that a variety of techniques are effective despite the paucity of research documenting treatment outcome. Graduates of physical therapy programs have been encouraged to write case reports and to conduct research to confirm or reject the claims of clinical success. As a former member of the Research Advisory Panel of the Foundation for Physical Therapy, as past president of the APTA's Section on Research, and as an Editorial Board member for *Physical Therapy*, I have had the opportunity to watch the recent exponential growth of physical therapy researchers, who are responding to the need for a body of experimentally confirmed knowledge. As you read this book, take note of how many persons have contributed to theory related to electrical stimulation therapy in the recent past. Although there remains significant work to be done, we should point with pride to the studies that have evaluated the clinical success of treatment as well as to the basic science research that has begun to explore some of the mechanisms responsible for treatment success. With the wealth of new information, an educator can now suggest stimulus parameters for optimal muscle "strengthening" that are derived from a conceptual framework in addition to empirical information. The student of electrophysiologic methods can now go to the literature and find studies that support or refute claims of clinical success.

Andrew Robinson, Lynn Snyder-Mackler, and the contributing authors are to be commended for not preparing a cookbook on treatment techniques. This text instead encourages the reader to ask why a treatment might work and to learn basic principles of evaluation and treatment that can be utilized regardless of the machine available. The critical reviews of the contemporary literature and the excellent summaries should enhance the critical thinking skills of the consumer, allow the reader to recognize the deficits in our present-day knowledge base, and inspire additional research.

Rebecca L. Craik, PhD, PT
Department of Physical Therapy
Beaver College
Glenside, Pennsylvania

Preface

When we wrote the first edition of this textbook in the late 1980s, electrotherapy had just gone through a decade-and-a-half of exponential explosion in its use. Portable electrical stimulators, TENS units, were everywhere and "Russian stimulation" was all the rage. The past five years have brought relatively few innovations in the field of electrotherapy given the plethora of devices that were brought to market in the preceding 15 years. The late 1980s and early 1990s, however, finally began to bring us a body of literature on the effectiveness of electrotherapeutic interventions. On the instrumentation side, innovations were largely confined to those that made devices easier to use.

We have made some changes in this edition in response to comments from students, their teachers, and clinicians. We have added a chapter entitled Electromyographic Biofeedback to Improve Voluntary Motor Control, written by a leading expert in the field, Stuart Binder-Macleod, PhD, PT. Although it is not electrotherapy, technically speaking, biofeedback is often taught in conjunction with electrotherapy in clinical electrophysiology courses. This topic is also a logical extension of our inclusion of clinical electrophysiologic assessment. Another change from the first edition is an expansion of the chapter on iontophoresis, by Charles Ciccone, PhD, PT. Dr. Ciccone has written extensively in the area of pharmacology and physical therapy and brings that perspective to bear on the subject of iontophoresis.

In the area of neuromuscular electrical stimulation (NMES), there have been rich additions to the literature in the past five years. For this reason, we have divided the old NMES chapter in two. The first chapter (Chapter 4) deals primarily with augmentation of muscle strength. The second (Chapter 5) covers the topics of NMES for control of posture and movement.

The edition includes some new features. We have added study questions to each chapter. They are provided to reinforce the material presented. The book has undergone a major renovation with respect to the illustrations, which have all been substantively redrawn. An appendix that covers peripheral neuroanatomy has been added. We hope that these changes bring a clearer, more user-friendly orientation to the text, which will aid in its use in the classroom and clinic.

We have maintained the character of the original text in that it is organized by treatment outcome. We believe that this is the strongest component of this book—that it reflects the way clinicians use electrotherapy in practice. We have continued to give clinical examples, via the inclusion of case studies, in each of the chapters on applications. The studies have been updated to exhibit changes in today's research and clinical practice.

We would like to acknowledge the contributions of our colleagues at Ithaca College and the University of Delaware, as well as technical assistance from Patrick Williams and Jill Hartzler. Once again, we must thank our colleagues who provided us with copies of in-press manuscripts so that this text would be up-to-date at the time of publication. Editors John Butler and Linda Napora were remarkably patient with us as we labored over this edition, and they deserve kudos as well.

Acknowledgments

We would like to thank many individuals for their advice and assistance in the preparation of the second edition of this text. First, we want to acknowledge the efforts of our contributors. Their writing has both improved and expanded the content of this second edition making the text more valuable to the readers of the first edition and expanding the utility of the text to more practitioners interested in the use of electrotherapeutic and electrophysiologic assessment procedures. Our appreciation goes out to Dr. Rebecca Craik of Beaver College who, in spite of a busy professional schedule, agreed to review the text and write the foreword to the second edition. Our special thanks to J. Chris Castel of PTI in Topeka, Kansas, who kindly consented to review Chapter 2, on instrumentation. His suggestions significantly improved the content of that chapter and made the discussion on electrical stimulators much more user-friendly. We extend our gratitude to the many educators who use this text for their constructive suggestions on how the book could be improved in the second writing. We also appreciate the efforts of three members of Williams & Wilkins's group: John Butler for advocating the production of the second edition, Linda Napora for encouraging the authors and editors to prepare chapters, and Anne Stewart Seitz for coordinating the text's production. Last, we wish to recognize our students and colleagues at Ithaca College and the University of Delaware who supported our efforts and provided important feedback as we discussed much of the material contained within the text in our classroom and laboratory presentations.

Contributors

Stuart Binder-Macleod, PhD, PT
Department of Physical Therapy
University of Delaware
Newark, Delaware

Charles D. Ciccone, PhD, PT
Department of Physical Therapy
School of Health Sciences and Human Performance
Ithaca College
Ithaca, New York

Anthony Delitto, PhD, PT
Department of Physical Therapy
School of Health and Rehabilitation Science
University of Pittsburgh
Pittsburgh, Pennsylvania

LCDR Robert Kellogg, MS, PT, ECS
United States Navy *and*
Department of Exercise and Sports Science
University of Florida
Gainesville, Florida

David J. Mayer, PhD
Department of Anesthesiology
Medical College of Virginia
Virginia Commonwealth University
Richmond, Virginia

Donald D. Price, PhD
Department of Anesthesiology
Medical College of Virginia
Virginia Commonwealth University
Richmond, Virginia

Andrew J. Robinson, PhD, PT
Department of Physical Therapy
School of Health Sciences and Human Performance
Ithaca College
Ithaca, New York

Lynn Snyder-Mackler, ScD, PT
Department of Physical Therapy
University of Delaware
Newark, Delaware

Contents

Chapter 1

Basic Concepts in Electricity and Contemporary Terminology in Electrotherapy

Andrew J. Robinson

$\bigwedge\!\bigwedge\!\bigwedge$ **E**lectricity is one of the basic forms of energy in the science of physics and can produce significant effects on biological tissues. One purpose of this chapter is to briefly review necessary concepts of electricity and electromagnetism as they relate to therapeutic electrical stimulation. Equations describing electrical phenomena are kept to a minimum, and frequent analogies are used to allow the reader to visualize what may be happening in human tissues as electrical stimulation is applied. The focus of the chapter is the conceptualization of relevant electrical phenomena rather than memorization. Another purpose of this chapter is to discuss the terminology used to qualitatively and quantitatively describe the electrical currents employed in clinical applications. These terms will be used throughout the text to ensure the clear, unambiguous, and consistent communication of the technical details of stimulation procedures.

Fundamental Concepts of Electricity

Electric charge

A discussion of electricity and electromagnetism must begin with the concept of electric charge. **Electric charge** (or simply "charge") is a fundamental physical property that is difficult for many to understand. Some physics instructors try to clarify the concept by saying that charge is a basic property like "mass" or "time." Although true, this approach does not always provide an individual with more insight into the nature of charge.

The problem in attempting to explain charge is that charge is operationally defined. That is, one cannot ever "see" a charge, but one can see through experimentation how charge manifests itself. For example, in high school or college physics classes your instructor may have demonstrated how rubbing cloth on amber (a yellowish fossil resin) would allow the amber to attract lightweight substances such as bits of paper. Ancient scientists described this property of amber as static electricity, which is nothing more than a manifestation of the electromagnetic force of attraction exerted by the "charged" particles within the amber. It is now known that the amber becomes charged by the exchange of electrons in it with those in the atoms of the cloth as they are rubbed together. As a result, both the cloth and the amber show the ability to attract or repel a variety of other charged objects.

Charge is the property of matter that is the basis of electromagnetic force. Sophisticated experiments in recent years have characterized the properties of electric charge (1). These studies have shown that two types of electric charge, positive and negative, exist in nature. At the simplest level, charge is carried by the electrons (negative charge) and the protons (positive charge) of atoms. Like charges repel each other, and opposite charges attract each other. Charge can be transferred from one object to another (charges may be separated), but charge can be neither created nor destroyed.

The concept of electric charge is not limited to the subatomic level of matter. An electrically neutral atom is one that contains an equal number of protons and electrons. If an atom of an element loses electrons without changing the number of protons in the nucleus, it becomes positively charged. If an atom gains electrons, it becomes negatively charged. Atoms of elements with either an excess or a deficiency of electrons are called **ions**. Atoms that are positively charged are referred to as **cations,** and negatively charged atoms are called **anions**.

Objects or substances may also become electrically charged. Consider the charges on the terminals of a simple dry cell battery. As a consequence of the chemical reactions taking place within the battery, one metal terminal (the **cathode**) gains electrons and becomes *negatively* charged while the other metal terminal (the **anode**) loses electrons and becomes *positively* charged. The anode and cathode of a battery are sometimes referred to as the poles of the battery. The term **polarity** is used to indicate the relative charge (positive or negative) of the terminals or essential leads of an electrical circuit at any one moment in time.

The force exerted between two electrical charges can be determined experimentally and is expressed in coulombs (C). The coulomb force *(F)* between two stationary charges, (q_1) and (q_2), is proportional to the magnitude and sign of the charges and is inversely proportional to the square of the distance *(r)* between them, as expressed by Coulomb's law:

$$F \propto (q_1 \cdot q_2)/r^2.$$

The law simply states that the larger the respective charges or the closer the two charges, the larger will be the attractive (or repulsive) force between them. The coulomb forces of electrons and protons are equal in magnitude but opposite in sign. The coulomb force for a single electron is 1.6×10^{-19}C. Consequently, to produce a charge of 1 C requires the presence of 6.24×10^{18} electrons.

Electric Field

The electric force of charged particles is carried to other charged particles by the **electric field** (E) that each charge creates around itself. Charges transmit force through an electric field in a manner analogous to the way Earth's force of gravity is transmitted via gravitational fields. The characteristics of electric fields created between two oppositely charged substances and two substances of the same charge are illustrated in Figure 1.1.

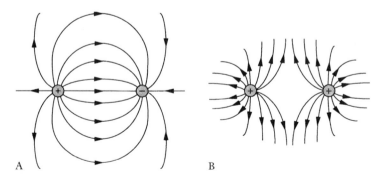

Figure 1.1 Electric field lines around oppositely charged particles **(A),** and two like charges **(B).** Configuration of the field lines reflects the attraction of oppositely charged particles and repulsion between similarly charged particles.

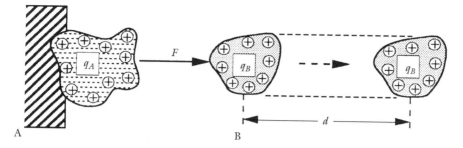

Figure 1.2 The effect of approximating two similarly charged objects (having charges q_A and q_B): **A.** Fixed position. **B.** The coulomb force *(F)* of repulsion between the two objects will tend to separate the objects by some distance *(d)*.

Voltage

To understand the concept of voltage, consider the situation diagrammed in Figure 1.2. Large charged substance *A* is brought close to small similarly charged substance *B*. As the two charged masses are approximated, the repulsive coulomb force of *A* is transmitted through *A*'s electric field and is "felt" by *B*—raising *B*'s electrical potential energy *(PE)*. If free to move, *B* will shift to a new position some distance *(d)* away from its original position. When *B* is moved, substance *A* has done work that amounts to the product of the average coulomb force applied to *B* and the distance moved by *B*; that is,

$$W = F \cdot d.$$

As substance *B* moves, the potential energy initially gained by interaction with *A* is lost in doing the work. Thus,

$$W = \Delta PE.$$

Since the work done is directly proportional to the charge on *B* and since the change in potential energy is directly proportional to the charge on *B*, the voltage is defined as

$$V = \Delta PE / q_B.$$

The **voltage** is the change in electrical potential energy between two points in an electric field per unit of charge and is synonymous with the term **electrical potential difference**.

From a more practical standpoint, voltage represents the driving force that makes charged particles move and is often referred to as the **electromotive force**, or **EMF**. Voltages are produced when oppositely charged substances are separated, when like-charged substances are approximated, or when charged particles within a system are not evenly distributed.

The standard unit for voltage is the volt (V). One volt is equal to a 1-J (joule) change in energy per coulomb of charge:

$$1\,V = 1\,J / 1\,C.$$

Voltages used in electrotherapeutic applications may be as small as the millivolt (mV, 10^{-3} V, thousandths of a volt) range or as high as several hundred volts (applied over an extremely short time).

Conductors and insulators

Charged particles such as electrons in metals or ions in solution will tend to move or change position by virtue of their interactions with other charged particles. In other words, charged particles will tend to move in matter when electrical potential differences exist. For charged particles to move when subjected to a voltage, they must be free to do so. Those substances in which charged particles readily move when placed in an electric field are called **conductors**. Metals such as copper are good conductors. The atoms of metals tend to give up electrons from their outer orbital shell quite readily when placed in an electric field. If a negatively charged substance is brought very near to one end of a long metal wire, electrons closest to the substance will be displaced along the wire away from the mass of similar charge. Biological tissues contain charged particles in solution, in the form of ions such as sodium (Na^+), potassium (K^+), or chloride (Cl^-). Human tissues are conductors because the ions there are free to move when exposed to electromotive forces. The ability of ions to move in human tissues varies from tissue to tissue. Muscle and nerve are good conductors, whereas skin and fat are poor conductors.

In contrast to substances that allow easy movement of charged particles in an electric field, **insulators** are substances that do not tend to allow free movement of ions or electrons. Rubber and many plastics are good insulators.

Electrical current

The study of electric charges in motion is of greater importance to the understanding of therapeutic electrical stimulation than is the examination of the properties of charges at rest. The movement of charged particles through a conductor in response to an applied electric field is called **current** *(I)*. The conduction of electrical charge through matter from one point to another is the transfer of energy that brings about physiological changes during the clinical application of electrical stimulation.

Producing electrical current requires *(a)* the presence of freely movable charged particles in some substance and *(b)* the application of a driving force to move the particles. In metal circuits, electrons are the movable charged particles, whereas in biological systems, ions in body fluids (electrolytic solutions) are the conductive media. The forces that induce current in these media are the applied voltages. The magnitude of current induced in a conductive medium is directly proportional to the magnitude of the applied voltage:

$$\text{Current} \propto \text{Voltage} \qquad (I \propto V).$$

Current is strictly defined as the amount of charge *(q)* moving past a plane in the conductor per unit time *(t)*, or

$$I = \Delta q / \Delta t.$$

The standard unit of measurement for current is the ampere (A), which is equal to the movement of 1 C of charge past a point in 1 sec. Currents used in electrotherapeutic applications are very small and are generally measured in milliamperes (mA, 10^{-3} amperes, thousandths of an ampere) or in microamperes (μA, 10^{-6} amperes, millionths of an ampere). Often analogies are drawn between electrical currents in conductors and the movements of liquid (fluid current) in hydraulic systems, in order to gain an intuitive sense of electric current.

Resistance and conductance

The magnitude of charge flow is determined not only by the size of the driving force (voltage) but also by the relative ease with which electrons or ions are allowed to move through the conductor. This characteristic of conductors may be described in two ways. The property of conductors called **resistance** *(R)* describes the relative opposition to movement of charged particles in a conductor. Conversely, the property called **conductance** *(G)* describes the relative ease with which charged particles move in a medium. For metals, resistance is dependent on the cross-sectional area *(A)*, length *(l)*, and resistivity *(ρ)* of the conductor by the formula

$$R = \rho \cdot (l/A).$$

The standard unit of resistance is the ohm (Ω). The magnitude of the current induced in a conductor is inversely proportional to the resistance of the conductor:

$$I \propto 1/R.$$

An alternative way to describe the ability of charged particles to move in conductors, conductance, is inversely related to resistance:

$$\text{Resistance} = 1/\text{conductance}.$$

The standard unit of conductance is the mho (in the International System [SI], it is the siemens).

The resistance of electrical conductors is analogous to the opposition to fluid movement that occurs in hydraulic systems. Just as the resistance to fluid movement increases as the diameter of pipe decreases (or the length of pipe is increased), the resistance to electrical current increases as the diameter of the conductor decreases (or the length of conductor increases).

Ohm's law

The relationship between those factors, voltage and resistance, which determines the magnitude of current *(I)*, is expressed in **Ohm's law**:

$$I = V/R \qquad \text{or} \qquad V = I \times R.$$

Ohm's law simply states that the current induced in a conductor increases as the applied driving force *(V)* is increased or as the opposition to charge movement *(r)* is decreased. Alternatively, Ohm's law may be expressed in terms of conductance rather than resistance:

$$I = V \times G \qquad \text{or} \qquad V = I/G.$$

Capacitance and impedance

In order to understand current in biological tissues, two other electrical concepts must also be introduced. **Capacitance** is the property of a system of conductors and insulators that allows the system to store charge. Currents produced in biological tissues are influenced not only by tissue resistance but also by the tissue capacitance.

An electrical circuit device, the capacitor, is made up of two thin metal plates separated by an insulator (or *dielectric*) (Fig. 1.3*A*). If a fixed voltage is applied across the capacitor, current does not pass through the device because of the presence of the insulating material. However, the potential difference between the two plates of the capacitor exerts a force on the molecules within the insulator, raising the potential energy within these molecules (Fig. 1.3*B*).

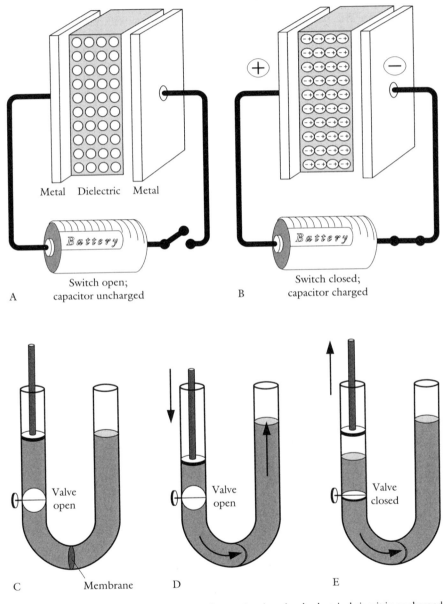

Figure 1.3 Diagrammatic representation of a capacitor in a simple electrical circuit in uncharged (**A**) and charged (**B**) states. A capacitor stores electric energy by the deformation of dielectric molecules. A hydraulic analog of an uncharged capacitor (**C**), charging capacitor (**D**), and charged capacitor with charging force removed (**E**). Energy is stored in the deformation of an impermeable elastic membrane.

If the applied voltage is removed, the stored energy (electrical potential difference across the capacitor) remains until the capacitor is discharged through some conductive pathway.

A capacitor stores electrical energy in a manner similar to that for an elastic impermeable membrane placed in a hydraulic system. Consider the situation illustrated in Figure 1.3*C*, where a thin rubber *membrane* is placed in the base of an inelastic tube. A piston is used to produce a driving force on the fluid— no fluid actually passes through the membrane (no current is produced); the driving force causes the membrane to distend (Fig. 1.3*D*). The membrane stores energy by virtue of its distended shape. If the *valve* in the tube is closed and then the piston pressure is released, the membrane will remain in the distended, energy-storing position until the valve is reopened (Fig. 1.3*E*). If the valve is reopened, the recoil of the membrane will produce a movement of the fluid (current) that will continue until the membrane returns to its original, rest position. Thus the elastic membrane in the hydraulic circuit stores energy that induces a fluid current just as a capacitor stores electrical energy that induces an electrical current.

Notice that the membrane in the tube blocks fluid flow (current) through the tube when a constant, unidirectional piston pressure is applied, just as a capacitor blocks electrical direct current when a constant voltage is applied. Although capacitive systems tend to block direct currents, they tend to allow alternating currents to pass. For a system at a particular capacitance, the higher the frequency of alternating current, the better the current will be passed through the system.

The capacitance of a capacitor or any similarly constructed system of conductors and insulators is expressed in farads (F); 1 F is the magnitude of capacitance as 1 C of charge is stored when 1 V of potential difference is applied.

The term **impedance** *(Z)* describes the opposition to alternating currents like the term *resistance* describes the opposition to direct currents. Impedance takes into account both the capacitive and resistive opposition to the movement of charged particles. When dealing with clinical electrical stimulation, it is more appropriate to express the opposition to current in terms of impedance because human tissues are better modeled as complex resistor and capacitor (R–C) networks. Since impedance is dependent on the capacitive nature of biological tissues, its magnitude is dependent on the frequency of applied stimulation. In general, the higher the frequency of stimulation, the lower will be the impedance of tissues. The standard unit of impedance is the ohm.

Language of Electrotherapeutic Currents

Traditional and commercial designations of currents

Electrical currents have been used for therapeutic purposes for hundreds of years. With the development of different forms of electrical generators during this century, the types of electrical currents employed in therapeutic applications have proliferated. The introduction to the health care market of many different types of stimulators producing different forms of electrical currents has been accompanied by substantial confusion in communication regarding the characteristics of the currents generated. Before 1990, no system had been developed to standardize descriptions of electrical currents used in electrotherapy.

Characterization of electrotherapeutic currents was often driven by historical developments or by the commercial sector. Figure 1.4 shows a number of the various types of currents traditionally employed in electrotherapy and their traditional designations. Figure 1.5 illustrates several commercially designated current (or voltage) waveforms.

Differentiation among these traditional and commercial types of currents was often based on only a single current characteristic such as the amplitude of voltage or the frequency of stimulation. Such unidimensional distinctions led to dichotomous designations—such as "low-volt vs. high-volt" or "low-frequency vs. medium-frequency" stimulators—that persist even today. In the mid-1980s, the Section on Clinical Electrophysiology (SCE) of the American Physical Therapy Association recognized that such arbitrary differentiation of electrotherapeutic currents along with the proliferation of commercial des-

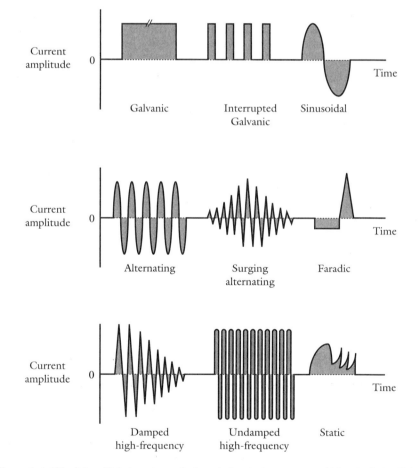

Figure 1.4 "Traditional" designations of selected electrical currents used historically in clinical practice. Each graph shows changes in *current amplitude* over *time*.

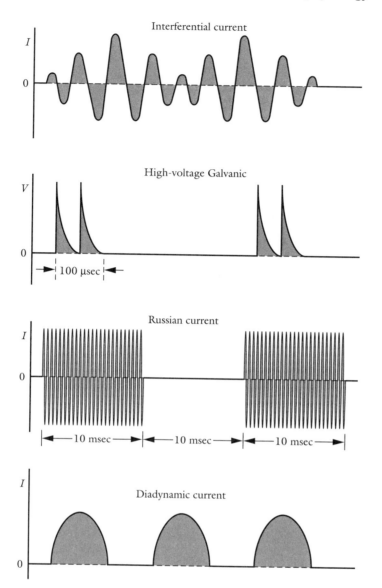

Figure 1.5 "Commercial" designations of selected electrical currents available from certain contemporary stimulators. Graphs show either changes in current amplitude over time or changes in voltage amplitude over time.

ignations of currents fostered confusion in communication regarding elec-trotherapy. In an attempt to alleviate the problem, the SCE developed a monograph addressing terminology in electrotherapy; it provides guidelines for qualitative and quantitative descriptions of electrotherapeutic currents (2). Although the remainder of this chapter presents much of the content of that SCE monograph, an appreciation for the traditional and commer-cial designations of electrotherapeutic currents is still needed since most of the literature published through the 1980s used traditional or commercial terminology.

Contemporary designations for electrotherapeutic currents

Types of Electrotherapeutic Currents

Electrical currents used in contemporary clinical electrotherapy may generally be divided into three types: direct current, alternating current, and pulsed (pulsatile) current. In the following segment of this chapter, we will differentiate among these types of current based on their qualitative and quantitative characteristics.

Direct Current

The continuous or uninterrupted unidirectional flow of charged particles is defined as **direct current** (DC). This form of current has traditionally been referred to as "galvanic" current; however, this is no longer the preferred term. Direct current in a simple electronic circuit is produced by a fixed-magnitude voltage applied to a conductor with a fixed resistance (Fig. 1.6*A*). The source of the fixed electromotive force (EMF) is the *battery*, where chemical reactions produce an excess of electrons on one pole *(cathode)* and a deficiency in electrons on the opposite pole *(anode)*. The opposition to *current* in the circuit is represented as a *resistor*. When the *switch* in the circuit is closed, electrons flow from the area of high concentration *(cathode)* to the area of low concentration *(anode)*. This flow, which is impeded by the resistance of the wire, will continue until the charge difference between terminals is eliminated—when the chemical reactions within the battery can no longer provide free electrons to the negative terminal. Although the movement of charged particles in this circuit is from negative to positive terminals, current *(I)* is, by convention, specified as moving from positive to negative terminals. The current that flows through this circuit is represented in Figure 1.6*C*, a graph of *current amplitude* over *time*.

The movement of electrons in this simple circuit is analogous to the movement of water molecules in a simple hydraulic circuit (Fig. 1.6*B*). The driving force in this fluid model is represented as the pressure difference created by the *pump* and is analogous to the voltage difference across the battery. The water molecules are analogous to the free electrons in the electrical circuit. The hydraulic *resistance* (opposition to the flow of water) is represented primarily by the narrowing of the tubing halfway through the circuit and is analogous to the resistance of the wire in our simple electrical circuit. The fluid will flow in the circuit as long as the pump maintains a pressure difference, just as electron flow will continue as long as the battery maintains an electrical potential difference. The volume of fluid that passes by a point in the fluid circuit per unit of time (current) will remain constant as long as the pressure gradient is kept constant and the geometry of the tubing is maintained. A drop in either the pressure gradient or the diameter of the tube would reduce fluid flow just as a drop in voltage or an increase in circuit resistance would reduce electron flow.

Direct current induced in an electrolytic aqueous solution containing both positively and negatively charged ions (cations and anions respectively) is associated with the movement of these two types of ions in opposite directions. Figure 1.7 illustrates the ionic movements in an electrolytic solution when

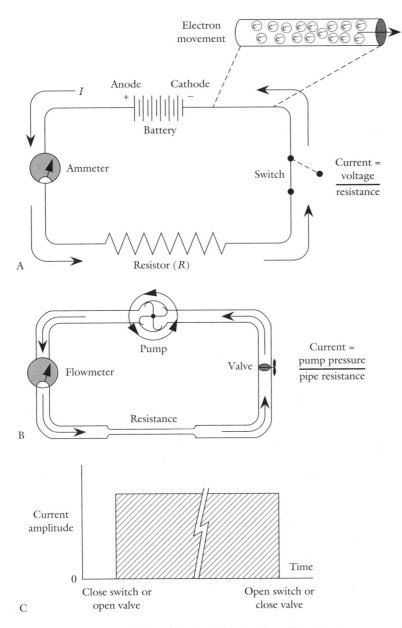

Figure 1.6 Diagram of a simple electrical circuit showing the unidirectional movement of electrons in response to a constant driving force (**A**). Hydraulic analog of a simple electrical circuit showing unidirectional movement of fluid in response to constant pressure produced by a pump (**B**). Graphical representation of direct current on a *current amplitude*-vs.-*time* plot (**C**).

the solution is exposed to a constant-voltage electric field. As can be seen in the figure, anions move toward the positively charged electrode *(anode)*, and cations migrate toward the negatively charged electrode *(cathode)*. The movement of each type of ion in the solution occurs at a fixed rate as long as the battery voltage is constant. The migration of ions or electrically charged

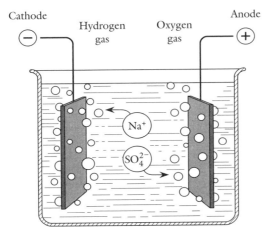

Figure 1.7 An example of ionic movements in a solution with negatively charged ions (anions) moving toward the *anode* and positively charged ions (cations) moving toward the *cathode* when exposed to a fixed electromagnetic field. In aqueous solutions (water as solvent), *hydrogen gas* is liberated from the *cathode* and *oxygen* is released from the *anode*.

molecules according to their charge when exposed to a fixed EMF is called *electrophoresis* and is the basis of *iontophoresis,* a therapeutic technique used to drive electrically charged medications through the skin (see Chapter 9). Figure 1.7 also illustrates the liberation of gases near the electrodes, which often accompanies DC effects on electrolytic solutions. In this case, a reduction reaction occurs at the cathode to produce *hydrogen gas* (H_2), and an oxidation reaction occurs at the anode to yield *oxygen gas* (O_2). The use of electric energy to produce such chemical reactions is called *electrolysis.*

Alternating Current

Alternating current (AC) is defined as the continuous or uninterrupted bidirectional flow of charged particles. To produce this type of current, the voltage applied across a simple circuit oscillates in magnitude and the polarity of the applied voltage is periodically reversed. Electrons in the circuit first move in one direction. When the electric field is reversed the electrons move back toward their original position. An alternating current may be produced by rotating a fixed voltage source in the circuit as illustrated in Figure 1.8*A.* The alternating current that flows through this circuit is represented in Figure 1.8*C,* a graph of *current amplitude* over *time.* Alternating currents are characterized by the frequency *(f)* of oscillations and the amplitude of the electron or ionic movement. AC frequency is expressed in hertz (Hz) or in cycles per second (cps). The reciprocal of frequency (1/*f*) defines a value, known as the *period,* that is the time between the beginning of one cycle of oscillation and the beginning of the next cycle.

A better understanding of alternating currents can be attained if we once again examine the force and flows in a fluid-filled system. Alternating electrical current is analogous to fluid in a closed system moving first in one direction and then back in the opposite direction. Consequently, for many types of alternating current there is no net movement of charged particles when the alternating electric field is withdrawn. For fluid to move back and forth in a hydraulic system, the pressure gradient must first be in one direction, then fall for an instant to zero, and finally reverse direction. In the case of the hydraulic pump in Figure 1.8*B,* the pump rotates back and forth much like the agitator in a washing machine. The oscillating pressure produced by the pump produces a back and forth movement of the fluid.

Alternating voltages applied to electrolytic solutions (as opposed to a metal conductor) produce cyclical movements in anions and cations in the solution. For some period of time, these ions "see" and "feel" an anode and cathode oriented in one fashion, and then the polarity of the electrodes in the solution

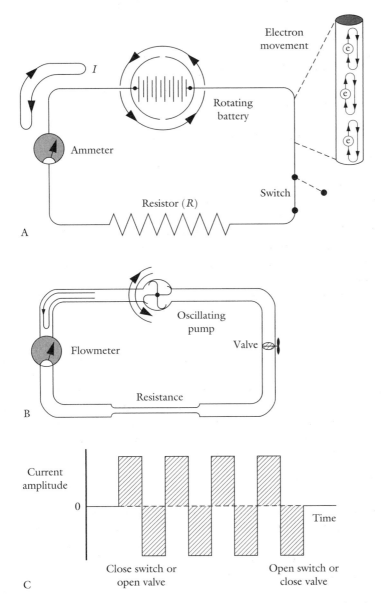

Figure 1.8 Simple electrical circuit in which a *battery* rotates at constant speed and regularly changes the direction of the driving force (voltage) acting on the electrons in the conductor **(A)**. Note the back-and-forth movement of electrons. Hydraulic analog circuit of the electrical circuit in **(A)** illustrating the back-and-forth movement of the pump, which produces the alternating movement of fluid within the system **(B)**. Graphical representation of the alternating current is produced in **(A)** on a *current amplitude*-vs.-*time* plot **(C)**.

switches. Consequently the ions in solution move back and forth in solution just as electrons move back and forth in metals when exposed to an alternating voltage.

Alternating currents are used in a number of electrotherapeutic applications. The most common contemporary use of AC clinically is in interferential electrical stimulation, where two circuits each producing sinusoidal AC are applied simultaneously to a patient for the management of problems such as pain (see Chapter 7).

Pulsed Current

Pulsed current (pulsatile, or interrupted, current) is defined as the uni- or bidirectional flow of charged particles that periodically ceases for a finite period of time. A description of this type of current may not be found in basic physics textbooks, but the term is important because it describes the most commonly used form of current in clinical applications of electrical stimulation. Physicists and engineers might refer to pulsed current as either *interrupted DC* or *interrupted AC*.

Pulsed current is characterized by the features of an elemental unit of this type of current, called a **pulse**. A single pulse is defined as an isolated electrical event separated by a finite time from the next event. That is, a single pulse represents a finite period of charged particle movement.

If a fixed voltage is applied to a simple resistive electrical circuit as shown in Figure 1.6A, a unidirectional current will be induced in the conductor. If the circuit is periodically interrupted by opening and closing a switch in the circuit, the electron movement produced will start and stop in synchrony with the closing and opening of the switch. The current produced is intermittent and in one direction only and is referred to as *monophasic pulsed current*.

In a similar manner, if an alternating voltage is applied to the simple electrical circuit as shown in Figure 1.8 and the circuit is interrupted on completion of each cycle of the alternating voltage, the electrons in the conductors will briefly move back and forth, stop, and then begin to oscillate again. The current produced is intermittent and the charged particle movement is bidirectional. Such a current is called a *biphasic pulsed current*. The changes in amplitude of the biphasic pulsed current for each pulse are determined by the changes in the amplitude of the applied voltage.

Descriptive Characteristics of Pulsed or Alternating Current Waveforms

The qualitative and quantitative features of current pulses (or a single cycle of AC) are most easily understood by examining graphically the current amplitude changes that occur over time. The shape of the visual representation of a single pulse or AC cycle on a current vs. time (or voltage vs. time) plot is called the **waveform**. An enormous number of types of pulsed current or AC waveforms can be generated in conductors. Some examples of waveforms produced by commercially available clinical electrical stimulators are illustrated in Figures 1.4 and 1.5. A single pulse or cycle of AC may be characterized by its amplitude- and time-dependent characteristics as well as a number of other descriptive features (Table 1.1).

Table 1.1
Descriptive Characteristics of Pulsed and Alternating Current Waveforms

Characteristic	Common Designations
Number of phases	Monophasic
	Biphasic
	Triphasic
	Polyphasic
Symmetry of phases	Symmetric
	Asymmetric
Balance of phase charge	Balanced
	Unbalanced
Waveform or phase shape	Rectangular
	Square
	Triangular
	Sawtooth
	Sinusoidal
	Exponential

Number of Phases in a Waveform

The term **phase** refers to unidirectional current flow on a current/time plot. A pulse that deviates from the zero-current line (baseline) in only one direction, like that shown in Figure 1.9*A*, is referred to as **monophasic.** Such a pulse could be produced by intermittently interrupting a constant voltage source applied to a conductor. In a monophasic pulse, charged particles in the conductive medium move briefly in one direction, according to their charge, then stop.

A pulse that deviates from baseline first in one direction and then in the opposite direction is called **biphasic** (Fig. 1.9*A*). This type of pulse can be produced by intermittently interrupting an alternating voltage source applied to an electrical circuit. In a biphasic pulse, charged particles move first in one direction and then move back in the opposite direction.

Waveforms with three phases are called **triphasic,** and those with more than three, **polyphasic.** Some commercially produced waveforms that have been referred to by other authors as polyphasic may in fact be an uninterrupted series of biphasic waveforms when reduced to the simplest common electrical event.

Symmetry in Biphasic Waveforms

For biphasic pulses or AC cycles, the manner in which charges move back and forth may or may not be the same. If the way in which current amplitude varies over time for the first phase of a biphasic waveform is identical in nature but opposite in direction to that in the second phase, the biphasic waveform is described as **symmetrical** (Fig. 1.9*B*). That is, a waveform is described as symmetrical if the first phase is the mirror image of the second phase of a biphasic pulse or single cycle of AC. On the other hand, a waveform is referred to as **asymmetrical** if the way in which current amplitude varies in the first phase of a biphasic pulse is not the mirror image of the second phase (Fig. 1.9*B*).

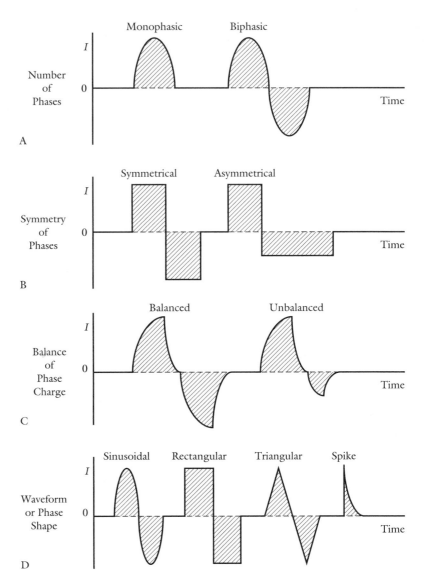

Figure 1.9 Characteristics of pulsed or alternating current waveforms.

Charge Balance in Biphasic Waveforms

For symmetrical biphasic waveforms, the total amount of current for one phase is equal to the absolute value of the total current flowing in the second phase. This condition may or may not be true for asymmetrical biphasic waveforms. If for an asymmetrical biphasic waveform the time integral for current in the first phase is not equal in magnitude to the time integral in the second phase, then the waveform is called **unbalanced**. More simply stated, if the area under the first phase of a biphasic waveform is not the same as the area under the second phase, the waveform is called unbalanced. If the area under the first phase of a biphasic waveform is equal to the area under the second phase, the waveform is described as **balanced**. Examples of balanced and unbalanced biphasic waveforms are shown in Figure 1.9*C*. From a clinical perspective, the use of unbalanced waveforms may result in noticeable differences in the sensation of stimulation under surface electrodes.

Waveforms

A very common descriptive approach to the characterization of pulsed and AC waveforms is the use of terms to denote the geometric shape of the pulse or cycle phases as they appear on the graph of current (or voltage) vs. time. Shape designations frequently encountered in the professional and commercial literature include **rectangular, square, triangular, sawtooth,** and **spike**. Alternatively, shapes can be ascribed based on the mathematical function that would give rise to a graph (or portion thereof) of similar shape. Two examples of such designations are waveforms based on **sinusoidal** or **exponential** changes in current (or voltage) over time. Figure 1.9*D* illustrates several of the common waveforms.

Combining Qualitative Terms to Describe Pulsed or Alternating Currents

The descriptive terms defined above are of limited value unless a system is developed to link these terms in a consistent manner. Figure 1.10 shows an organizational chart that can be used to assign qualitative descriptions to pulsed current or AC waveforms. From the examination of waveforms, one first decides which type of current is displayed. Next, the number of waveform phases is determined, followed by the symmetry and balance of charge for biphasic waveforms. Finally, a shape designation may be assigned either to the entire pulse or very often to the first phase of biphasic pulses.

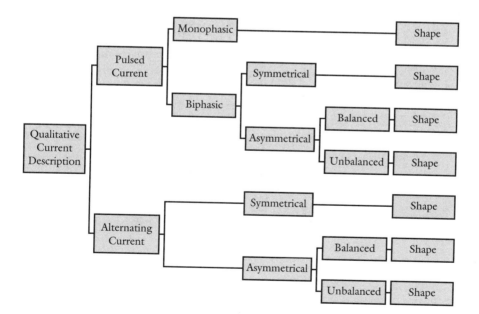

Figure 1.10 Diagram of the system for combining descriptive current designations in the naming of alternating or pulsed current waveforms.

The naming of the current waveform then proceeds from right to left along the chart. Figure 1.11 shows several current waveforms and indicates the qualitative description of these currents using the system shown in Figure 1.10. Note that this proposed system for naming electrotherapeutic currents may not be sufficient for describing all possible types of currents, but it does allow therapists and other practitioners to consistently describe most of the currents used in contemporary practice.

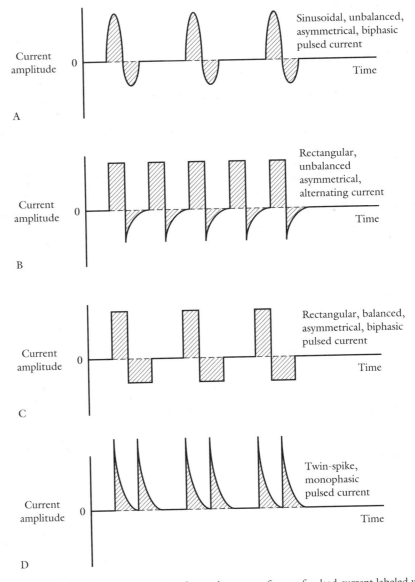

Figure 1.11 Graphical representation of several common forms of pulsed current labeled with appropriate "descriptive" designations. Waveforms represented in (**B**) and (**D**) have previously been called *high-voltage pulsed galvanic* and *faradic* respectively.

Table 1.2
Quantitative Characteristics of Pulsed and Alternating Currents

Amplitude-Dependent Characteristics
  Peak amplitude
  Peak-to-peak amplitude
  Root-mean-square amplitude
  Average amplitude

Time-Dependent Characteristics
  Phase duration
  Pulse duration
  Rise time
  Decay time
  Interpulse interval
  Intrapulse interval
  Period
  Frequency

Amplitude- and Time-Dependent Characteristics
  Phase charge
  Pulse charge

Quantitative Characteristics of Pulsed and Alternating Currents

Characteristics of Single Pulses

Pulsed current or AC waveforms may be quantitatively characterized by their amplitude- and time-dependent features (Table 1.2). Amplitude is a measure of the magnitude of current with reference to the zero-current baseline at any one moment in time on a current-vs.-time graph. Alternatively, the amplitude may be a measure of the driving force (voltage) applied to induce a current when a waveform is plotted as a voltage-vs.-time graph (Fig. 1.12A). Amplitude-dependent properties of current pulses (or voltage pulses) can be characterized by the measurement of the following.

Peak amplitude: the maximum current (voltage) reached in a monophasic pulse or for each phase of a biphasic pulse.

Peak-to-peak amplitude: the maximum current (voltage) measured from the peak of the first phase to the peak of the second phase of a biphasic pulse.

Of these two methods of measuring current (voltage) amplitude, peak amplitude of each phase is recommended. Other ways to describe current amplitude, such as **root-mean-square** (RMS, or effective) amplitude or **average current** per unit time, depend on the particular waveform examined. For instance, the RMS value for a pure sinusoidal waveform equals about 70% of the peak amplitude value, whereas the average current for the same waveform is about 64% of the peak value. Illustrations of these measures of current amplitude are given in Figure 1.12. Average current and RMS current measures take pulse shape into account and may more accurately reflect the stimulating power of the waveform than peak amplitude measures.

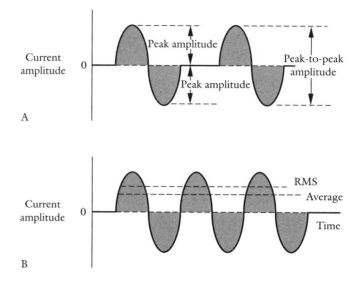

Figure 1.12 Sinusoidal AC waveforms and their amplitude-dependent characteristics. Amplitudes may be expressed as either *peak* amplitudes for each phase or *peak-to-peak* pulse amplitude **(A)**. Alternatively, root-mean-square *(RMS)* or *average* amplitudes can be used to describe the magnitude of currents or voltages **(B)**.

The amplitude of currents applied using clinical stimulators is sometimes referred to as the *intensity of stimulation*. Therefore, controls on clinical generators that regulate the amplitude of induced current (voltage) are often labeled "Intensity." Since the term *intensity* is also frequently used to describe pulse charge, it is recommended that the term intensity *not be used* at all to describe amplitude characteristics of pulsed current or AC waveforms.

A variety of time-dependent characteristics are used to quantify current pulses (Fig. 1.13). Time-dependent pulse characteristics of currents include the following.

Phase duration: the elapsed time between the beginning and the end of one phase.

Pulse duration: the elapsed time between the beginning and the end of all phases in a single pulse; on clinical stimulators the pulse duration is often incorrectly labeled "Pulse Width."

Period: the elapsed time from a reference point on a pulse waveform or cycle of AC to the identical point on the next successive pulse; the reciprocal of frequency (Period $= 1/f$). For pulsed current, the period is equal to the pulse duration plus the interpulse interval.

Interphase interval: the elapsed time between two successive phases of a pulse; also known as the **intrapulse interval**.

Rise time: the time for the leading edge of the phase to increase in amplitude from the zero current baseline to peak amplitude of one phase.

Decay time: the time for the trailing edge of the phase to return to the zero current baseline from the peak or maximum amplitude of the phase.

These time-dependent characteristics of pulses are generally expressed in microseconds (μsec, μs, 10^{-6} sec, millionths of a second) or milliseconds (msec, ms, 10^{-3} sec, thousandths of a second) when dealing with applications of pulsatile currents in clinical electrotherapy.

One of the most important quantitative characteristics of pulses from a physiologic standpoint is the charge carried by an individual pulse or phase of a pulse. The **phase charge** is defined as the time integral of current for a single phase. That is, the phase charge is represented by the area under a single phase waveform (Fig. 1.13*B*). As such, the phase charge is determined by both the amplitude of the phase and the duration of the phase. The magnitude of the phase charge will provide an indication of the relative influence a pulse will have in producing changes in biological systems. The **pulse charge** of a single pulse is the time integral for the current waveform over the entire pulse (Fig. 1.13*B*). For a typical biphasic pulse, the pulse charge is the sum of the area under each phase. For monophasic waveforms the pulse charge and phase charge are equal. Phase and pulse charges are expressed in coulombs, and pulse charges commonly found in clinical stimulation fall into the microcoulomb (μC, 10^{-6}, millionths of a coulomb) range.

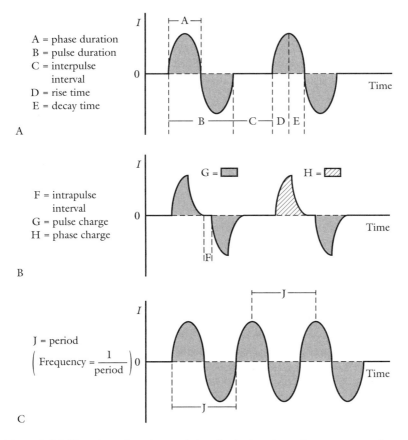

Figure 1.13 Time-dependent characteristics of pulsed or alternating current waveforms.

Characteristics of a Series of Pulses

In addition to those terms used to quantify the features of individual pulses, a number of important terms are used to describe a series of pulses, the usual manner in which electrical currents are induced in biological tissues for their therapeutic effects. Among those terms are the following two.

Interpulse interval: the time between the end of one pulse and the beginning of the next pulse in a series; the time between successive pulses (Fig. 1.13*A*).

Frequency *(f)*:   the number of pulses per unit time for pulsed current expressed as pulses per second (pps); the number of cycles of AC per second expressed in cycles per second (cps) or hertz (Hz); often on clinical stimulators the frequency of stimulation control is labeled "Rate."

Since voltage and current are directly proportional, many of the terms used to describe the amplitude- and time-dependent characteristics of currents may also be used to describe the voltage pulse features that induce these current waveforms.

Current Modulations

Amplitude and Duration Modulations

In the use of electrical stimulation for management of patient problems, current amplitude- and time-related characteristics are often varied in a prescribed fashion. Changes in current characteristics may be sequential, intermittent, or variable in nature and are referred to as **modulations**. Several of the quantitative characteristics of pulsed current and AC are modulated in selected clinical applications. Variations in the peak amplitude of a series of pulses are called **amplitude modulations** (Fig. 1.14*A*). Regular changes in the time over which each pulse in a series acts are referred to as **pulse** or **phase duration modulations** (Fig. 1.14*B*). **Frequency modulations** consist of cyclic variations in the number of pulses applied per unit time (Fig. 1.14*C*). The illustrations of modulations shown in Figure 1.14 occur in a systematic fashion. Modulations in amplitude, pulse duration, or frequency can also be provided randomly.

Another modulation encountered rather frequently in clinical electrical stimulation is **ramp (surge) modulation**. Ramp modulations are characterized by an increase (ramp up) or decrease (ramp down) of pulse amplitude, pulse duration, or both, over time. In the past, ramp modulations have been referred to as *rise time* and *fall time*. However, these two terms are now used to describe single pulse characteristics, not the variations in features of a series of pulses.

Timing modulations

A continuous, repetitive series of pulses (series of pulses at a fixed frequency) or a segment of AC is called a **train** (Fig. 1.15, *A* and *B*). Systematic variations in the pattern of delivery of a series of current pulses are referred to as **timing modulations**. Several terms are now recognized for describing timing modulations. They include the following.

Burst: a finite series of pulses; a finite interval of alternating current delivered at a specified frequency over a specified time interval (Fig. 1.15, *C* and *D*). The time interval over which the finite series of pulses or AC cycles is delivered is called the **burst duration**. The time period between

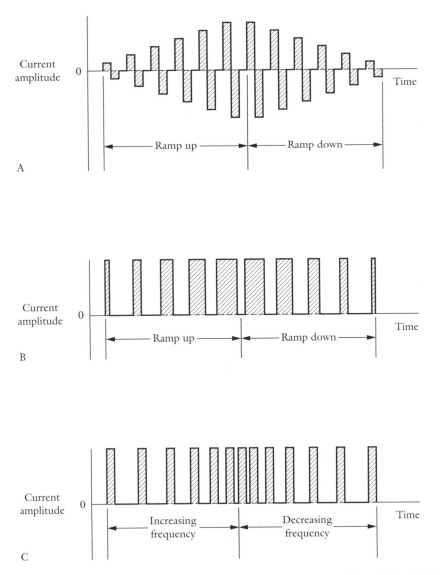

Figure 1.14 Examples of automatic modulations of stimulation characteristics: **A.** Amplitude modulation. **B.** Pulse duration modulation. **C.** Frequency modulation.

bursts is called the **interburst interval**. In contemporary clinical applications of such burst modulations, the burst duration and interburst interval are usually on the order of a few milliseconds.

In some forms of electrotherapy, trains of pulses, trains of AC, or series of bursts are applied to patients without any interruption for the entire treatment period. Such a pattern of stimulation is often described as a **continuous mode** of stimulation. In many other approaches, trains of pulses, trains of AC cycles, or series of bursts are often applied to individuals for times ranging from a few seconds to a minute or more, followed by comparable periods of no stimulation before stimulation is resumed. That is, trains or

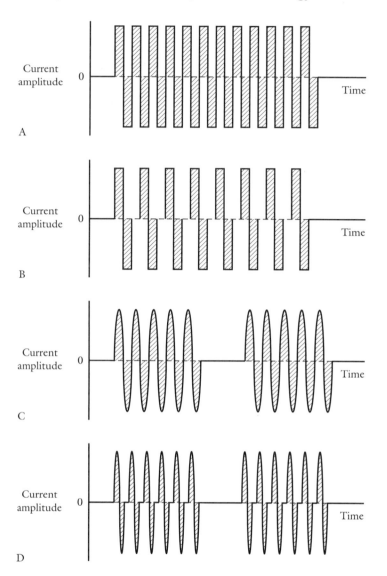

Figure 1.15 Examples of stimulation "trains" and "burst" modulations: **A.** Continuous train of rectangular, symmetric, biphasic AC waveforms. **B.** Continuous train of rectangular, symmetric, biphasic pulsed current waveforms. **C.** Burst-modulated, sinusoidal AC waveforms. **D.** Burst-modulated, sinusoidal pulsed current waveforms.

series of bursts are intermittently or regularly interrupted. Such patterns of stimulation are quantitatively characterized by two time intervals, called the *on time* and the *off time*, defined in the following.

On time: the time during which a train of pulses, trains of AC, or a series of bursts is delivered in a therapeutic application.

Off time: the time between trains of pulses, trains of AC, or a series of bursts.

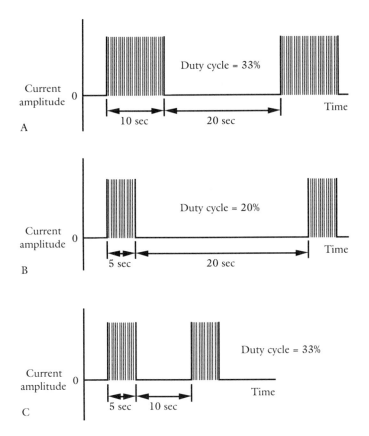

Figure 1.16 Examples of on times and off times of stimulation and the concept of *duty cycle*, with 2-pps monophasic pulsed currents at fixed amplitude: **A**. 10-sec on time and 20-sec off time. **B**. 5-sec on time and 20-sec off time. **C**. 5-sec on time and 10-sec off time.

A closely associated characterization of the interrupted patterns of stimulation used in many clinical applications is embodied in the concept of the **duty cycle**. The duty cycle of stimulation is the ratio of on time to the sum of on time plus off time multiplied by 100, expressed as a percentage (Fig. 1.16).

$$\text{Duty cycle} = \frac{\text{on time}}{(\text{on time} + \text{off time})} \times 100\%$$

For example, if the on time equals 10 sec and the off time equals 30 sec, the *duty cycle* for such a pattern of stimulation would be 25% (Fig. 1.16A). A very different pattern of stimulation with an on time of 5 sec and an off time of 10 sec yields the same 25% duty cycle (Fig. 1.16B). For this reason and because in some cases the duty cycle has erroneously been equated with the simple ratio of on time divided by off time, confusion has arisen from the use of the term *duty cycle*. For clear documentation of stimulation patterns, specific on times and off times of stimulation should be specified rather than using the duty cycle or on/off ratios.

Summary

This chapter has presented fundamental concepts in electricity and standardized terminology associated with the application of electrotherapeutic currents. The review of basic electrical concepts was included to refresh the reader's memory of physical entities and principles that form a foundation for understanding the electrical and chemical events associated with clinical applications of electricity. The standardized qualitative and quantitative terminology was presented to facilitate clear communication between researchers, clinicians, students, and manufacturers involved in the use and development of clinical electrotherapy. To this end, the standardized terminology is used throughout the remaining chapters. The various quantitative characteristics defined in this chapter represent the features of electrical stimulation that must be either selected or regulated by therapists or other practitioners in order to safely and effectively employ electrotherapy to achieve therapeutic outcomes.

Study Questions

For answers, see Appendix B.

1. The driving force that makes charged particles move is called ___(a)___, ___(b)___, or ___(c)___.

2. The movement of charged particles in a conductor is called a _____.

3. The opposition to the movement of charged particles in an electrical circuit is called _____.

4. The opposition to the movement of ions in biological systems is called _____.

5. The negative pole of a battery or electrical circuit is called the ___(a)___, and the positive pole is called the ___(b)___.

6. Positively charged ions are called ___(a)___, and negatively charged ions are called ___(b)___.

7. Anions are attracted to the *[cathode/anode]* and repelled from the *[cathode/anode]*.

8. Ohm's law describes the relationship between ___(a)___, ___(b)___, and ___(c)___.

9. The larger the voltage applied to an electrical circuit, the larger will be the _____ produced in the circuit.

10. For a fixed applied voltage, if the impedance of tissue is decreased, the magnitude of the current will _____.

11. The three types of current used in contemporary electrotherapy are ___(a)___, ___(b)___, and ___(c)___.

12. The shape of the visual representation of currents on a current amplitude-vs.-time graph is called a _____.

13. Give the standard unit of measurement for the following:

    a. current _____

    b. electromotive force _____

    c. resistance _____

d. capacitance _____

e. impedance _____

f. conductance _____

g. pulse current frequency _____

h. AC frequency _____

i. peak amplitude (current) _____

j. pulse duration _____

k. phase charge _____

l. on time/off time _____

14. Use the flowchart in Figure 1.10 to assign qualitative descriptions to the currents diagramed below.

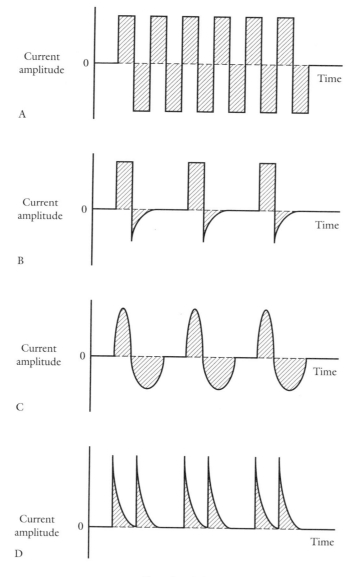

Question 14

15. Draw the following types of currents on current amplitude-vs.-time graphs.

    a. Triangular, symmetrical, biphasic pulsed current

    b. Rectangular, balanced, asymmetrical, biphasic pulsed current

    c. Sinusoidal, unbalanced, asymmetrical alternating current

    d. Square, monophasic pulsed current

16. For the pulsed current waveforms shown, provide labels for the lettered amplitude- and time-dependent characteristics.

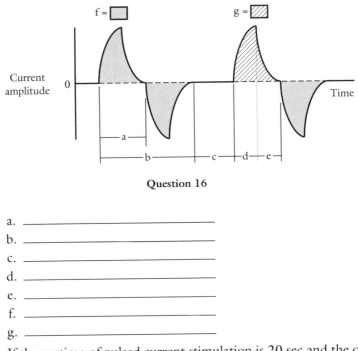

Question 16

    a. _____

    b. _____

    c. _____

    d. _____

    e. _____

    f. _____

    g. _____

17. If the on time of pulsed current stimulation is 20 sec and the off time is 60 sec, the duty cycle of stimulation is _____.

References

1. Urone PP. Physics with Health Science Applications. New York: Harper and Row, 1986:264-343.

2. American Physical Therapy Association. Electrotherapeutic Terminology in Physical Therapy. Section on Clinical Electrophysiology, Alexandria, VA: American Physical Therapy Assoc., 1990.

Chapter 2

Instrumentation for Electrotherapy

Andrew J. Robinson

A wide variety of devices are commercially available for electrotherapeutic applications. Instrumentation in electrotherapy is rapidly proliferating because of advances in engineering, developments in electrotherapy research, and a widened scope of problems managed by electrotherapy. A clear understanding of the components and operating principles of electrical stimulation devices is essential for safe and effective clinical applications. This first portion of this chapter describes (a) design features and general principles for operating electrical stimulators, (b) typical controls to regulate parameters of stimulation, (c) common types and features of commercially available stimulators, and (d) types of electrodes and general considerations in the selection and application of electrodes. The objective of this chapter is to provide a description of electrotherapeutic stimulators valuable to the practitioner rather than a detailed presentation suitable only to the biomedical engineer. Discussion of commercial classes of stimulators is not intended to include specific descriptions of each manufacturer's stimulator features. Instead, the intention is to present the characteristics and features of devices generally marketed within a particular class. Such information may also be of value to those responsible for the acquisition of electrical stimulators for clinical applications. In the last portion of the chapter, issues related to electrical safety in the selection and clinical utilization of stimulators are discussed.

Design Features of Stimulators for Electrotherapeutic Applications

Electrotherapeutic devices are used to induce electrical currents in body tissues. When electrotherapy has been determined to be an appropriate therapeutic intervention, the first question faced by the clinician is, What stimulator do I use to achieve the desired outcome? The choice of a stimulator for a particular application is determined by a number of design-related factors, including the following.

1. Does the stimulator produce the specific parameters of electrical stimulation required to bring about the desired effect? Are the waveform, output frequency range, output channel timing, etc., sufficient for the particular application?
2. Does the stimulator have sufficient maximum output amplitude to achieve the desired effect?
3. Does the stimulator have a sufficient number of output channels for the application?
4. Does the stimulator have appropriate controls to make any necessary adjustments of stimulation parameters during treatment?
5. Can stimulation parameters be adjusted while stimulation is being provided, or does the instrument need to be turned off?
6. Is a battery-operated stimulator appropriate for the application, or should a line-powered stimulator be selected?
7. Is the stimulator preprogrammed, or should the stimulation parameters be adjusted separately for each type of application?
8. Are important safety features included in the stimulator design?

Questions such as these related to stimulator design are important considerations in the planning of an electrotherapy program. Clinicians who employ electrotherapy need to be aware of the different design features of stimulators as well as the potential advantages and disadvantages of each. If the clinician has a clear understanding of the output characteristics and control features of a stimulator, he or she need not understand the operation of internal circuitry or discrete electronic components (e.g., oscillators, transformers, rectifiers) to be able to safely and effectively use the stimulator.

Portable, battery-operated versus line-powered stimulator designs

Depending on the stimulation characteristics applied, that is, the magnitude and duration of the stimulation required, the power requirements of stimulation devices may be small or large. Those therapeutic interventions that require relatively high stimulator output for long periods dictate the use of standard, line-powered (115 V, sinusoidal AC) devices. One advantage of line-powered stimulators is that stimulation parameters should remain at set levels for as long as treatment is provided, given no interruption in the power supply. The main disadvantage of the line-powered devices is that they can only be applied when patients can remain stationary during stimulation.

When an electrotherapeutic intervention does not require relatively high device output and/or prolonged periods of stimulation, battery-powered (1.5 V to 9.0 V) stimulators are often selected. Battery-operated stimulators in general have much lower peak output amplitudes than line-powered units. Portable, battery-driven devices have the advantage of allowing stimulation while a patient is moving, as in the stimulation of muscle during ambulation or stimulation for pain control while a patient is on the job. The performance of portable stimulators is dependent on battery life, which is in turn determined by the particular stimulation procedure employed. Suffice it to say that the higher the amplitude requirements or the more frequently the stimulator is used, the greater the power consumption and the shorter will be the battery life. In most uses of battery-operated therapeutic stimulators, batteries should be either recharged (nickel-cadmium batteries) or replaced (alkaline batteries) frequently because the power drain is substantial even in short-term applications and an undesirable drop in the output amplitude may result as treatment is continued. Carbon batteries are generally not suitable for use in portable stimulators because of their short operating life and the potential for leakage of corrosive chemicals, which may damage the stimulator.

Analog control versus digital control stimulator designs

Analog control stimulators are those that allow the adjustment of stimulation parameters primarily through the use of rotary knobs or dials. Since these stimulator controls are mechanical in nature (e.g., potentiometers), analog-controlled stimulators may also be referred to as "hardware controlled." Digital control stimulators on the other hand are those that allow adjustment through the use of push buttons or pressure-sensitive switches. In general, devices with digital control design include some form of visual display (e.g., an LCD) that allows one to select from a menu of available stimulation parameters. Today's digital design stimulators are generally microprocessor-based and are also called "software-controlled." Figure 2.1 illustrates two portable neuromuscular stimulators, one with more traditional analog controls and a newer design with digital control.

Both analog and digital instruments have their advantages and disadvantages. The primary advantage of analog controls is the ability to continuously adjust a treatment parameter, thus allowing the knowledgeable user to make fine adjustments in stimulation characteristics while the treatment is occurring. In contrast, digital controls often allow only incremental adjustment of parameters. This disadvantage can be minimized if the size of the increments from one setting to the next are made as small as possible. For example, frequency adjustment in increments of 1 pps (pulse per second) as opposed to increments of 10 pps would enhance the user's ability to precisely control the pattern of stimulation for a variety of applications. In addition, modification of stimulation characteristics with digital devices often requires the user to scroll through menu options. This control approach is time-consuming if changes are desired during the course of treatment. Analog controls may be more "user-friendly" in the hands of those who were nurtured on analog-controlled radios and televisions, whereas those who have been regularly exposed to digitally controlled electronics (e.g., computer games) may be more comfortable with digital stimulators. Some digital devices have a memory capability that allows the storage of a set of stimulation parameters for later use. Preprogrammed patterns of stimulation (protocols) may be useful to clinicians who are less familiar with the electrotherapy literature; such preprogrammed patterns make stimulator use easy. The use of preprogrammed protocols assumes that the stimulator manufacturer has selected stimulation characteristics that have been shown to be effective and safe for a particular application. Analog-controlled stimulators usually do not have preprogrammed memory although some would argue that the positions in which the knobs and switches are left constitutes memory—at least a memory of the last treatment. In general, stimulators with primarily analog controls require the user to be more knowledgeable regarding electrotherapy and to be aware of the expected consequences of stimulation parameter adjustment.

Some contemporary stimulators incorporate both analog control and digital control technology in their design and hence take advantage of the benefits of each approach.

Controls to Regulate the Characteristics and Pattern of Electrical Stimulation

Although an understanding of the function of the internal components of electrical stimulators may be of interest to the user, this knowledge is not critical for one to be able to employ electrical stimulators in the clinic. In contrast, a knowledge of the types of current produced, the waveform characteristics, and how the stimulation parameters are regulated is essential for the safe and effective application of electrotherapy. The specific output parameters that can be adjusted, selected, or set by users of electrical stimulators vary from device to device. The parameters are adjusted with analog or digital controls, usually located on one main control panel of the stimulator. In

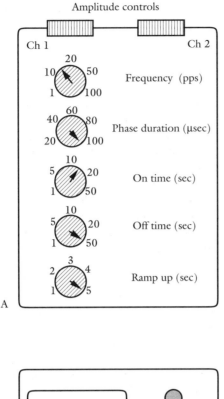

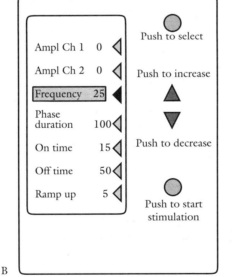

Figure 2.1 Control panel of portable stimulators with analog controls (**A**) and digital controls (**B**).

some cases (e.g., portable stimulators) these controls are located in covered or hidden compartments. Such a design reduces the chance that the patient will inadvertently or intentionally change the stimulus parameters and cause an unintended pattern of stimulation to be applied. Controls located in hidden compartments are often initially set and/or adjusted only by the health care professional responsible for implementing the stimulation.

No industry standard exists for the types of stimulator controls that every type of stimulator should possess. The stimulator controls common to many units currently used in clinical practice are discussed below. The focus of the following discussion is on analog as opposed to digital controls.

Waveform selection controls

The proliferation of commercially available stimulation devices has been accompanied by the development of numerous types of current (or voltage) waveforms that can be used in clinical electrotherapy. Most devices designed before the early 1980s included waveform generators capable of producing only one type of waveform from a single instrument. Today, however, many commercially available stimulators are capable of producing several distinct waveforms. One of the first steps in applying stimulation clinically is the selection of a suitable waveform. This is achieved in some stimulators by a control generically referred to as a **waveform selector.** Waveform selectors may be rotary switches, pressure-sensitive switches, or other types of controls labeled with small diagrams of the available waveforms (Fig. 2.2). In some microprocessor-based stimulators, waveforms can be selected from a menu on the display screen. The choice of the "most appropriate" waveform for a particular outcome should be based on findings or recommendations from the scientific or professional literature. Information on waveform selection for specific applications will be found in subsequent chapters.

Amplitude controls

The **amplitude control(s)** on stimulators allows the user to adjust the output voltage from the output amplifier circuit. On direct current generators the amplitude controls regulate the magnitude of current and generally do not allow the current to exceed 5 mA (milliamperes) at maximum settings. On AC or pulsed current generators, amplitude controls allow the practitioner to adjust the peak amplitude or peak-to-peak amplitude of AC cycles or pulses (Fig. 2.3A). Depending on the specific stimulator, either output current or voltage is regulated, and amplitudes usually do not exceed peak values of 100–200 mA or 500 V, respectively.

Analog amplitude controls are variable resistors, or potentiometers. Ideally, these potentiometers should be linear in nature. In a linear potentiometer, every increment of rotation of the potentiometer dial produces an equal change in stimulator output amplitude (Fig. 2.3B). For instance, a one-quarter turn of the dial would change the resistance by 25% and correspondingly would alter the output amplitude by 25%. Commonly, however, the amplitude controls are nonlinear (1, 2). In such cases, a one-quarter turn of the dial may produce only a small (e.g., 5%) rise in amplitude in the initial range of rotation, whereas a one-quarter turn at the end of the range of rotation might produce a much larger (e.g., 50% or greater) increase in output amplitude. The user who is unaware of this common characteristic of amplitude controls may inadvertently deliver a patient a startling stimulus during the course of amplitude adjustment.

Another feature of amplitude controls that varies among stimulators is the number of revolutions of the dial required to vary the amplitude from 0 to 100% of available output. Most often amplitude controls vary output over

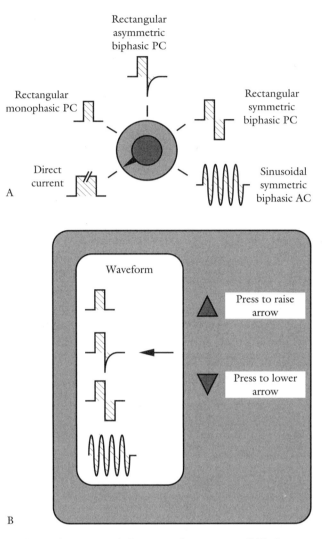

Figure 2.2 Waveform selector control illustrating the options available for treatment applications: **A.** Analog control. **B.** Digital control.

the range of a single clockwise turn (actually about $\frac{3}{4}$ of a revolution). In other cases, amplitude controls must be rotated several complete revolutions in order to vary output from minimum to maximum levels.

In some instances, the amplitude controls have a switch built-in at the beginning of the range of rotation. Such amplitude dials (Fig. 2.3C) must be switched to the "Off" position before the stimulator can be turned on in subsequent sessions. This is a valuable safety feature that is designed to prevent the output circuits of the stimulator from being "powered up" when the amplitude control is left at a level other than the zero output. This design of output controls is intended to prevent an unexpected stimulation to the patient.

On digital or microprocessor-based stimulators, output amplitude is adjusted in incremental steps by depression of pressure-sensitive switches.

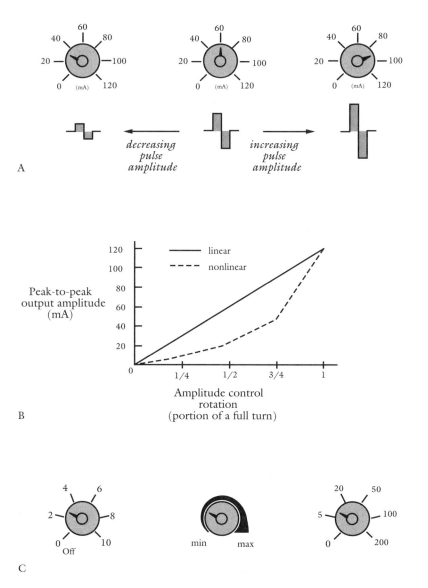

Figure 2.3 Amplitude controls: **A.** The effect of changes in dial setting on pulse amplitude. **B.** The relationship between the amplitude control setting and peak-to-peak pulse amplitude. **C.** Examples of the amplitude control labeling used on commercial stimulators.

Amplitude controls are labeled "Intensity" or "Voltage," and numbers ranging from 1 to 10 indicate the lowest to highest level of stimulation output. Users of stimulation devices should be aware that the numerical labels often bear little relation to the actual or relative amplitude of stimulation being applied to an individual. In other cases, controls may have no numerical labeling related to output amplitude or may have labeling that reflects the nonlinear characteristics of the control (Fig. 2.3C). Precise information on relative output changes in current or voltage as amplitude controls are adjusted may be obtained only by observing the analog or digital meters available on some devices or by monitoring stimulator output using an oscilloscope. No industry standards exist for monitoring or displaying the amplitude of either monophasic or biphasic current waveforms in stimulators. Furthermore, the absolute amplitude of stimulation does not bear a consistent relationship to evoked response (e.g., muscle contraction level) from subject to subject or from session to session in the same subject. In actual clinical stimulation sessions, one of the best indicators of the amplitude is the reliable patient's subjective perception of the stimulation, as discussed in Chapter 3.

The relationship between the amplitude control setting and the actual output of the stimulator should be periodically checked to ensure proper functioning. Adjustments of analog amplitude controls should not produce large voltage transients (unexpected voltage fluctuations) that may prove to be uncomfortable to patients. Any malfunctioning of amplitude controls should be immediately corrected by either replacement or cleaning of the controls.

Ideally, independent amplitude controls should be available for each output channel. On some devices, however, only one amplitude control is present, for changing amplitude on two output channels simultaneously. In some cases, the single output amplitude control for "two channels" is accompanied by a dial called a **balance control.** Adjustment of the balance control shifts the relative amplitude of stimulation from one channel to another (Fig. 2.4). The balance control raises the amplitude on one channel and simultaneously decreases it on the other. Such a design is more difficult to

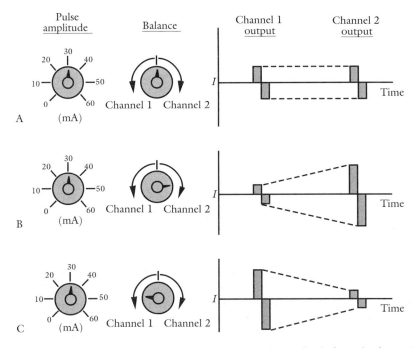

Figure 2.4 The effects of balance control adjustment on pulse amplitude for each of two stimulation channels. **A.** Output amplitude balanced between two channels. **B.** Output amplitude increased on *channel 2* while decreased on *channel 1*. **C.** Output amplitude increased on *channel 1* while decreased on *channel 2.*

use than one having independent amplitude controls on each channel. When the balance control is initially set to equally divide the stimulator output and the amplitude control is gradually advanced, the amplitude increases by the same amount on each output channel. If amplitude control is set at a particular level and the balance control is adjusted, the amplitude of stimulation increases on one output channel and simultaneously decreases on the other. Such an arrangement can be problematic when, in a two-channel application, the desired amplitude is reached on the first channel but is insufficient on the second. In order to reach desired stimulation levels on each channel, the balance control is adjusted to raise the relative amount of stimulation to the second channel. This change decreases the stimulation of the first channel, which must then be compensated for by another amplitude adjustment. A number of such adjustments may be required before the desired level of stimulation is reached on each channel.

The maximum output amplitude of electrical stimulators has been used as a basis for categorizing these devices. The common differentiation of stimulators based on output amplitudes is "low volt" and "high volt." Low-volt devices are those that produce peak amplitudes of less than 100–150 V, whereas high-volt devices produce peak amplitudes of over 150 V. Differentiation among stimulators based on output amplitudes alone is not recommended because such a system does not take into account many of the other important characteristics of stimulation, such as the output frequencies or the phase (or pulse) durations. In addition, the so-called high-volt devices are capable of producing "low-voltage" output when the amplitudes are adjusted within the low range of the device capability.

Phase duration and pulse duration controls

Phase duration controls adjust the duration of a single phase of biphasic pulses, usually the first phase of the pulse. **Pulse duration controls** regulate the total duration of individual waveforms in pulsed current stimulators (Fig. 2.5). Such controls are common on portable, battery-operated stimulators designed for pain reduction, and they are becoming more common in recent designs of 60-Hz AC-powered devices for neuromuscular stimulation applications. Phase and pulse duration controls are not found on most AC stimulators because the cycle duration is determined by the frequency of stimulation.

Phase and pulse duration controls are often labeled "Pulse width" or "Width." Such designations are not recommended since phase and pulse durations are measured in units of time (generally milli- or microseconds), not units of length. A numerical-scale labeling of analog phase (or pulse) duration controls in equal increments from 1 to 10 is common. Like that for analog amplitude controls, such labeling may not reflect relative changes in phase (or pulse) duration as the duration dial is rotated. Similarly, the adjustment of phase duration dials may not produce linear changes in phase duration (1). In some stimulator designs, the phase duration control is a rotary switch rather than a smoothly rotating potentiometer. Often in such cases, the phase duration control will allow the user to select the specific phase or pulse duration shown on the dial label. In digital stimulator designs, phase durations can be

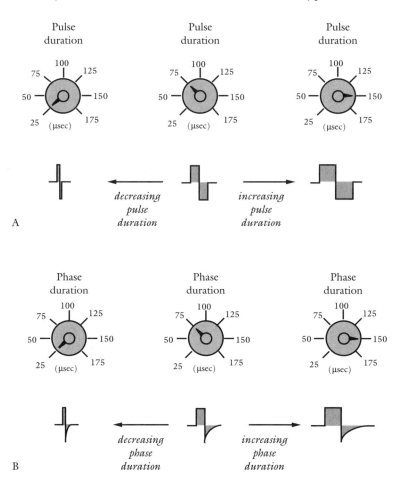

Figure 2.5 A. Pulse duration controls showing the effect of changes in dial setting on the pulse duration. **B.** Phase duration controls showing the effect of changes in dial setting on the duration of the first phase of a rectangular, asymmetric, biphasic pulse.

selected from screen menu options or can be adjusted in preset increments by the use of pressure-sensitive switches.

On most commercially available pulsed current stimulators, one phase (or pulse) duration control adjusts this parameter for all output channels. This configuration appears to be adequate for most contemporary electrotherapy protocols.

The adjustment of amplitude and phase or pulse duration controls changes the phase and/or pulse charge of each waveform applied. Increasing either the amplitude, phase (or pulse) duration, or both increases the net charge per stimulus and induces a larger current in biological tissues. That is, the stimulus is "stronger" as either amplitude or phase duration is increased. Since increasing either amplitude or phase duration increases the phase charge, these controls regulate what has been traditionally referred to as the *intensity* or *strength* of stimulation. As discussed in Chapter 3, these two parameters of stimulation determine the number of nerve or muscle fibers activated in response to stimulation.

Frequency controls

Frequency controls are provided on electrical stimulators to allow users to set the number of pulses (or AC cycles) delivered through each channel per second (Fig. 2.6), the number of bursts applied per second (Fig. 2.7), or the frequency of pulses within a burst (Fig. 2.8). Frequency controls on stimulators are usually labeled simply "Rate" or "Pulse rate," "Burst rate," etc., depending on which parameter of stimulation is being regulated. On many commercially available stimulators, more than one type of frequency control is present.

As with pulse amplitude and phase duration controls, rotating the frequency dial to a particular setting may not reflect the actual pulse or burst frequency delivered. Similarly, as the frequency dial is adjusted, incremental changes in the position of the dial may not reflect equivalent changes in frequency. This more imprecise type of frequency control is present on some portable stimulators designed for protocols to reduce pain (transcutaneous electrical nerve stimulation [TENS] devices), where an exact frequency of activation is not as critical to achieving effective stimulation (1). The portable stimulators designed for the activation of innervated muscles (referred to as functional electrical stimulation [FES] devices) usually have much more accurately labeled frequency controls. On microprocessor-based stimulator designs, the frequency of stimulation is often selected by controls linked to a digital frequency display, which provides an accurate indication of the frequency selected.

The frequency of pulsed stimulation commonly ranges from 1 to 1000 pps (pulses per second), and burst frequency generally ranges from 1 to 100 bursts per second. The frequency of pulses within bursts (carrier frequency) or AC frequencies may range from 1 to 10,000 pulses or cycles per second.

The output frequency characteristics of various stimulators have been used as a basis for their classification (3): stimulators capable of producing 1 to 1000 pps have been commonly referred to as **"low-frequency"** stimulators,

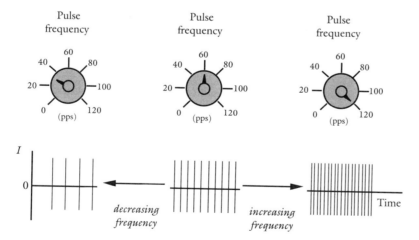

Figure 2.6 Pulse frequency controls showing the effect of changes in dial setting on the frequency of pulses produced.

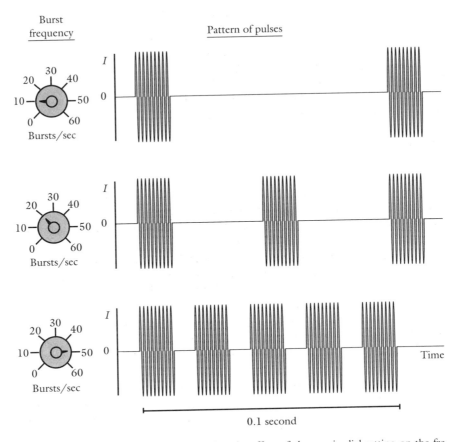

Figure 2.7 Burst frequency controls showing the effect of changes in dial setting on the frequency of bursts produced.

those producing 1000 to 10,000 pps have been called **"medium-frequency,"** and those producing pulses at frequencies in excess of 10,000 pps have been designated as **"high-frequency."** Such a classification scheme is commonly used in literature originating in Europe and Canada as well as in commercial circles. For this reason, those interested in electrotherapy should be aware of the meaning of these frequency designations. However, the differentiation among stimulators on the basis of frequency characteristics alone is not recommended. Some stimulators are capable of producing AC or pulsed currents at frequencies that cross the frequency boundaries just described and hence cannot be accurately categorized by a single designation.

On time and off time controls

When intermittent stimulation is required for a particular therapy, controls are necessary to set the duration of stimulation and the duration of "rest" between the periods of stimulation. The controls associated with setting these periods of stimulation and rest are called **duty cycle** or **cycle time** controls. In general, duty cycle controls are labeled "On time" and "Off time;" they will adjust periods of stimulation and rest from 1 to 60 sec. Many devices designed for neuromuscular stimulation include on time and off time controls, whereas many portable electrical stimulators designed for pain control do not. Figure 2.9 shows how on time and off time control settings regulate the pattern of a pulsed current.

Some stimulators allow the timing of stimulation on a channel to be manually controlled by the use of a remote switch. Some remote switches begin stimulation when the switch is depressed and stop stimulation when the switch is released; others operate in the opposite way. This timing-control approach is often found on devices used in neuromuscular stimulation applications, especially those used to produce contractions of muscle during functional activities.

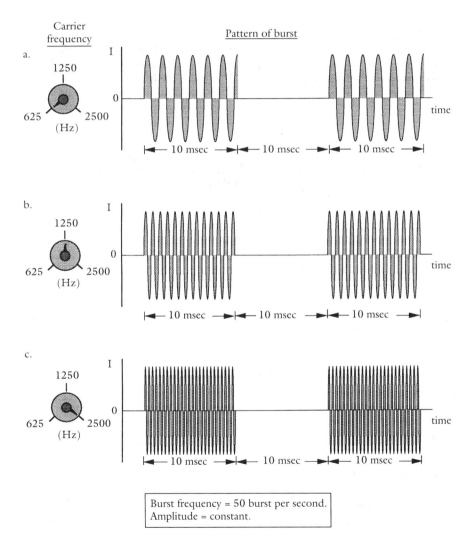

Burst frequency = 50 burst per second.
Amplitude = constant.

Figure 2.8 Carrier frequency controls showing the effect of changes in dial setting on the frequency of AC within each burst.

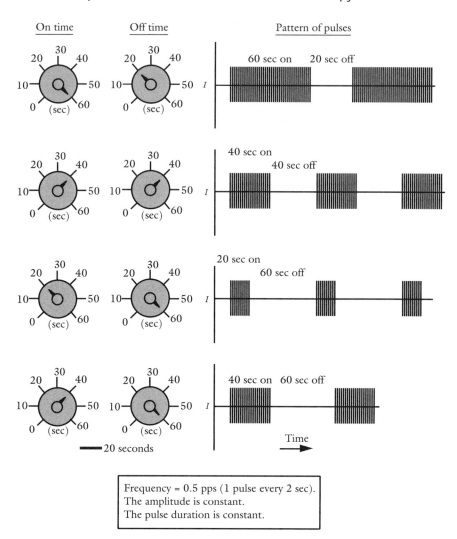

Figure 2.9 On time and off time controls showing the effect of changes in dial settings on the pattern of pulses produced.

Ramp up and ramp down controls

In order to automatically increase or decrease the phase or pulse charge in a pattern of stimulation, **ramp modulation controls** are included in several types of commercially available stimulators. These controls allow the therapist to set the number of seconds over which the amplitude (or phase/pulse durations) will gradually increase (or decrease) to (or from) the maximum value set by the amplitude control (Fig. 2.10). Ramp modulations on the leading and trailing ends of a train of pulses provide a more comfortable onset and cessation of stimulation in a variety of applications, especially when very high levels of stimulation are required for a particular therapeutic goal. In neuromuscular stimulation applications, the inclusion of a ramp up time on the leading edge of a train of pulses or AC allows for the gradual recruitment of motor nerve fibers and hence the gradual increase in muscle fiber contraction, which results in the smooth increase in muscle force output. The gradual onset of muscle stimulation produces contractions that more closely mimic those produced in functional activities during voluntary muscle activation, and the gradual onset is more comfortable for the individual receiving the stimulation. Ramp controls are usually not available on portable stimulators designed specifically for pain control. Depending on the instrument used, rise time and fall time controls may be labeled "Slope time," "Surge time," "Ramp up," or simply "Ramp."

Although ramp modulations are commonly associated with the automatic regulation of either amplitude or phase duration characteristics, they may also be employed to automatically vary the frequency of stimulation.

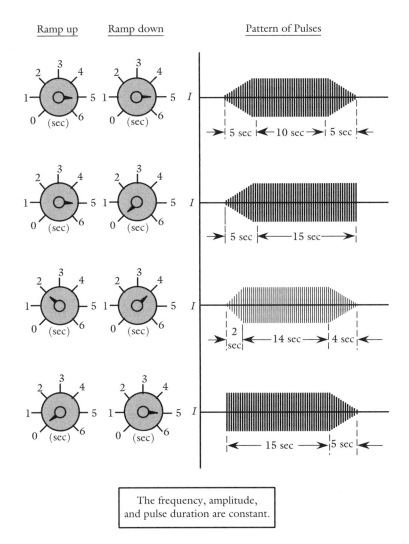

The frequency, amplitude, and pulse duration are constant.

Figure 2.10 Ramp controls showing the effect of changes in dial setting on the time over which the amplitude of pulses is automatically increased (ramp up) or decreased (ramp down) during a 20-sec train of pulses.

Programmed stimulation pattern controls

Programmed stimulation controls are available on some stimulation devices. Generally in analog control devices, these controls are simply slide or rotary switches that allow the health care professional to select from the available stimulation patterns. In digitally controlled stimulators, preprogrammed stimulation can be selected from a menu display. If the option for preprogrammed stimulation is chosen, many of the time-dependent and/or amplitude-dependent characteristics of the stimulation pattern are automatically set or modulated. In some portable stimulators, designed for pain control, such preprogrammed patterns of stimulation can labeled "Burst mode," "Modulated mode," or even simply "Pain." In other devices, designed primarily for the activation of muscle, preprogrammed patterns of stimulation are labeled "Muscle stimulation." The option to select a predetermined set of stimulation parameters can be very useful when a patient is taught to use stimulation at home. For example, such preprogrammed parameters are often employed in the application of transcutaneous nerve stimulation for pain, for which parameters such as pulse amplitude, pulse duration, or both are automatically modulated. On the other hand, the choices offered by preprogrammed controls may limit the utility of a device for a wide range of applications especially if the user is unable to program stimulation patterns tailored to a specific application on a specific patient population.

Preprogrammed stimulus options should be clearly described in the technical literature of a stimulator in order to determine if the pattern of stimulation is appropriate for a particular application. No uniform industry standards exist for the patterns of stimulation produced in the various "modes" of stimulation. The manufacturer's operating manual should be consulted for a description of precisely what stimulation characteristics are automatically modulated in specific predetermined "modes" of stimulation.

Output channel selection controls

The signal from a single stimulator output amplifier is generally referred to as an *output channel*. Many stimulators with more than one channel allow the user to select the timing of stimulation through each channel. There are four common ways for stimulation to be coordinated when two channels are applied in a single treatment approach. First, the stimulator may be set so that both channels provide uninterrupted stimulation (no off time) for the entire treatment period. This pattern of two-channel stimulation is usually called the **continuous,** or **constant, stimulation mode** (Fig. 2.11*A*). Second, many devices may turn stimulation on and off to both channels at the same time (Fig. 2.11*B*). This mode of stimulation is called **synchronous, simultaneous,** or **interrupted mode.** The *synchronous* and *simultaneous* designations for this pattern of two-channel stimulation are the less-confusing designations. The third way in which the timing of stimulation is regulated in two-channel applications is to alternate stimulation between the two channels. That is, one channel provides stimulation while the second channel is off and vice versa (Fig. 2.11*C* and *D*). This pattern of stimulation on two channels is generally referred to as **alternate** or **reciprocal mode.**

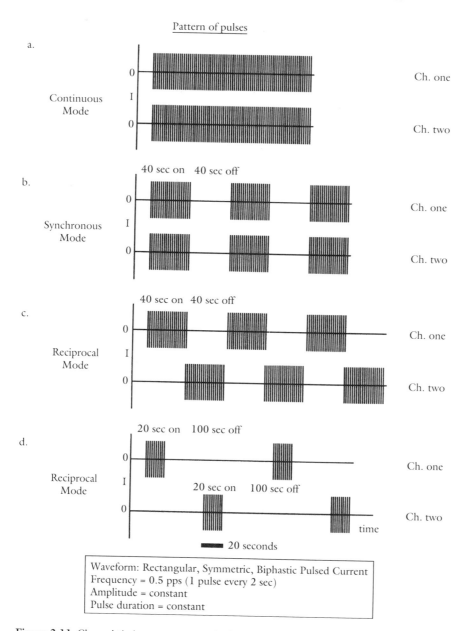

Pattern of pulses

a.

Continuous
Mode

0   Ch. one
I
0   Ch. two

b.

Synchronous
Mode

40 sec on   40 sec off

0   Ch. one
I
0   Ch. two

c.

Reciprocal
Mode

40 sec on   40 sec off

0   Ch. one
I
0   Ch. two

d.

Reciprocal
Mode

20 sec on   100 sec off

0   Ch. one
I

20 sec on   100 sec off

0   Ch. two
time

■■■ 20 seconds

Waveform: Rectangular, Symmetric, Biphastic Pulsed Current
Frequency = 0.5 pps (1 pulse every 2 sec)
Amplitude = constant
Pulse duration = constant

Figure 2.11 Channel-timing output controls showing three of the common patterns of stimulation: **A.** Continuous mode with both channels producing uninterrupted trains of pulses. **B.** Synchronous mode with both channels producing the same pattern of pulses simultaneously. **C** and **D.** Reciprocal mode with each channel producing alternating brief trains of pulses.

Finally, a few stimulators provide a stimulation pattern on two channels called **delayed mode.** In this mode, stimulation is started on one channel, and the stimulation is delayed on a second channel for periods ranging from one to several seconds. Such control over the timing of stimulation output can be of value in the electrical stimulation of skeletal muscles, when contraction in one muscle group should precede contraction in a second muscle group in order to produce a functional limb movement.

When the option to select on times and off times is included in a stimulator, each channel should be linked to a **channel output indicator.** These small lights simply inform the user about when stimulation is being applied through each channel. Channel output indicators allow the user to know at what time during a stimulation program the amplitude of stimulation may be adjusted.

The ability of the health care provider to select the timing of stimulation between two channels is a desirable feature for a stimulator because it allows a broader range of treatment applications than would otherwise be possible. The reader is cautioned that the labeling of controls that regulate the timing of two-channel stimulation has no industrywide standard. The operator's manual of a specific instrument should be consulted for the user to learn how timing patterns between two channels can be controlled.

Treatment timer

Many commercially available stimulators include a timing device, called the *treatment timer,* that allows the user to set the time period over which a pattern of stimulation will be provided for a single treatment session. Timers on many stimulators permit the adjustment of treatment times of up to 60 min long, and some include timers that extend the treatment time up to 99 min. Some timers will trigger an audible signal when the time has expired. Treatment timers automatically turn stimulation off after the selected treatment time has expired. Devices with electromechanical timers as opposed to electronic timers should be periodically checked to ensure that the timer is functioning properly.

Certain types of devices, such as portable stimulators designed for pain control, do not contain treatment timers and will run continuously until turned off or until the batteries powering the unit discharge to the point where they no longer drive the electronics.

Electrode Systems for Electrotherapy

An electrode is a conductive material that serves as the interface between a stimulator and the patient's tissues. Electrodes are connected to stimulators by insulated wires called *electrode leads, cables,* or *cords.* In most applications, electrodes are attached to the skin (surface electrodes). In other applications, electrodes have been designed to be implanted near tissues such as peripheral nerve or bone (invasive or indwelling electrodes) or in body cavities (internal electrodes). The discussion in this segment of the chapter is restricted to surface electrodes.

The material from which the electrodes are fabricated, electrode sizes and shapes, the location of the electrodes with respect to relevant tissues, and the orientation of the electrodes with respect to each other, all need to be considered in the development of an electrotherapy plan of care.

Types of surface electrodes

Stimulating electrodes used in today's conventional electrotherapy are usually made of a polymer or electrically conductive, carbon-impregnated, silicon rubber. Some electrodes provided with selected stimulators are made of metals such as stainless steel or aluminum foil. Surface-stimulating electrodes require the use of a coupling medium to provide a lower-resistance pathway for the flow of current from stimulator to tissue. For the flexible conductive rubber electrodes, the coupling medium may be an electrolytic paste, gel, cream, or liquid. Some commercially available electrodes are coated with a self-adhesive conductive polymer that serves as the coupling agent. In the case of metal electrodes, sponges soaked with tap water (not distilled water) are most commonly used to provide a pathway for current. The coupling media serve to decrease the impedance at the interface between the electrode and the skin. Skin impedance may be further reduced by proper cleaning of the skin with a mild, hypoallergenic soap or by mild skin abrasion. A number of the commonly available types of surface electrodes are illustrated in Figure 2.12.

Electrodes are available for repeated use, short-term use (2–15 days), or for single use. The flexible, conductive rubber electrodes provided by manu-

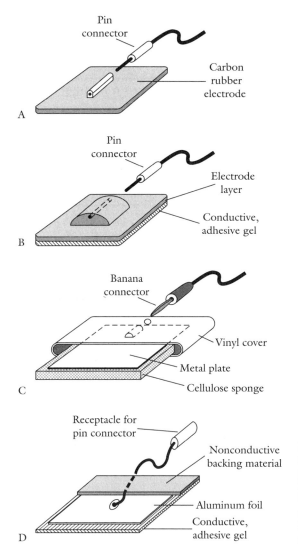

Figure 2.12 Diagrams of common surface electrodes: **A.** Simple conductive rubber electrode. **B.** Conductive rubber electrode bonded to a conductive, adhesive gel. **C.** Metal/sponge electrode. **D.** Foil/conductive adhesive electrode.

facturers with most stimulators are fabricated to provide long-term, repeated use and will generally maintain their conductive properties for many months. After a period of time, however, the conductive rubber electrodes will begin to deteriorate and will not provide uniform current flow over their entire surface. This change in an electrode's ability to uniformly conduct will be accompanied by variations in current density beneath the electrodes and may result in uncomfortably high levels of stimulation in small spots (hot spots), often perceived by patients as a burning sensation. For this reason, flexible, conductive rubber electrodes should be replaced regularly (e.g., every three to six months). To maintain carbon rubber or similar electrodes, they should be cleaned with a mild soap solution after every application.

Flexible conductive rubber/gel and metal/sponge electrode systems should be securely attached to the skin with elastic straps to ensure uniform contact with the skin. Some manufacturers provide custom-cut adhesive bandages to secure carbon rubber electrodes. For active users of electrotherapy, the adhesive bandages may provide better long-term stabilization of the electrodes.

Disposable electrodes available commercially are designed for short-term use on a single patient. Disposable electrodes are often fabricated with a thin carbon rubber or foil layer and a conductive medium, which acts as the coupling agent. In some cases, the gel also acts as an adhesive, and in other cases, an adhesive bandage surrounds the electrode and secures it to the skin. The foil/adhesive gel electrodes have electrical resistances on the order of 10 to 100 Ω, whereas carbon rubber/gel systems have resistance values of approximately 1000 Ω. Some self-adhering electrodes designed for single or short-term use have impedances in excess of several thousand ohms (3).

Different commercially available electrodes made from what appear to be similar materials may not be similar with respect to their ability to conduct (3). Unfortunately, the significance of these findings for clinical electrotherapeutic applications remains to be determined.

Significance of size and shape of surface electrodes

The required contact area of the stimulating electrodes depends in part on the area of excitable tissues to be stimulated. An electrode that is too large or of the wrong shape may cause the current to spread to excitable structures other than the nerve or muscle of interest. Electrodes are available in a wide range of sizes and shapes in order to accommodate the spectrum of electrotherapeutic procedures common to contemporary practice.

With uniform electrode conductivity, the *current density* (amount of current per unit of conduction area) is inversely proportional to the electrode contact area. Therefore, as electrode contact area decreases, current density increases. This means that if the same electrical voltage is applied first across a pair of small electrodes and then across a pair of large electrodes, the amplitude of stimulation will feel greater beneath the smaller pair. If one small and one large electrode are used in a single application, the stimulation will generally be perceived as greater beneath the small electrode.

In the documentation of electrotherapeutic procedures, it is recommended that the type, size, and shape of electrodes be clearly specified and that changes made in any of these electrode characteristics also be noted in the medical record.

Designation of electrode location in therapeutic applications

The placement of electrodes in electrotherapy is critical to achieving benefits. For this reason, the exact location should be clearly recorded. To accurately describe electrode placement, measurements should be made from the center of electrodes to known anatomical landmarks or structures (e.g., 5 cm proximal to the superior edge of the patella over the rectus femoris or 10 cm distal to the axillary fold in the midaxillary line). These measurements should be recorded after initial treatments, and changes in electrode position should be documented in a similar fashion in the medical record. This quantitative approach is the most clear method to describe electrode location.

Historically, electrode placements have been described in relation to the relative position of the electrodes employed. Three terms—*monopolar, bipolar, and quadripolar*—have been regularly used to describe electrode orientations (4).

In the simplest **monopolar electrode orientation,** a single electrode is placed over the "target area" or over the tissue where the greatest effect is desired (Fig. 2.13*A*). The second electrode of the stimulating circuit is placed at some distance well away from the target area in order to complete the electrical circuit. The electrode placed in the target area is sometimes called the *stimulating*, or *active*, electrode, whereas the electrode placed distant from the target area is often referred to as the *dispersive, indifferent*, or *reference* electrode. Referring to the second electrode, away from the target area, as the "ground electrode" is inappropriate because this electrode is not actually connected to earth ground. In the monopolar orientation, the electrode in the target region is generally smaller than the electrode away from the target area. As a result, the smaller "stimulating" electrode commonly elicits activity in excitable tissue before the larger "dispersive" electrode does because the current density beneath the smaller electrode is greater at any particular amplitude of stimulation. The use of the term *active* to describe the smaller electrode is not recommended because it implies to some that the larger electrode is "inactive" and will not elicit any type of physiologic response. Those who have regularly used electrical stimulation in the clinic will attest that this implication is not always true. The large contact area of the larger electrode minimizes the underlying current density, and hence excitation of nerve and muscle by this electrode is less likely to occur. Referring to the large electrode as either "indifferent" or "reference" is also misleading and has no particular value.

In some situations, the single electrode placed in the target region is replaced by two electrodes connected to the same single lead (Figure 2.13*B*). Even though this electrode-placement configuration uses two electrodes in the target area, both are connected to a single output lead and hence this pattern of placement is also referred to as *monopolar.*

In the simplest **bipolar electrode orientation,** both surface electrodes from a stimulator channel are placed over the target area (Fig. 2.14*A*). This

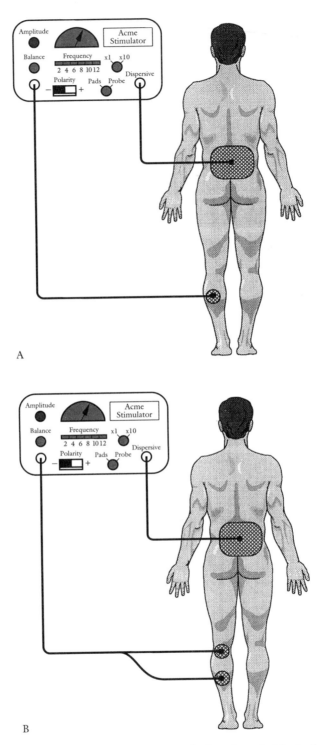

Figure 2.13 Monopolar electrode configurations: **A.** One electrode (the "active" electrode) in the target region (calf) and a second electrode away from target region. **B.** Two electrodes bifurcated from one channel lead (pole) in the target region (posterior calf) and second electrode away from the target region.

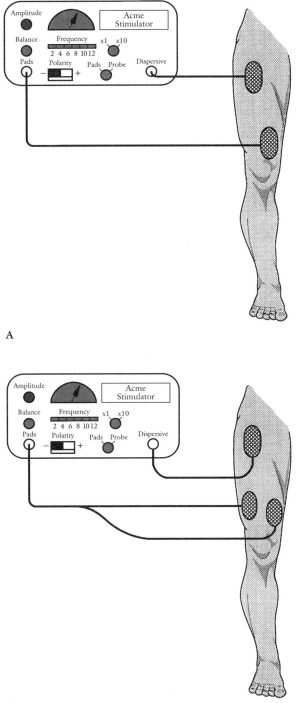

Figure 2.14 Bipolar electrode configurations: **A.** Simple bipolar arrangement with both electrodes from a single stimulator channel in the target region (anterior thigh). **B.** One channel lead bifurcated with two electrodes over lower target region and second lead of the channel with single electrode in target region.

pattern of electrode placement provides current that is more limited to the excitable tissue of interest. The electrodes used in the bipolar technique are usually equal in size. In such cases, the relative ability of each electrode to activate nerve or muscle will be equal when symmetric biphasic waveforms are applied. In some instances, the bipolar technique may require that one of the two electrodes be smaller in order to obtain the desired response; in such a case, the smaller of the two electrodes will be relatively more effective in the activation of excitable tissue.

Another bipolar configuration of electrode placement that uses more than two electrodes originating from a single channel is shown in Figure 2.14B.

In the **quadripolar electrode configuration,** two electrodes from two separate stimulating circuits are positioned in the primary target area. Two general quadripolar configurations of electrode placement are shown in Figure 2.15. In Figure 2.15A, the four electrodes are positioned such that currents induced by the respective circuits intersect, interact, or interfere with each other. This type of electrode placement has been used in conjunction with the interferential stimulation technique described later in this chapter. In Figure 2.15B, two electrodes from each of two stimulation channels are placed in the target region (low back) in a manner that would not result in an intersection of the currents induced by each channel. This electrode orientation from two channels, shown in Figure 2.15B, might be described as bipolar for channel 1 in the right low back and bipolar for channel 2 in the left low back. In describing electrode locations, the term *quadripolar* simply refers to the placement of four electrodes from two stimulation channels in the target area, whereas the term *interferential* refers to a particular relative orientation of the two sets of electrodes that employs the quadripolar electrode configuration.

Since the terms *monopolar, bipolar, and quadripolar* can each be used to represent various combinations of electrode placement, these descriptive terms should be accompanied by an explicit description of each electrode's location, and the electrode types and dimensions should be clearly specified.

Special-purpose electrodes

Handheld Probe Electrode

Although simple surface electrodes are the most commonly used electrodes in clinical electrotherapy, a variety of special-purpose electrodes are available. One of the more frequently used special electrodes is the **handheld probe electrode** (Fig. 2.16). Modern versions of these electrodes generally have two controls included in their design: an amplitude control and a switch to control the timing of stimulation. Other parameters of treatment are set on the stimulator to which the electrode is attached. Handheld electrodes were originally developed for use in classical electrophysiologic tests such as reaction of degeneration (RD) or strength-duration (S-D) testing. Handheld probe electrodes are now commonly used for locating motor points, for the stimulation of very small muscles, or for the stimulation of very small points such as in electroacupuncture.

Electrodes for Iontophoresis

In traditional iontophoresis (the electrophoresis of charged medications), the treatment electrodes used were either aluminum or tin foils placed over

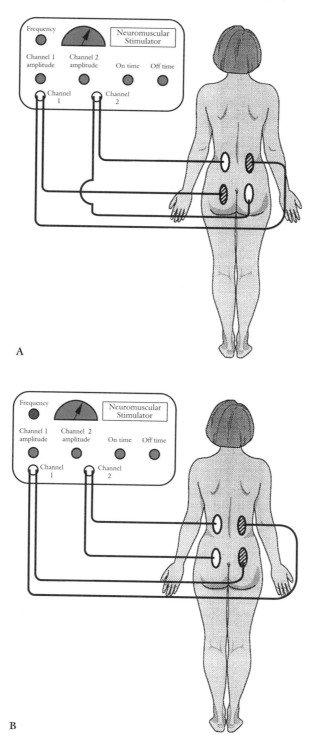

Figure 2.15 Quadripolar electrode configurations in low back target region; **A.** four electrodes from two stimulation channels arranged such that currents produced by channel 1 (shaded electrodes) intersect with currents produced by electrodes of channel 2. **B.** Four electrodes from two channels arranged such that currents produced by the channel 1 electrodes (shaded) do not intersect with currents produced by channel two electrodes.

an absorbent pad (Fig. 2.17*A*). The pad contained either an aqueous solution of the medication or a simple electrolytic coupling agent. When the pad contained only the coupling agent, the medication was placed on the skin surface beneath the pad and metal electrode. This traditional iontophoresis electrode system was often difficult to use and even contact of the electrode with the skin surface was hard to achieve. In recent years, manufacturers of stimulators designed for iontophoresis have developed special electrodes that hold or store the medication to be used. The three basic designs provide either *(a)* a plastic reservoir for medication injection (Fig. 2.17*B*), *(b)* the incorporation of the medication into the conductive gel adhesive (Fig. 2.17*C*), or *(c)* an absorbent reservoir pad (Fig. 2.17*C*).

Commercially available iontophoresis electrodes come in a number of sizes and shapes. Iontophoresis electrodes may be designed for single- or multiple-session use. Medication-containing iontophoresis electrodes should not be used on multiple patients.

Electrodes for Internal Applications

A number of unique electrodes have been designed for electrotherapy requiring stimulation within body cavities. Rectal and vaginal electrodes have been developed for the electrical activation of musculature associated with the control of urination, defecation, and ejaculation. Other electrodes have

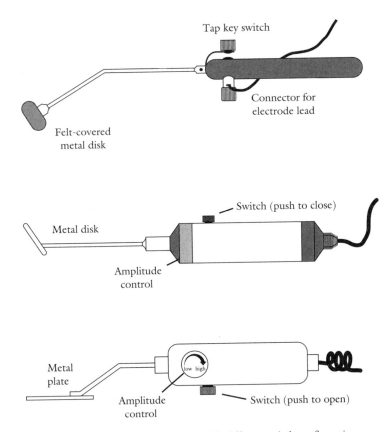

Figure 2.16 Handheld electrodes with different switch configurations.

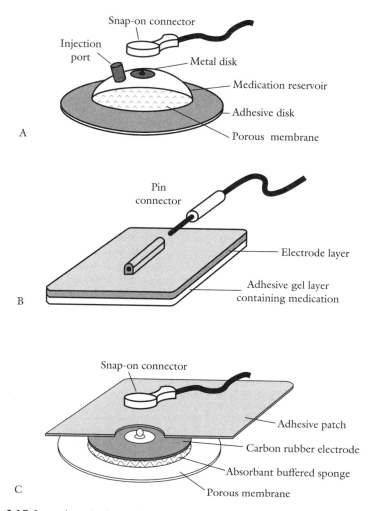

Figure 2.17 Iontophoresis electrodes: **A.** Reservoir electrode (no longer commercially available). **B.** Conductive rubber electrode with medication contained in adhesive gel. **C.** Buffered sponge electrode with medication contained within sponge.

been designed for intraoral stimulation programs for the management of disorders such as temporomandibular joint syndrome. Principles and procedures related to the application of these special electrodes are addressed in subsequent chapters.

Types of Stimulators

Electrical stimulators for therapeutic and diagnostic applications can be generically described by their external power source (battery versus line power) and by the nature of electrical stimulus that is applied to the patient (constant current versus constant voltage). In contrast, commercially available stimulators are usually categorized based on the characteristics of the stimulus waveform or some characteristic related to a stimulation technique associated with the particular stimulator.

Constant current versus constant voltage stimulators

Constant (or regulated) current instruments provide current that flows at a constant amplitude within a specified range of impedances. By Ohm's law, the voltage output of the device varies to maintain the current at a constant level as the tissue impedance changes. Constant (or regulated) voltage instruments provide a constant amplitude voltage within a specified range of impedances. The current flow varies inversely with impedance to maintain the voltage output of the device.

Each type of stimulator has potential advantages and disadvantages in clinical applications. During treatment at a particular amplitude setting, constant current stimulators will automatically reduce the driving voltage when electrode contact improves or when electrical transmission increases, thereby maintaining the desired level of stimulation. In cases where electrode contact or transmission is reduced, constant current stimulators will automatically increase the driving voltage, which may result in skin burns as current density rises to very high levels. For this reason, well-designed constant current stimulators often feature a voltage limit, which cannot be exceeded regardless of an electrode/tissue impedance increase.

Constant voltage stimulators have advantages and disadvantages similar to those for constant current devices. In cases where electrode contact or electrical transmission is reduced, constant voltage stimulators will automatically reduce the current produced and thereby lessen the chance of skin burns resulting from increased current density. Conversely, if electrode/tissue contact improves and transmission is better with a constant voltage device, induced currents could significantly increase, resulting in an undesirably high level of stimulation.

Commercially available stimulators

TENS Devices

The resurgence of interest in electrotherapy occurred in the early 1970s with the widespread development and marketing of small (about $2.5 \times 4 \times 1$ inches), lightweight (< 200 g), portable stimulators called **TENS (transcutaneous electrical nerve stimulation)** units. The development of these compact stimulators was made possible by two developments—the postulation of the gate-control theory of pain control and the miniaturization of electronic components. Also contributing to TENS development was the ever-increasing incidence of pain related to cumulative trauma disorders (e.g., carpal tunnel syndrome) and the searching done by health professionals for pain control solutions.

Most devices marketed specifically for pain control applications have common features related to output currents and controls. In general, TENS devices produce a rectangular, asymmetric, biphasic pulsed current. In some early models, the output waveforms were unbalanced in phase charge, but today most stimulation waveforms are balanced.

Today's TENS devices are typically two-channel stimulators with independent amplitude controls for each channel. Pulse duration (also called *pulse width*) controls are routinely present and allow the user to vary the pulse duration from low values (20–50 μsec) to high values (250–600 μsec). A pulse frequency (also called *rate*) control is standard for TENS units and allows adjustment from a low frequency of 2 pulses per second (pps) to maximum frequencies ranging from 125 to 200 pps. The pulse duration and pulse frequency are most often the same value on each output channel. Recommendations regarding the stimulation characteristics of TENS devices can be found in the American National Standard for Transcutaneous Electrical Nerve Stimulators, published by the Association for the Advancement of Medical Instrumentation (AAMI) (5).

The feature that distinguishes TENS devices from many other types is the presence of several predetermined stimulation modulation options (Fig. 2.18) These may include systematic modulations to the pulse frequency, pulse duration, both pulse frequency and pulse duration, burst modulation, and amplitude modulation. In each of these modulation modes, the stimulus parameter(s) modulated is systematically lowered from and raised to the

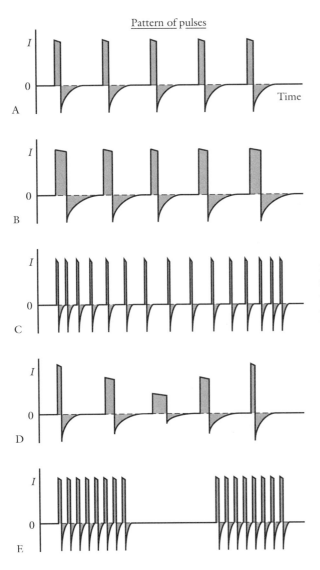

Figure 2.18 Patterns of stimulation commonly produced by TENS devices: A. Normal mode of continuous train of pulses. B. Pulse duration mode with automatic modulation of pulse duration to preset level. C. Frequency modulation with automatic increase and decrease in frequency to preset level. D. Strength duration modulation with automatic reductions of pulse amplitude and increases in pulse duration followed by increases in pulse amplitude accompanied by decreases in pulse duration. E. Burst modulation with regular bursts of pulses.

maximum setting. Modulation modes were included in the design of TENS devices when clinicians recognized that for certain patterns of stimulation (low amplitude, continuous pulse trains) patients quickly lost the ability to perceive the stimulation. By providing modulation to one or more parameters of stimulation, this reduction in the perception of stimulation is often avoided.

Although the acronym *TENS* is presently associated with electrical stimulation for pain control, the reader should be aware that most contemporary electrotherapy for problems other than pain involves transcutaneous stimulation of peripheral nerve fibers. Many clinicians when faced with the management of pain in a patient preferentially choose TENS devices even though many other types of clinical stimulators may provide a pattern of stimulation analogous to those generated by the commercially available, portable TENS devices.

High-Voltage Pulsed Current Stimulators

Following the rapid introduction of TENS devices, in the mid1970s a second commercial class of constant voltage stimulators began to appear in large numbers in clinics across the United States. This class, the high-voltage pulsed galvanic stimulators (also called *high-volt pulsed current* [HVPC] stimulators) were actually developed in the mid-1940s; they generate a twin-spike, monophasic pulsed current waveform (Fig. 2.19*A*) with peak spike ampli-

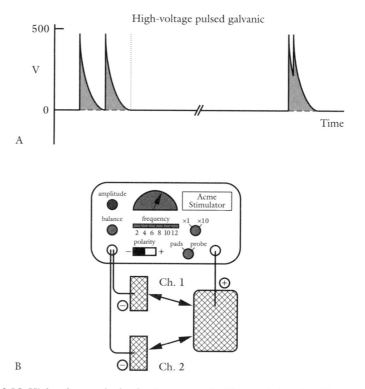

Figure 2.19 High-voltage pulsed galvanic current. **A.** Characteristic twin-spike monophasic waveform with longer and shorter interval between spikes respectively. **B.** Schematic diagram of a typical configuration of output channels and electrodes associated with some commercially available stimulators.

tudes of up to 500 V and pulse durations of about 50–200 μsec at frequencies ranging from 1 to approximately 120 twin-spike pulses per second. In early models, the ranges of available settings for on time and off time were often limited. Designs have been changed over the years so that today many HVPC stimulators offer much more flexibility in the selection of treatment parameters. Since these stimulators produce monophasic waveforms, the output polarity of electrode leads does not change during stimulation. Most HVPC devices allow the user to select and manually switch the polarity of the output leads.

The typical twin-spike, monophasic pulsed current stimulator includes a control that adjusts the time between the beginning of the first spike waveform and the beginning of the second waveform. This time between the two spikes may be reduced such that the two waveforms overlap. During stimulation, as the twin spikes progressively overlap, the stimulation is perceived as being stronger.

One characteristic of "high-volt" stimulators is the design of the output channel leads. On many of the stimulators in this class, two output leads at one polarity are each used in conjunction with a single lead of opposite polarity (Fig. 2.19B). Such an arrangement of output leads actually reflects the presence of a single channel of stimulation. However, some HVPC devices that employ this lead arrangement are labeled as having two channels.

High-voltage pulsed stimulators are now available in both line-operated and battery-operated models, with the major difference being that the peak output voltages are about 100–200 V lower in the battery-operated designs.

Neuromuscular Electrical Stimulators

In the latter 1970s after the rapid proliferation of TENS units for pain control, interest in electrotherapy was boosted by research reports from the Soviet Union that suggested that regular electrical activation of muscle was more effective than exercise in strengthening skeletal muscle in elite athletes. This research resulted in improvements in the development and design of a class of electrical stimulators for **neuromuscular electrical stimulation (NMES).**

Both pulsed and alternating currents are used in NMES devices. The current originally used by the Soviet researchers was a 2500-Hz, sinusoidal, symmetric alternating current that was burst-modulated every 10 msec to provide 50 bursts per second. This form of stimulation has been promoted commercially as "Russian stimulation" (Fig. 2.20A). Today's NMES devices use a variety of waveforms. Most stimulators marketed for NMES produce either *(a)* the burst-modulated, 2500-Hz AC just described, *(b)* a rectangular, balanced, symmetric, biphasic pulsed current (Fig. 2.20B), or *(c)* a rectangular, balanced, asymmetric biphasic pulsed current (Fig. 2.20C). To date, no single waveform has been found to be superior for all NMES applications in all patient populations. Research has demonstrated that individuals may have a preference for particular waveforms used in NMES devices; selection of the waveform should be based on the ability to evoke the desired level of contraction as well as patient tolerance of the procedure.

Both portable, battery-operated units and line-powered stimulators are available for NMES applications. In general, portable NMES units have lower maximum power output than line-powered ones. For this reason, portable

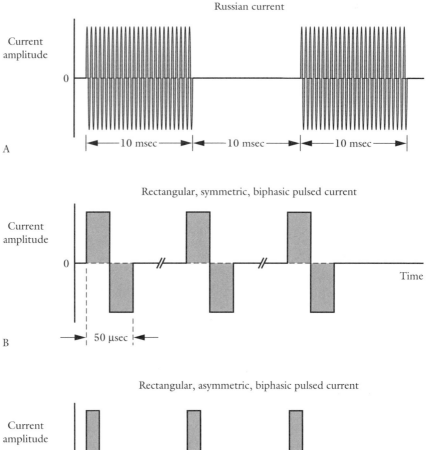

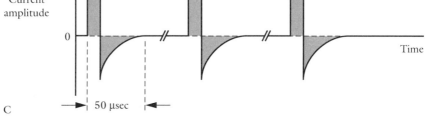

Figure 2.20 Examples of currents used with commercially available stimulators designed for neuromuscular electrical stimulation applications.

stimulators may not have enough capacity to maximally activate large muscle groups such as the quadriceps. In addition, portable NMES stimulators, like portable TENS devices, have performance characteristics that are limited by battery life; therefore, batteries should be replaced or recharged frequently. Portable stimulators have a distinct advantage over line-powered models in applications where stimulation evokes functional muscle contraction, such as stimulation of the tibialis anterior and peroneal muscles during the swing phase of gate.

NMES devices usually have two output channels, but some models are available with four channels. Stimulators designed for activation of normally innervated muscle usually have independent amplitude controls for each output channel. Controls are commonly available to provide stimulation on all channels simultaneously (synchronous mode), alternately (reciprocal mode), or such that one or two channels are providing stimulation while the other output channels are off and vice versa. Some NMES units have a "delayed mode," which initiates stimulation on one channel and then begins stimulation on a second channel after a brief delay of from one to several seconds. Usually the output frequency and phase (or pulse) duration controls are common to each channel. Frequency modulation and phase (or pulse) duration modulation are generally not included in NMES stimulators.

The three other features that distinguish stimulators designed for NMES are on time/off time controls, ramp controls, and "Set output" controls. On time/off time controls allow the user to control the number of seconds of muscle contraction (on time) and the number of seconds of rest (off time). The ramp control adjusts the number of seconds of the on time over which the amplitude gradually reaches the maximum setting (ramp up) or decreases from the maximum setting (ramp down) to zero output. Alternatively, the timing of stimulation can be controlled on many NMES units by a manually operated remote switch (trigger). Such switches are particularly useful in cases where a functional muscle contraction is desired. *Set output* controls allow the user to adjust the amplitude of stimulation to the desired level before switching the device into an on time/off time sequence.

Most line-powered NMES units contain analog or digital meters that reflect some aspect of the amplitude of stimulation. Metering should be available for each output channel and should reflect the actual current applied during stimulation (delivered current). Observation of the meters during the application of stimulation allows the user to know when stimulation begins on an output channel and when stimulation has reached the maximum set value. This information is valuable since the output amplitude should not be adjusted during the ramp up or during the off time especially when the maximum tolerated levels of stimulation are being reached. On portable NMES units, no output amplitude meter is present. Instead, small lights are used to indicate when stimulation begins and ends. When using such portable NMES instruments, the user must estimate when peak amplitudes of stimulation have been reached before further amplitude adjustment is made. Some NMES units will include both "Channel on" lights as well as output metering.

A final feature common to well-designed NMES devices is the safety switch, which allows the stimulator to be shut off in the event of an unexpected response to stimulation. Often such switches are provided to patients during treatment because they are likely to become aware of an unwanted response well before the clinician who is operating the device.

Recommendations on performance standards and safety requirements for neuromuscular electrical stimulators are being developed by the Association for the Advancement of Medical Instrumentation (Neuromuscular Electrical Stimulator and Electrical Muscle Stimulator Devices, AAMI, 3330 Washington Boulevard, Suite 400, Arlington, VA, 22201-4598).

Interferential Stimulators

In the early-to-mid-1980s, another class of stimulators was introduced to the American market, the interferential stimulators. These stimulators are generally characterized as two-channel stimulators producing sinusoidal, symmetrical, alternating currents at frequencies of several thousand cycles per second (e.g., 2000–5000 Hz) on each channel. In clinical applications of the original versions of these devices, both channels are used simultaneously with AC frequencies set at slightly different values and electrodes are oriented in a quadripolar orientation as shown in Figure 2.15*A*. In this or similar electrode placements, the ionic movement in the tissues results from the electromotive forces produced by both channels. In other words, the current that would be produced by one channel interacts with ("interferes with") the current produced by the second channel and produces a net ionic movement different from that produced by either channel alone. Proponents of this form of electrotherapy propose that the resultant current (called an *interference current*) is similar to an amplitude-modulated, sinusoidal alternating current with *beat*, frequency, which is equal to the difference in AC frequency on the two chan-

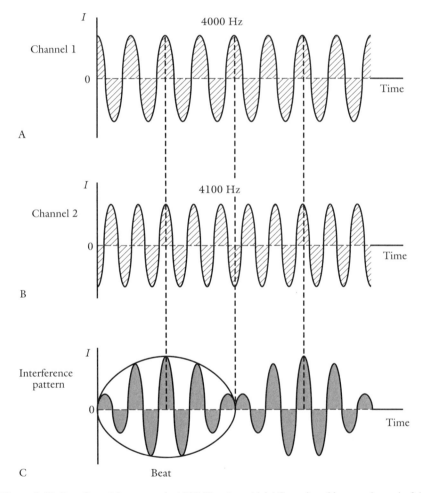

Figure 2.21 Interferential current: **A.** 4000-Hz, sinusoidal AC produced by one channel of the stimulator. **B.** 4100-Hz sinusoidal AC produced by second channel of stimulator. **C.** Amplitude-modulated, sinusoidal AC produced by the "summation" of currents from channels 1 and 2 at point of intersection given a homogeneous medium.

nels (Fig. 2.21) (6). Such summation of effects to produce an interference current assumes a perfectly homogeneous conductor. Human tissues are not homogeneous in conduction of electrical currents, and the proposed summation of effects across two channels is unlikely to occur precisely as described. Since the frequencies of stimulation using the interferential technique are an order of magnitude higher than those used with pulsed current procedures, proponents claim that tissue impedance is reduced and that currents are induced in deeper tissue levels than those induced by devices generating much lower frequencies. Such claims have been challenged in the literature (7), and the relative effectiveness of interferential stimulation remains to be demonstrated in the clinical and scientific literature. Nonetheless, interferential therapy has gained a solid foothold in clinics for the management of a wide range of disorders.

An alternative approach to the design of interferential stimulators is to use two channels of stimulation and to have the AC carrier frequency on each channel be the same (e.g., 5000 Hz) and in phase. This carrier frequency is then burst-modulated (1–250 bps). In addition, the output of each channel may then be amplitude-modulated. This approach is referred to as *full-field technique* and is purported to increase the volume of tissue exposed to the induced currents.

The types of controls on interferential stimulators vary. Amplitude controls range from a single dial for controlling both channels to independent controls, on each channel. In general, controls are available to select the difference in frequency between the two channels (the beat, or interference, frequency). On many units, options are available to continuously modulate the beat frequency. Most instruments contain a metering system for output amplitude. Early interferential stimulator designs often did not incorporate on time and off time controls, which meant that such devices would not be ideal for NMES applications. Recent interferential stimulator models have included on time and off time controls and ramp modulation controls to make them more suitable for NMES applications.

Diadynamic Stimulators

Stimulators classified as diadynamic devices are characterized by the production of waveforms derived from the electronic processing of sinusoidal, symmetric, alternating current. The two main types of processing of the AC currents are half-wave rectification and full-wave rectification. Figure 2.22*A* illustrates the source AC and the alterations produced by half- and full-wave rectification. Briefly, half-wave rectification simply eliminates the second half of each AC cycle and produces a monophasic pulsed current with a pulse duration equal to the interpulse interval and a frequency equal to that of the original AC. This form of diadynamic current (Figure 2.22*B*) has been traditionally referred to as *single phase* or *monophasé fixe (MF)*. Full-wave rectification of sinusoidal AC produces a monophasic pulsed current with no interpulse interval and at twice the original AC frequency. This form of current (Figure 2.22*C*) has been called *double-phase* or *diphasé fixe (DF)*. Devices that produce these two basic forms of monophasic pulsed currents also amplitude- and/or time-modulate the basic waveforms to produce the two other types of diadynamic currents shown in Figure 2.22, *D* and *E*. Diadynamic stimulators may be unfamiliar to many readers in the United States,

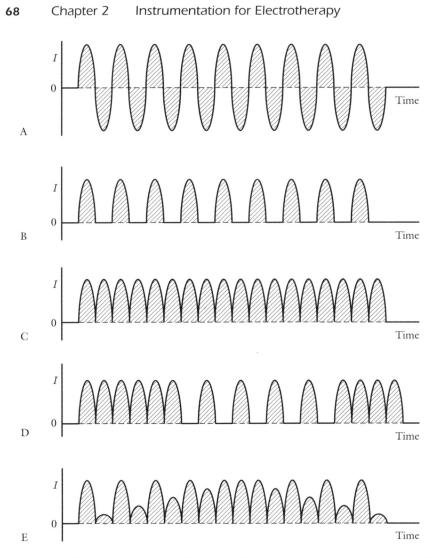

Figure 2.22 Diadynamic current: **A.** Sinusoidal AC, which serves as precursor current. **B.** Monophasé fixe, from half-wave rectification of source current. **C.** Diphasé fixe, from full-wave rectification of source current. **D.** Courtes periodes, from alternating periods of full-wave rectification followed by half-wave rectification. **E.** Longues periodes, amplitude modulation of full-wave rectified source current.

but in fact these stimulators have been in use in Canada and in European countries for quite some time.

Iontophoresis Stimulators

Iontophoresis, the delivery of medication to subcutaneous tissue by electrical currents, has been used clinically for decades. The devices used for this procedure produce direct current (DC). The fundamental stimulator features required to use this technique are an amplitude control, output current ammeter, and power switch. Historically, iontophoresis has been administered using line-powered electrical stimulators that converted the AC power to DC. Very often these devices were designed to perform traditional electrodiagnostic procedures and could produce peak DC currents of 30 mA or more. Since iontophoresis treatments use no more than 5 mA, these devices were not particularly well suited for this electrotherapeutic application.

In recent years, a number of manufacturers have developed portable, battery-operated stimulators designed solely for iontophoresis treatments. These devices are usually powered by a 9-V battery and have maximum output settings of 5 mA. Most are equipped with an accurate ammeter, which provides the user with a reliable indication of the amplitude of the current used on each output channel. Other features include a treatment timer and a low-battery indicator, which alerts the user to replace the battery. Some models are designed with a control system that allows the user to select the dosage of current expressed in milliampere-minutes. This feature may represent an improvement over previous designs that simply included an ammeter to detect the magnitude of current, since the expression of dosage in milliampere-minutes reflects the total charge transfer during treatment. Additionally, new iontophoresis devices also include a sensing circuit that continuously monitors the output circuit impedance and activates an alarm when electrode contact is inadequate to achieve ion transfer. Some iontophoresis units have a switch that allows the polarity (positive or negative) to be chosen for the lead connected to the electrode containing the charged medication. The relatively new iontophoresis devices are designed to be used with the specially designed drug containment electrodes described previously.

Microcurrent Stimulators

The most recent class of stimulators to appear on the commercial market are referred to as *microcurrent stimulators.* As the name indicates, these devices produce very low amplitude currents (8). The output of this class of instruments is so small that few are capable of producing 1 mA of total current. For this reason, the electrical output of such constant current devices is not generally great enough to activate electrically excitable tissues—nerve and muscle. The normal mode of application of the microcurrent devices is at levels that do not activate even cutaneous sensory nerve fibers, and as a result, patients do not have any perception of the tingling sensation so commonly associated with electrotherapeutic procedures. This form of electrotherapy has been referred to as *subliminal stimulation.* No industry-wide standard has been developed for what types of currents are produced by devices marketed in this class. Figure 2.23 shows an example of the output waveform produced by some microcurrent devices. The individual waveforms are characteristically rectangular, monophasic pulses that periodically reverse polarity. A number of parameter controls related to the delivery of these pulse patterns are typically included. Amplitude controls allow the adjustment of peak amplitude from 0 to 600 μA. Some microcurrent units provide for an automatic amplitude ramp for the series of pulses delivered. Frequency controls generally allow one to set the frequency of monophasic pulses from 0.1 pps to approximately 1000 pps. One or two channels of stimulation are commonly available on either line-powered or battery-operated units.

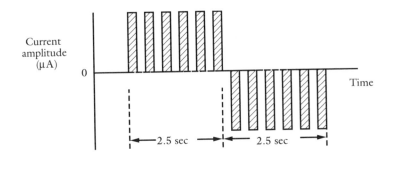

Pulse frequency: 1 – 990 pps

Figure 2.23 Example of one type of current produced by some microcurrent stimulators: rectangular, monophasic pulsed current with regular reversal of electrode polarity.

Point Stimulators

Point stimulators are devices designed to provide noxious-level stimulation to trigger points or acupuncture points. These stimulators typically have one output channel with one lead connected to a conventional surface electrode and the second lead connected to a handheld electrode with a small ($\approx$ 1 mm^2) metal tip. Devices in this class consist of two primary components—a stimulator and a point locator. These may produce either rectangular, monophasic pulsed current or rectangular, symmetrical, biphasic pulsed current. Controls are usually available for the adjustment of pulse amplitude, phase duration, and frequency of pulses. The upper range of phase duration is in excess of 100 msec, a range that is much greater than that found on most other classes of stimulators; the range enhances the ability to produce noxious stimulation through the point electrode. Treatment timers on some models have limited ranges ($\leq$ 1 min) because treatment times associated with the stimulation of individual trigger or acupuncture points are short.

The point locator associated with point stimulators is a metering system that is used to identify points on the skin with either high conductance or low resistance—a finding thought to be characteristic of trigger points and acupuncture points.

Safety Considerations for Electrotherapy

Electrical hazards may develop with electrotherapeutic stimulators or may occur when these devices are not used in a safe manner; stimulators can be hazardous to both patients and providers (9, 10, 11). This segment of the chapter describes common types of electrical hazards that may be associated with the application of electrotherapeutic devices, and it outlines some steps to minimize these hazards.

Sources of electrical hazards

Electrical hazards occur when a person comes into contact with a current-carrying conductor(s). A current-carrying conductor is usually thought of as the metal wires that supply electrical power to home or worksite receptacles or the wires within the power-supply cords to electrical devices. Breaks in the insulation on such cables or wires through misuse, normal wear and tear, or deterioration with age expose the conductor. The contact of body tissues with such bare wires may result in shocks with potentially serious consequences.

Less-obvious types of potential current-carrying conductors include the metal cases, or other external conductive components on line-powered instruments, that can carry current as a result of contact (electrical short) with internal conductors, incorrect electrical connections within the device, or other factors. Such currents are commonly referred to as leakage currents. Low levels of leakage current (< 1 mA) are routinely found on most line-powered instruments. Normally this current does not present a problem because instrument cases are connected to a wire (safety ground) that provides a low-resistance pathway to ground (Fig. 2.24A). If this pathway does not exist, as when using a line-powered device with a two-prong plug (or "cheater" adapter) or if the wire to the safety ground is broken for any reason, leakage currents can flow to ground through an individual who is touching the device—resulting in electrical shock (Fig. 2.24B). Further, care should be taken when an individual is connected to two pieces of equipment simultaneously. If the ground connections in the two devices are separate, current may flow from the higher- to the lower-voltage machine through the patient—also producing a potentially harmful shock.

The passage of current to ground through pathways other than the power-supply circuit is also called a *ground fault*. An additional approach used to protect against the development of significant ground faults or leakage currents is the use of electrical devices called **ground fault interrupters** (GFIs, or sometimes ground fault circuit interrupters, GFCIs). A GFI monitors the amount of current going to and returning from a line-powered device. When the amounts of outgoing and returning currents differ by more than 3–5 mA, the GFI opens a switch and instantly cuts the current to the device. GFIs are designed to trip in 1/40th of a second. There are two types of GFIs in power-supply circuits (Fig. 2.25): a circuit breaker GFI, located in a circuit breaker box, and a receptacle GFI, located within a wall outlet. It is important to note that ordinary circuit breakers are not suitable substitutes for GFI devices on power-supply circuits to areas where line-powered stimulators are used. This has particular significance in those situations where line-powered devices are used in home treatment. Battery-operated stimulators may be an appropriate alternative for many home treatment programs unless a receptacle GFI is properly installed.

Electrical hazards may also be the result of a misapplication of therapeutic currents. Whether an electrical hazard exists depends on the magnitude of the available current. In physiological systems, high current levels cause damage regardless of the particular type of current.

Potential adverse effects of 60-Hz alternating current

Currents as low as 1 mA RMS (60-Hz AC applied for 1 sec) are perceived as a tingling sensation. The same current applied for 1 sec at 16 mA RMS causes muscles to contract so strongly that an individual cannot overcome electrically induced contraction through the volitional contraction of antagonistic muscle

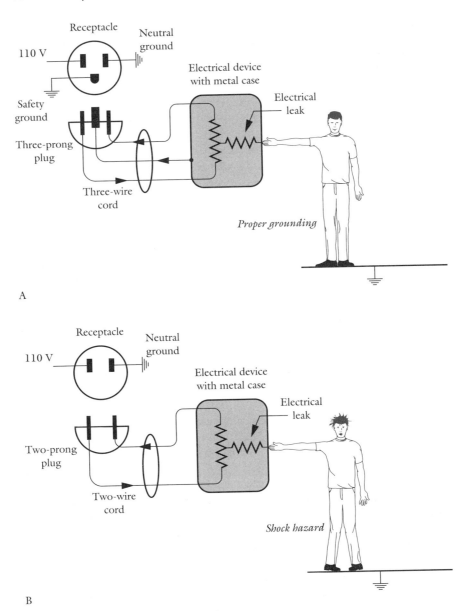

Figure 2.24 A. Pathway to ground of a leakage current in a device with a *three-prong plug*. **B**. Pathway to ground of a leakage current in a device with a *two-prong plug*.

groups. Higher levels of electricity can cause tissue damage, respiratory arrest, and cardiac arrest. More than 80 mA RMS of current can cause ventricular fibrillation as well as skeletal muscle contractions that are so rapid and forceful that involuntary jerking can pull the person away from the electrical contact. A startling reaction caused by higher-level currents can result in secondary accidents (e.g., falling).

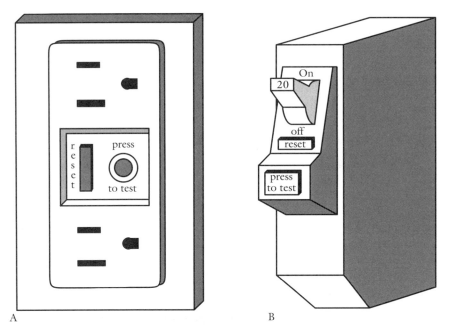

Figure 2.25 Ground fault–interrupt devices: **A.** GFI outlet. **B.** GFI circuit breaker.

Potential adverse effects of direct current at electrotherapeutic levels

Tissue damage, especially to the skin, can be caused by clinical uses of direct current even at low amplitudes. Skin damage may result from electrolytic reactions that occur when DC passes through the skin. Acidic reactions can occur beneath the positive electrode (anode), and alkaline reactions can occur beneath the negative electrode (cathode) used in DC treatment procedures (12). If the buildup of acids or alkaline by-products is sufficient, skin damage in the form of blistering burns may result. Additionally, coagulation of microcirculation (capillary circulation) may occur beneath the anode of a DC circuit and result in tissue necrosis secondary to ischemia. Although DC clinical applications to the skin have the potential to raise tissue temperature, adverse skin responses are not thought to be thermal burns.

The extent of damage that can result from DC is determined by the amplitude of current applied, the time over which the current is allowed to flow, and the tissue impedance. For these reasons, DC electrotherapeutic procedures are generally limited to applications of current at 5.0 mA or less for periods of 15 minutes or less, and steps are taken to reduce skin impedance prior to treatment. Electrode sizes are selected so that current densities do not exceed 0.1 to 0.5 mA per square centimeter of electrode surface area. For these treatment parameters, the patient should not report any sensation beneath the electrodes. A report of a tingling sensation as continuous DC is applied should signal the health care provider to reduce the amplitude and reexamine the treatment setup to insure uniform electrode contact. Close observation of the skin is required in each patient receiving DC stimulation in order to avoid adverse reactions. This inspection is particularly important due to the anesthetic effect that may occur beneath the electrodes.

Although some forms of pulsed monophasic currents have been reported to have chemical effects similar to those of DC, very short duration, monophasic pulsed currents do not apparently produce harmful electrochemical effects on the skin (12).

Steps to minimize electrical shock hazards

Health care providers using electrical stimulators are responsible for ensuring that such devices do not harm their patients. This segment of the chapter addresses safety considerations relating to three fundamental questions. What are the safety considerations related to the *equipment* chosen for a clinical procedure? What are the safety considerations with respect to the *environment* in which electrotherapy is applied? What are the safety considerations related to the *actual use of electrical stimulators?* The lists of recommendations do not represent all possible steps that can be taken to reduce the shock hazards in clinical electrotherapy.

Suggestions for equipment selection

One of the dilemmas facing the user of electrical stimulators is how to decide which device(s) to choose for electrotherapeutic applications. One feature to look for in North America is the approval by a nationally recognized testing laboratory such as Underwriters' Laboratories (UL), Canadian Standards Association (CSA), or Electronic Testing Laboratories (ETL). In general, these organizations apply a standard set of tests to electronic equipment, examining the device performance. For instance, depending on the type of equipment, leakage currents must be less than 50–100 μA in order to receive approval. The approval assigned by these organizations is for the design of the instrument, not for each stimulator manufactured, and therefore does not guarantee performance as instruments age or in those subjected to heavy use. Safety certification for portable battery-operated devices is currently not required.

NMES devices should meet the AAMI performance standards. Such standards have been developed for TENS devices and are being developed for NMES devices. The standards include the presence of a safety interlock that prevents voltage transients in the patient circuit when a device is powered up or down and an output interlock that does not allow the device to be energized unless the amplitude control is first returned to the minimum setting. Instruments for electrotherapy should also be designed such that supply-voltage fluctuations do not result in marked increases in stimulator output while in use. In addition, new stimulators should be shielded in such a manner as to protect from interference from diathermies and other forms of electromagnetic interference.

Forms have been developed to assist in the decision-making process for selection of electrotherapeutic devices (13).

Suggestions for equipment inspection and maintenance

All types of electrical stimulators should be subjected to periodic inspection and preventive maintenance by a qualified biomedical engineer. No standard exists for frequency of inspection for all types of electrical stimulators. Frequency of inspection is generally dictated by frequency of use. Minimally, electrical stimulators should receive an annual inspection. For devices used heavily, inspection should be performed every other month. Any incident such as dropping a stimulator, liquid spills on or near a stimulator, or unexpected electrical shocks to a patient or health care provider should trigger an inspection of the device before use is resumed. New equipment should be inspected prior to use.

Routine equipment inspections include verification that the output characteristics are consistent with standards provided by the manufacturer and consistent with existing performance standards for particular classes of stimulators.

Preventive maintenance is essential. Any component (switch, dial, meter, lights, connector, lead, electrode, etc.) not found to be in original working order should be replaced or repaired immediately by qualified personnel. Some manufacturers supply service contracts for such equipment maintenance.

Records of the results of all safety inspections and routine maintenance should be stored and be made available during subsequent inspections.

Although safety inspections are normally performed by qualified biomedical personnel, knowledgeable users of electrical stimulators should be aware of the elements of the inspection because they are in the best position to recognize potential problems with the structure or operation of these instruments. One of the first signs of a developing problem is the intermittent failure of a component that often occurs during clinical use but may not arise at the time of inspection. Components that should be checked by the user or qualified personnel include the following.

1. Proper operation of all switches, dials, push buttons, pressure-sensitive switches, or other types of stimulator controls. Dirt in controls or broken controls interfere with stimulator output. Adjustment of malfunctioning controls during a treatment may result in an abrupt change in a stimulation parameter and possibly an uncomfortable stimulus to a patient.
2. Proper function of meters and auditory and visual signals.
3. Loose connections at stimulator or electrode interface.
4. Frayed or broken wires, broken lead insulation, broken power plugs.
5. Worn, old, or otherwise malfunctioning electrodes.

Environmental factors and safety in electrotherapy

Users of electrical stimulators must also be aware of a variety of factors related to the environment in which theses devices are to be used. These environmental factors include the following.

1. Line-powered stimulators should be connected to a power-supply circuit protected by a ground fault–interrupt (GFI) outlet or GFI circuit breaker. As discussed previously, GFIs are designed to shut off the supply circuit if the current provided to the stimulator differs from the current returning from the device by 3–5 mA. GFI outlets or GFI circuit breakers should be tested monthly to ensure proper operation. A line-powered stimulator need not be connected to a GFI-protected power circuit if the stimulator is double-isolated and meets the design standards of Underwriters' Laboratories Standard 544. This feature is particularly important for line-powered stimulators to be used in the home setting since many circuits will not be protected by GFI devices.

2. All stimulators should be isolated from the voltage transients (surges) that may occur during power outages or shifts to emergency power. In hospitals, isolation transformers or other types of surge protection devices are generally included in power supply lines. Such devices may not be available in private homes or in private practice settings. In these cases, a biomedical engineer should be consulted to determine the method of surge protection.

3. Three-prong wall outlets should be tested annually to ensure a secure fit with plugs (pull tension: minimum 4 oz.) and proper grounding. Before initial use of any new supply circuit, the polarity of the circuit should be tested.

4. Use of line-powered electrical stimulators should be avoided in areas where water may accumulate on floors or where individuals can come into contact with grounded conductors such as water pipes, radiators, or cases of other grounded instruments.

Suggestions for safe clinical use of electrical stimulators

Once a stimulator has been shown to function according to specification and the power-supply system is safe, the third set of safety considerations relates to the operators' appropriate use of the devices in patient treatment. The following procedures are suggested to minimize risks and undesired responses to patient and clinician.

1. Be sure that the prospective patient does not have any contraindications to the specific electrotherapeutic application considered. Refer to the specific listings of contraindications in subsequent chapters, professional literature, and manufacturer operating instructions.

2. Never place line-powered stimulators next to radiator pipes, water pipes, or any other ground pathway within reach of the patient or clinician.

3. To prevent interference or the production of transients in the patient circuit, do not use stimulators within 3 meters of short-wave diathermy and microwave devices. Some stimulators may be properly shielded to prevent such interactions.

4. Read and understand the operating manual for the stimulator. If the manual is missing, request a replacement copy. If the precise function

of the various components and controls is not clearly specified in the operating manual, the device should not be used until a clear understanding of its operation is obtained from the manufacturer.

5. Apply power to the stimulator before connecting the electrodes to the patient. Never use "cheater" plugs, frayed plugs or cords, or extension cords. Do not remove the ground prong of the power plug to accommodate a two-prong outlet receptacle.

6. Use only the power switch to turn the power on and off.

7. Never disconnect a machine from the wall power outlet with the power turned on.

8. Always turn the output amplitude controls to zero before applying electrodes to the patient.

9. Do not apply or remove surface electrodes while currents are applied to the patient circuit. Use electrodes designed for use with the particular stimulator.

10. Gradually increase stimulation amplitude after all other parameters of stimulation have been set.

11. Do not make major incremental adjustments to stimulation parameters while applying maximally tolerable stimulation.

12. Do not adjust the amplitude of stimulation during the off time.

13. Adjust only one parameter of stimulation at a time.

14. Reduce the output amplitude of stimulation before turning the power off.

15. Only one individual should make changes in stimulation parameters during an electrotherapeutic procedure.

16. Never pull the plug from the socket by pulling on the power cord.

17. Always explain the treatment procedure to the patient; describe what sensations the patient is to expect during stimulation as well as those sensations that are not desired or anticipated.

18. Do not use electrotherapeutic approaches on individuals who are unable or unwilling to clearly communicate.

19. Do not use battery-operated stimulators with a recharger plugged in unless line leakage current is less than 10 μA.

Problems encountered with the use of any type of biomedical instrument should be reported to the Medical Device and Laboratory Product Problems Reporting Program (US Pharmacopeia, 12601 Twinbrook Parkway, Rockville, MD 20852) at (800) 638-6725.

Summary

This chapter has described the design features of electrotherapeutic devices, generic stimulator controls, the classes of electrotherapeutic stimulators, and issues related to their safe use. The chapter was intended to provide an introduction to the common features of stimulators to enable the reader to better understand how to operate these machines safely and effectively. This information should enable individuals to make informed decisions regarding instrument selection. Safety considerations were outlined to emphasize that contemporary clinical electrotherapy as defined by the content of this text is not without potential harmful effects. With the information provided here and in subsequent chapters, the informed practitioner should be able to provide patients with electrotherapeutic treatments that can be both safe and of lasting benefit.

Study Questions

For answers, see Appendix B.

1. Describe the two types of power supplies used in electrotherapeutic stimulators.

2. The component of an electrical stimulator that determines the magnitude of the current (or voltage) produced is called the _____ .

3. A number of amplitude- and time-dependent characteristics of stimulation waveforms are controlled or modulated by electronic devices called _____ .

4. Define the function of the following generic stimulator controls:
   a. Waveform selector
   b. Ramp up and ramp down controls
   c. On time and off time controls
   d. Phase duration control
   e. Pulse frequency control
   f. Burst frequency control
   g. Amplitude control
   h. Output channel controls

5. The figure shows the controls on a commercially available neuromuscular stimulator. Describe the function of each labeled control.

6. What is a surface electrode for electrotherapy? How do the various types of surface electrodes differ?

7. What is the minimum number of electrodes associated with a single output channel of a stimulator?

8. What is meant by monopolar electrode orientation? bipolar electrode orientation? quadripolar electrode orientation? Draw a diagram illustrating examples of these types of electrode placements.

9. In a tabular format, compare and contrast the common characteristics of a typical TENS device with a typical portable NMES unit.

10. What type of electrical current is produced by common interferential stimulators?

11. How are iontophoresis devices unique in comparison to other commercial classes of stimulators?

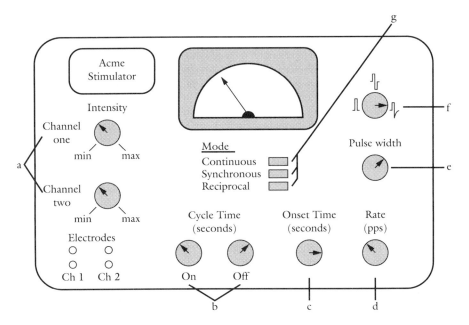

Question 5. Front control panel of a hypothetical electrical stimulator.

12. What is the difference between a two-prong and a three-prong power-supply plug? Which should be used on medical devices and why?

13. What is a ground fault? Describe the function of a ground fault–interrupt device.

14. What components of electrical stimulators should be routinely inspected by the user to screen the device for electrical safety?

References

1. Witters DM, Lapp AK, Hinckley, SM. A descriptive study of transcutaneous electrical nerve stimulation devices and their electrical output characteristics. J Clin Electrophysiol 1991;3:9-16.

2. Campbell JA. A critical appraisal of the electrical output characteristics of ten transcutaneous electrical nerve stimulators. Clin Phys Physiol Meas 1982;1:141-144.

3. Nolan MF. Conductive differences in electrodes used with transcutaneous electrical nerve stimulation devices. Phys Ther 1991;71:746-751.

4. American Physical Therapy Association. Electrotherapeutic terminology in physical therapy. Section on Clinical Electrophysiology, Alexandria, VA: American Physical Therapy Assoc., 1990.

5. Association for the Advancement of Medical Instrumentation. American national standard for transcutaneous electrical nerve stimulators. ANSI/AAMI NS4-1985. Arlington, VA: Assoc. for the Advancement of Medical Instrumentation, 1985.

6. Hansjuergens A, May H-U. Traditional and modern aspects of electrotherapy. 2nd ed., [monograph] Nemectron Medical, Inc., 1984.

7. Alon G. Principles of electrical stimulation. Nelson RM and Currier DP, eds. Norwalk, CT: Appleton and Lange, 1991:48, 58-61.

8. Picker RI. Current-low-volt pulsed microamp stimulation Part 1. Clin Management Phys Ther 1989;9:10-14.

9. Arledge RL. Prevention of electrical shock hazards in physical therapy. Phys Ther 1978;58:1215-1217.

10. Berger WH. Electrical shock hazards in the physical therapy department. Clin Management Phys Ther 1985;5:26-31.

11. Roth HH, Teltscher ES, Kane IM. Electrical safety in health care facilities. New York: Academic Press, 1975.

12. Newton RA, Karselis TC. Skin pH following high voltage pulsed galvanic stimulation. Phys Ther 1983;63:1593-1596.

13. Nolan TP. Choosing electrotherapeutic devices. 1993;1(7):43-49,92. Correction. 1993;1(9):11.

Chapter 3

Physiology of
Muscle and
Nerve

Andrew J. Robinson

$\Lambda\Lambda$ **E**lectrical stimulation (ES) has many therapeutic applications. Currently and historically, clinical electrical stimulation has been used primarily to activate electrically excitable tissues—muscle and nerve. The appropriate use of ES in therapeutic applications requires that individuals possess a clear understanding of the basic structure and function of these tissues, including the mechanisms of their activation by the central nervous system (CNS) and by electrical currents. The primary objective of this chapter is to review the physiology of muscle and nerve as it applies to electrical stimulation. The effects of electrical stimulation on other tissues (such as skin and bone) are considered in Chapter 8.

Electrical Excitability of Muscle and Nerve

Membrane structure and the resting membrane potential

Structurally, the cell membranes of muscle and nerve cells are similar to those of other cells, consisting of a phospholipid bilayer containing a variety of protein molecules. The proteins in nerve and muscle cell membranes, however, serve several special functions that set nerve and muscle apart from many other body tissue types. Membrane proteins may serve as *(a)* **receptor proteins**, which act as binding sites for neurotransmitters or neuromodulators, *(b)* **channel proteins**, which under certain conditions form a pore through the membrane for the movement of ions such as sodium, potassium, chloride, or calcium and *(c)* **transport proteins**, which bind and transfer substances such as sodium and potassium through the membrane, often against concentration gradients (Fig. 3.1). Each of these types of proteins is thought to undergo a change in shape (conformational change) as they perform their respective functions.

By virtue of its structure, the excitable cell membrane functions as a barrier to the movement of substances between the intracellular and extracellular spaces. Substances such as oxygen, carbon dioxide, and water move readily through the phospholipid bilayer. In contrast, ions do not move readily through the membrane, and there exists a differential permeability of the excitable cell membrane to different electrically charged particles. In the resting state, when the nervous system is not activating excitable tissues, excitable membranes are *(a)* readily permeable to potassium ions, *(b)* slightly permeable to sodium ions, and *(c)* impermeable to a number of large negatively charged proteins and phosphates (anions). Since a large number of anions are trapped within the cell and the membrane is readily permeable to potassium, the positively charged potassium ions are drawn into the cell by electromotive forces as the resting state is being established.

Another important property of excitable membranes in the resting state is that they use cellular energy in the form of adenosine triphosphate (ATP) to actively transport sodium and potassium across the membrane. This active **sodium–potassium pump** moves sodium ions out of the cell while moving potassium ions in. An important feature of the sodium–potassium pump is that it is capable of moving the ions against the electrochemical forces tending to oppose their movement in the resting state.

As a consequence of both the selective permeability of the membrane and the membrane pumping mechanisms, the distribution of charged particles

83

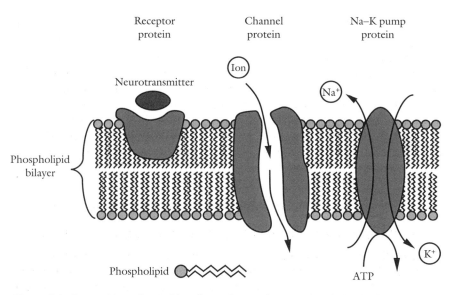

Figure 3.1 Composition of excitable cell membrane, showing phospholipid bilayer and three general types of membrane-bound proteins.

across the membrane of electrically excitable cells is not uniform (Fig. 3.2). The concentration of sodium (Na^+) is higher in the fluid surrounding the cells, and the concentration of anions and potassium (K^+) is higher inside. The resulting separation of electrically charged particles gives rise to an electrical potential difference across the membrane. This transmembrane potential difference or voltage is referred to as the **resting membrane potential** (**RMP**), and hence the excitable cell is described as **polarized**. The magnitude of the resting membrane potential is approximately -90 mV for muscle and is slightly less, at about -75 mV, for peripheral nerve fibers. The absolute RMP values for muscle and nerve of different species vary. As long as the membranes of these cells remain intact and ATP can be supplied to the Na^+–K^+ pump, the resting membrane potential of these cells remains stable. By virtue of the concentration differences of sodium and potassium across the membrane and the relative negative charge within the resting membrane, both the chemical and electrical forces acting on sodium are directed into the cell as the net electrochemical force acting on potassium is directed out of the cell.

Action potentials

Muscle or nerve cells are said to be unique because of specialized cell membrane properties that enable these cells to initiate and propagate action potentials. What are action potentials and why do they occur?

In contrast to the membranes of nonexcitable cells, muscle and nerve cell membranes may quickly and dramatically alter their permeability to ions in response to chemical, electrical, thermal, or mechanical interventions. When an appropriate "stimulus" is applied to a small region of the excitable cell (e.g., peripheral nerve fiber), the membrane permeability to sodium ions transiently increases. Since both the concentration and electrical forces for sodium are directed inward, sodium ions move into the cell and the transmembrane potential is reduced (approaches zero). The reduction in transmembrane potential, referred to as **depolarization** (Fig. 3.3*A*), is gradual at first as a small amount of sodium leaks into the cell. When the transmembrane potential reaches a critical voltage level called the **threshold**, voltage-sensitive sodium and potassium channels in the membrane undergo a conformational change and open (Fig. 3.3*B*). Permeability to sodium abruptly increases as sodium gates are thrown open, whereas permeability to potassium increases slowly since the potassium gates open gradually. Because both the electrical and chemical forces acting on sodium are initially directed into the cell, sodium rushes in when sodium channels open. As a result, the transmembrane potential rapidly depolarizes as positively charged sodium ions are added to the negatively charged cell interior. If sodium channels remained open for very long, sodium ions would continue to enter the cell until the electrical and chemical forces acting on sodium ions were balanced—the sodium equilibrium potential—which would occur at approximately +60 mV. In fact, sodium influx actually stops at a point when the transmembrane potential reaches about +35 mV because at this level the sodium channels close (sodium inactivation gates close, Fig. 3.3*B*). At this voltage, the membrane again becomes relatively impermeable to sodium. At the time when sodium gates become closed, the potassium gates are approaching their fully open position. Since the inside of the cell is now positively charged and only a few potassium ions have left the cell, both the electrical and chemical forces acting on the intracellular potassium ions strongly force potassium ions out of the cell's interior. As the potassium gates fully open, potassium ions rush out of the cell—thereby removing positive charges from within—and the transmembrane potential becomes progressively more negative. Potassium efflux from the cell quickly makes the transmembrane potential negative again, in a process called **repolarization**.

The forces moving potassium out of the cell are so great and the potassium gates remain open long enough that the membrane is repolarized slightly past

Intracellular fluid		Extracellular fluid	
sodium	10 mEq/L	....140 mEq/L	
potassium	140 mEq/L	..... 4 mEq/L	
calcium	0.0001 mEq/L	.... 2 mEq/L	
chloride	4 mEq/L	....100 mEq/L	
phosphates	75 mEq/L	...... 4 mEq/L	
glucose	10 mg/dL	.... 90 mg/dL	
proteins	40 mEq/L	...... 5 mEq/L	
amino acids	200 mg/dL	..... 30 mg/dL	

Figure 3.2 Intracellular and extracellular concentration differences for selected ions and molecules across the membrane of excitable cells.

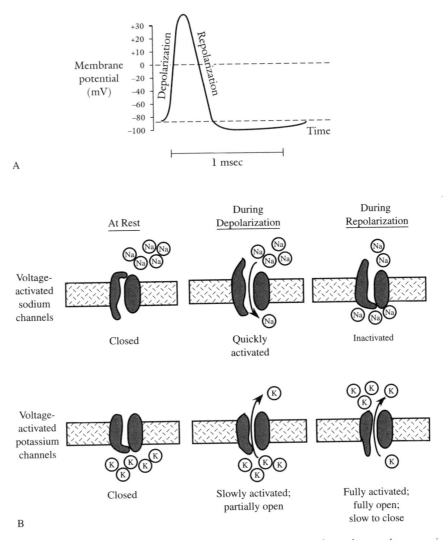

Figure 3.3 A. Intracellular action potential: the transmembrane voltage changes that occur in excitable cells during "excitation." **B.** Diagrammatic representation of changes in the voltage-gate sodium and potassium channels during the action potential.

the resting membrane potential, about 10 to 20 mV below it. This **hyper-polarization** of the membrane proceeds to a point at which no net electrochemical forces act on potassium; that point is the potassium equilibrium potential. At that time, potassium gates continue to close, and passive diffusion of ions across the membrane then quickly restores the membrane potential to its original level.

The transmembrane voltage changes that occur in response to excitable cell stimulation are collectively called an **action potential (AP)** and are generally completed within 1 msec. These voltage changes occur in response to each stimulus sufficient to raise the transmembrane potential to threshold level. The voltage changes of the AP are simply a reflection of the sodium and potassium currents that occur over time across the membrane in response to the opening of the sodium and potassium gates. If an excitatory stimulus is large enough to depolarize the membrane to its threshold level, the transmembrane voltage changes that occur by virtue of ion movement happen in precisely the same manner for each stimulus applied. Because the amplitude of voltage changes in response to stimulation is constant from stimulus to stimulus, the AP is described as "all or none" in character. The APs that occur in nerve and muscle cells in response to either normal nervous system activation, the application of sensory stimuli, or electrical stimulation are for the purposes of this text, identical.

In monitoring the normal function of peripheral nerve and muscle, single, isolated APs are rarely produced. Usually, volitional activation of nerve and muscle produces series of APs in these cells. The frequency of APs produced in nerve and muscle during normal activities rarely exceeds 60 per second, and ordinary discharge range from 5 to 15 impulses per second. In contrast, during electrical stimulation of excitable cells, the AP frequency mirrors the stimulation frequency up to a maximum rate of near 1000 pps. This maximum-possible rate of AP generation is determined by the period following the onset of the potential during which the fiber is absolutely or relatively inexcitable (called the *absolute* or *relative refractory period*).

Action Potential Propagation

Although the opening of sodium and potassium gates and resultant action potential have been discussed as though they occur at one segment of the membrane, the channel opening and associated voltage changes spread to adjacent portions of the nerve or muscle membrane and trigger the same sequence of ionic movements across the membrane. In this manner, the action potential is said to propagate, or be transmitted, along the surface of the cell. The resultant transmembrane voltage changes (action potential) monitored in one portion of a nerve axon or muscle cell will be identical to those recorded in any other region (Fig. 3.4).

The rate of AP propagation (conduction velocity) along membranes is not the same for all excitable cells. In muscle fibers and nerve fibers of different diameters, the speed of AP propagation increases as the diameter of the fiber increases. The differences in AP conduction velocity in excitable cells of different sizes are due to differences in the passive electrical properties (cable properties) among the various types of fibers. The two most important passive electrical properties with respect to the transmission of ionic currents in nerve and muscle fibers are the membrane resistance (r_m) and the internal (intracellular) resistance (r_i).

In response to electrical or chemical activation, current spreads within the fiber and becomes attenuated (reduced) with distance. This attenuation occurs because ionic movement longitudinally along an axon is opposed by r_i and because some ions leak out of the fiber through its membrane. The higher the r_i or the lower the r_m, the less will be the spread of longitudinal currents along the fiber and hence the lower will be the propagation speed of AP currents. Small-diameter fibers have relatively high internal resistance as compared to large-diameter fibers and consequently have lower conduction velocities than large fibers.

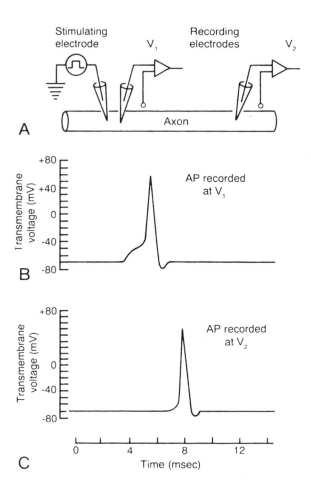

Figure 3.4 A. Experimental setup for examining action potential propagation. B. Action potential recorded at V_1. C. Action potential recorded at V_2. (Redrawn with permission from Kuffler SW, Nicholls JG. From neuron to brain: A cellular approach to the function of the nervous system. Sunderland, MA: Sinauer Associates, Inc., 1976.)

The transmission of APs along the membrane of both muscle or unmyelinated peripheral nerve fibers occurs in a continuous manner involving every segment of the membrane. Many types of peripheral nerve fibers, however, are surrounded by an insulating (nonconductive) material known as *myelin*. This myelin layer is interrupted at intervals *(nodes of Ranvier)* of about 1 mm, exposing small segments of the axon membrane to the extracellular fluids. In myelinated axons, only those portions of the axon membrane in contact with the extracellular fluid at the nodes of Ranvier undergo the permeability changes and ionic currents associated with the AP. Ionic currents do not occur in those segments of the membrane covered by myelin because the membrane resistance in those regions is extremely high; those areas of membrane have relatively few voltage-gated channel proteins. When an AP is initiated by electrical stimulation at one node, ionic currents spread longitudinally until they reach the adjacent node. As a result, current attenuation is less and impulse propagation is more rapid than with unmyelinated excitable cells.

The AP currents moving along the axon membrane in myelinated axons can be viewed as moving in a stepwise manner from one node to the next. This form of AP propagation in myelinated axons is called **saltatory conduction**, a term derived from the Latin verb *saltare*, which means "to leap." The conduction velocity of myelinated nerve fibers is much higher than that of either unmyelinated axons or muscle fibers because myelinization reduces the amount of membrane involved in the ionic movements of the action potential. Hence myelinization effectively reduces the amount of axon membrane through which ionic currents must flow.

Composition of Peripheral Nerves

Peripheral nerve fibers

Peripheral nerves link the spinal cord to structures in the extremities and trunk such as muscle and skin. The primary components of peripheral nerves, such as the median or sciatic, are the nerve fibers (axons). Nerve fibers are merely long, thin projections from nerve cell bodies that lie in or near the spinal cord or brainstem. Most peripheral nerves contain axons from three general types of nerve cells: motoneurons, first-order sensory neurons, and autonomic neurons (Fig. 3.5). Because most nerves contain each of these components, most peripheral nerves are referred to as *mixed*. Some peripheral nerves contain almost entirely axons originating from sensory cells and hence are called *sensory* nerves. No peripheral nerves contain only motor or autonomic fibers.

The cell bodies of motoneurons lie in either the ventral horn regions of the spinal cord or in brainstem motor nuclei of cranial nerves. The multipolar cell bodies of spinal motoneurons project a single axon through the ventral roots of the spinal cord and along the normally well established routes of peripheral nerves. All motoneuron axons terminate in muscle. Functionally, motoneurons can be divided into two groups: **alpha motoneurons**, the cells that innervate ordinary striated skeletal muscle fibers and **gamma motoneurons**, smaller multipolar cells that innervate the special types of muscle fibers contained within a sensory receptor called the *muscle spindle*.

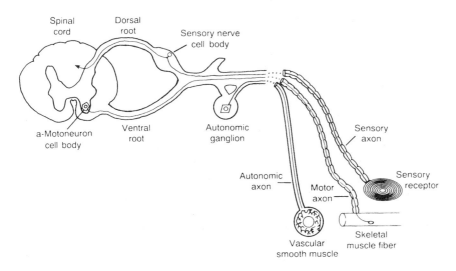

Figure 3.5 Spinal cord cross-section, showing the dorsal and ventral roots and types of axons (sensory, motor, autonomic) that contribute to the formation of peripheral nerve.

The cell bodies of first-order sensory nerve cells lie in the dorsal root ganglia of the spinal cord or in selected cranial nerve sensory nuclei. These sensory neurons are either unipolar or bipolar cells that project one axon toward the spinal cord and a second axon toward the periphery. The peripheral process of sensory neurons is the segment contained within the peripheral nerve. The peripheral axons of about half of the dorsal root cells terminate as free nerve endings in structures such as skin, muscle, and joint capsules. The remaining peripheral sensory axons end within some specialized type of sensory receptor such as the Pacinian corpuscles, the Golgi tendon organs, or the muscle spindles that are designed to respond best to one type of sensory stimulus.

The cell bodies of the autonomic nerve fibers in peripheral nerves are found in the chain ganglia that lie parallel to the spinal cord or in ganglia of the cranial nerves. The axons of autonomic neurons in peripheral nerves terminate in such places as the sweat glands and the smooth muscle in the walls of blood vessels.

Under normal conditions of central nervous system (CNS) activation, the motoneurons and autonomic neurons send APs away from the spinal cord and toward the periphery. Conversely, sensory cells of all types normally initiate APs in the periphery in response to stimuli and transmit the information to the spinal cord, whence it can be relayed to higher brain centers. With these primary functions noted, peripheral nerves can be viewed as simply AP transmission pathways. Action potentials are carried along these "two-way streets," and the transmission of signals in one direction along sensory fibers is not impeded by transmission in the opposite direction along motor or autonomic fibers. The propagation of APs along axons in the "normal" or "physiologic" direction in a particular type of nerve fiber is referred to as **orthodromic propagation**. Orthodromic transmission for motor and autonomic fibers is away from the spinal cord, whereas orthodromic transmission of sensory fibers is toward the spinal cord. Although peripheral axons normally transmit APs either away from or toward the CNS, all types of peripheral nerve fibers are capable of transmitting APs in both directions from the point of initiation. For instance, this can occur when electrical stimulation is applied at some point along the course of peripheral fibers. The transmission of APs in the direction opposite the physiological direction is called **antidromic propagation**. Antidromic AP transmission in sensory cells, for example, is directed away from the CNS, whereas antidromic propagation in motoneurons is toward the CNS.

The types of fibers in peripheral nerve differ in size and structure. In general, the larger the cell body of origin, the larger the diameter of its projecting axon and the higher the AP conduction velocity. The differences in conduction velocity among the various types of fibers in peripheral nerve have formed the basis for the development of two commonly used nerve fiber classification schemes (Table 3.1). The scheme developed by Gasser and colleagues (1, 2) is used to describe all types of axons (motor, sensory, and autonomic) in peripheral nerves, and the Lloyd scheme (3) is restricted to the description of only sensory peripheral nerve fibers. Table 3.1 indicates the distributions of fiber diameters and conduction velocities for axons of specified type. The fiber diameters and conduction velocities were determined in animal nerves and generally exceed those seen in human peripheral nerve fibers. In spite of

Table 3.1
Peripheral Nerve Axon Classification Schemes

Scheme 1[a]	Scheme 2[b]	Diameter (μm)	CV (m/sec)	Type of nerve fiber
A_α	Ia	12–20	72–120	Muscle spindle primary afferent
	Ib	12–20	72–120	Golgi tendon organ afferent
		12–20	72–120	Skeletal muscle efferent
A_β	II	6–12	36–72	Touch-pressure receptor afferent
		5–12	20–72	Muscle spindle secondary afferent
A_γ		2–8	12–48	Muscle spindle efferent
A_δ	III	1–5	6–30	Pain-temperature afferent
B		< 3	2–18	Preganglionic autonomic efferent
C	IV	< 1	< 2	Pain-temperature afferent
		< 1	< 2	Postganglionic autonomic efferent

[a]Gasser scheme: all peripheral nerve fibers
[b]Lloyd scheme: sensory fibers only

this, the relative differences in conduction speed of axons are believed to be similar in human nerves. Large-diameter fibers innervating muscle spindle, Golgi tendon organ, and skeletal muscle conduct much more rapidly than the small-diameter myelinated and unmyelinated pain afferents.

Peripheral nerve as a composite of many tissues

Although the axons of sensory, motor, and autonomic fibers constitute the primary component of peripheral nerves, many other important types of tissue are contained within nerves. Structures that support peripheral nerve fiber function include Schwann cells, connective tissues, and blood vessels.

All peripheral nerve fibers are surrounded by Schwann cells; they provide several important functions in peripheral nerve. As indicated in the discussion of action potential propagation, approximately one-half of the peripheral axons in the average nerve are myelinated. Myelination, the segmented covering of peripheral axons, serves to increase the velocity of AP conduction. In addition, Schwann cells insulate fibers from each other—which may explain why APs that are propagating distally in motor and autonomic fibers do not interfere with sensory APs traveling centrally.

Peripheral nerve axons are also surrounded by three layers of connective tissue that serve several important roles in relation to nerve transmission (Fig. 3.6). The **epineurium** is a dense sheath of connective tissue that surrounds the entire peripheral nerve and projects inward to enclose and separate bundles of nerve fibers. One role of the epineurium is to protect peripheral axons from compressive forces applied to the nerve. The **perineurium** is a thin sheath of flattened cells layered concentrically around bundles of nerve fiber (fascicles). These perineurial sheaths maintain intrafascicular pressure, act as a bidirectional diffusion barrier, and supply much of the elasticity and tensile strength (passive resistance to stretch) of peripheral nerves. The innermost layer of connective tissue is the **endoneurium**, which is a fine connective tissue matrix surrounding every nerve fiber. The endoneurial layer contains Schwann cells and endoneurial fluid, which make up the extracellular environment of each nerve fiber.

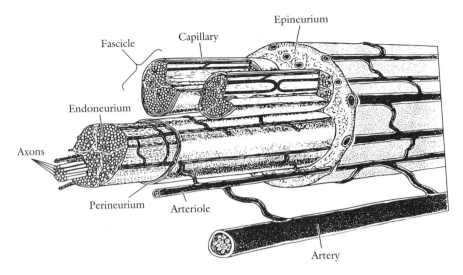

Figure 3.6 Diagram of the components of peripheral nerve: connective tissue layers (epineurium, perineurium, and endoneurium), neural elements (axons contained within nerve fascicles), and vascular (arteries, arterioles, and capillaries). (Reprinted with permission. Lundburg G, Dahlin L. The pathophysiology of nerve compressions. Hand Clinics 1992; 8(2):219.

Peripheral nerves have a well-developed vascular supply (Fig. 3.6). Longitudinally oriented arterioles and venules are present in the epineurium. These vessels regulate the blood flow to and from the peripheral nerve via sympathetic innervation of the smooth muscle contained within their walls. At regular intervals, epineurial arterioles branch toward the center of the nerve to serve one or more nerve fascicles. Smaller diameter, longitudinally oriented arterioles and venules are also found lying between cells of the perineurium. Branches arise and pass obliquely to form capillary beds within the fascicles. The capillaries of peripheral nerves are relatively large in diameter (approx. twice the size of muscle capillaries) and are contained mostly in the endoneurium within fascicles. The intrafascicular capillary beds are characterized by extensive U-loop anastomoses between capillaries, which provide several alternative pathways for blood flow to capillaries that serve the individual nerve fibers.

Skeletal Muscle Structure

Muscle cell structure

The primary component of striated skeletal muscle is the **muscle fiber**. Muscle fibers are cylindrical, elongated cells that have diameters commonly ranging from 50 to 200 μm and lengths that may reach many centimeters. These threadlike cells provide muscle with the electrical and contractile properties that are of special interest to those who study electrotherapy and electrodiagnosis. Occupying most of the interior of the muscle cell are bundles of specialized proteins called **myofibrils** (Fig. 3.7). These cylindrical elements are about 1 to 2 μm in diameter and are found throughout the entire length of the muscle fiber. A single, large muscle fiber may contain as many as several thousand myofibrils. When a longitudinal section of muscle is viewed with a light microscope, a myofibril has a banded appearance, with alternating light and dark areas. This "striated" appearance extends along the entire length of the myofibril. On closer inspection, with an electron microscope, the structure of the myofibril becomes clearer (Fig. 3.7D). The repeating pattern of the myofibril is due to the arrangement of a system of contractile proteins, the thick and thin **myofilaments**, contained within the myofibril.

The repeating unit of the myofibril containing light and dark areas is called the **sarcomere** (Fig. 3.7E). A single sarcomere extends between structures known as Z lines. Thick myofilaments are located in the center of each sarcomere, and thin myofilaments extend from each Z line toward the center of

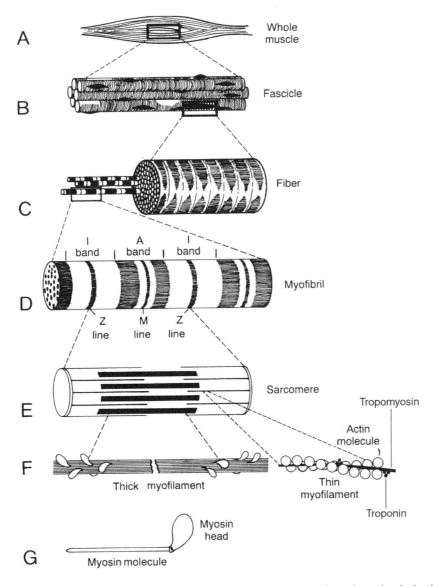

Figure 3.7 Structure organization of skeletal muscle from whole muscle to the molecular level. (Redrawn with permission from Bloom W, Fawcett DW. A textbook of histology, Philadelphia: W.B. Saunders, 1975.)

the sarcomere and overlap the ends of the thick filaments. The dark bands (A bands) of muscle are made up of thick filaments and overlapping thin filaments. The light bands (I bands) contain only thin myofilaments extending from the centrally located Z line.

Biochemical analysis has revealed that the thick filaments are composed of the contractile protein **myosin**. This protein has a globular head connected to a long, thin tail. Approximately two hundred myosin molecules are arranged to make up a single thick filament (Fig. 3.7*F*). Thin filaments are primarily composed of the contractile protein **actin**. Actin molecules are spherical, and the thin filament is made up of two long chains of actin molecules twisted together much like a twisted strand of pearls. Included on the thin filament are two other proteins, **troponin** and **tropomyosin**, which are regulatory proteins that control the interaction between the contractile proteins, actin and myosin. Troponin binds free calcium ions and controls the position of tropomyosin. Tropomyosin is a long, thin protein that, by virtue of its position on the thin filaments, interferes with actin–myosin coupling, in the resting state.

Surrounding each myofibril in a sleevelike fashion is a tubular network called the **sarcoplasmic reticulum (SR)** (Fig. 3.8). The SR is a modified en-

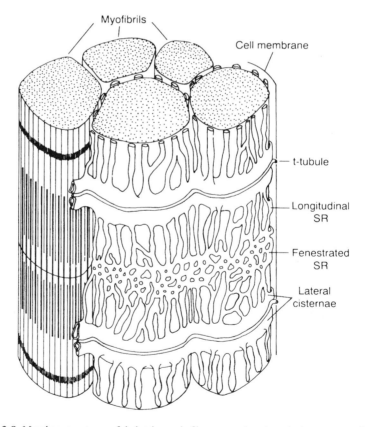

Figure 3.8 Membrane systems of skeletal muscle fiber: sacroplasmic reticulum surrounding myofibrils and t-tubules extending centrally from the surface. (Redrawn with permission from Ham AW, Cormack DH. Histology. Philadelphia: J.B. Lippincott, 1987.)

doplasmic reticulum containing a high concentration of free calcium ions. The SR is divided into units along the length of the muscle fiber, and each unit is made up of three components. The central portion of each SR segment is called the **fenestrated SR**. The tubular network extending from the central region is called the **longitudinal SR**. The longitudinal SR connects to larger cavities called the **lateral** or **terminal cisternae**. The lateral cisternae from adjacent segments of the SR wrap around the **t-tubules**, which are invaginations of the surface of the cell membrane.

Just beneath the surface of the cell membrane of each muscle fiber lie the elongated or ovoid **nuclei** of the cell. The nuclei lie along the length of the muscle fiber and are the storage sites for the genetic instructions for cellular activities such as the production of muscle protein. Since the muscle fibers are so very long, it may be that many nuclei are required along their length in order to efficiently regulate protein synthesis and degradation within the cell.

Another set of internal structures that are located along the entire length of the muscle fiber are the **mitochondria**. Muscle fiber mitochondria, also known as *sarcosomes*, are rather evenly distributed throughout the volume of the muscle fiber and are commonly found lying between myofibrils. They contain the enzymes associated with controlling the oxidative metabolism of muscle fuels such as free fatty acids and glucose. Consequently, sarcosomes are the primary location for ATP production; ATP is required to supply the energy for numerous intracellular processes, including protein synthesis, membrane pumps, and muscle contraction.

The last major internal component of muscle cells, which surrounds all of the intracellular structures, is the **cytoplasm**. This intracellular fluid (also called the **sarcoplasm**) is mostly water but contains a variety of other substances that are critical to the normal function of the muscle cell. The sarcoplasm is the storage site for muscle fuels including free fatty acids, glucose, and a number of nucleosides like ATP. All of the enzymes of glycolysis are located in the cytoplasm, and hence the cytoplasm is the site at which glycogen and glucose are anaerobically metabolized to produce ATP. The sarcoplasm is also rich in amino acids, which are the basic building blocks of proteins. Lysosomes are suspended in the cytoplasm for the handling of damaged intracellular components. Finally, the muscle cell fluid contains myoglobin, an oxygen-binding and transport protein, which is essential for the maintenance of oxidative metabolic processes.

The cell membrane of the muscle fiber is called the **sarcolemma**. The membrane serves to regulate those substances that are allowed to enter and leave the cell, to maintain the muscle fiber's resting membrane potential, and to propagate muscle action potentials as described previously. The specialized region of the muscle fiber membrane where action potentials are normally initiated is called the **motor endplate** (Fig. 3.9). At the motor endplate, the axon of the nerve cell innervating the muscle fiber, the alpha motoneuron, terminates. This region of the muscle cell contains specialized receptor proteins for the neuromuscular junction transmitter acetylcholine. When acetylcholine is released from the motoneuron terminal and binds to the receptors at the motor endplate, the cell membrane in the region becomes more permeable to sodium ions. So many sodium ions leak into the cell that a muscle action

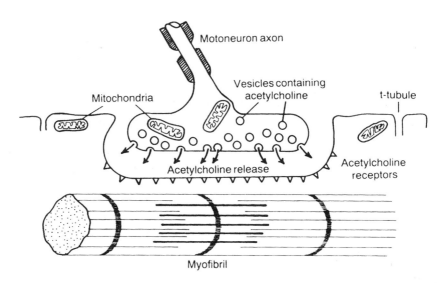

Figure 3.9 Diagrammatic representation of the neuromuscular junction. (Redrawn with permission from Vander et al. Human physiology: The mechanisms of body function. New York: McGraw-Hill, 1990.)

potential is initiated each time the motoneuron to the muscle is activated. The normal neuromuscular junction is therefore often described as an effective synapse.

Another special feature of the sarcolemma is that at regular intervals the cell membrane is invaginated to form channel-like structures called **t-tubules**, which dive toward the center of the cell (Fig. 3.8). The lumen of each t-tubule is continuous with the extracellular fluid, and the t-tubules provide a pathway for action potential propagation to the center of the muscle fiber.

Muscle as a composite of many tissues

In a discussion of muscle's structural features, it is common to focus primarily on the organization of muscle fiber groups and on the morphological features of individual muscle fibers. Although this is certainly valuable as a basis for understanding, another often overlooked perspective is that of the whole skeletal muscle as a composite of many different types of tissue. Certainly the primary component of muscle is the muscle fiber or muscle cell. Fiber contraction gives whole muscle the ability to generate and sustain force used to either move or stabilize the skeleton. Several other types of tissue, however, are present within whole skeletal muscle that may significantly influence the muscle's ability to generate or transmit force.

One important component of whole muscle is connective tissue. The connective tissue within whole muscle is subdivided into three main segments. First, the **epimysium** is the dense layer of connective tissue that surrounds the entire muscle. Second, the **perimysium** is the layer of connective tissue that extends from the epimysium and surrounds groups of muscle fibers, called *fascicles*. Finally, the **endomysium** is the fine, lacy, delicate layer of connective tissue—extending from the perimysium and composed of collagen fibers and fibroblasts—that invests closely each muscle fiber. These three layers of connective tissue account for much of muscle's resistance to passive stretch. Changes in either the character or amount of connective tissue in muscle can dramatically alter the way in which the contraction force is transferred to the skeletal system.

Another important secondary component is skeletal muscle's vasculature. As they penetrate the epimysial connective tissue layer, large arteries rapidly branch into complex arteriolar networks in the perimysial layer. The arterioles in muscle are the main blood vessels regulating the muscle circulation, and they branch to form elaborate capillary beds. In skeletal muscle, each fiber is served by at least one and usually several capillaries that lie next to the cell membrane in the endomysial layer. At the capillary–muscle cell interface, exchange of fuels and metabolic substrates and by-products occurs. Muscle's capillary networks condense to form venules, which drain into larger veins, which in turn exit muscle at the point where the arterial supply entered. Muscle's circulatory network is vital to the normal function of this tissue because muscle is one of the body's largest consumers of oxygen and energy. Without an efficiently operating vascular supply, muscle rapidly loses its ability to sustain contraction, as reflected in the weakness and low fatigue resistance often seen in individuals with circulatory disorders of muscle.

A third component of whole muscles consists of nervous tissue. Skeletal muscles contain both myelinated and unmyelinated nerve fibers. Myelinated nerve fibers innervate not only skeletal muscle fibers but also numerous types of sensory receptors that monitor muscle activity. Sensory receptors include spindles that monitor muscle length, rate of the change of muscle length, and under certain conditions muscle force. At the musculotendinous junction, Golgi tendon organs are receptors that monitor the active tension generated by muscle during contraction. In addition, muscle contains numerous free nerve endings that are the terminations of either myelinated or unmyelinated nerve fibers associated with the detection of muscle pain. Other nerve fibers contained within muscle include the larger myelinated axons of alpha motoneurons, which when activated cause the skeletal muscle fibers to contract, and the smaller-size axons of the gamma motoneuron system, which regulate the sensitivity of muscle spindles. The remaining nervous system structures in muscle are the fibers of the autonomic nervous system that are primarily involved in the control of muscle circulation by the regulation of the vascular smooth muscle in the walls of arterioles and venules.

All of the neural elements within whole muscle play important roles. The alpha motoneuron axon endings in muscle are essential to initiate muscle contraction, as will be detailed in the next segment of this chapter. The sensory receptors in muscle and the sensory afferent fibers that transmit their information to the brain and spinal cord provide valuable information to the central nervous system regarding muscle position and the state of contraction. Such sensory feedback is essential to the regulation of muscle contraction and hence to the control of posture and movement. Dysfunction of whole muscle's neural components may lead to a reduction in muscle's ability to produce levels of contraction appropriate to the demands of our routine activities.

Physiology of Muscle Contraction
Initiation of contraction

The series of electrical and chemical events that describes how skeletal muscle is brought to contraction is referred to as *excitation–contraction coupling*. To activate muscle, the central nervous system first initiates action potentials in the axons of alpha motoneurons (Fig. 3.10). Once initiated, the nerve action potential passes very rapidly along the peripheral axons and finally sweeps over the membrane of the motor nerve terminals. As the nerve action potentials invade the terminals, acetylcholine—the transmitter substance at the neuro-muscular junction (NMJ)—is released into the fluid in the region of the motor endplates through a process called **exocytosis**. By diffusion, the acetylcholine moves across the gap between the nerve and muscle membranes. When the transmitter reaches the muscle cell membrane, it is rapidly bound to special-ized receptor proteins in the endplate region of the membrane. Immediately on the binding of acetylcholine, the membrane in the region of the receptors increases its permeability to sodium ions by the opening of sodium channel proteins. Since both electrical and chemical gradients for sodium ions are di-rected inward, sodium ions move into the muscle cell and the transmembrane

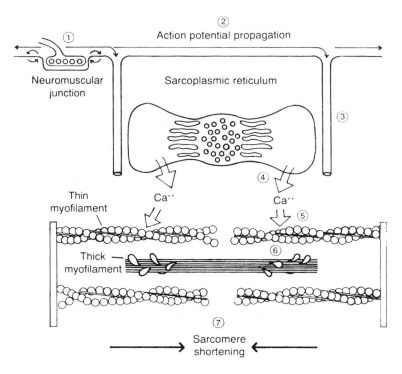

Figure 3.10 Diagrammatic representation of the sequence of events leading to muscular con-traction: *1*, Neuromuscular junction activation and initiation of muscle action potential. *2*, Ac-tion potential propagation along the muscle membrane. *3*, Action potential conducting along t-tubules. *4*, Calcium release from the sarcoplasmic reticulum. *5*, Calcium binding to troponin and subsequent movement of tropomyosin. *6*, Cross-bridge formation between actin and myosin. *7*, Sarcomere shortening. (Redrawn with permission from Vander et al. Human physiology: The mechanisms of body function. New York: McGraw-Hill, 1990.)

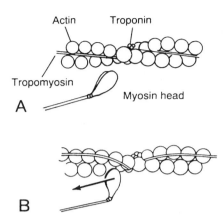

Actin   Troponin

Tropomyosin

Myosin head

A

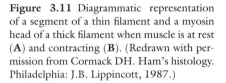

B

Figure 3.11 Diagrammatic representation of a segment of a thin filament and a myosin head of a thick filament when muscle is at rest (**A**) and contracting (**B**). (Redrawn with permission from Cormack DH. Ham's histology. Philadelphia: J.B. Lippincott, 1987.)

muscle potential is reduced (becomes less negative) as positive charges enter the muscle cells. This reduction in the transmembrane muscle potential can be recorded as an **endplate potential**. In the normal muscle, the endplate potential produced in response to the transmitter released from the motor nerve terminal is always great enough to increase the sodium permeability of the muscle fiber membrane just outside the NMJ, thus triggering a **muscle action potential**. Sodium ions rush into the muscle followed by potassium ions flowing out of the cell in a regenerative process that quickly sweeps over the surface of the entire muscle fiber (see previous discussion on resting membrane potential and action potential for further details).

As the processes of the muscle action potential pass over the membrane, they encounter the invaginations of the surface membranes called *t-tubules*. The action potential currents invade the t-tubules and are carried along the membrane toward the interior of the cell. As the depolarization is carried along the t-tubule, it triggers the opening of voltage-sensitive calcium channels and hence an increase in the permeability of the lateral cisternae of sarcoplasmic reticulum to calcium ions. Since the sarcoplasmic reticulum contains a high concentration of calcium ions in comparison to the muscle cytoplasm, calcium rapidly diffuses into the cytoplasm. Once released into the sarcoplasm, calcium diffuses away from the sarcoplasmic reticulum into the region of the thick and thin filaments.

On the surface of the thin filaments lies the regulatory protein troponin, which has a very high affinity for calcium ions. Any free calcium in the cytoplasm rapidly binds to any open calcium-binding site on the troponin molecule. When calcium binds to the troponin molecules on the thin filaments, a conformational (structural or shape) change of troponin is produced. Since troponin is attached to another protein, tropomyosin, as the troponin changes shape, it pulls on the tropomyosin and causes an alteration in the position of tropomyosin on the thin filament (Fig. 3.11). Normally, tropomyosin is located in a position on the thin filament covering the site on each actin molecule, which has a high affinity or attraction for the heads of the myosin molecules extending from the thick filaments. As troponin draws the tropomyosin away from these binding sites, actin molecules and myosin heads quickly attach. This attachment of myosin heads to the binding sites on actin (cross-bridge formation) immediately triggers a rapid change in shape of the myosin molecules. In effect, the myosin heads swivel and, in so doing, draw the thin filaments toward the center of the sarcomeres. This interaction between actin and myosin, which occurs at literally thousands of sites, is the process that produces the force of muscle contraction.

Cessation of contraction

The interaction of actin and myosin at any available site will continue as long as sufficient amounts of ATP are available to fuel the swiveling of myosin and as long as the binding sites on the actin molecules remain unblocked. In response to a single surface muscle action potential, a pulse of calcium is released. Some of the calcium binds to troponin, cross-bridges between actin and myosin are rapidly formed, and muscle force generation begins. Almost as soon as it is released from the lateral cisternae, calcium is actively transported back into the sarcoplasmic reticulum (SR) by ATP-dependent pumps that lie in the wall of the longitudinal cisternae. As the concentration of free calcium falls in the cytoplasm, other bound calcium ions dissociate from the troponin. The troponin thus returns to its original shape and the tropomyosin falls back into its rest position, where it blocks actin-binding sites for myosin. Very rapidly, the number of cross-bridges that can form is reduced and the force of contraction falls. When all actin-binding sites are blocked by tropomyosin, no thick and thin interaction is possible and hence no contraction force is generated.

Factors influencing muscle force production

Frequency of Activation

Figure 3.12 shows the force produced within a segment of muscle in response to activation by a single action potential when the fibers are held at a constant length. The force produced reaches its maximum value almost instantaneously and then gradually falls back to zero. The force pattern generated is a reflection of the very rapid formation of many actin and myosin cross-bridges within the muscle followed by the gradual reduction in the number of cross-bridges as calcium is gradually resequestered into the sarcoplasmic reticulum

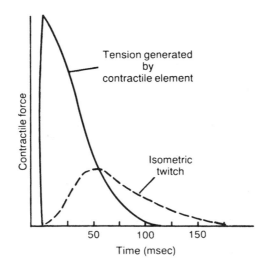

Figure 3.12 Contractile force over time in response to a single action potential. Tension generated by thick and thin filament interaction with the muscle (calculated) is transformed into the isometric twitch response (measured) by muscle's series elastic elements. (Redrawn with permission from Vander et al. Human physiology: The mechanisms of body function. New York: McGraw-Hill, 1990.)

and myosin-binding sites on actin are covered by the tropomyosin regulatory proteins.

Although in theory this force profile is produced within the muscle segment, the force that can be measured experimentally in response to a single activation is very different in character. Figure 3.12 also shows the isometric force measured from a segment of muscle in response to a single stimulus. In contrast to the force produced within the muscle segment, the force actually measured from the segment rises gradually to an amplitude that is less than that produced by the internal contractile apparatus. This force measured over time in response to a single activation is called the **isometric twitch**. For mammalian skeletal muscles, the peak in twitch tension is reached about 20 to 100 msec after the force begins to rise.

Why is the form of the isometric twitch different than that predicted for the internal contractile apparatus of muscle? The answer is found by examining the way in which the contractile machinery is harnessed to skeletal system. As discussed previously, the contractile proteins of muscle that produce its force are encircled by elements such as the sarcoplasmic reticulum, the muscle cell membrane, and the layers of connective tissue throughout the muscle. The layers of connective tissue within the muscle segment in turn merge at each end of the muscle to form the tendons of muscle. Each of these structures (and possibly others) are elastic in nature. That is, they possess a springlike quality such that applied forces stretch them, but when the force ends, they return to their original size and shape. The force produced within a muscle segment is modified (reduced in amplitude and delayed in time) by the presence of these muscle components—which may account for the form of the measured isometric twitch.

By way of further explanation, the contractile and elastic elements of muscle can be modeled as shown in Figure 3.13. The force produced by the contractile elements (CE, i.e., the system of thick and thin filaments) of muscle must be transmitted through the springlike tissue elements, called the *series elastic components (SEC)* of muscle, before it may be measured in the real world. When the contractile elements of muscle are activated, some of the force produced is used to stretch out the SEC while the remainder is transferred through the skeletal system. The amount of force used to stretch out the SEC is equal to the difference in the peak of the internal force produced and the peak in the isometric twitch tension. The time difference between the contractile element peak force and the peak in measured isometric twitch is due to the fact that some time is required to stretch the SEC. A comparison of the contractile element and twitch forces reveals that twitch tension is a poor indicator of the muscle segment's true ability to generate force.

Thus far, the consequences of muscle activation by only a single, isolated stimulus has been described. During the natural activation of muscle, however, a series of muscle action potentials is usually produced on the muscle cell membranes. How does such a train of action potentials influence the force produced by the muscle?

In response to the first muscle action potential of a train, the contractile element of muscle is activated and the force of a twitchlike contraction is produced. If a second action potential is delivered before all of the calcium released by the first action potential is resequestered, more calcium from the

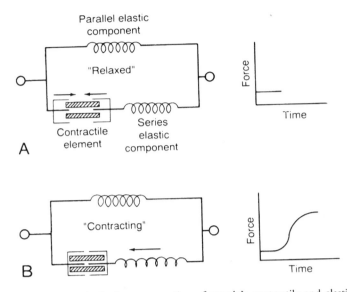

Figure 3.13 Diagrammatic single representation of muscle's contractile and elastic elements (**A**) at rest and (**B**) in response to a single action potential under isometric conditions. Note how contractile activity stretches the series elastic components but does not influence the parallel elastic component length.

SR is released and actin–myosin cross-bridge cycling is allowed to continue. As a result, the force measured from the activated muscle segment rises again. Since the SEC stretched in response to the first AP and it does not have time to fully relax before the second AP, less of the tension generated in response to the second activation is required to stretch out the SEC. Consequently, more tension is transmitted through the springlike harness and measured in the outside world. Subsequent APs in the train produce even further increases in the measured force output of the activated tissue. The profile of force output that appears as oscillations in the tension record is called **unfused tetanus** (Fig. 3.14*A*). Unfused tetanic contractions are generally elicited in response to activation of the muscle at frequencies from approximately 3 to 20 stimuli per second depending on the particular type of muscle activated.

As the frequency of skeletal muscle activation is increased, the isometric force that can be measured from muscle also increases. In addition, the oscillations in force gradually disappear, and the contraction becomes progressively smoother. Smooth, strong contractions produced in response to the very rapid delivery (>30/sec) of action potentials to muscle is a state called **fused tetanus**. Maximum fused tetanic contractions are better indicators of muscle's contractile capabilities than twitch contractions.

Number of Fibers Activated

A second important factor associated with the regulation of muscle force is the number of muscle fibers activated in a contraction. As the percentage of total muscle fibers activated by motor nerve fibers is increased, the amount of force produced by the muscle at a fixed frequency of activation is increased (Fig. 3.14*B*). The increase in force as greater numbers of muscle fibers are recruited into contraction results simply from increasing the total number of actin–myosin cross-bridges formed.

A third factor that determines the active tension that a muscle can produce is the muscle length. The length of muscle at which the maximum number of cross-bridges between thick and thin filaments can be formed is called **rest length**. For many human limb muscles in situ, muscle length can be altered by ±20% of rest length (Fig. 3.15*A*). For example, when the elbow is fully extended, the biceps will be stretched to a length 20% greater than that in midrange. Conversely, if the elbow is fully flexed, the biceps length will be about 20% less than rest length. Muscle generates the maximum amount of tension in the ±10% of rest length range. In biceps for example, the maximum tension is generated by the muscle in the middle 50% of the elbow's range of motion. If the biceps is either lengthened or shortened such that the elbow position is out of the middle 50% of ROM, biceps tension-generating capacity is reduced.

Up to this point, muscle contraction has been discussed only under isometric conditions, when the muscle is held at a constant length. If during activation the muscle is allowed to change length, the velocity of shortening becomes a determinant of the amount of force that the muscle can generate. Figure 3.15*B* shows the force–velocity relationship typically determined for striated muscle. The graph reveals that the highest forces are produced by muscle under isometric conditions when the velocity of shortening is zero and progressively declines as the speed of shortening is increased.

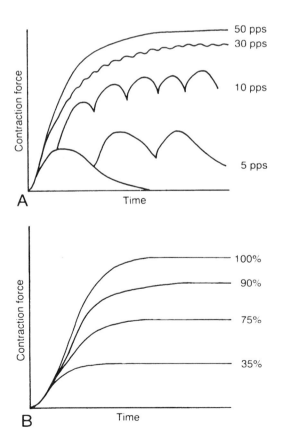

Figure 3.14 Contraction forces recorded from muscle in response to (**A**) changes in the frequence of activation (rate coding) of a fixed portion of the muscle and (**B**) changes in the percentage of muscle fibers activated (recruitment) at a tetanic frequency of activation.

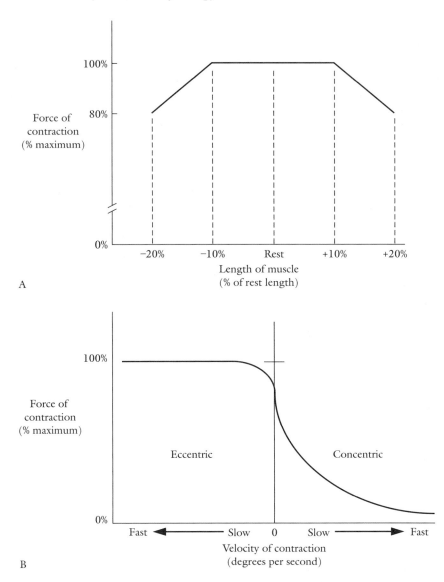

Figure 3.15 Contraction forces produced by muscle as a function of (**A**) muscle length and (**B**) velocity of concentric isotonic contraction.

Nonhomogeneous Nature of Skeletal Muscle

Skeletal muscle fiber characteristics

In the previous description of the structural and contractile characteristics, skeletal muscle was treated as if all muscles where identical. In fact, it has been known for several hundred years that several different types of skeletal muscle exist and that the differentiation of skeletal muscle appears to be related to how the muscles are used in normal activity. A classical comparison is made between the muscles of the birds of flight and those of domestic fowl.

As you may know, the breast musculature of chickens is pale in comparison to the dark leg muscles. These barnyard animals rarely use the breast muscles, which move their wings, but use the leg muscles the majority of their waking hours while foraging. In contrast, the breast musculature of migratory birds is dark while leg muscles are paler in appearance. Such observational findings led early scientists to suspect that there were two types of muscle fibers, light and dark, that made up skeletal muscles. With the advent of microscopic and histochemical techniques in the early and middle part of this century, we learned that most mammalian skeletal muscles are composed of not two but rather three types of muscle fibers.

Over the years, two major classification schemes have been developed to describe the different types of fibers within mammalian striated muscles (Table 3.2). The percentages of these fibers that make up muscle varies from muscle to muscle and from species to species for analogous muscles. In humans, the muscles of the extremities are frequently composed of half fast-twitch muscle fibers and half slow-twitch fibers.

Basic functional element of the neuromuscular system: the motor unit

From a structural perspective, skeletal muscle is composed of single muscle fibers of three different types. From a functional perspective, however, muscle contraction does not occur by the isolated activation of single muscle fibers. During normal activities, the force of muscle contraction is produced by the activation of muscle fiber groups acting in concert in response to a stimulus provided by the single nerve cell serving each fiber in the group. A motoneuron along with the group of muscle fibers it innnervates is called a **motor unit** (Fig. 3.16). The cell body of the multipolar neuron of the motor unit is located in the ventral horn of the spinal cord. A large-diameter, myelinated axon projects from the cell body, passes through the ventral root of the spinal

Table 3.2
Classification Schemes and Characteristics of Skeletal Muscle Fibers

| | Muscle Fiber Type | | |
Characteristic	IIB FG[a]	IIA FOG[b]	I SO[c]
Contraction speed	Fast	Fast	Slow
Myofibrillar ATPase	High	High	Low
Fiber diameter	Large	Medium	Small
Capillary supply	Sparse	Rich	Rich
Oxidative enzyme activity	Low	Med-High	High
Mitochondrial content	Low	High	High
Glycolytic enzyme activity	High	High	Low
Glycogen content	High	High	Low
Myoglobin content	Low	High	High

[a]FG = fast twitch, glycolytic
[b]FOG = fast twitch, oxidative, glycolytic
[c]SO = slow twitch, oxidative

cord and subsequently along peripheral nerves until it reaches the target muscle. As the axon enters muscle it divides many times into fine branches, which in turn innervate single muscle fibers. The muscle fibers of a single motor unit are usually distributed over a rather large portion of the cross-sectional area of the muscle, and all of the fibers of a motor unit are of the same histochemical type. The number of muscle fibers in motor units ranges from about 200 to 300 in hand muscles to over 1000 fibers per motor unit in the large muscles of the lower extremity.

Over the past two decades, the properties of single motor units in skeletal muscle have been extensively examined. Nearly all motor units in the muscles studied thus far can be classified into one of three major types. Table 3.3 shows some of the common characteristics of the three major classes of motor units. The presence of varying proportions of motor units of different type provides muscle with the ability to respond appropriately to postural and locomotive demands. Muscles that are required to produce moderate-to-low levels of

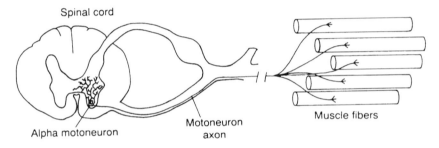

Figure 3.16 Basic functional element of the neuromuscular system, the motor unit.

Table 3.3
Classification and Characteristics of Motor Units in Skeletal Muscles

Motor Unit Type	FF[a]	FR[b]	S[c]
Muscle Fiber Type	IIb	IIA	I
	FG	FOG	SO
Characteristic			
Contraction speed	Fast	Fast	Slow
Twitch contraction time	Short	Short	Long
Resistance to fatigue	Low	High	V High
Tetanic tension	High	Intermed	Low
Number fibers/unit	Large	Intermed	Small
Frequency of use	Low	Intermed	High
Recruitment order	Last	Intermed	First
Size of unit cell body	Large	Intermed	Small

[a]FF = fast twitch, fatigable
[b]FR = fast twitch, fatigue-resistant
[c]S = slow twitch, very fatigue-resistant

tension over long periods of time contain a higher percentage of fatigue-resistant muscle fibers. Muscles required to produce quick, high force levels for brief intervals contain a higher percentage of strong, fast-twitch fatiguable units.

Control of Force Generation in Volitional Contraction

Studies of voluntary contraction in humans have demonstrated that two main processes are employed by the central nervous system to regulate the force output of skeletal muscle. To initiate contraction in a particular muscle, the CNS must first excite alpha motoneurons, which innervate the muscle. The number of motoneurons, and hence the number of motor units activated, is one primary determinant of the level of muscle contraction produced. **Recruitment** is the term used to describe the process of increasing the number of motor units brought to excitation in order to produce increasing levels of muscle contraction. Precisely which motor units are recruited is determined by both descending and reflex inputs to alpha motoneurons. If excitatory inputs to the motoneurons sufficiently overwhelm inhibitory inputs to these cells, action potentials will be produced in the neuron's axon that will evoke contraction.

One question that faced neuroscientists for years was, How does the CNS know which motoneurons to activate in order to produce a particular level of contraction? Evidence now exists that indicates that motoneurons are recruited in most contractions in an orderly sequence. The CNS command to begin muscle contraction first activates the smallest (highest internal resistance) alpha motoneurons. If more force is required to properly perform an activity, the command signals from the CNS are increased and progressively larger (lower internal resistance) motoneurons are activated. The recruitment order of motor units into both reflex and voluntary contractions is dependent on the size of the motoneuron cell body. The small motoneurons require less synaptic current to excite them sufficiently to produce action potentials. As the size of motoneurons increases, greater amounts of synaptic current are required to excite these cells. The size-dependent recruitment of motor units into contractions is now commonly referred to as the **"size principle."** Since the size of the alpha motoneuron is related to the type of muscle fibers innervated by the neuron (Table 3.3), recruitment of motor units in contractions will generally follow from type S motor units to type FR units and finally to type FF units as the level of contraction increases. Those motor units designed to generate tension for relatively long periods without substantial fatigue (types S and FR) are therefore used most in volitional contractions. Type FF units, which are capable of producing high tension levels for very short periods are used only occasionally, in high-force-level contractions.

The second way in which the nervous system regulates muscle tension generation is by controlling the rate of motor unit discharge during recruitment. This control process is commonly referred to as **rate coding**. If higher levels of tension are required as a contraction is proceeding, the CNS may simply increase the discharge rate of the motor units already active in the contraction.

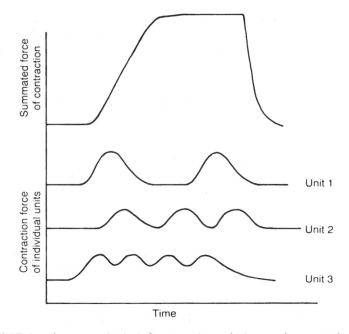

Figure 3.17 Asynchronous activation of motor units results in smooth summated contraction of whole muscle.

As described previously, increasing the firing rate of motor units increases their force output by summation of contraction. Discharge rates of motor units in human voluntary contractions rarely exceed 30 pps. During voluntary muscle contraction, motor units are recruited in an asynchronous manner (Fig. 3.17). That is, motor units are not all activated at the same instant in time. This asynchronous activation has the effect of smoothing the tension output of the whole muscle even when the individual firing frequencies of motor units may be well below the level at which fused tetanus would occur for the individual units involved.

Discharge frequencies of motor units recruited in voluntary contraction are also not all the same. Some units may be discharging at low steady frequencies while others may discharge irregularly at even lower frequencies.

The relative importance of recruitment and rate coding in the regulation of muscle tension has not been clearly established. Some evidence suggests that in the muscles of the hand, which often produce delicate, controlled movements, rate coding may be the predominant mechanism by which force is regulated. In contrast, in the large muscles of the lower extremity, recruitment may be the dominant mechanism regulating the level of contraction.

Clinical Electrical Stimulation of Nerve and Muscle

In the intact peripheral neuromuscular system, the application of electrical stimulation can induce action potentials in nerve and muscle that are indistinguishable from those evoked by the normal action of the nervous system. Furthermore, action potentials evoked by electrical stimulation in peripheral alpha motoneuron axons elicit contraction of skeletal muscle that appears to the uninformed to be identical to voluntary contraction. In fact, muscle contraction in response to electrical stimulation is very different from that produced by normal physiological mechanisms. One of the objectives of the remaining segments of this chapter is to identify the similarities and differences between the nature of contraction in response to electrical stimulation and normal physiologic activation.

Electrical stimulation may also activate peripheral sensory nerve fibers and the nerve fibers of the autonomic nervous system. The electrical stimulation of sensory fibers gives rise to the characteristic sensations associated with such stimulation and has been found useful in the management of a variety of pain syndromes—but it does not produce patterns of sensory action potentials like those generated under physiological activation.

Activation of excitable tissues with electrical stimulation

How does the induction of current in biological tissues bring about the activation of nerve or muscle? Figure 3.18 shows two electrodes of a stimulator

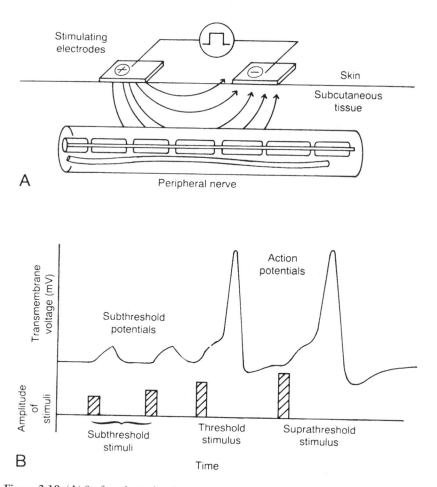

Figure 3.18 (**A**) Surface electrodes placed near a peripheral nerve. (**B**) Transmembrane voltage changes in response to stimuli of gradually increasing amplitude.

circuit applied to the skin directly over a small branch of a peripheral nerve. As stimulation is initiated, one electrode briefly contains an excess of negative charge and the other electrode is deficient in negative charge. Ions in the region will migrate toward or away from these electrodes according to their charge. The pattern of current induced by the movement of these ions is represented by the lines connecting the electrodes. Some of the ionic movement occurs in the extracellular fluid, and some of the current passes through the nerve membrane. The net effect of this current is a slight depolarization of the nerve membrane. If the brief current induced across the membrane is very small, the change in transmembrane potential will rapidly return to resting membrane potential. If the current induced across the membrane is large enough, an action potential will be evoked and propagated along the membrane. This evoked action potential is identical to that produced along the fiber membrane in response to physiological activation.

Action potentials evoked in peripheral nerve are transmitted along the nerve fiber in both directions from the point of initiation. There is conduction in the direction of "normal" propagation (orthodromic) and action potential propagation in the opposite direction (antidromic). Antidromic conduction is toward the periphery in sensory nerve fibers and toward the CNS in motor nerve and autonomic fibers.

Stimulus characteristics for activation of excitable tissues

The current induced in biologic tissues must be of sufficient amplitude (strength) and duration to bring excitable cells to a critical transmembrane voltage, called **threshold**, in order to evoke an action potential. For a particular single excitable cell, there exists a family of stimulus strength–duration (S–D) combinations that can bring the cell to the threshold of depolarization. Plotting the S–D combinations that are sufficient to activate an excitable cell, and connecting the points, yields a **strength–duration curve** for the threshold of that type of cell (Fig. 3.19). Any stimulus S–D combination that falls below or to the left of this line will not initiate an action potential and are described as *subthreshold*. Those stimuli with S–D characteristics above or to the right of the curve are called *suprathreshold* and are always adequate to activate the fiber.

For a particular type of excitable tissue, the *minimum amplitude* of a stimulus that will activate the tissue at very long stimulus durations is called the **rheobase**. The *minimum duration* of a stimulus at twice the rheobase that is just sufficient to activate the excitable tissue is called the **chronaxie**. In the past, electrical stimulators were used clinically to determine the rheobase and chronaxie values for the peripheral neuromuscular system. The reliability of such techniques is poor. Today, sophisticated electrodiagnostic procedures such as nerve conduction velocity studies and electromyography are commonly used to assess peripheral nerve and muscle integrity.

Peripheral nerve fibers are inherently more excitable by electrical stimulation than muscle fibers. Consequently, stimuli with S–D characteristics sufficient to activate peripheral nerve fibers will not be strong enough to activate isolated muscle fibers. Stimuli must have much larger amplitudes and longer durations to initiate action currents in denervated muscle fibers. Figure 3.20 shows the S–D curves for three type A peripheral nerve fibers and isolated (denervated) muscle fibers.

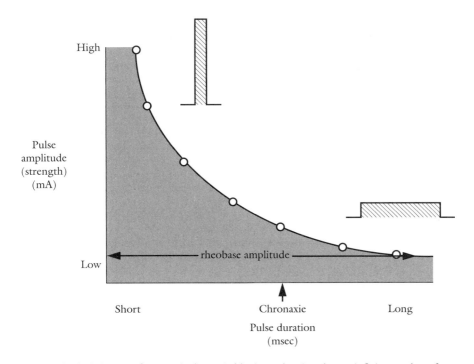

Figure 3.19 S–D curve for a particular excitable tissue showing that an infinite number of amplitude and duration combinations are sufficient to the excite.

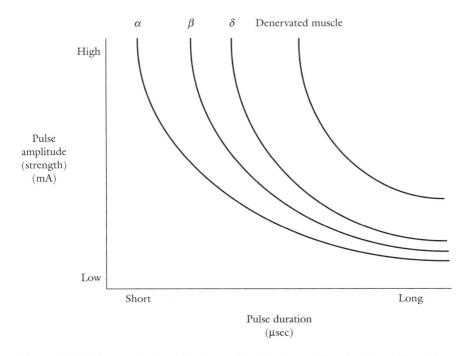

Figure 3.20 S–D curves for three size classes of peripheral nerve fibers (α, β, and δ) and denervated muscle fibers.

Since nerve fibers contained within a peripheral nerve are not of identical diameter and internal resistance, the relative excitability of these fibers by electrical stimulation varies. When a mixed peripheral nerve is *directly* stimulated, those fibers with the largest diameter and lowest internal resistance are the most easily excited. When electrical currents are produced in peripheral nerve, the A_α group of nerve fibers are those most likely to be activated. To activate group A_β, A_δ, or group C fibers requires stimuli of progressively larger amplitude and/or duration. The process by which increasing numbers of nerve fibers are activated by progressively increasing the amplitude and/or duration of the stimulus is called *fiber recruitment*. For isolated motor nerve stimulation, the pattern of recruitment tends to be in an order from largest-diameter (lowest internal resistance) fiber to smallest-diameter fiber (highest internal resistance) in peripheral nerve, which is a pattern opposite to that occurring during volitional activation.

The effects of inducing currents on nerve and muscle fibers depend not only on the inherent excitability of these tissues but also on their location with respect to the electrodes used to transfer the current. In general, the closer the excitable tissue to the electrodes, the more likely it is to be activated by the current. At a fixed intensity of stimulation, small-diameter axons very close to the stimulating electrodes may be activated before large-diameter fibers located further away (Fig. 3.21). As a result, A_β touch-pressure sensory nerve fibers near the skin are commonly recruited before the more inherently excitable A_α motor and sensory fibers. Since the distance between the electrode(s) and target tissue influences the precise nature of the activation, it is important to position the electrodes as close as possible to the target tissue to achieve optimal activation.

Because both the inherent excitability of nerve fibers and the location influence the order of nerve activation, when ES is applied over a mixed peripheral nerve—containing motor, touch-pressure sensory, and pain sensory fibers—ES will normally evoke a pins and needles sensory response before either motor or painful responses (Fig. 3.22). If the amplitude or duration of stimuli are increased sufficiently, motor responses (muscle contractions) will be produced and superimposed on the sensory stimulation. If stimuli amplitudes or durations are increased even more, ES may evoke a painful response, which occurs simultaneously with the sensory and motor responses. In most clinical electrotherapy applications, the pins and needles sensation produced by the stimulation of cutaneous touch-pressure afferents will precede motor or painful responses. If stimulation is applied in a region where no alpha motoneuron axons or skeletal muscle fibers exist (e.g., over bony prominences), an increase in stimulation amplitude may produce a painful response in the absence of muscle contraction.

Another important consideration in applying ES to activate muscle and nerve is the polarity of the electrodes. This is especially important in the use of monophasic pulsatile currents or asymmetrical biphasic currents. With monophasic waveforms of fixed amplitude and duration, for example, the negatively charged electrode (cathode) is usually slightly more effective in activating excitable tissue than the positively charged electrode (anode).

In the past, when the initiation and cessation of a single pulse were not electronically controlled, pulse durations were relatively long. The application of fixed-amplitude current to motor nerves in this fashion revealed that con-

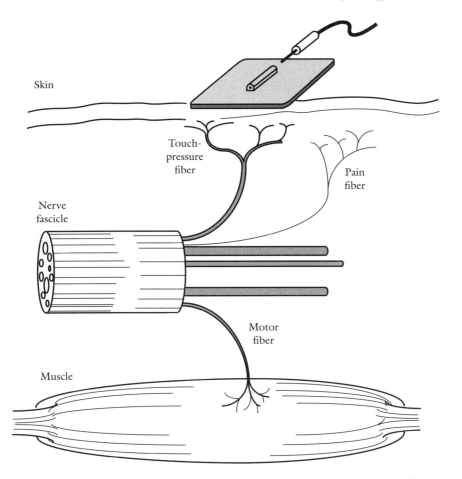

Figure 3.21 Diagrammatic representation of the location of excitable tissue with respect to surface electrodes. Note that sensory fibers are closer than motor fibers.

tractions could occur in response to both closing the stimulating circuit and opening it. In addition, contractions could be elicited by either cathodal or anodal stimulation. Overall, the response to the cathode upon closing the circuit (cathode closing current, CCC) was the most effective approach to neuromuscular stimulation. The contraction response evoked by the anode upon closing the circuit (anode closing current, ACC) was the next strongest, followed by the contraction beneath the anode upon opening the circuit (anode opening current, AOC) and the cathode upon opening the circuit (cathode opening current, COC) respectively. The relationship between anodal and cathodal opening and closing currents has been called *Erb's polar formula*. With the very short duration pulses used today in electronically controlled stimulation, contractions are not elicited by opening currents during neuromuscular stimulation.

Clinical responses to nerve and muscle stimulation

Sensory-Level Stimulation

When electrodes are placed on the surface of the skin and the amplitude of electrical stimulation is gradually increased, three general forms of response

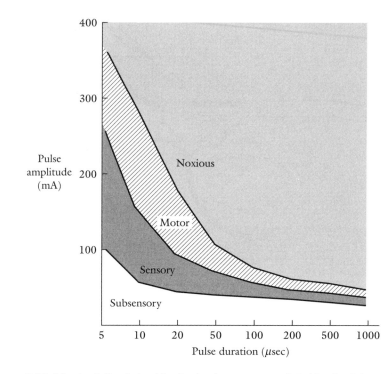

Figure 3.22 Stimulus S–D relationships for the three common clinical levels of electrical stimulation.

are consistently noted in normal individuals. At relatively low amplitudes, the first response reported by subjects is sensory. Commonly, subjects identify a pins and needles sensation if the frequency is greater than a few pulses per second or a "tapping" sensation if the frequency is set at 1 to 5 pps. This **sensory-level stimulation** is not uncomfortable to most subjects and results from the excitation of those sensory nerve fibers that lie in or near the skin in the vicinity of the electrodes, where the current density is greatest. As the stimulation amplitude is gradually increased, the sensation becomes stronger and often spreads to the region between the electrodes and deeper into tissues. As can be noted from Figure 3.22, sensory-level stimulation can be achieved by the use of many different stimulus amplitude and duration combinations. Whether or not sensory fibers are activated is independent of the stimulation frequency selected.

If sensory-level stimulation at frequencies greater than about 15 pps is maintained for prolonged periods of time, the subject will generally note a gradual diminution in the ability to sense the stimulation, a phenomenon called **adaptation**. Adaptation may be reduced by intermittently interrupting the stimulation or by varying either its amplitude or frequency.

Motor-Level Stimulation

As the intensity of stimulation is gradually increased, the tingling sensation felt by subjects increases as progressively greater numbers of sensory nerve fibers are recruited. In addition, the activation threshold of alpha motoneuron axons lying in peripheral nerves innervating skeletal muscle is soon reached (motor-level stimulation). The excitation of alpha motoneuron axons in the intact nervous system produces muscular contraction. At low levels of motor stimulation, the contraction may only be detected by palpation of innervated muscles. As the amplitude (or pulse duration) is increased, the contraction soon becomes strong enough to produce visible joint movement.

The form and strength of muscular contraction produced by ES are determined by the same two processes involved in voluntary contraction. There is progressive activation of greater numbers of alpha motoneuron axons as the amplitude of stimulation is increased, i.e., recruitment. Motoneuron recruitment increases the number of muscle fibers activated (Fig. 3.23) and hence increases the force output of the stimulated muscle.

Compared to volitional recruitment, however, recruitment order in ES tends to be reversed. In voluntary contraction, motor units are recruited from smallest to largest as the requirements for force are increased. In stimulated contraction, recruitment tends to occur from largest to smallest as the stimulation strength is gradually increased (4). Stimulated contraction occurs by an activation of type FF (fast twitch, fatigable) motor units first, followed by type

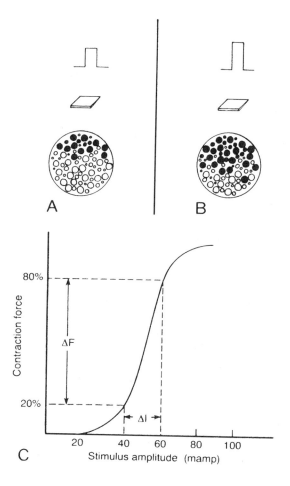

Figure 3.23 (**A** and **B**) Pattern of peripheral axon recruitment in response to stimuli of increasing amplitude. (**C**) Relationship between amplitude of stimulation and force of muscular contraction. (Redrawn with permission from Benton LA, Baker LL, et al. Functional electric stimulation—A practical clinical guide. Downey, CA: The Professional Staff Association of the Rancho Los Amigos Hospital, 1981.)

FR (fast twitch, fatigue-resistant) units, and ending with type S (slow twitch, fatigue-resistant) units. This reversed order of motor unit recruitment in electrically induced contractions is not as stable as that for voluntary contraction. For example, if the axons of fatigue-resistant units are located significantly closer to the stimulating electrodes than axons for type FF units, these units may be recruited before the fatigable type.

Recruitment in stimulated contractions is controlled by varying either the magnitude or the duration of each stimulus or some combination of the two. Increasing either amplitude or pulse duration increases the phase charge of each stimulus, increasing the activation "punch" in each waveform.

Figure 3.23C shows the relationship between the amplitude of fixed-duration stimuli delivered at a tetanic frequency and the resultant contraction force of activated muscle. The graph shows that after the motor threshold is exceeded, very small increases in stimulation amplitude produce relatively large increases in the force of muscular contraction as recruitment increases rapidly. During the use of ES to activate muscle, users must exercise caution in the adjustment of stimulation amplitude to avoid undesirable marked increases in the level of muscle contraction.

On initial exposure to ES for muscle activation, patients frequently exhibit a low tolerance to the stimulation. As a consequence, evoked muscle contraction may not be sufficient to produce either physiologic changes or appropriate joint movements. Repeated exposure to ES is generally accompanied by an increase in patient tolerance and associated increases in the force of evoked contraction. Increases in stimulation amplitude may be possible not only from session to session but also between individual contractions within the same treatment session.

The second primary mechanism regulating the force of contraction in voluntary contraction, rate coding, is also a primary mechanism in regulating stimulated contractions. If the stimulation frequency is low, in a range of 1–5 pps, elicited contractions will be twitchlike in character. If the frequency of stimulation is gradually increased, partially fused contractions are formed followed by fused tetanic contractions as the stimulation frequency rises above a range of 25–30 pps. As in voluntary contraction, as the frequency of motor unit activation rises, the force output of muscle rises.

Another key difference between voluntary and electrically induced contractions is the synchronous activation of all motor units in stimulated contractions. The initiation of thick and thin filament interaction occurs in all recruited fibers at about the same time.

It should be readily apparent by now that electrically elicited contractions are not truly physiological contractions. In spite of this fact, electrical stimulation can be used clinically to produce physiological changes within muscle for strengthening and movement, which allow individuals to function better in their daily activities. Both isometric and isotonic contractions can be produced by the electrical stimulation of muscle. Isometric contractions can be produced by physically blocking the movement about a joint or by simultaneously stimulating both the agonists and antagonists acting on it. Isotonic movements can be evoked by ES when the joint on which the stimulated muscle acts is not stablized. A common application of an isotonic movement produced by stimulation is ankle dorsiflexion during ambulation to assist in the swing phase of gait. Such stimulation is frequently triggered by some special switching device placed beneath the sole of the foot. In eliciting isotonic contractions, the user must realize that very strong isotonic contractions to the fully shortened position of a muscle may produce severe cramping in the belly of the stimulated muscle, and hence such contractions should be avoided in most therapeutic applications.

The application of an uninterrupted train of pulses at an amplitude high enough to elicit muscle contraction will very rapidly induce muscle fatigue. To combat this potential problem in clinical applications, stimulator trains are commonly interrupted at prescribed intervals. When the stimulation is adjusted to a 1:1 on time to off time ratio (e.g., 10 sec on, 10 sec off), muscle tends to fatigue quite rapidly. In contrast, 1:5 ratios (e.g., 10 sec on, 50 sec off) tend to significantly reduce muscle fatigue.

Noxious-Level Stimulation

The final level of clinical ES is that which is painful to the subject and is referred to as **noxious-level stimulation**. In this situation, the intensity of stimulation is so high that many A_δ and C nerve fibers, which normally carry signals associated with painful stimuli, are activated. Action potentials produced in first-order nociceptive afferents in turn activate the spinothalamic pathways leading to the somatosensory cortex and result in the perception of pain. Since the level of stimulation required to activate the small-diameter pain afferents is generally very high, strong muscular contractions often accompany this form of stimulation if motor axons are located near enough to the stimulation electrodes. If noxious stimulation is carried out in a region where no motor axons or muscle fibers are present (e.g., over a bony prominence), a painful response may be elicited without the production of any muscular contraction.

The precise stimulation parameters that evoke a painful response will vary from patient to patient. Other factors such as electrode placement, stimulation waveform, and type or location of sensory fibers in the region of stimulation will also influence the level of stimulation that will evoke a noxious response. Details regarding these issues will be addressed in subsequent chapters.

Summary

This chapter has reviewed the essential features of nerve and muscle morphology and physiology in order to provide a foundation for understanding the clinical application of neuromuscular electrical stimulation. In addition, volitional and stimulation-induced muscle contraction properties have been compared and contrasted.

Specific techniques commonly employed in clinical settings for the management of selected disorders will be presented in subsequent chapters. More detailed information on the structure and function of nerve and muscle can be found in comprehensive references and reviews. Astute users of electrotherapy must remain sensitive to the fact that results of ongoing research and clinical studies must be integrated into their knowledge. Such new information may prompt significant changes in the application of electrical stimulation to obtain optimal therapeutic benefit.

Study Questions

1. a) What are the two main ions associated with the establishment of the resting membrane potential? b) What is their relative distribution across the membrane of excitable cells?

2. What components of excitable cell membranes are responsible for the change in membrane permeability when nervous tissue is excited to produce an action potential?

3. What is the sodium–potassium pump?

4. Draw a graph of the voltage changes occurring at one point in an excitable membrane when an action potential is produced. Be sure to accurately label axes.

5. What are the three functional types of axons in a mixed peripheral nerve?

6. What is the difference between orthodromic and antidromic propagation in peripheral nerve axons?

7. What are the connective tissue layers in peripheral nerve? Name one major function of each layer.

8. The primary component of striated skeletal muscle is the ___(a)___ . Bundles of contractile proteins are called ___(b)___ . The myofilament are composed of the contractile proteins, ___(c)___ and ___(d)___ , along with the two regulatory proteins, ___(e)___ and ___(f)___ .

9. Describe the excitation–contraction coupling process in a stepwise manner.

10. The two primary factors regulating the force of skeletal muscle contraction in either voluntary or stimulated contractions are ___(a)___ and ___(b)___ .

11. Define the motor unit. Describe the differences between the three different types of motor units.

12. Two factors that determine the order of activation of peripheral nerve fibers in response to electrical stimulation are ___(a)___ and ___(b)___ .

13. (a) Define sensory-level stimulation.

    (b) Define motor-level stimulation.

    (c) Define noxious-level stimulation.

14. To increase the recruitment of motor nerve fibers in NMES, either the ___(a)___ or the ___(b)___ of the stimuli may be increased.

15. What is the function of the sarcoplasmic reticulum?

References

1. Erlanger J, Gasser HS. Electrical signs of nervous activity. Philadelphia: University of Pennsylvania Press, 1938.

2. Gasser HS, Grundfest H. Average diameters in relation to spike dimensions and conduction velocity in mammalian fibers. Am J Physiol 1939;127:393.

3. Lloyd DPC, Chang HT. Afferent fibers in muscle nerves. J Neurophysiol 1948;11:199-207.

4. Trimble MH, Enoka RM. Mechanisms underlying the training effects associated with neuromuscular electrical stimulation. Phys Ther 1991;71(4):273-282.

Suggested Readings

Cormack DH. *Ham's Histology, 9th ed*. Philadelphia: JB Lippincott, 1987.

Nicholls JG, Martin AR, Wallace BG. *From neuron to brain*. Sunderland, MA: Sinauer Associates, Inc., 1992.

Mountcastle VB, ed. *Medical physiology: Volume 1, 14th ed*. St. Louis, MO: C V Mosby, 1980.

Vander AJ, Sherman JH, Luciano DS. *Human physiology: The mechanisms of body function, 6th ed*. New York: McGraw-Hill, 1994.

Chapter 4

Electrical Stimulation of Muscle: Techniques and Applications

Anthony Delitto,
Lynn Snyder-Mackler,
and Andrew J. Robinson

Although electrical stimulation of muscle has long been used therapeutically, two factors precipitated a renewed interest in the effects of electrical stimulation on innervated muscle. The first factor was the development of new, more versatile types of stimulators for pain control; stimulators were therefore more readily available for **neuromuscular electrical stimulation (NMES).** A second factor that increased interest in NMES was reports in the mid-1970s on the effectiveness of NMES training programs to promote strength development in elite athletes and healthy individuals. Contemporary work on NMES effects on skeletal muscle focuses not only on its effects on healthy muscle, but also on its usefulness in the management of weak muscles.

The purposes of this chapter are to: (a) describe general adaptations of skeletal muscle to changes in patterns of activation, (b) describe procedures used to assess muscle strength and endurance using NMES, (c) review studies examining the effectiveness of NMES for strengthening healthy and weak muscle, (d) outline the general clinical considerations and procedures for NMES for strengthening muscle, (e) describe common features of NMES devices, and (f) outline conditions considered to be precautions or contraindications to the use of NMES. In addition, case studies will be presented at the end of this chapter to illustrate how NMES for strengthening can be integrated into a comprehensive treatment plan for patients with muscle weakness.

Muscle Plasticity in Response to Electrical Stimulation

Skeletal muscle characteristics are not immutable (1). In response to changes in the use of muscle, its structural, biochemical, and physiological characteristics adapt to meet the imposed demands more appropriately.

Adaptations to prolonged, low force level activity

Muscles that are required to perform at relatively low force levels for prolonged periods of time in day-to-day activities develop an enhanced capability to provide ATP from energy stores to fuel repetitive muscle contraction. This "endurance" muscle has an enhanced capability to oxidatively metabolize muscle fats, carbohydrates, and protein. Frequent low-level use of skeletal muscle increases the activity of oxidative metabolic enzymes within muscle and is associated with an increase in the number of mitochondria—the primary location of these enzymes. The increase in muscle's oxidative enzymes in response to use is accompanied by an increase in the content of the oxygen transport protein, myoglobin, and a rise in the number of capillaries bringing oxygen to the muscle fibers. These changes characterize the adaptive muscle response not only to voluntary endurance training programs, but also to the changes in use brought about by chronic low-level electrical stimulation of muscle (2).

Adaptations to intermittent, high force level activity

When muscle is activated so that the majority of muscle fibers are recruited, high forces are generated. The strength of these contractions requires high energy; as a result, the energy stores used to fuel these contractions are very rapidly depleted. Consequently, this form of contraction can be sustained for only very short periods of time before fatigue begins to set in. In spite of the rapid onset of fatigue, this pattern of muscle contraction is associated with changes in the muscle that lead to increases in muscle strength.

Histochemical, biochemical, and physiological studies reveal that the major adaptation of muscle to high force level contractions is an increase in the content of muscle's contractile proteins, actin and myosin. As the amount of contractile protein in muscle fibers is increased, the number of cross-bridges that are formed increases. The force produced by the activation of muscle is directly proportional to the number of cross-bridges that form, which accounts for the marked increase in muscle's ability to generate tension in response to high-level activation. Both volitional activation and electrical stimulation are capable of inducing a rise in the amount of muscle fiber contractile protein.

Assessment of strength and endurance

Neuromuscular electrical stimulation (NMES) can be used to augment the strength of either healthy or injured muscle. Strength is defined as the maximal force or torque a muscle or muscle group can generate at a specified velocity (3). Muscle endurance is the reciprocal of muscle fatigue, which has been defined as a decrease in the force-generating capacity of a muscle after recent activation (4, 5). This section will review electrical tests to assess muscle strength and endurance and the results of experimental studies on (*a*) levels of torque produced in muscle in response to NMES, (*b*) comfort, (*c*) the strengthening effect of NMES training in normal and patient populations, and (*d*) muscle fatigue in response to NMES.

Muscle strength is typically assessed as **maximum voluntary contraction (MVC)** force or torque. Isometric, isotonic, isokinetic, eccentric, or concentric contractions can be measured. In our experience, determination of MVC is best done by targeting combined with a superimposed electrical stimulus. A superimposed pulse or a burst of supermaximal electrical stimulation to evaluate the force-generating capacity of a muscle has been used as a research tool for many years. The patient contracts volitionally and a supramaximal stimulus (100 pulses per second [pps], 600 μsec duration) is superimposed on the contraction. If the patient is not activating the muscle completely, an increase in force as a result of the stimulus is recorded. If the patient is contracting fully, no augmentation of force is noted in response to stimulation, or a slight decrease in force is seen (6). Recently, this technique has been used to evaluate muscles with weaknesses attributed to a significant central nervous system component such as a failure of voluntary activation or "reflex inhibition" (7, 8, 9). The potential of this technique for generating useful clinical information is great. In fact, the failure to use this technique may result in a gross underestimation of the torque-generating capability of a muscle (10). Shih recently demonstrated the magnitude of this underestimation in the quadriceps muscles of a group of healthy subjects and a group with weak quadriceps. The underestimation averaged 20% in both groups. Errors in clinical judgment can arise from the failure to accurately measure strength. Presumably, the wrong clinical strategy can be implemented as a result of this erroneous information. A volitional strength test does not determine whether the measurement reflects muscle atrophy, motivational factors, or an inability to activate morphologically normal motor units (so-called reflex inhibition). The appropriate clinical strategies to treat each of these causes of low force output differ substantially. In the first case, a program of volitional strengthening exercise is most common. The second case, motivation, requires only training in fully activating the muscle, as there is no true weakness or inhibition. In the third case, volitional exercise would not help and another strategy, such as the use of electromyographic (EMG) biofeedback, might be appropriate.

Snyder-Mackler and colleagues recently used the burst superimposition technique to evaluate the presence of reflex inhibition in a group of patients with anterior cruciate ligament (ACL) rupture (9). Here, too, the burst superimposition technique resulted in a higher measured torque value. Training patients to produce an MVC was not difficult, and all were trained within a single test session. No patient required more than four trials; most required only two to learn to maximally activate the quadriceps. Snyder-Mackler and colleagues demonstrated that some patients do have a failure of voluntary activation after ACL rupture. If the burst superimposition technique had not been used, the weakness could have been attributed to atrophy rather than reflex inhibition. These patients should not respond well to the typical treatment prescribed for weakness: a volitional exercise program.

Volitionally and electrically elicited fatigue tests have been used to quantify muscle fatigue. A potential disadvantage of tests in which volitional contraction forces are used to assess the amount of fatigue is that it may not be possible to isolate the site of fatigue. If subjects are not well motivated or have a disorder that affects central drive (e.g., following a cerebral vascular accident), measured losses in force generation may not reflect actual failure of the contraction within muscle. However, clinically, volitional effort may be most representative of the clinical entity being evaluated.

Many investigators have attempted to use electrically elicited fatigue tests as a clinical tool to sustain force output through repeated contraction (11, 12). Before reviewing the specifics of this literature, it is important to point out the limitations with regard to clinical fatigue tests and the clinical decisions that can be made based on the results of these tests. Operationally defining fatigue and endurance is an elusive task whether using electrically elicited or volitional contractions. Physiologically, fatigue has a very strict definition related to the oxidative capacity of a muscle: an increase in the aerobic capacity (VO_2). Clinicians, however, whether they use volitional or electrical tests to evaluate changes in endurance capabilities in response to treatments, may actually be attempting to quantify something other than physiological fatigue. Does the clinical observation that muscles appear to fatigue more quickly indicate that the involved muscles are truly less endurant or are the observed phenomena due solely to muscle weakness? Since the involved muscles are weaker, they must be contracting at a higher percentage of their capacity to perform typical functional tasks, such as ambulation. Thus, even if both involved and uninvolved muscles fatigue at the same rate, the weaker muscles will be able to maintain functional force levels for shorter periods of time and will appear to the clinician to fatigue sooner.

Electrically elicited fatigue tests have provided new insights into muscle physiology. The physiological differences between electrical and volitional activation (see Chapter 3), and the facts that the stimulus to the muscle can be standardized and is not affected by central factors such as motivation and change in effort, make these tests attractive, at least in theory. Snyder-Mackler and Binder-Macleod have used a modification of a fatigue test, originally described by Burke and colleagues to categorize cat single motor units, to assess muscle performance in patients following surgical repair of their anterior cruciate ligaments. A 600-μsec monophasic pulse is delivered at a train frequency of 100 pps for one-third of a second each second for 3 minutes (13). McDonnell and colleagues (11) investigated the use of electrical stimulation to assess the fatigue characteristics of muscle using a commonly available electrical stimulator. Using a 2500-Hz alternating current interrupted at 50 bps with a 5 sec on : 2 sec off duty cycle, these workers recorded the torque decrement over 50 electrically elicited contractions of the quadriceps femoris of healthy subjects. From an initial value of 60% of maximal voluntary isometric torque, declines of 65% were noted. This measure of electrically elicited fatigue was reproducible in repeated testing of healthy subjects. Although physiological insights, such as contractile characteristics and muscle fiber composition, are readily available from these results, whether useful clinical information can be gained from these tests has yet to be determined.

Torque-Generating Capabilities of NMES

Voluntary exercise to augment strength is based on the overload principle: training at high contraction intensities (greater than 75% of maximal) for a low number of repetitions (usually fewer than 10). In an effort to establish analogous paradigms for NMES strength training, investigators have compared maximal electrically elicited torque and **maximal voluntary isometric torque (MVIT)**.

Kramer and others (14) compared the torque-generating capabilities of three different forms of stimulation: ES 1 (1-μsec pulse duration (PD), asymmetrical alternating current delivered at 100 Hz), ES 2 (200-μsec PD, asymmetrical, biphasic pulsed current delivered at 100 pps), and ES 3 (2-μsec PD, monophasic pulsed current delivered at 45 pps). The MVIT of the quadriceps femoris was recorded for each machine. The electrically elicited torques produced were 93% of MVIT for ES 1, 67% of MVIT for ES 2, and 53% of MVIT for ES 3. These investigators also found that superimposition of NMES onto voluntary isometric contractions produced no appreciable increase in torque.

Walmsley and coworkers (15) performed a similar study comparing the torque-generating capabilities of four different stimulators: ES 1 (2500-Hz alternating current interrupted at 10-msec intervals to produce 50 bursts/sec), ES 2 (250-μsec PD, 4000-Hz alternating current modulated at 75 beats/min), ES 3 (400-μsec PD, asymmetrical, biphasic pulsed current delivered at 50 pps), and ES 4 (200-μsec PD, monophasic pulsed current delivered at 60 pps). The electrically elicited torques produced in the quadriceps femoris were 87% of MVIT for ES 1, 46% of MVIT for ES 2, 84% of MVIT for ES 3, and 68% of MVIT for ES 4. Superimposition of voluntary contraction on NMES contraction increased torque levels for all stimulators but in no instance exceeded MVIT.

DeDomenico and Strauss (16) tested the ability of a variety of stimulators to produce torque in the quadriceps femoris of healthy females. The average evoked isometric contraction ranged from 47 to 74% of MVIT and an interclass correlation demonstrated a direct relation between pulse duration and torque-generating capability. These investigators noted, however, that there was tremendous individual variation among subjects.

Snyder-Mackler and colleagues (17) compared the isometric torque-generating capabilities of three stimulators: ES 1 (2500-Hz alternating current interrupted at 10-msec intervals to produce 50 bursts/sec), ES 2 (250-µsec, 4000-Hz alternating current modulated at 50 beats/min), and ES 3 (400-µsec, symmetrical, biphasic pulsed current delivered at 50 pps). The electrically elicited torques produced in the quadriceps femoris expressed as percentages of MVIT were 68% for ES 1, 45% for ES 2, and 61% for ES 3. These investigators also noted tremendous intersubject variation.

Locicero studied the isotonic torque-generating ability of a 2500-Hz alternating current modulated at 50 bursts/sec at three different velocities (0, 60, and 240 degrees/sec). The stimulator was able to generate an average of 93 to 104% of the MVC of the quadriceps femoris muscles of 20 women and 10 men (18).

Delitto and his colleagues evaluated the torque-producing capability of 2500-Hz current interrupted at 75 bursts/sec in healthy college students while studying discomfort with stimulation (19). Similar to Snyder-Mackler and coworkers (17), they found that subjects who participated were able to generate an average of 70% of their MVIT.

From the levels of contraction evoked in these and other studies, one can conclude that many forms of NMES are capable of producing torque comparable to those volitional contraction levels that induce strengthening on repeated application. But can muscles be activated at levels greater than MVC? Delitto and colleagues (20) reported average training contraction intensities of 112% of the quadriceps MVIT in a single-subject study of an elite weight lifter, the only study to date where training contraction intensities in excess of 100% of MVIT have been demonstrated. It should be noted, however, that this subject is precisely the type of subject described by Kots.

Enoka has made the argument that the MVC does not represent the maximum force that a muscle can exert. Rather, the MVC represents a relative maximum, one that can be achieved under the constrained conditions of the test. Most studies that have examined motor unit discharge during high-force contraction have reported that neither the instantaneous nor the average discharge rate approach the frequencies that are necessary to electrically elicit the maximum force (21, 22, 23). This argument supports the possibility of the kind of activation described by Kots.

Dose–Response Relationship in NMES for Strengthening

One assumption implicit in the studies of torque-generating capability is that stronger stimulation will result in greater strengthening effects. In a recent study of 52 patients after anterior cruciate ligament reconstruction, Snyder-Mackler and coworkers (24) described a dose–response relationship for NMES. They found that the greater the electrically elicited training contraction force, the greater was the quadriceps femoris muscle recovery after anterior cruciate ligament reconstruction. Training contraction intensity was linearly related to quadriceps femoris muscle strength only for training contraction intensities above 10% of the uninvolved maximum voluntary contraction force. This suggests that there is a threshold training contraction intensity for strengthening of the quadriceps femoris muscle group using this technique. They also found that there was no correlation between the electrically elicited training contraction force and quadriceps femoris muscle recovery when portable stimulators were used. The portable stimulators were set to deliver a 300-μsec phase duration stimulus at 50 pps and a maximum current amplitude of 100 mA. The average training contraction intensity for the portable stimulation group was 8.9% at an average milliamperage level of 83 mA (range 64–100 mA), which is 83% of the capacity of the stimulator. Portable stimulators were not capable of producing average training contraction intensities above the 10% threshold in the quadriceps femoris muscles of the patients in this study.

Comfort and Electrical Stimulation

One classic approach to studying experimental pain methodologies is to use electrical stimulation. Subject discomfort is often the limiting factor in using electrical stimulation in clinical settings, especially when high contractile forces are sought for strength training regimens. There have been numerous studies that have investigated discomfort associated with electrical stimulation as administered clinically (19, 25, 26, 27). The methodologies used in these studies were visual analog scales, categorical scales, forced choice techniques, and magnitude estimation.

Most studies have focused on subject preference for various current characteristics, studying electrical stimulation as though discomfort was solely a function of current form. Delitto and Rose (26) evaluated the relative comfort of three different waveforms (triangular, sinusoidal, and square), delivered at 2500 Hz and interrupted at 10-msec intervals to produce 50 bursts/sec, and found that "individual preferences" existed for each current form. A similar conclusion was reached by Bowman and Baker (25) when comparing 2500-Hz sinusoidal current delivered in 10-msec bursts at 50 bursts/sec to "single pulsed" current (symmetrical biphasic current, 300 μsec pulse duration) when using a forced choice paradigm. In a study comparing identical current forms, Grimby and Wigerstadt-Lossing (27) concluded that subjects preferred the single-pulsed currents.

Two limitations are apparent in the previous studies. First, the contractile force elicited differed substantially, with Delitto and Rose (26) eliciting contractile forces in excess of 60% of MVC and Bowman and Baker (25) eliciting contractile forces of as little as 20 foot-pounds, the latter most likely less than 10% of MVC. Second and more importantly, the discomfort associated with electrical stimulation is studied as though it is a sole function of current form, as if one particular current will stimulate the maximal number of motor neurons and the minimal number of nociceptors. This is not consistent with most pain theories, which make it very clear that discomfort and pain are not solely a function of nociceptor input and instead are moderated by cognitive and behavioral factors.

Delitto and colleagues (19) attempted to address these limitations by studying cognitive behavioral influences of subject tolerances to electrically elicited contractions at contractile forces that are reported to have training (e.g., strengthening) effects as well as electrical stimulation in which there were no contractile forces (nociceptive input only). The results of this study suggested that three variables are important when considering discomfort associated with electrical stimulation: the subject's preferred coping style, whether the stimulus caused a muscle contraction, and whether the subject was judging the intensity or the unpleasantness of the stimulation. Therefore, they suggest that clinicians be aware that individual differences exist in preferred coping styles when a subject is faced with an aversive event (e.g., electrical stimulation), which may include "monitoring" (information-seeking) or "blunting" (information-avoiding) behaviors. The clinician should attempt to match the environment to the subject's coping style. For example, if a subject prefers a blunting strategy, providing a quiet environment with some type of distraction (e.g., music with headphones) may improve their tolerance to electrical stimulation. In addition, the clinician should be aware that in addition to the nociceptive input, the electrically elicited contractile force contributes uniquely to the discomfort associated with electrical stimulation; that is, the greater the force, the more intense is the discomfort irrespective of current intensity.

NMES Strength Training Studies

Two issues must be kept in mind when reviewing the literature related to the strengthening effect of electrical stimulation: subject selection (patients or healthy college students) and the type of stimulator used (battery-powered or line-current driven).

There are numerous studies in which the general purpose was to compare electrical stimulation to volitional exercise. Taken as a whole, the results of these studies vary. The picture becomes more distinct when the studies are separated according to the characteristics of the populations from which each study obtained samples.

NMES-Induced strengthening in healthy subjects

Massey and coworkers (28) examined the strengthening effect of NMES in four groups of United States Marine Corps recruits: group A (electrical stimulation alone), group B (progressive resistive exercise), group C (isometric exercise), and group D (unexercised control). A monophasic, pulsed current of unknown pulse duration was delivered at 1000 pulses per second (pps). The middle deltoid, pectoralis major, trapezius, biceps, triceps, wrist flexor, and wrist extensor muscles were stimulated bilaterally in group A at the **maximally tolerated contraction (MTC)** level. All groups trained three times per week for 9 weeks. Isometric shoulder, elbow, and grip strengths were used as dependent measures. Training with NMES produced strength gains equivalent to those of volitional exercise (groups B and C). All three experimental groups demonstrated increased strength versus unexercised controls. This study is one of the few that has shown similar increases in strength for both.

Currier and associates (29) compared simultaneous NMES and voluntary isometric exercise to voluntary isometric exercise alone and to an unexercised control group. A monophasic pulsed current of unspecified pulse duration was delivered at 25 pps. The quadriceps femoris muscle was stimulated to MTC. Training sessions were five times per week for 2 weeks. Both experimental groups increased their isometric knee extension torque when compared to the control. Increases in knee extension torque were comparable for both experimental groups.

Halbach and Straus (30) compared NMES to isokinetic exercise for muscle strengthening in only six subjects. A monophasic pulsed current of 500-μsec duration was delivered at 50 pps to the quadriceps femoris to elicit maximal tolerated isometric contractions (MTICs) in training. The voluntary exercise group was trained isokinetically. Training occurred five times per week for 3 weeks. NMES was found to be inferior to isokinetic exercise for increasing strength. However, the voluntary exercise group was trained in the same fashion as all groups were tested, whereas the NMES group trained isometrically, which may have affected the validity of the dependent measure.

Eriksson and coworkers (31) compared quadriceps NMES (using a 500-μsec monophasic pulse delivered at 200 pps at 15 sec on and 15 sec off) to isometric exercise in eight subjects. MTICs and isometric exercise training occurred three to four times per week for 4 weeks. Increases in isometric strength were equivalent for both groups. Using a multitude of t-tests, the researchers reported biochemical changes after stimulation similar to those seen in intense voluntary exercise.

Romero and others (32) compared NMES to an unexercised control group. The quadriceps femoris was stimulated to MTIC by a 250-μsec half-cycle duration, alternating current delivered at 2000 Hz. Training sessions occurred two times per week for 5 weeks. Using a number of t-tests in their analyses, the authors reported significant increases in the muscle stimulation group in a few of the dependent measures.

NMES using a 2500-Hz, sinusoidal alternating current interrupted for 10 msec at 10-msec intervals (resulting in 50 10-msec bursts/sec) became the focus of a great deal of attention beginning in 1977. These current characteristics were described by Soviet researcher Yakov Kots during a symposium on NMES (1977, unpublished) and were purported to produce intense muscle contractions (110–130% of MVIC) with no discomfort in elite athletes. Training for 3–4 weeks reportedly produced 30–40% strength gains as well as functional gains (in vertical jump and other measures). Endurance was reported to be increased after 6–8 weeks of training. No data were presented (33).

Spurred by these reports, Kramer and Semple (34) examined the effects of isometric exercise and muscle stimulation of the quadriceps femoris. Due to the unavailability of a machine capable of producing the current characteristics described above, these investigators used a symmetrical, rectangular alternating current of unspecified pulse duration and frequency to stimulate the muscle to MTIC. Using similar training regimens, NMES alone, NMES plus isometric exercise, and isometric exercise alone were compared to unexercised controls. All experimental groups produced significant gains in isometric strength when compared with the control group, but there was no difference among the experimental groups.

In the late 1970s a stimulator was developed that produced the current characteristics described by Kots. This device was used to provide training sessions of 15 10-sec contractions followed by 50 sec of rest. Using this device, Currier and Mann (35) compared isometric exercise, NMES, simultaneous NMES, and isometric exercise with an unexercised control group in the quadriceps femoris. These authors were the first to report the amplitude of the stimulated contraction as a percentage of the MVIT, producing a standard that was analogous to that used in the voluntary exercise and strengthening literature. In this study, current amplitude was increased to 60% of pretest MVIT. Training occurred three times per week for 5 weeks. Stimulation produced significant gains in isometric strength in quadriceps when compared to controls, but not when compared to the isometric exercise group. No additional benefit was noted with simultaneous NMES and isometric exercise. In addition, no increase in isokinetic torque-producing capability was found in any of the groups.

Laughman and associates (36) subsequently published the results of a similar study. Average NMES intensity was 33% MVIT, whereas the voluntary exercise group averaged 78% MVIT. Quadriceps training consisted of five sessions per week for 5 weeks. Both the NMES and exercise groups demonstrated significant increases in MVIT of the quadriceps femoris muscle when compared with unexercised controls, but, again, no difference was found between the two experimental groups.

Selkowitz (37) compared NMES to unexercised controls and continuously monitored training intensity, which averaged 91% of pretest MVIT. Training occurred seven days per week for 4 weeks. The NMES group increased quadriceps MVIT by 18%, suggesting that the pretest MVITs may have been artificially low. Training intensities and strength gains in the experimental group were positively and significantly correlated.

McMiken and coworkers (38) compared NMES to the quadriceps femoris (using a 100-μsec monophasic pulse delivered at 75 pps) to voluntary isometric exercise. NMES training intensities were approximately 80% of MVIT. Subjects trained for 10 sessions in a 3-week period. Muscle stimulation produced significant increases in MVIT when compared to pretest values, but there was no difference between the two experimental groups.

Mohr and colleagues (39) compared NMES to isometric exercise and unexercised controls using a 20-μsec pulse duration, monophasic pulsed current at 50 pps. Subjects trained five times per week for 3 weeks. No significant increase in MVIT was noted in the quadriceps femoris of the NMES group when compared with unexercised controls. This type of current may not produce contractions of sufficient intensity to elicit a strengthening effect because of its extremely short pulse duration.

Stefanovska and Vodovnik (40) compared a 2500-Hz alternating current burst-modulated 25 times per sec to a 300-μsec, monophasic pulsed current delivered at 25 pps to an unexercised control group. Quadriceps training intensities were only 5% of MVIT. Subjects trained five times per week for 3 weeks. Their training paradigm included MVIT measurements for the NMES groups prior to and following each training session. Although MVIT increases were noted for both NMES groups when compared to unexercised controls, the relatively low NMES training intensities compared to the high amplitude of multiple daily MVICs might indicate that the MVIT increases in the NMES groups were attributable to the voluntary exercise of the MVIT measurements.

Wolf and coworkers (41) compared an exercise group, a simultaneous NMES and exercise group, and an unexercised control group. The stimulator used in this study delivered a 300-μsec, monophasic pulse at 75 pps. Intensity was set at MTIC of the quadriceps, bilaterally. A full body squat was the training exercise. Dependent measures included force measurements from a computerized squat machine and two functional measures, 25-yard-dash time and vertical jump. Again, NMES and exercise was comparable to exercise alone for quadriceps strengthening. Of particular interest in this study is the carryover to functional tasks of the increase in strength. In both experimental groups there was an increase in vertical jump and decrease in 25-yard-dash time.

Alon and coworkers (42) examined the effect of NMES alone or combined with volitional exercise on the strength of the abdominal musculature. Subjects participated in NMES or voluntary training three times per week for 4 consecutive weeks. Measurements of abdominal strength were taken at 1-week intervals over the course of the study. NMES to the abdominal musculature was provided with symmetrical biphasic waveforms at maximal tolerable intensities with 5 sec on time and 5 sec off time initially. Contract/relax cycles were increased to 7.5, 10, and 12.5 sec over the remaining weeks of training. The number of contraction repetitions was increased by 20% weekly from baseline voluntary repetition levels determined in pretest measures. NMES combined with stimulation produced the greatest increases in muscle strength followed by stimulation alone.

NMES strengthening in patient populations

William and Street (43), in a study of 20 patients with quadriceps femoris atrophy, used 20 min of NMES per session for 13 sessions and found "recovery of normal quadriceps function in the majority of cases." The current characteristics were not described and no control group was used.

Johnson and coworkers (44) studied the effects of electrical stimulation on quadriceps MVIT in 40 patients with chondromalacia patellae. NMES of unspecified current characteristics was delivered at 65 Hz, three times per week for 6 weeks. Each session consisted of 10–15 min of NMES. A 36% increase in MVIT was noted. No control group was employed.

Eriksson and Haggmark (45), in their study on healthy subjects cited earlier, used the identical regimen to compare NMES and isometric exercise in eight patients who had undergone knee ligament surgery. They reported less observable atrophy of the thigh, marked functional improvement (using an ordered rating scale), and an increase in oxidative activity in the NMES group as compared to the isometric exercise group.

Godfrey and colleagues (46) compared isometric exercise to NMES in 35 patients referred for "strengthening" who had recently either undergone knee surgery or sustained a knee injury. A 60-Hz alternating current was used to produce MTICs of the quadriceps femoris. The isometric exercise group trained at 75% of MVIT. Patients trained 5 days per week for 3 weeks, 10–15 min per day. The dependent measure was peak isokinetic torque measured at several velocities. Increase in peak torque at the slowest isokinetic speed (3 rpm) was significantly higher in the NMES group than in the exercise group. An aggregate score developed from the isokinetic measures at the three tested speeds (3, 10, 25 rpm) was also significantly higher in the NMES group than in the exercise group.

Gould and coworkers (47) compared NMES to isometric exercise in 20 patients who had undergone open joint meniscectomy. A 100-μsec, monophasic pulsed current delivered at 35 pps was used to produce MTICs of the involved quadriceps femoris. Training occurred daily for 2 weeks. The percent of MVIT (compared to the unoperated limb) was significantly greater for the NMES group.

Grove-Lainey and associates (48) compared isometric exercise alone to NMES plus isometric exercise in a crossover design study on the quadriceps femoris of postoperative knee patients. The stimulator produced a 250-μsec alternating current modulated at 100 bps. Training consisted of 22 sessions of MTICs over a 6-week period. Although there was no significant difference reported between the groups at the end of the 6-week period, the authors concluded that the MVIT changes (posttest minus pretest scores) tended to be greater in the 3rd to 6th week in the NMES plus exercise group when compared to exercise alone.

Morrissey and coworkers (49) compared NMES to unexercised patients following anterior cruciate ligament (ACL) reconstruction. The portable stimulator used in this study produced a 350-μsec, monophasic pulsed current delivered at 50 pps at a duty cycle of 10 sec : 50 sec. Training consisted of 6 hours of cycled maximally tolerated muscle stimulation per day. The results showed a less pronounced decrease in quadriceps MVIT and in thigh circumference in the muscle stimulation group at the end of a 6-week immobilization period when compared to unexercised controls. Six weeks later (12 weeks postsurgery, 6 weeks after immobilization and NMES had ceased), all differences between the groups had disappeared.

Singer and others (50) examined the effect of three different stimulators on patients with long-term knee pathology and quadriceps atrophy. The stimulators delivered pulsed current with pulse durations ranging from 75 to 350 μsec and frequencies of 50 and 100 pps. Patients were stimulated to MTIC 15 min per day, 7 days per week for 4 weeks. An aggregate increase in MVIT for the 15 subjects was 22% over pretest values.

Sisk and coworkers (51) investigated the effect of NMES plus isometric exercise versus isometric exercise alone on the quadriceps femoris strength of a group of 22 patients immobilized after ACL reconstruction. The stimulator produced a 300-μsec PD, balanced, symmetrical, biphasic pulsed current at 40 pps and a timing cycle of 10 sec on : 30 sec off. Training was at MTIC, 8 hours per day, 7 days per week, for 6 weeks. No difference in MVIT was noted between the groups.

Currier and coworkers (52) compared NMES and NMES combined with with PEMF (pulsed electromagnetic frequencies) superimposed on volitional quadriceps/hamstring contraction with the knee in full extension. They compared thigh girth before surgery to that after 6 weeks of treatment three times per week with 10 contractions per session. Although using pulsed electromagnetic frequencies is an intriguing approach to strength training, the study by Currier and coworkers represents the only study done on patients with this technique. The emerging technology and cumbersome procedure makes this option less attractive as compared to electrical stimulation.

Delitto and colleagues (53) compared high-intensity electrical stimulation to volitional exercise in the early postoperative period of patients undergoing ACL surgery. Twenty patients were randomly and independently assigned to an NMES and to a volitional exercise group. All quadriceps activity was accompanied by simultaneous hamstring contraction to counter anterior translation forces at the knee. Their posttest design used a dependent measure of isometric strength measured at 65 degrees of flexion within the 4–6 week period after surgery. The results of this study clearly show a better response to high-intensity electrical stimulation than volitional training in the early postoperative period (Fig. 4.1).

Snyder-Mackler and coworkers (54) conducted a similar study, in that the postoperative ACL reconstructions were randomly assigned to an exercise or an electrical stimulation group and a posttest design was employed. A 2500-Hz alternating current, burst-modulated at 75 bps was used to deliver 15 contractions per session, three times per week. Their dependent measures were isokinetic strength (90 and 270 degrees per sec) as well as a more functional quadriceps strength measure that they operationalized as knee excursion during stance.

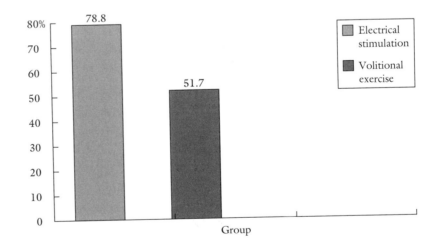

Figure 4.1 A comparison of isometric quadriceps femoris muscle torque expressed as a percentage of the MVIC of the uninvolved quadriceps for electrical stimulation and volitional exercise groups.

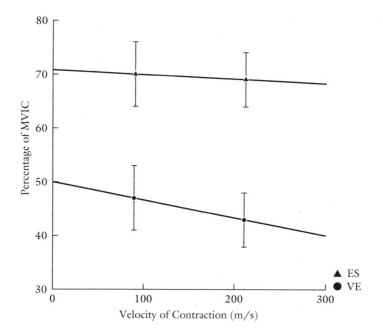

Figure 4.2 Graph showing isokinetic torque of the quadriceps femoris as a function of contraction velocity. *ES*, electrical stimulation group and *VE*, volitional exercise group. (Reprinted with permission from Snyder-Mackler L, Ladin Z, Schepsis A, et al. Electrical stimulation of the thigh muscles after reconstruction of the anterior cruciate ligament. Effects of electrically elicited contractions of the quadriceps femoris and hamstring muscles on gait and strength of the thigh muscles. J Bone Joint Surg [(Am)] 1991;73:1025–1036.)

The isokinetic results for the quadriceps femoris musculature are illustrated in Figure 4.2. Again, high-intensity stimulation showed a clear benefit over volitional training when co-contractions were employed. In addition, the study showed that patients receiving high-intensity NMES had significantly better temporal gait measures. Finally, the Snyder-Mackler and coworkers (54) study showed a significant correlation between quadriceps strength and the amount of knee flexion excursion during stance phase.

Most recently, Snyder-Mackler, Delitto, and colleagues (55) randomly assigned 110 subjects immediately following ACL reconstruction to one of four groups: *(a)* high-level electrical stimulation, *(b)* high-level volitional exercise, *(c)* low-level electrical stimulation (portable stimulator), and *(d)* combined high- and low-level electrical stimulation. After 4 weeks, quadriceps femoris muscle strength and stance phase knee kinematics were assessed. Quadriceps strengths averaged over 70% of the uninvolved quadriceps for the two high-level electrical stimulation groups (high-level electrical stimulation and combined), 57% for the high-level volitional exercise, and 51% for the low-level electrical stimulation group. Knee joint kinematics were again in this study directly and significantly correlated with quadriceps strength.

Starring (56) published a case report of a patient with persistent low back pain. A 2500-Hz alternating current with a burst frequency of 75 bps and a 15 sec : 50 sec on/off ratio was used. Electrodes were placed bilaterally over the erector spinae in the lumbar region. After 6 weeks of treatment the patient was asymptomatic. No measures of lumbar extensor muscle strength were taken, but the author notes that substantial electrically elicited contractions were achieved, necessitating stabilization with straps over the pelvis to prevent anterior pelvic tilt.

Kahanovitz and coworkers (57) compared NMES to an exercised and an unexercised control group in a study of the erector spinae muscles in patients with low back pain. The stimulator produced a 425-μsec PD symmetrical, biphasic pulsed current. Training consisted of daily 20-min sessions of either exercise or maximally tolerated electrical contraction for 4 weeks. The increase in isokinetic strength was significantly greater in the NMES group than in either the exercise or control group.

Karmel-Ross (58) reported on a case series of patients with spina bifida. The quadricep femoris muscles were stimulated during standing or walking for 30 min each day. Stimulation intensity was adjusted to patient tolerance. A battery-powered stimulator using a pulse duration of 347 μsec and a frequency of 35 pps was used. On/off times ranged from 8 sec on : 24 sec off at the beginning of the study to 8 sec on : 8 sec off at the completion of the study. Dependent measures included quadriceps torque and timed functional tasks (level walking, stair ascending, and stair descending). Four of the five subjects improved at the functional tasks over the treatment period, and only two (the two oldest subjects) had improved quadriceps torque production. Karmel-Ross suggested that the muscle testing in the youngest subjects was not reliable.

These studies represent new avenues of application of NMES for strengthening. Controlled randomized trials in these and other populations still need to be undertaken.

NMES for Strengthening Using Portable versus Clinical Stimulators

Several recent studies suggest that the use of portable (battery-driven) electrical stimulators may not be as effective in producing strength gains in muscle as training with clinical (60-Hz, AC-driven) stimulators. Sisk and coworkers (51) found no difference between NMES plus exercise and exercise alone in strengthening the quadriceps following reconstructive knee surgery. Morrissey and associates (49) found NMES using a portable stimulator better than nothing (unexercised controls) for quadriceps strengthening during immobilization, but this apparent effect was only temporary. Wigerstad-Lossing and colleagues (59) compared electrical stimulation and exercise to exercise alone in a sample of ACL patients in the early postoperative period. Twenty-three subjects participated. Patients in the electrical stimulation group simultaneously performed volitional exercise. The results of this study showed a beneficial effect when adding electrical stimulation to the treatment regimen. Draper and coworkers (60) compared EMG biofeedback and portable NMES used in conjunction with quadriceps setting and straight leg raise exercises, three times per day for 30 min per session, for 4 weeks. There was no control group or comparison group in which volitional exercise alone was used. The biofeedback group recovered to 46% of the contralateral quadriceps MVIC, whereas the portable electrical stimulation group recovered to 38%. A comparison of these results with those of Delitto and colleagues (53) and Snyder-Mackler and coworkers (54), who used high-intensity NMES, is illustrated in Figure 4.3. As can be seen in the graph, the magnitude of the experimental effect of the studies using high-intensity NMES is much greater than that of those using the portable stimulation.

Since portable stimulators have current characteristics comparable to those of clinical stimulators and may produce contractions with sufficient amplitude for strengthening, the differing results are puzzling.

Snyder-Mackler, Delitto, and colleagues (24) established a dose–response curve for electrical stimulation regimens designed to improve quadriceps femoris muscle recovery in patients after ACL reconstruction. They analyzed data from a subsample ($N = 52$) of patients receiving electrical stimulation who were involved in a large ($N = 110$), multicenter, randomized clinical trial investigating treatment strategies designed to enhance quadriceps femoris recovery. Training contraction forces were monitored by logging the electrically elicited knee extension torque and expressing this as a percentage of the uninvolved quadriceps maximal voluntary contraction force. After 4 weeks of training, isometric muscle performance was assessed and a dose–response curve was generated. A significant, linear correlation was found between training contraction force and quadriceps recovery. Subjects training with console, clinical generators trained at higher contraction forces than those with portable, battery-operated generators; such training resulted in higher quadriceps femoris recovery and improved gait. The dose–response curve is illustrated in Figure 4.4. These results support the use of clinical, high-output electrical stimulation and do not support the use of low-output or battery-powered stimulators when the goal is quadriceps femoris recovery in the early phases of rehabilitation after ACL surgery.

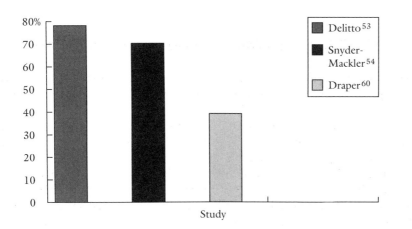

Figure 4.3 A comparison of quadriceps femoris muscle torque expressed as a percentage of the MVIC of the uninvolved quadriceps for electrical stimulation groups in the three cited studies.

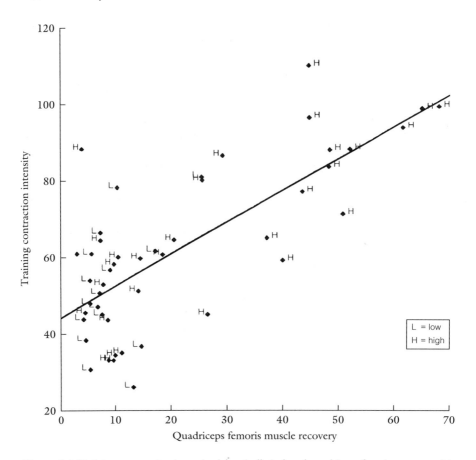

Figure 4.4 Training contraction intensity (electrically induced quadriceps force) versus quadriceps femoris muscle torque expressed as a percentage of the MVIC of the uninvolved quadriceps. (L = low, H = high).

NMES Effects on Muscle Endurance and Fatigue

Few studies have been performed to date to investigate the effects of NMES training on the ability of muscle to sustain force over time with either repeated or sustained voluntary contractions.

Alon and coworkers (61) found that abdominal musculature stimulation training designed to enhance muscle strength did not significantly influence the endurance characteristics of these trunk muscles.

Hartsell (62) examined the effect of NMES training on the volitional endurance characteristics of the quadriceps in normal subjects. NMES training consisted of 10 daily maximally tolerated contractions (10-sec contraction/50-sec rest each), five days per week for a period of 6 weeks. "Square waves" with a pulse duration of 2 msec were applied at a frequency of 65 pps. Quadriceps endurance was found to be increased following the stimulation program, but the small increases were not significantly greater than those achieved through exercise alone.

One of the key problems in determining the effects of NMES on muscle fatigue is the fact that NMES training programs examined to date have been based on volitional strength training programs. No clinical studies are available that have applied the principles of volitional endurance training (low-amplitude contractions, high number of repetitions) to NMES training for improving muscle endurance.

That NMES can induce muscle fatigue is well known. The precise relationship between fatigue and stimulated contraction duration and rest has not been clearly established for most muscles. Packman-Braun (63) has performed a study examining the effects of 1 : 2, 1 : 4, and 1 : 6 (5 sec on : 5 sec off, 5 sec on : 15 sec off, 5 sec on : 25 sec off, respectively) duty cycles on the development of fatigue in the wrist extensor musculature of hemiparetic patients. She found that 1 : 2, 1 : 4, and 1 : 6 timing cycles were progressively less fatiguing. These findings are in general agreement with those of Benton and associates (64) presented in the early 1980s.

Binder-Macleod (65) has recently demonstrated that contraction force and frequency both directly affect fatigue, but their effects are independent of one another. In order to achieve the high force levels necessary for strengthening, frequencies higher than critical fusion frequency (tetany) are required. The higher the frequency, the more fatiguing is the contraction. High contraction intensities also provoke fatigue. The manipulation of these variables to minimize fatigue is critical for functional electrical stimulation. The use of electrical stimulation to augment muscle strength, however, allows for much more flexibility in manipulating on/off times to minimize fatigue while using high frequencies and contractile forces to maximize strength gains.

Conclusions Regarding NMES and Strengthening

The basic question in the muscle stimulation literature is, how does NMES of muscle compare to voluntary exercise or voluntary exercise in combination with NMES in the ability to alter muscle performance? After reviewing the literature on the use of NMES in healthy subjects with no apparent muscle weakness, the following conclusions can be drawn.

1. Increase in strength usually occurs in the NMES group compared to an unexercised control group.
2. No difference generally exists between the NMES group and a voluntary exercise group with similar regimens. Each group has significant increases in muscle strength compared to unexercised controls. Electrical stimulation has strengthening benefits at lower training contraction intensities.
3. No added benefit is attributed to simultaneous NMES and voluntary exercise over either alone.
4. Electrically elicited contractions of the quadriceps femoris muscle in the range of 80–100% of MVIT are possible; in some cases those in excess of 100% MVIT are also possible.
5. Evidence exists that certain muscle stimulation regimens produce greater strength gains than voluntary exercise in studies of patient populations.

6. A positive correlation exists between training contraction intensity and strength gains in persons with muscle weakness.

7. A positive correlation exists between phase charge and torque-generating capability in patients.

Methodological Concerns with Studies of NMES

Caution should be exercised in reading and interpreting studies comparing NMES and voluntary exercise for strengthening. Many studies lack sophistication in design, data sampling, analysis, and/or documentation. Some major flaws to be aware of when reviewing articles on NMES for strengthening include:

1. Use of change scores: pre- minus post-measurements (change scores) may lack reliability (66).

2. Failure to accurately describe methods: many studies do not completely report important experimental details, such as training parameters and current characteristics.

3. Inappropriate statistical analysis: use of multiple t-tests within one study without appropriate adjustment in the α level. For example, one of the studies reviewed above employed up to 20 independent t-tests. With an α level of 0.05, 1 of the 20 t-tests will be significant by chance alone.

4. Generalizability of results: improvements in functional status cannot be inferred from improvements in isometric or isokinetic strength.

5. Specificity of results: results in one population cannot be extended to another group (e.g., healthy vs. patients, young vs. old). In addition, results obtained from NMES experimentation on one muscle (e.g., quadriceps femoris) may not be validly applied to other muscles.

The results of a study exhibiting any of these flaws should not be automatically dismissed. Clinical research is difficult to perform without breaking some methodological rules. An awareness of possible methodological shortcomings should facilitate more discriminating review.

Clinical Considerations and Procedures for NMES Strengthening

A review of the literature concerning NMES strengthening and torque generation yields a number of general guidelines for clinical applications (Table 4.1). NMES strengthening procedures are based on several analogous volitional exercise programs for strengthening. In voluntary exercise programs, tetanic muscle contractions are produced at a relatively high percentage of the maximum voluntary torque levels (overload principle). In NMES for strengthening, stimulated contractions should be tetanic and at the highest tolerated torque level. The peak contraction torque ranges used for the quadriceps commonly exceed 60% of MVIT. Maximum torque levels for other lower extremity musculature may be comparable, and maximum torque levels for upper extremity and trunk musculature remain to be established.

Table 4.1
Characteristics of NMES Strengthening Programs

Type of current	Pulsatile or burst-modulated AC
Amplitude of stimulation	Maximum tolerable
Phase duration	20–1000 μsec
Waveform	Subject preference
Frequency of stimulation	30–75 pulses or bursts per sec
On/off times	10–15 sec on: 50–120 sec off
Type of contractions	Isometric
Number of contractions per session	10 at maximum tolerable intensity
Frequency of sessions	3 times per week

Snyder-Mackler, Delitto, and coworkers have demonstrated a threshold contraction intensity for quadriceps strengthening (55). Although thresholds have not been described for other muscles, it is essential that the clinician consider two factors related to this when choosing a stimulator. The stimulator must be able to cause a contraction of a sufficient intensity to strengthen the muscle. This necessitates measurement of electrically elicited contraction intensity, not current amplitude or pulse charge. Additionally, the stimulator must have sufficient reserve current amplitude to continue to present a training stimulus to the muscle as it gets stronger.

Recommended Stimulator Features and Controls for NMES for Strengthening

A stimulator for NMES for strengthening should be designed to allow a high degree of freedom in adjusting current characteristics. Preset parameters or programs for stimulation may be somewhat limiting and do not allow the adjustment of current characteristics for each individual. A stimulator with considerable flexibility may be a better investment in the long run as more is learned about optimal stimulation characteristics for strengthening. Flexibility in stimulus adjustment places additional responsibilities on users who need to know the consequences of changing each stimulator setting in order to provide safe and effective treatment. Recommendations on the design and performance characteristics of stimulators intended for NMES applications are available from the Association for the Advancement of Medical Instrumentation (AAMI).

Output channels and amplitude controls

A stimulator for NMES applications generally has at least two channels for stimulation. In a stimulator with two or more channels, controls are often provided to set either simultaneous stimulation from channels or alternating stimulation (both channels active at the same time). This is important, for example, during synchronous activation of two synergystic muscle groups to produce two-joint movement or synchronous activation of two antagonistic muscle groups (co-contraction) to minimize joint movement by producing

an isometric contraction. In contrast, alternating or reciprocal stimulation between channels is useful in providing alternating patterns of active movement
(e.g., flexion followed by extension) across a joint.

Each channel on NMES devices should have independent amplitude controls. Controls should also be present to adjust stimulus pulse or phase duration, frequency of stimulation, ramp modulations, and on time/off time.
Analog design of amplitude controls allows the user to carefully adjust stimulation to evoke maximal contraction.

Pulse amplitude and pulse/phase duration controls

The pulse amplitude control and the phase (or pulse) duration control together regulate the charge of each pulse and so determine the number of peripheral nerve fibers recruited with each stimulus. Analog amplitude controls,
those that allow continuous amplitude adjustment over the available range,
are preferable to incremental amplitude controls for NMES applications. Analog amplitude controls allow the user to make very small adjustments and
hence provide fine control over the level of evoked muscle contraction.

Phase duration or pulse duration controls may or may not be provided on
commercially available stimulators designed for NMES applications. Optimal
pulse durations for NMES have not been definitively established. Investigators have suggested that the optimal pulse duration likely lies between 50
and 1000 μsec. Since the question of optimal pulse duration is unresolved,
the contemporary "ideal" stimulator should provide pulse duration control
in the 20- to 1000-μsec range.

Each output channel of NMES units should have channel output indicators
that allow the user to know when stimulation is being applied through the
channel.

Frequency controls

Continuous adjustment of rate from 1 pulse (or burst) per sec to about 80–
100 pps allows a user to closely regulate the rate at which muscle is activated.
Smooth tetanic muscular contractions will be ensured in normally activated
muscle when the stimulation frequency is set at more than 50 pps. Stimulation at this frequency is rapid enough to elicit smooth contraction in even the
most rapidly contracting motor units. Although 50-pps stimulation ensures
a smooth tetanic contraction, some workers employ stimulation at lower frequencies (e.g., 30–35 pps) in clinical applications and have reported positive
results.

Theoretically, as the frequency of stimulation is increased, the opposition to
current flow (impedance) by the tissue falls. This has led to the development
of stimulators employing bursts of stimulation with carrier frequencies in the
2000- to 4000-Hz range. Such devices were expected to be able to produce
higher levels of muscle contraction with less patient discomfort than more
traditionally available stimulators producing 1–100 pps stimulation. Research
studies published to date have not substantiated this claim.

On time/off time controls

On time/off time controls of NMES stimulators are necessary because continuous or uninterrupted stimulated skeletal muscle contraction leads to very

rapid muscle fatigue or force failure. "On times" of muscle stimulation for many applications are usually set to 10 or 15 sec. "Off times" are generally adjustable up to about 1 or 2 min and have been most thoroughly examined at about 60 sec.

Ramp modulation controls

Ramp modulation controls are included in neuromuscular electrical stimulators so that the phase or pulse charge of each stimulus can be gradually increased or decreased. The gradual rise in stimulus charge over several seconds allows the gradual recruitment of nerve fibers and a more comfortable initiation of contraction for the subject. This type of modulation also allows more effective use of NMES in neurologically impaired patients. A declining ramp at the end of a contraction allows a smooth gradual drop in the force produced by muscle.

Treatment duration timer

A treatment duration timer is usually included in NMES devices to provide control of the total duration of stimulation and automatic shutoff of the unit at the desired time. In general, treatment durations are adjustable up to 60 min. Many stimulators have a safety switch that can be used by the patient to shut off the stimulation if it becomes uncomfortable. Such safety switches override the treatment timer.

To achieve tetanic contractions, pulsatile or burst-modulated alternating currents have been the most common forms of current successfully used at frequencies ranging from 30 to 80 pulses or bursts per second. Lower stimulation frequencies may not provide smooth tetanic contraction in all muscles. Stimulators must provide a range of available on and off times to mitigate the effects of fatigue.

Electrode Considerations in NMES for Strengthening

The selection and application of electrodes may influence the level of contraction evoked in response to NMES. Electrode sizes should suit the muscle(s) to be activated. If electrodes are too large, current may spread to antagonistic muscle groups. If electrodes are too small, current density may be so high that subject tolerance is exceeded before sufficient levels of contraction are reached for strengthening. No available evidence suggests that one type of commercially available electrode or coupling agent is superior to others for strengthening applications.

Proper electrode placement is of paramount importance in NMES. Illustrations of common electrode placements are shown in Figure 4.5. In general, both electrodes from a single stimulator circuit are placed over the muscle(s) to be activated. In some cases, four electrodes from two separate stimulator channels may be required to produce contraction levels sufficient to strengthen muscle. Commonly, motor points of the muscle to be activated are first located. The clinician should position electrodes over the belly of the muscle to be strengthened and carefully adjust their position as motor

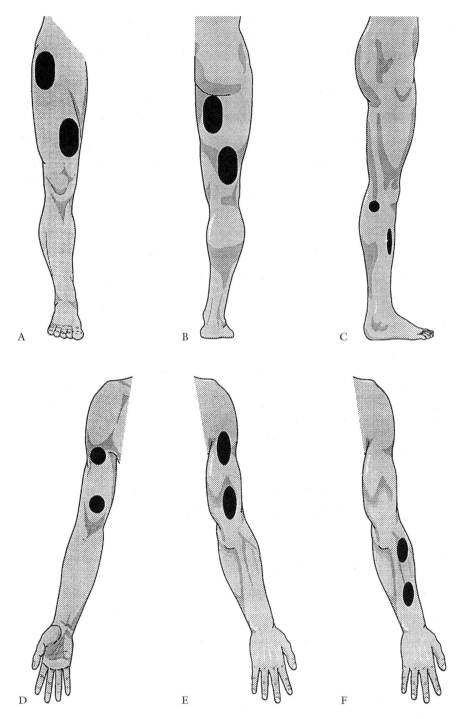

Figure 4.5 Common electrode placement sites for NMES of **A,** the quadriceps, **B,** the hamstrings, **C,** the tibialis anterior and peroneals, **D,** the biceps brachii, **E,** triceps brachii, **F,** wrist extensors, and **G,** lower back extensors.

threshold level stimulation is applied to determine the site of optimal stimulation. When electrode position has been established, all electrodes should be firmly secured with elastic wraps to ensure uniform electrical contact with the skin and minimal electrode movement during stimulation.

Application Principles for NMES for Strengthening

Before beginning the first stimulation session, the clinician should explain clearly what sensations patients are likely to experience during NMES. The patient should also be provided with a cutoff safety switch to stop stimulation if tolerance is exceeded. In addition, firm and proper patient stabilization should be provided. Minimal motion should occur in body segments proximal and distal to the region stimulated and in the joint influenced by the stimulated contraction. Many commercially available dynamometers and their attachments afford excellent positioning and stabilization for NMES applications.

Once the majority of stimulation parameters have been selected, electrodes secured, and subject stabilized, the initial stimulation session begins. The objective for muscle strengthening applications is to achieve the maximum tolerable level of muscle contraction. As treatment is initiated, the amplitude of stimulation should be increased gradually until motor threshold is reached and exceeded. In the first stimulation session, the amplitude of stimulation should be increased from contraction to contraction to subject tolerance.

During early treatment sessions, some patients will not tolerate stimulation at amplitudes sufficient to produce muscle strengthening. In such patients a 5- to 7-day program of stimulation designed to increase subject tolerance should be implemented. In these sessions, low amplitudes of stimulation are administered to produce threshold (just visible) contraction for periods of less than 10 sec. This allows the patient to adapt to the feeling of the electrical current. After a rest period of 10–20 sec, another 10-sec contraction is attempted at a higher stimulus amplitude. In this fashion, the current amplitude is slowly increased until the desired level of contraction is obtained. Throughout this process, the ongoing encouragement and reassurance of the clinician is essential in order to achieve levels of stimulation sufficient to produce muscle strengthening. If adequate torque levels are not produced in response to NMES using a particular set of current parameters, other combinations of stimulation parameters can be tried by the clinician in attempts to improve the torque response to NMES.

Precautions and Contraindications for NMES

The precautions or contraindications for NMES strengthening and endurance training include the following.

1. NMES over the thoracic region because current may interfere with the function of vital internal organs, including the heart.
2. NMES in the thoracic region of patients with demand cardiac pacemakers because the current may interfere with pacemaker activity and may lead to asystole or ventricular fibrillation.

3. NMES in regions of phrenic nerve or urinary bladder stimulators because current may interfere with the normal operation of these devices.

4. NMES over the carotid sinus because current may interfere with the normal regulation of blood pressure and cardiac contractility and may produce bradycardia or cardiac arrythmia.

5. NMES in hypertensive or hypotensive patients because autonomic responses may adversely affect control of blood pressure.

6. NMES in areas of peripheral vascular disorders, such as venous thrombosis or thrombophlebitis, because of the risk of releasing emboli.

7. NMES in regions of neoplasm or infection because muscular and circulatory effects may aggravate these conditions.

8. NMES on the trunk of pregnant females because of the risk of inducing uterine contractions that may influence the developing fetus.

9. NMES in close proximity to diathermy devices because of the potential for loss of control of stimulation parameters.

10. NMES in areas of excessive adipose tissue, as in obese patients, because levels of stimulation required to activate muscle in such patients may produce adverse autonomic reactions.

11. NMES in patients who are unable to provide clear feedback regarding the level of stimulation, such as infants, senile subjects, or individuals with mental disorders.

Careful observation by the clinician is required in all applications of electrical stimulation. Detrimental responses to stimulation often occur rapidly and require quick reaction to avoid serious injury. Patients should not be allowed to use NMES independently until the clinician is confident that they have been trained adequately in the safe use of the device and are aware of the signs and symptoms of adverse reactions that may occur.

Case Studies

The following case studies serve as examples of the ways in which NMES can be used to strengthen muscle. The electrotherapeutic treatment plan is described in detail, but additional treatment is only outlined.

Case 1

A 24-year-old female is referred to physical therapy 1 week after surgery for reconstruction of her left anterior cruciate ligament.

Examination: She is wearing a postoperative knee orthosis on the left and ambulates with the orthotic locked in full extension. Examination of the knee reveals no overt signs of inflammation and well-healed surgical incisions. Left thigh girth is 6 cm less than right. Comparison of MVIT of left and right knee extensors and flexors shows significant weakness on the left using a burst superimposition technique.

Assessment: Well-healed postsurgical knee with demonstrable atrophy and weakness as compared to contralateral extremity.

Plan: Therapy will include remedial exercise and NMES to augment strength of the thigh musculature by either co-contraction of both hamstrings and quadriceps, monitoring knee extension torque, or NMES of the quadriceps with the patient stabilized in knee flexion greater than 45°.

Detailed Electrotherapeutic Plan:

Mode of stimulation: Cycled NMES

Type of stimulator: Clinical

Electrodes and electrode placements: two equal-size electrodes placed on the quadriceps (Figure 4.5*A*); knee extension must be blocked at 45–65°.

Duration and frequency of treatment: three times per week, 10 contractions per session. Current amplitude is increased with each contraction to maximally tolerated contraction magnitude above 60% of the quadriceps femoris MVIC.

Rationale: It is generally recognized that ligament protection must be observed by the therapist while exercising or electrically stimulating the quadriceps femoris muscle in patients with ACL pathology. Forceful knee extension through the range from 45° of flexion to full extension may be accompanied by anterior translational forces that are deleterious to the reconstructed ACL and secondary supportive structures of the knee. The clinician can safely stimulate the quadriceps femoris in the way listed above. NMES can be used at any time during the rehabilitation phase after ACL surgery, but is superior to voluntary exercise in increasing isometric strength of knee extensors. Electrode placement is designed to minimize compressive forces on the patellofemoral joint and decrease the likelihood of other lower extremity musculature being stimulated simultaneously. A clinical stimulator is recommended because portable stimulators have not been shown to be effective for this type of NMES. The effectiveness of the program should be assessed weekly. Isometric knee extension and flexion torques should be measured, prior to stimulation, with the knee in 60° of flexion.

Case 2

A 45-year-old man with a recent history of lumbar instability and pain in the right leg is referred for physical therapy.

Examination: Examination reveals a slight loss of the lumbar lordosis and no lateral shift. Movement tests into extension and right pelvic translocation produce centralization of the patient's symptoms; flexion peripheralizes them. Returning from a flexed position, he uses his hands to climb up his thighs in order to extend. Sacroiliac torsion, compression, and distraction tests are negative. The Oswestry score is 45/100. The patient states that his symptoms are easily evoked with changes in position. Sometimes sitting evokes his leg pain and sometimes standing and walking. This is borne out over several days of examination and treatment when he presents with lateral shifts alternating from left to right.

Assessment: This patient has symptoms that are consistent with those in the immobilization principle category described by Delitto, Erhard, and Bowling (67).

Plan:

1. Stabilization exercise
2. Brace immobilization
3. NMES

Detailed Electrotherapeutic Plan: The patient is placed prone on a treatment table and stabilized across the pelvis with a belt. This is the exception to isometric contraction and measuring contraction force principles because of the difficulty in stabilizing the patient in a dynamometer for back muscle strength measurement. Electrodes are placed as in Figure 4.5G, bilaterally over the area of the multifidus muscle.

Type of stimulator: Clinical NMES, burst-modulated AC

Burst frequency: 50–75 bps

On time: 10 sec

Off time: 50 sec

Muscle contraction force: To tolerance, but at least sufficient to cause an anterior pelvic tilt

Treatment duration: 10–15 contractions per session; 2–3 times per week

Rationale: Intermittent, high-force, electrically elicited contractions have been shown to augment muscle strength. Use of small electrodes over the deep back extensors should permit more isolated strengthening of these muscles.

Summary

This chapter has reviewed the essential features of neuromuscular electrical stimulators, the clinical application of NMES to augment muscle strength, the stimulation parameters and techniques employed in NMES, and precautions and contraindications for NMES applications. Advances in NMES applications and techniques are occurring at a faster pace than ever before. Therapists using NMES and other electrical techniques should strive to stay abreast of new developments in the field. If possible, clinically active therapists are encouraged to add to the knowledge base in NMES through the careful design and implementation of clinical research studies. Through the coordinated efforts of laboratory and clinical researchers, both the breadth and depth of understanding and applications will increase. The patients we serve will benefit most from these efforts.

Study Questions

For answers, see Appendix B.

1. Given a patient with weakened right knee extension secondary to prolonged cast immobilization and a neuromuscular stimulator that produces rectangular, symmetric, biphasic pulsed current, describe the parameters of stimulation and a training program that may strengthen knee extension.

Stimulation Characteristics:

amplitude:

pulse/phase duration:

frequency of stimulation:

on time/off time of stimulation:

NMES Training Program Characteristics:

number of contractions per training session:

number of training sessions per week:

2. What is (are) the major adaptation(s) in skeletal muscle in response to:

(a) high-amplitude (>30% MVIC), low-repetition (10–15 contractions) daily NMES?

(b) low-amplitude (<30% MVIC), high-repetition (5 sets of 10 contractions each) daily NMES?

3. One pattern of NMES, developed by the Soviet researcher Kots, has been extensively studied with respect to NMES for strengthening. Completely describe the parameters of stimulation developed by Kots and outline the training regimen employed by this researcher to improve strength in elite athletes.

4. What range of stimulation frequencies are used for electrical stimulation to augment muscle strength?

5. Lengthening the off cycle of NMES is a good strategy for decreasing _____ during NMES for strengthening.

6. How can NMES be used to test the strength of a muscle?

7. What are the essential characteristics of an electrical stimulator to be used for NMES?

Pulse duration:

Frequency:

Amplitude:

8. Electrically elicited training contraction force is _____ related to strength augmentation.

9. The relationship between phase charge and electrically elicited muscle contraction force is _____.

10. List three contraindications to NMES.

References

1. Rose SJ, Rothstein JM. Muscle mutability. Part 1. General concepts and adaptations to altered patterns of use. Phys Ther 1982;62:1751-1830.

2. Salmons S, Henriksson J. The adaptive response of skeletal muscle to increased use. Muscle and Nerve 1981;4:94-105.

3. Knuttgen HG, Kraemer WJ. Terminology and measurement in exercise performance. J Appl Sport Sci Res 1987;1:1-10.

4. Bigland-Ritchie B, Woods JJ. Changes in muscle contractile properties and neural control during muscular fatigue. Muscle & Nerve 1984;7:691-699.

5. Vollstad NK, Sejersted OM. Biochemical correlates of fatigue. A brief review. Eur J Appl Physiol 1988;57:336-347.

6. Snyder-Mackler L, Binder-Macleod SA, Williams P. Fatigability of the human quadriceps femoris muscle following anterior cruciate ligament reconstruction. Med Sci Sports Exer 1993;25:783-789.

7. Jones DW, Jones DA, and Newham DJ. Chronic knee effusion and aspiration: the effect on quadriceps inhibition. Brit J Rheumatol 1987;26:370-374.

8. Newham DJ, Hurley MV, and Jones DJ. Ligamentous knee injury and muscle inhibition. J Orthop Rheumatol 1989;2:163-173.

9. Snyder-Mackler L, De Luca PF, Williams PR, Eastlack ME, Bartolozzi AR. Reflex inhibition of the quadriceps femoris muscle after injury or reconstruction of the anterior cruciate ligament. J Bone Joint Surg 1994;76-A:555-560.

10. Shih Y-F. Can burst superinposition predict maximal voluntary force? University of Pittsburgh Masters Thesis 1994.

11. McDonnell MK, Delitto A, Sinacore DR, Rose SJ. Electrically elicited fatigue test of the quadriceps femoris muscle. Phys Ther 1987;67:941-945.

12. Binder-Macleod SA, Snyder-Mackler L. Muscle fatigue: clinical implications for fatigue assessment and neuromuscular electrical stimulation. Phys Ther 1993;73:902-910.

13. Burke RE, Levine DN, Tsairis P, Zajac FE. Physiological types and histochemical profiles in motor units of the cat gastrocnemius. J Physiol 1973;234:723-748.

14. Kramer J, Lindsay D, Magee D, Mendryk S, Wall T. Comparison of voluntary and electrical stimulation contraction torques. J Orthop Sports Phys Ther 1984;5:324-331.

15. Walmsley RP, Letts G, Vooys J. A comparison of torque generated by knee extension with a maximal voluntary muscle contraction vis-a-vis electrical stimulation. J Orthop Sports Phys Ther 1984;6:10-17.

16. DeDomenico G, Strauss GR. Maximum torque production in the quadriceps femoris muscle group using a variety of electrical stimulators. Austral J Physiother 1986;32:51-56.

17. Snyder-Mackler L, Garrett M, Roberts M. A comparison of torque generating capabilities of three different electrical stimulating currents. J Orthop Sports Phys Ther 1989;11:297-301.

18. Locicero R. The effect of electrical stimulation on isometric and isokinetic knee extension torque: interaction of the Kinestim electrical stimulator and the Cybex II+. J Orthop Sports Phys Ther 1991;13:143-148.

19. Delitto A, Strube MJ, Shulman AD, Minor SD. A study of discomfort with electrical stimulation. Phys Ther 1992;72:410-421.

20. Delitto A, Brown M, Strube MJ, Rose SJ, Lehman RC. Electrical stimulation of quadriceps femoris in an elite weight lifter: a single subject experiment. Int J Sports Med 1989;10:187-191.

21. Enoka RM and Fuglevand AJ. Neuromuscular basis of the maximum force capacity of a muscle. In: MD Grabiner, ed. Current issues in biomechanics. Champaign, IL: Human Kinetics, 1993:215-235.

22. Enoka RM, Stuart DG. Neurobiology of muscle fatigue. J Appl Physiol 1992;72:1631-1648.

23. Enoka RM. Personal communication, 1993.

24. Snyder-Mackler L, Delitto A, Stralka SW, Bailey S. Electrically elicited training contraction intensities and strength recovery in the quadriceps femoris muscle after anterior cruciate ligament reconstruction. Phys Ther (in press).

25. Bowman BR, Baker LL. Effects of waveform parameters on comfort during transcutaneous neuromuscular electrical stimulation. Ann Biomed Eng 1985;13:59-74.

26. Delitto A, Rose SJ. Comparative comfort of three waveforms used in electrically eliciting quadriceps femoris contractions. Phys Ther 1986;66:1704-1707.

27. Grimby G, Wigerstad-Lossing I. Comparison of high- and low-frequency muscle stimulators. Arch Phys Med Rehabil 1989;70:835-838.

28. Massey BH, Nelson RC, Sharkey BC, Comden T. Effects of high frequency electrical stimulation on the size and strength of skeletal muscle. J Sports Med Phys Fit 1965;5:136-144.

29. Currier DP, Lehman J, Lightfoot P. Electrical stimulation in exercise of the quadriceps femoris muscle. Phys Ther 1979;59:1508-1512.

30. Halbach JW, Strauss D. Comparison of electro-myo stimulation to isokinetic power of the knee extensor mechanism. J Orthop Sports Phys Ther 1980;2:20-24.

31. Eriksson E, Haggmark T, Kiessling KH, Karlsson J. Effects of electrical stimulation on human skeletal muscle. Int J Sports Med 1981;2:18-22.

32. Romero JA, Sandford TL, Schroeder RV, Fahey TD. The effects of electrical stimulation of normal quadriceps on strength and girth. Med Sci Sports Exer 1982;14:194-197.

33. Kramer JF, Mendryk SW. Electrical stimulation as a strength improvement technique: a review. J Orthop Sports Phys Ther 1982;4:91-98.

34. Kramer JF, Semple JE. Comparison of selected strengthening techniques for normal quadriceps. Physiother Can 1983;35:300-304.

35. Currier DP, Mann R. Muscular strength development by electrical stimulation in healthy individuals. Phys Ther 1983;63:915-921.

36. Laughman RK, Youdas JW, Garrett TR. Strength changes in the normal quadriceps femoris muscle as a result of electrical stimulation. Phys Ther 1983;63:494-499.

37. Selkowitz DM. Improvement in isometric strength of the quadriceps femoris muscle after training with electrical stimulation. Phys Ther 1985;65:186-196.

38. McMiken DF, Todd-Smith M, Thompson C. Strengthening of human quadriceps muscles by cutaneous electrical stimulation. Scand J Rehab Med 1983;15:25-28.

39. Mohr T, Carlson B, Sulentic C, Landry R. Comparison of isometric exercise and high volt galvanic stimulation on quadriceps femoris muscle strength. Phys Ther 1985;65:606-612.

40. Stefanovska A, Vovovnik L. Change in muscle force following electrical stimulation. Scand J Rehab Med 1985;17:141-146.

41. Wolf SL, Gideon BA, Saar D, et al. The effect of muscle stimulation during resistive training on performance parameters. Am J Sports Med 1986;14:18-23.

42. Alon G, McCombe SA, Koutsantonis S, Stumphauzer LJ, Burgwin KC, Parent MM, Bosworth RA. Comparison of the effects of electrical stimulation and exercise on abdominal musculature. J Orthop Sports Phys Ther 1987;8:567-573.

43. William JCP, Street M. Sequential faradism in quadriceps rehabilitation. Physiotherapy 1976;62:252-254.

44. Johnson DH, Thurston P, Ashcroft PI. The Russian technique of faradism in the treatment of chondromalacia patellae. Physiother Can 1977;29:266-288.

45. Eriksson E, Haggmark T. Comparison of isometric muscle training and electrical stimulation supplementing isometric muscle training in the recovery after major knee ligament surgery. Am J Sports Med 1979;7:169-171.

46. Godfrey CM, Jayawardena H, Quance TA, Welch P. Comparison of electrostimulation and isometric exercise in strengthening the quadriceps muscle. Physiother Can 1979;31:265-267.

47. Gould N, Donnermeyer D, Gammon GG, Pope M et al. Transcutaneous muscle stimulation to retard disuse atrophy after open meniscectomy. Clin Orthop Rel Res 1983;178:190-197.

48. Grove-Lainey C, Walmsley RP, Andrew GM. Effectiveness of exercise alone versus exercise plus electrical stimulation in strengthening the quadriceps muscle. Physiother Can 1983;35:5-11.

49. Morrissey MC, Brewster CE, Shields CL, Brown M. The effects of electrical stimulation on the quadriceps during postoperative knee immobilization. Am J Sports Med 1985;13:40-45.

50. Singer KP, Gow PJ, Otway WF, Williams M. A comparison of electrical muscle stimulation isometric, isotonic and isokinetic strength training programmes. NZ J Sports Med 1983;11:61-63.

51. Sisk TD, Stralka SW, Deering MB, Griffin JW. Effects of electrical stimulation on quadriceps strength after reconstructive surgery of the anterior cruciate ligament. Am J Sports Med 1987;15:215-219.

52. Currier D, Ray JM, Nyland J, Rooney J, Noteboom JT, Kellogg R. Effects of electrical and electromagnetic stimulation after anterior cruciate ligament reconstruction. J Orthop Sports Phys Ther 1993;17:177-184.

53. Delitto A, Rose SJ, Lehman R, et al. Electrical stimulation versus voluntary exercise in strengthening the thigh musculature in patients after anterior cruciate ligament surgery. Phys Ther 1988;68:660-663.

54. Snyder-Mackler L, Ladin Z, Schepsis A, et al. Electrical stimulation of the thigh muscles after reconstruction of the anterior cruciate ligament. Effects of electrically elicited contractions of the quadriceps femoris and hamstring muscles on gait and strength of the thigh muscles. J Bone Joint Surg (Am) 1991;73:1025-1036.

55. Snyder-Mackler L, Delitto A, Bailey S, Stralka SW. Quadriceps femoris muscle strength and functional recovery after anterior cruciate ligament reconstruction: a prospective randomized clinical trial of electrical stimulation. J Bone Joint Surgery (in press).

56. Starring D. The use of electrical stimulation and exercise for strengthening lumbar musculature: a case study. J Orthop Sports Phys Ther 1991;14:61-64.

57. Kahanovitz N, Nordin M, Verderame R, Yabut S, et al. Normal trunk muscle strength and endurance in women and the effect of exercises and electrical stimulation, Part 2: Comparative analysis of electrical stimulation and exercises to increase trunk muscle strength and endurance. Spine 1987;12:112-118.

58. Karmel-Ross K, Cooperman D, Van Doren C. The effect of electrical stimulation on quadriceps femoris muscle torque in children with spina bifida. Phys Ther 1992;72:723-730.

59. Wigerstad-Lossing I, Grimby G, Jonsson T, et al. Effects of electrical muscle stimulation combined with voluntary contractions after knee ligament surgery. Med Sci Sports Exerc 1988;20:93-98.

60. Draper V, Ballard L. Electrical stimulation versus electromyographic biofeedback in the recovery of quadriceps femoris muscle function following anterior cruciate ligament surgery. Phys Ther 1991;71:455-464.

61. Alon G, McCombe SA, Koutsantonis S, Stumphauzer LJ, Burgwin KC, Parent MM, Bosworth RA. Comparison of the effects of electrical stimulation and exercise on abdominal musculature. J Orthop Sports Phys Ther 1987;8:567-573.

62. Hartsell HD. Electrical muscle stimulation and isometric effects on selected quadriceps parameters. J Orthop Sports Phys Ther 1987;8:203-209.

63. Packman-Braun R. Relationship between functional electrical stimulation duty cycle and fatigue in wrist extensor muscles of patients with hemiparesis. Phys Ther 1988;68(1):51-56.

64. Benton LA, Baker LL, Bowman BR, Waters RL. Functional electrical stimulation: a practical clinical guide. Downey CA: Professional Staff Association of the Rancho Los Amigos Hospital, 1981:65-71.

65. Binder-Macleod SA, Halden HE, Jungles KA. Force frequency relationship and fatiguability of human muscle: effects of stimulation intensity. Med Sci Sports Exer (in press).

66. Cohen J, Cohen P. Applied multiple regression correlation for the behavioral sciences. 2nd ed. Hillsdale, NJ: Lawrence Erlbaum Associates, 1983;72-73, 413-423.

67. Delitto A, Erhard RE, Bowling RW. A treatment based classification system for low back syndrome: identifying and staging patients for conservative treatment. Phys Ther (in press).

Chapter 5

Neuromuscular Electrical Stimulation for Control of Posture and Movement

Andrew J. Robinson

$\bigwedge\bigwedge\bigvee$ **E**lectrical stimulation for the activation of skeletal muscle is a therapeutic technique that has been used in physical medicine for over half a century. Through the early 1960s the use of electrical stimulation focused mainly on the management of denervation atrophy of skeletal muscle. Few studies were available on the use of electrical stimulation to activate nerve and normally innervated muscle via the peripheral nerve. A renewed interest in the effects of nerve stimulation and normally innervated muscle (neuromuscular electrical stimulation, NMES) gradually developed with the resurgence in the use of electrical stimulation for pain control and the development of new, more sophisticated types of stimulators.

The spark that ignited widespread interest in NMES did not come until the mid-1970s. In 1976 at the Olympic Games, in Montreal, Soviet athletes were observed using NMES in conjunction with voluntary exercise as a strength training technique. In 1977 the Russian researcher Kots, who had developed the stimulation technique, reported that NMES could produce muscle strength gains in elite athletes that are 30 to 40% greater than those achieved through exercise alone. These dramatic strength gains were achieved by eliciting muscle contractions that were 10 to 30% greater than those achieved by maximum voluntary muscle contraction. Western researchers quickly recognized the potential of such a technique and soon initiated studies designed to verify Kots's reports.

Although the results of western studies have not confirmed all of Kots's original findings, they did support the contention that NMES could strengthen normally innervated muscle. Contemporary work in this area focuses not only on NMES effects on normal muscle but also on its usefulness in the management of a variety of muscle disorders commonly encountered in patients in the clinical setting. The plasticity of skeletal muscle in response to altered patterns of activity (1, 2) as described in the previous chapter forms the basis of the clinical application of NMES for therapeutic effects.

The purposes of this chapter are to *(a)* identify the essential features of electrical stimulators required for NMES, *(b)* describe the important parameters of stimulation techniques used to elicit contraction in muscle, *(c)* identify and describe the various clinical indications and applications for sensory- and motor-level electrical stimulation, *(d)* identify the clinical precautions or contraindications of NMES, and *(e)* present case studies to illustrate NMES. The review of clinical applications will include ES for maintenance of range of motion, spasticity control, orthotic substitution, muscle facilitation/reeducation, functional electrical stimulation, and denervated muscle. NMES for increasing the strength and endurance of muscle is presented in the previous chapter.

General Considerations for Neuromuscular Electrical Stimulation (NMES)

The first step in clinical NMES is the selection of an appropriate stimulator. Recently a wide range of electrical stimulators has been introduced to the market. The clinician charged with making the choice of stimulators is deluged with product information on "high-voltage pulsed galvanic," "interferential,"

"Russian," and other types of stimulators; such information frequently serves more to confuse than to clarify the choice. The objective of this segment of the chapter is to provide the reader with essential information on the features of electrical stimulators that are important for the majority of clinical NMES applications.

Current waveform selection and NMES

What is the "best" current to employ in NMES? A synthesis of the results of several studies monitoring the force produced in response to stimulation, the subject's comfort level during stimulation, or the preference for a particular waveform during stimulation does not provide a final answer to this question but does provide insight into answering it. Relative tolerance to NMES procedures can be assessed by determining a subject's preferences or relative comfort or by determining the force of muscular contraction in response to maximally tolerable levels of stimulation.

Subject Preference/Comfort

As any clinician who has used electrical stimulation (ES) knows, one of the most important aspects of stimulation that can affect the potential outcome of treatment is the patient's tolerance to the procedure. Limited tolerance of stimulation may render NMES procedures ineffective. Delitto and Rose (3) assessed subject comfort in response to ES using three different symmetrical biphasic waveforms (sinusoidal, triangular, and rectangular). These three stimulation waveforms were used to electrically activate the quadriceps muscle group in normal subjects. The amplitude of stimulation was gradually increased during stimulation sessions to amplitudes that evoked similar magnitudes of knee extension torque. Subjects were then asked to indicate their level of comfort on a visual analog scale. No waveform was found to be "most" comfortable in evoking comparable levels of contraction. In a similar study, Bowman and Baker (4) compared symmetrical and asymmetrical biphasic waveforms applied to evoke contractions in the quadriceps muscles at approximately 20% of maximum voluntary levels. Their results suggested that the majority of subjects preferred symmetrical biphasic waveforms over asymmetrical biphasic waveforms in both constant voltage and constant current stimulators.

Both of these studies addressed the issue of tolerance to NMES (relative subject comfort or discomfort) as a function of only the stimulation current characteristics. The clinical implication of these reports was that several different forms of stimulation should be applied to a patient using trial and error to select the "best" type of current for the individual, in order to maximize muscle contraction and minimize discomfort.

A more recent report by Delitto and coworkers (5) has examined whether subject tolerance to NMES depends on other factors besides type of current, such as the individual's behavioral style of coping with stressful situations. In this study, a Miller Behavioral Style Scale questionnaire was used to establish coping styles of potential subjects for an NMES tolerance study. Based on the MBSS responses, two groups of subjects were selected. One group consisted of individuals who actively solicit information when faced with a potentially stressful situation ("the monitors"). The second group consisted of individuals who avoid information in a stressful situation and prefer some form of distraction during a stressful event ("the blunters"). The researchers hypothesized that NMES would be better tolerated by monitors. NMES was applied over the quadriceps muscle at 45%, 60%, 75%, and 90% of maximum tolerance levels using a conventional electrode placement. Pure sensory stimulation was also provided at 45%, 60%, 75%, and 90% of maximum tolerance by placement of electrodes medially and laterally over the knee joint, a location that would not give rise to muscle contraction. For both types of stimulation, a triangular, symmetric, biphasic, 2500-Hz AC was applied at 75 bursts per second. Participants were asked to rate the intensity and unpleasantness of the motor- and sensory-level stimulation at each intensity level with respect to a 20-mA reference current. The results showed that, although perceptions of stimulation intensity and unpleasantness varied between groups, the levels of evoked contraction (% maximum voluntary isometric contraction) did not significantly differ between groups.

Force of Contractions

The effectiveness of waveforms in eliciting contraction force depends on amplitude, duration, frequency, and waveform. The torque-generating capabilities of a particular pattern of stimulation cannot be inferred from any single current characteristic. To date, no single combination of these stimulation parameters has been shown to be optimal for eliciting maximal force of contraction for all subjects. A detailed review and analysis of the torque-generating capability of various forms of stimulation are contained in the previous chapter, on NMES used to augment strength.

Recent research indicates that many of the waveforms available on commercial stimulators may be effective in activating skeletal muscle when applied intelligently. Individuals may exhibit a waveform preference in NMES; consequently, the user should consider changing the waveform if the desired response is not achieved. The studies examining the question of waveform efficacy suggest that symmetrical biphasic waveforms and burst-modulated, sinusoidal AC waveforms may ultimately be the "most appropriate" for the electrical activation of normally innervated skeletal muscle. Very short duration monophasic waveforms may be found to be slightly less effective than symmetrical biphasic waveforms. Additional studies are required to substantiate these hypotheses.

Recommended stimulator features and controls for simple NMES applications

A stimulator for NMES applications should be designed to allow considerable flexibility in adjusting current characteristics. Such flexibility in stimulus adjustment places additional responsibilities on users, who need to know the consequences of changing each stimulator setting in order to provide safe and effective treatment. Recommendations on the design and performance characteristics of stimulators intended for NMES applications are available from the Association for the Advancement of Medical Instrumentation (AAMI) (6) and have been reviewed in more detail in Chapters 2 and 4. Basic requirements for most NMES applications include two output channels, pulse or phase duration controls (4, 7), frequency controls, independent amplitude controls, on time/off time controls, ramp modulation controls, and a timer.

Portable versus clinical stimulators for NMES

Both battery-operated, portable stimulators and 60-Hz AC, line-powered stimulators have been used for NMES. Some battery-operated stimulators may not provide enough output to bring skeletal muscle contraction to levels sufficient for therapeutic effects. In addition, some portable stimulators used at high-output levels drain battery power quickly and require frequent battery replacement or recharging. Another limitation of battery-operated units is, in general, the lack of any analog or digital output meter. As an advantage, portable stimulators do not tether patients to wall outlets, and consequently these devices are often the stimulator of choice for functional neuromuscular stimulation. In addition, newer battery designs have improved operating life and hence have increased the utility of portable NMES instruments. The reader is referred to Chapter 2 for more detail regarding the various types of commercially available stimulators.

Electrodes and electrode placement for NMES

The point on the surface of the skin overlying a muscle at which the smallest amount of current activates the muscle is called the **motor point**. Electrodes for NMES are frequently placed over these points in order to achieve maximal muscle stimulation with minimal current. The location of motor points for many muscles of the extremities and trunk is shown in Figure 5.1. In general, the motor point of a muscle is located over the belly of the muscle at or near the point at which the motor nerve enters the muscle. Motor threshold amplitude stimuli may be applied using small handheld electrodes to precisely localize the motor point for each muscle. In this procedure, single pulsatile stimuli are applied in the suspected region of the motor point. The amplitude of stimulation is adjusted to a level just sufficient to evoke a small muscle contraction. The position of the electrode is gradually shifted until the maximum muscle contraction is obtained at this fixed level of stimulation. Pad electrodes can then be positioned over this point for the remaining stimulation procedures.

Placement of electrodes over motor points may not always be desirable in order to achieve high levels of muscle contraction. Ferguson and coworkers (8) examined maximum tolerable quadriceps activation using a variety of electrode placements. They found that maximum torques (65% of maximum voluntary isometric contraction [MVIC]) were produced by bipolar, longitudinal placement of electrodes over the femoral triangle and either the vastus lateralis (VL) or vastus medialis (VM). The lowest evoked torques were found using a transverse orientation of electrodes—one electrode over VM and a similar electrode over VL. These findings have been verified in a similar study by Brooks and associates (9). NMES of the VL alone should not be performed without simultaneous activation of the VM muscle. Isolated activation of the VL may result in patellar subluxation. One subject withdrew from the Ferguson study (8) after experiencing the lateral subluxation of the patella. A similar patellar subluxation has been produced in the teaching

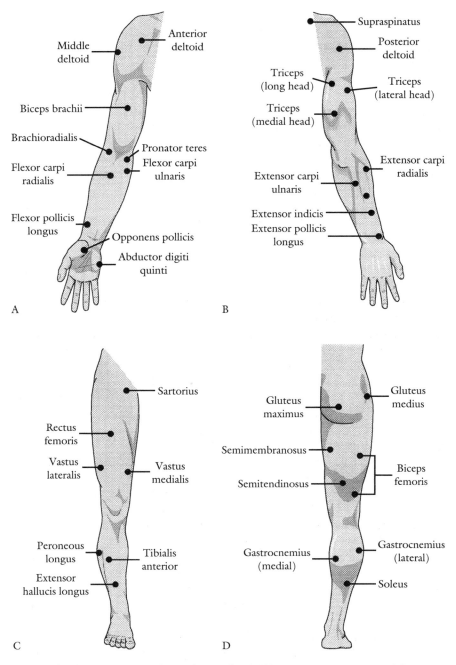

Figure 5.1 Approximate motor point for muscles on (**A**) anterior upper extremity, (**B**) posterior upper extremity, (**C**) anterior lower extremity, and (**D**) posterior lower extremity.

laboratory of the author during very low amplitudes of stimulation in a student who had a history of patellar subluxation on the opposite side during volitional contraction. The Ferguson study (8) has suggested that the placement of electrodes over the VL and femoral triangle for quadriceps activation should be reserved for subjects with a Q angle of less than 15°.

The bipolar placement (both electrodes over muscle) of electrodes appears to be used when large muscle groups are to be stimulated in NMES procedures. However, when smaller muscles are stimulated, a monopolar electrode placement is used to avoid an overflow of current to muscle the user does not want to activate.

A number of different types of electrodes are available for NMES. Conductive rubber electrodes and metal-sponge electrodes are available in a spectrum of suitable sizes. To date, no commercially available electrode has been shown to be superior for all NMES applications. In 1980 Nelson and coworkers (10) examined the "effectiveness" of four types of commercially available electrodes. These included felt-covered metal (63 cm^2), carbon-impregnated silicon rubber (54 cm^2), solvent-activated silver-impregnated tape (34 cm^2), and self-adhering pregelled (13 cm^2) electrodes. They found that for quadriceps NMES, the felt pad electrodes yielded the highest torque outputs (75% MVIC), whereas the pregelled electrodes yielded the lowest ($\sim$38% MVIC). Because of the marked differences in electrode sizes, these results may have been due more to electrode size than electrode construction.

Regardless of the type of electrode employed, each electrode must be securely attached to attain optimal stimulation. Electrically conductive gels or solutions used with conductive rubber electrodes must be evenly distributed over the surface of each electrode to ensure patient comfort during stimulation. All gel or liquid coupling agents are not equally safe or effective. The results of a Nelson and colleagues study (10) on electrode effectiveness revealed that salt-free electrically conductive gel used with carbon-impregnated silicon rubber electrodes maintained impedance during four days of continuous application, whereas the same type of electrodes coupled with a salt-based gel showed a 50% increase in impedance. Such increased impedances may impair the ability to achieve optimal stimulation when electrodes remain affixed in one location for extended periods of time.

The selection and application of electrodes may influence the level of contraction evoked in response to NMES. Electrode sizes should suit the muscle(s) to be activated. If electrodes are too large, current may spread to antagonistic muscle groups. If electrodes are too small, current density may be so high that subject tolerance is exceeded before desired levels of contraction are reached. Alon and coworkers (11) have suggested that the largest possible electrodes are used when muscular contraction is desired in an electrotherapeutic application because larger electrodes will produce stronger contractions without pain. Practitioners who use NMES should be aware that stimulation electrodes should be regularly replaced. This is especially true for siliconized rubber electrodes whose conductivity may change with repeated, regular use or with age. To avoid this problem of nonuniformity of conduction in siliconized rubber electrodes, some manufacturers have developed flexible foil electrodes bonded to a conductive, self-adhesive gel. Although these electrodes appear to offer a solution to the problem of conductivity changes in conductive rubber electrodes, the durability of these electrodes has yet to be established. Further research is needed regarding electrode effectiveness before claims of superior performance in NMES can be substantiated.

General application principles for NMES

Once a stimulator has been selected and electrodes have been chosen, and properly located and secured, the initial NMES session can begin. Before turning on the stimulator, the user should thoroughly explain the procedure to the subject. The explanation should include a description of the sensations and responses that may be produced as well as the goals of the stimulation program.

One objective for many NMES applications is to achieve high levels of stimulated muscle contraction. As treatment is initiated, the amplitude of stimulation should be gradually increased until motor threshold (the stimulation amplitude sufficient to initiate muscle contraction) is reached and then exceeded. In the first treatment session, the amplitude of stimulation may have to be increased from contraction to contraction in order to gradually increase subject tolerance (12) and increase the contraction force.

During initial treatment sessions, many patients will not tolerate stimulation at levels sufficient to produce the desired muscle contraction. In such patients, a 5- to 7-day program of stimulation designed to increase tolerance can be implemented. In these sessions, low amplitudes of stimulation are administered to produce intermittent, just-visible contractions for periods of less than 10 sec each. This allows the patient to adjust to the feeling of the electrical current. After a rest period of 10–20 sec, another 10-sec contraction is attempted at a higher amplitude. In this fashion, the current amplitude is slowly increased until the desired level of contraction is obtained. Throughout this process, the encouragement and reassurance of the health care provider is essential in order to achieve levels of stimulation sufficient to produce contractions that will achieve the therapeutic goal. If adequate contraction forces or movements are not produced in response to NMES using a particular set of current parameters, other combinations of parameters can be tried in an attempt to improve the response.

NMES for Restricted Joint Motion and Contractures

Effect of NMES on passive range of motion

Limitations in active and passive range of joint motion may result from a number of disorders. Orthopedic procedures such as immobilization of joints after fracture or severe sprains often result in tightening of joint capsule structures and the muscles that cross the affected joints. Neurological disorders such as spinal cord trauma, cerebrovascular accident, or traumatic head injury commonly lead to either lost or disordered control of skeletal muscle contraction and, subsequently, to joint and muscle contractures. Traditionally, stretching techniques used alone or in conjunction with thermal agents have been employed to prevent or resolve abnormal shortening of soft tissue structures. More recently a number of other manual procedures (e.g., myofascial release, joint mobilization) have also been used to improve range of motion. Recent studies have indicated, however, that NMES is an appropriate adjunctive therapy in the prevention and management of restricted joint movement.

In the mid-1970s Munsat and coworkers (13) published a report on the effect of NMES on long-term knee flexion contractures; they triggered renewed interest in the application of stimulation for range-of-motion (ROM) management. These workers reported that electrical stimulation of the quadriceps muscle 6 hours daily was effective in reducing knee flexion contractures in four out of five comatose patients. Each patient underwent surgical lengthening of the hamstrings prior to the initiation of NMES.

In 1979 workers at the Rancho Los Amigos Medical Center reported on the use of NMES in the management of wrist and finger flexion contractures in 16 hemiplegic patients (14). Treatment initially consisted of 15 minutes of NMES to the wrist and finger extensor muscles twice a day. As patient tolerance to stimulation increased, single sessions of electrical stimulation were increased in duration to 30 minutes and were provided three times per day, 7 days a week, for 4 weeks. Stimulation amplitude was increased to produce the "maximum comfortable extensor contraction" and elicited movement throughout the available ROM in wrist and finger joints. NMES was initiated in the hospital setting and continued at home. For those patients whose onset of hemiplegia was more than 4 months prior to initiation of treatment, NMES increased wrist, metacarpal-phalangeal, and proximal interphalangeal passive ROM on average 36°, 27°, and 17° respectively. These marked changes in passive ROM occurred in the absence of any other specific therapy designed to reduce flexion deformities. In addition, when NMES was discontinued in four patients, passive extension of wrist and finger joints gradually declined in spite of the self-application ROM exercises. This single study provides the best support to date for the use of NMES for stretching soft tissue structures that impair joint mobility.

Further research may reveal that the activation of contracted muscles across a joint with restricted passive movement is a more effective way of applying "stretching" forces to shortened structures than externally applied manual stretching techniques. NMES may be effective simply because the total stretch time often exceeds that of manual stretching techniques alone.

Effect of NMES on active range of motion

Electrical stimulation of innervated skeletal muscle may also be a valuable treatment technique to combat restrictions in *active* ROM (movement produced by volitional activation of muscle). Bowman and coworkers (15) have reported on the effectiveness of this NMES application in 30 hemiplegic patients. In their study, subjects demonstrated full passive ROM suggesting no significant soft tissue shortening in muscle or joint structures. Active ROM in these individuals was limited, however, to 5° to 30° of extension. Subjects in both the control and stimulated groups received "conventional" therapy consisting of passive ROM, active resistive exercise, traditional neuromuscular facilitation, and training in activities of daily living. Subjects in the "stimulated" group in addition received positional feedback stimulation training (PFST) consisting of resisted volitional contraction followed by NMES to the wrist and finger extensor muscles on the involved side using a custom-built feedback/stimulator device. PFST was provided for 30 minutes twice daily, 5 days a week for 4 weeks.

The position feedback device simply provided the subject with audio and visual feedback on wrist position and gave performance goals to help maintain subject motivation. The subjects in the stimulated group showed increases in active wrist extension that averaged 35°, whereas the control subjects' active wrist extension improved by only 8° on average. This improvement may have been due to the average 70 to 280% improvement in voluntary wrist extension strength that accompanied the improved ability to extend the wrist.

Winchester and coworkers (16) more recently reported on the effects on knee extension function of using PFST and twice-a-day NMES to the quadriceps in hemiparetic patients. All 40 patients in this study received the same traditional physical therapy for lower extremity rehabilitation. Those subjects assigned to the "stimulated" group exhibited increases in volitional quadriceps torque (673% over initial) and active ROM (33°) significantly higher than those achieved by unstimulated subjects, in the control group.

Whether the position feedback, the improved strength of volitional contraction, or the stimulation alone was critical to achieving these remarkable results is not known. Furthermore, measures of functional recovery were not used to determine its effect on the results. In spite of these shortcomings, these studies strongly support the continued use of NMES for the improvement of active ROM in neurologically involved patients. The efficacy of NMES applications for increasing the active and passive ROM of patients in orthopedic or other neuromuscular populations has yet to be demonstrated.

NMES application principles for improving range of motion

NMES for improving either passive or active ROM employs the same general techniques described previously in the discussion of muscle strengthening. A variety of appropriate stimulators are currently available on the market. Because this treatment program is frequently carried out at home by the patient or family, portable stimulators with rechargeable batteries and easy-to-operate controls are suitable. Many portable stimulators allow the clinician to set a number of stimulation parameters (such as on time/off time, phase/pulse duration, and ramp modulations) in a hidden compartment. The patient or family member then is faced with only the adjustment of stimulation amplitude and treatment time, simplifying the application. Close attention must be paid to training in electrode placement and skin care to achieve optimal stimulation and avoid possible adverse reactions (see Chapter 2). Although this form of NMES can be carried out independently by the patient or family in many cases, close supervision by the health care provider is recommended.

Documented use of NMES for ROM therapy employs contraction levels that are described as "fair-plus." That is, the amplitude of stimulation is adjusted to evoke a contraction that moves the joint through the full available ROM against a small amount of manual resistance (Fig. 5.2). The *fair-plus* designation is commonly associated with volitional contraction and may not be appropriate for quantifying muscle contraction force in response to NMES. The stimulated torque output as a percentage of maximal volitional torque in the muscle(s) of an uninvolved extremity may be more appropriate. However, the lack of sensitivity of some dynamometers to very low-level contractions may make this impossible. Although fair-plus levels of contraction may be sufficient for stretching contracted soft tissue, future research may show higher levels of evoked contraction (greater than fair-plus) to be more effective in achieving the best clinical results. If improvements in active ROM are indeed secondary to NMES-induced increases in muscle strength,

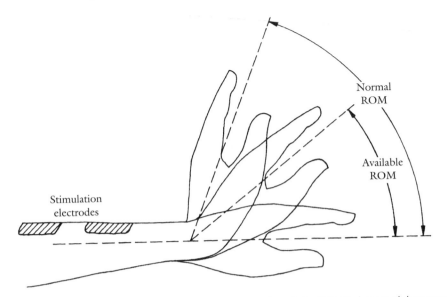

Figure 5.2 Wrist extension produced by NMES through available ROM to improve joint movement.

many of the principles and procedures discussed in Chapter 4 may be applicable to the treatment of active ROM problems. During the use of NMES for passive or active ROM limitations, caution must be exercised while evoking contractions to avoid abnormal compressive forces at joints, which may initiate inflammation.

NMES for Control of Spasticity

Skeletal muscle spasticity is a clinical phenomenon associated with damage to the central nervous system. Traumatic head injury, cerebrovascular accident, spinal cord disease, and a number of other disorders may all result in spasticity of limb and trunk musculature. Spasticity has been characterized as an abnormal muscle state in which phasic and tonic stretch reflexes are usually hyperactive, flexion reflexes are hyperactive, and dexterity and strength are decreased (17).

The precise nature of the neurological changes producing spasticity in various disorders has yet to be elucidated. In general, spasticity is thought to result from a disruption of the normal balance of neural inputs to the alpha motoneurons. Central nervous system disorders may result in an increase in the central or peripheral excitatory inputs to alpha motoneurons, a decrease in inhibitory inputs to alpha motoneurons, or some combination of these factors. The net effect of such input imbalances is increased alpha motoneuron excitability, increased muscle tone, and disordered motor control. In more mild cases, spastic muscles may not be spontaneously active but exhibit hyperreflexia, which resists either passive or active stretch of the muscle. In more severe cases, spastic muscles may be in a state of nearly constant contraction, which, if unabated, leads to marked loss of function, discomfort to the patient, and permanent joint contractures. The marked functional impairments associated with spasticity have led health care providers from many areas to search for clinical procedures to combat this problem. Electrotherapy is one procedure that has been used in the management of spasticity. Several different electrical stimulation approaches have been tried in attempts to manage spasticity in skeletal muscles.

Antagonist muscle or nerve stimulation to reduce spasticity

Electrical stimulation was used for the control of spasticity as far back as 1871 when Duchenne reported on the effects of the electrical activation of antagonists to spastic muscles (18). The first attempt in modern times to use this first therapeutic technique for spasticity control in humans was made by Levine and coworkers (17) in the early 1950s. In the study, they found that antagonists to spastic musculature were stimulated with uninterrupted "faradic" currents at a frequency of 100 pps using a monopolar electrode configuration. The electrode placed over the antagonist of the spastic muscle was positioned over the motor point for that muscle. Stimulation was applied at an amplitude sufficient to evoke a "maximum contraction." The spasticity of subjects in the clinical report was secondary to a number of disorders including hemiplegia and multiple sclerosis and in some cases was present in flexor and extensor musculature (e.g., biceps and triceps) simultaneously. Regardless of the underlying pathology, the authors reported a relaxation of the hypertonicity within several seconds of initiating stimulation as evidenced by the reduction in opposition to passive stretch of the spastic muscle. In several of the reported cases, the reduction in spasticity was accompanied by improvements in function associated with self-care activities, mobility, and posture.

In the late 1970s, Baker and coworkers (13) applied NMES to the wrist and finger extensor musculature of 16 hemiplegic patients with flexor spasticity in the forearm. Stimulation using rectangular monophasic pulses at a pulse duration of 200 μsec and frequency of 33 pps was applied using a 7 sec on/10 sec off timing cycle and a bipolar electrode setup. The pulse amplitude was increased to produce tetanic isotonic contractions with wrist and finger extension through the full available ROM. This pattern of stimulation was applied for 15 minutes twice each day in the initial sessions and was increased to 30-minute sessions three times per day, each day, as the patients' tolerance to the stimulation increased. After a 4-week program of stimulation, wrist and finger flexor spasticity in response to quick manual stretches were subjectively evaluated. Although the distributions of spasticity "grades" (severe, moderate, mild, and none) before and after treatment were not presented, the authors reported a reduction in wrist flexor spasticity immediately following stimulation that usually persisted for about thirty minutes.

Just three years later, Alfieri (19) also reported on the effects of antagonistic muscle stimulation for spasticity management in a group of 96 hemiplegic patients. For stimulation of muscles in the forearm and lower leg, trains of 0.5-msec-duration pulses at 50 pps were applied with a 2 sec on/2 sec off timing cycle. The pulse amplitude was increased exponentially as stimulation was applied and was adjusted to a level that did not produce any overflow and subsequent activation of the spastic muscles. For the quadriceps and deltoid muscles, square wave pulses (duration of 20–30 msec) were applied at a frequency of 0.5 Hz. For both patterns of stimulation, amplitudes were adjusted to evoke slight visible isometric contractions in each of 8 to 17 daily treatment sessions lasting on average 10 minutes. The effects of stimulation on muscle spasticity were evaluated "clinically" and by testing with the Ashworth scale (Table 5.1) (10 patients). The author states that spasticity was decreased in the muscles of 90% of patients for periods lasting from 10–15 minutes to 2 hours. In addition, subjective assessment suggested that spasticity remained reduced for 90% of those patients reexamined 4–16 weeks after termination of stimulation ($n = 64$).

Objective data on the effects of antagonist stimulation on spasticity can be found in the work of Carnstam's laboratory (20) reported in 1977. Seven subjects exhibiting spasticity in the plantar flexor muscle group(s) were examined. In the study, tibialis anterior was stimulated with 300-msec bursts of pulses (pulse duration of 0.5 msec, frequency of 30 pps) each second for a period of 10 minutes and evoked a submaximal contraction. Immediately before and after this stimulation, the force for the Achilles tendon reflex was measured. They found that in several subjects examined, the amplitude of the Achilles tendon reflex was reduced after stimulation, suggesting a drop in plantar flexion spasticity. A decade later, Petersen and Klemar (21) reported on the effects of motor-level stimulation of the tibialis anterior on plantar flexor spasticity in 22 patients. NMES was applied by placement of electrodes over the tibialis anterior motor point or the peroneal nerve. Monophasic pulsed currents at 33–50 pps and pulse duration of 250 μsec were applied with 2–3 sec on time and 10 sec off time for 30 minutes twice each day. The stimulation program was carried out for four weeks at which time both subjective and objective measures of hypertonicity were made. The subjective clinical evaluation performed just after cessation of stimulation revealed a decrease in passive

Table 5.1
Ashworth and Penn Scales for Spasticity Assessment

Ashworth muscle tone scale (resistance to manual stretch)	
Muscle tone score	Degree of muscle tone
1	Normal muscle tone
2	Slight increase in tone ("catch" with passive movement)
3	Moderate increase in tone (passive movement possible)
4	Marked increase in tone (passive movement difficult)
5	Severe increase in tone (rigid in flexion or extension)

Penn spasticity scale	
Spasm score	Frequency of spasm
0	No spasms
1	Mild spasm induced by stimulation
2	Infrequent full spasms less than 1/hr
3	Spasms >1/hr but <10/hr
4	Spasms >10/hr

Reprinted with the permission of *Paraplegia*. Adapted from Halstead LS, Seager SWJ. The effects of rectal probe electrostimulation on spinal cord injury spasticity. Paraplegia 1991;29:43-47.

resistance to stretch, reduced hypertonia, and decreased clonus. Subjective improvement was also reported by the majority of subjects, who described decreases or elimination of spasticity for 6–14 hours after each stimulation session. Objective measures of spasticity (number of flexor spasms per hour, T-reflex, and passive resistance to stretch) were unchanged after stimulation. Only the magnitude of the electrically elicited flexion reflex was significantly reduced. Apkarian and Naumann (22) also examined the effects of tibialis anterior stimulation on plantar flexor spasticity. They found that a brief burst of motor-level stimulation (minimal perceptible contraction) to the tibialis anterior preceding stretch reduced mechanically elicited stretch reflex responses of the plantar flexors in three of six patients.

How does using ES on antagonist muscles or nerves to spastic muscle relieve spasticity? The precise answer is as yet unknown. One plausible explanation was originally presented by Levine and coworkers (17). As stimulation is applied to the peripheral nerve leading to the antagonist muscle, the large-diameter Ia muscle spindle afferent fibers originating in the muscle are excited (Fig. 5.3). The action potentials generated in these fibers are transmitted to the spinal cord and excite spinal interneurons, which in turn inhibit the activity in the motoneurons to the spastic muscle. Although the activation of such a reciprocal inhibition pathway may lead to an immediate reduction of activity in the spastic muscle during stimulation, it does not explain why such stimulation may reduce spasticity and alpha motoneuron excitability for extended periods following the cessation of stimulation, a finding that has been repeatedly demonstrated in several of the investigations reviewed above. In addition, some evidence exists that indicates that antagonist nerve or muscle stimulation does not reduce spasticity by activation of Ia afferent reciprocal inhibition pathways. Apkarian and Naumann (22) varied the time at which a stimulus was delivered to the tibialis anterior prior to evoking a plantar flexor stretch reflex. If Ia afferent reciprocal inhibition pathways were responsible for

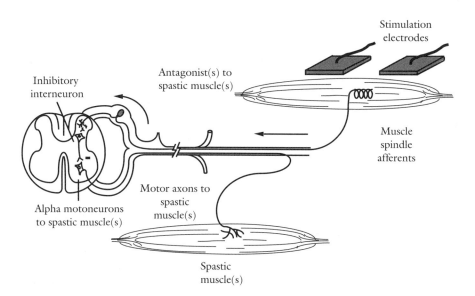

Figure 5.3 Proposed mechanism of action of antagonist stimulation to control spasticity. Stimulation of antagonist(s) Ia afferents reciprocally inhibits motoneurons to spastic muscle(s).

spastic stretch reflex inhibition, they estimated that the maximum inhibition would occur at about 40 msec after stimulation. They found, however, that the maximum inhibition occurred when stimulation preceded the stretch by about 160 msec. They hypothesized that antagonist motor nerve stimulation may reduce spasticity by the activation of multisynaptic spinal cord pathways associated with flexion reflex activation.

NMES of spastic muscle to reduce hypertonicity

A second stimulation approach examined for the reduction of spasticity is the use of NMES to electrically activate the spastic muscle(s). Lee and coworkers (23) were among the first to study this approach for spasticity management on 27 spinal cord injury patients. Bipolar electrode placement over the spastic muscle was followed by either continuous "faradic" or sinusoidal stimulation at frequencies ranging from 60 to 100 pps for faradic current or 60 to 350 Hz for sinusoidal current. The amplitude of stimulation was set initially and adjusted periodically to maintain "maximum contraction." In some cases, stimulation intensity was manually increased and decreased to elicit alternating contraction and relaxation in the spastic muscles; however, no on times and off times were specified. Total treatment time in a single session was 15 minutes. Assessment of hypertonicity before and after stimulation was made by four observers using an arbitrary five-point scale ("excellent" to "poor" muscle relaxation). The experimenters reported that sinusoidal stimulation produced greater reduction in level and longer duration of relaxation than did "faradic" tetanization. Another early study of the effects of spastic muscle stimulation on the level of hypertonicity was reported by Vogel and colleagues (24). In this study, continuous AC stimulation (2000 Hz) was applied to spastic quadriceps and plantar flexor muscles simultaneously with DC stimulation (40 V) through one electrode over the low back (anode) and plantar surface of the foot (cathode). The rationale for such simultaneous AC/DC stimulation was not discussed. Spasticity was assessed by examining passive resistance manual stretch using a four-point scale. Each of 15 patients with complete or incomplete traumatic spinal cord lesion experienced reductions in spasticity that persisted for 1–24 hours. Patients also reported "greater ease and comfort" in performing activities of daily living. Those patients with spasticity associated with multiple sclerosis could not tolerate the stimulation program and consequently did not show any improvement in spasticity following treatment.

Tetanic stimulation of spastic muscle was also examined by Bowman and Bajd (25) in 10 spinal cord injury patients. Slight diminution of spasticity was found in only 4 of the 10 subjects studied following repetitive stimulation.

More recently, Robinson and coworkers (26) attempted to quantitatively assess the effects of spastic muscle stimulation on post-stimulation spasticity by use of pendulum testing. The **pendulum test** of spastic quadriceps was developed by Bajd and Vodovnik (27) and is designed to quantify the hyperactive stretch reflex under standardized conditions (Fig. 5.4). In this test, the leg is first extended through full available passive range and then the foot is released. The leg is allowed to swing freely and the oscillations (knee flexions and extensions) produced by repeated stretch reflexes in the quadriceps are measured. An index of spasticity called the *R2n* is then calculated. An R2n greater than 1 indicates a normal limb, whereas R2n values less than

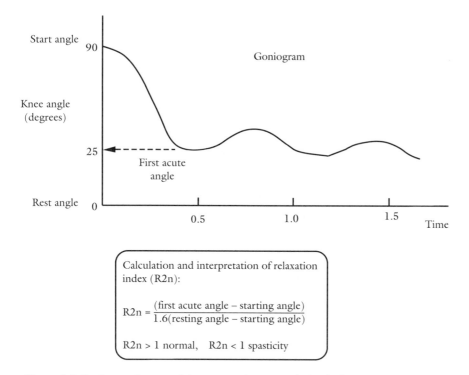

Figure 5.4 Goniogram from pendulum test used to assess the level of spasticity, calculation of the relaxation index, and interpretation of the relaxation index.

1 indicate muscle spasticity. R2n indices were determined before and after motor-level stimulation was applied alternately to the quadriceps bilaterally. Monophasic pulsed stimulation at 100-mA amplitude, 20-pps frequency, and 0.5-msec pulse duration were applied with a reciprocating 2.5-sec on time and off time for 20 minutes (10 minutes of stimulation to each quadriceps group). Immediately following stimulation, R2n values increased, indicative of a reduction in spasticity. The greatest reduction in spasticity was noted in those subjects initially showing the most spasticity. Testing was repeated 24 hours after the stimulation; it revealed that the immediate effect of NMES on spastic muscle did not persist.

The exact mechanism(s) by which the NMES of spastic muscle may reduce spasticity is unknown. Early investigators proposed that intense NMES of spastic muscle without interruption produced fatigue in the spastic muscle to a level sufficient to account for their results. The drop in peak torque measured by Robinson and coworkers (26) immediately following a 10-minute stimulation session supports this contention. The simple theory is that a fatigued spastic muscle produces a lesser contractile response to the abnormally high, spontaneous motoneuron input that is sent to muscle following injury or disease of the CNS.

More recently, other investigators (28) have hypothesized that NMES may reduce hypertonicity as a result of the effects of antidromically propagated action potentials evoked in the motoneuron axons to the spastic muscle. The action potentials propagated toward the spinal cord along motoneurons may not only invade the cell body of the motoneuron but may also pass along recurrent collateral axons, which are thought to synapse with spinal inhibitory interneurons called *Renshaw cells*. The antidromic activation of these interneurons in turn inhibits the activity of agonist and synergistic hyperactive motoneurons (Fig. 5.5). Such a hypothesis has yet to be experimentally tested and verified.

Combined agonist and antagonist NMES to control spasticity

A third approach to the use of ES in spasticity management has been to alternately activate spastic and antagonist muscles in a rhythmical manner. Vodovnik and coworkers (29) employed this technique in the early 1980s on seven spinal cord–injured patients with clinical evidence of thigh muscle spasticity. Bipolar electrode placements were made using four channels of two portable stimulators over the hamstrings and quadriceps of each leg. Isotonic contractions of quadriceps and hamstrings were elicited using asymmetrical biphasic pulses at a rate of 30 pps, pulse duration of 300 μsec, and an amplitude of 100 mA. The timing cycle for stimulation was 5 sec on and 5 sec off, with the hamstrings in one limb stimulated while the quadriceps in the opposite limb was activated. A **Wartenburg spasticity test (pendulum test)** (30) was used to assess the influence of the 30-minute stimulation program delivered over a period of 5 days. Because of variability in the tone of knee flexor and extensor muscles, subjects in the study were divided into three groups. Only two of the seven patients demonstrated consistent spasticity, and the qualitative results indicated that the hypertonia was reduced following the stimulation program. The results did not suggest that such a reciprocal pattern of activation is superior to either agonist or antagonist approaches alone.

Sensory-level stimulation in the control of spasticity

The ES approaches to spasticity management reviewed thus far have applied stimulation to either the spastic muscle, its antagonist muscle or nerve, or some combination of these approaches. In most of these applications, the

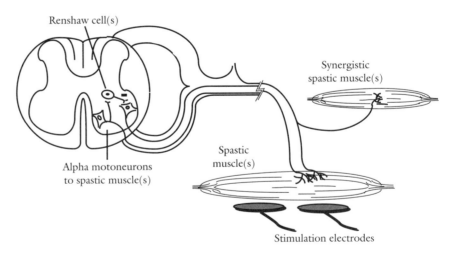

Figure 5.5 Proposed mechanism of action of agonist stimulation to control spasticity. Stimulation over the spastic muscle(s) antidromically activates motoneurons to the spastic muscles. Action potentials propagated toward the spinal cord activate inhibitory interneurons (Renshaw cells), which inhibit alpha motoneurons to the spastic muscle(s).

level of stimulation was sufficient in amplitude to evoke muscular contraction. Lower levels of stimulation activating only sensory fibers in peripheral nerves have also been studied for the control of spasticity.

In 1985 Bajd and coworkers (31, 32) placed electrodes over the L3,4 dermatome medially below the knee and laterally above the knee. *Sensory-level stimulation* was applied to this dermatome associated with the same level as the motor nerves to the spastic quadriceps of spinal cord–injured subjects using rectangular, monophasic pulses (duration = 0.3 msec) at a frequency of 100 pps for 20 minutes. Pendulum testing (26) of the quadriceps was performed before, just after, and 2 hours after the stimulation as a measure of hypertonicity in six spinal cord–injured patients with moderate spasticity in the quadriceps bilaterally. Three of the six patients demonstrated marked reductions in spasticity as revealed by the rise in the relaxation indices derived from pendulum testing. The reduction in muscle tone had not persisted by the test given 2 hours after stimulation.

More recently, Hui-Chan and Levin (33) reported on the sensory-level electrical stimulation of the nerve trunk innervating the antagonist of spastic plantar flexor muscles. In this study, sensory-level ES was applied to the common peroneal nerve innervating the tibialis anterior. The effects of stimulation were examined on measures of plantar flexor muscle spasticity in 10 hemiparetic patients. Stimulation was applied using a continuous train of rectangular, asymmetric, biphasic pulses at 99 pps, pulse duration of 0.125 msec, and an amplitude set to twice the sensory threshold level. This pattern of stimulation was applied continuously for 45 minutes. Spasticity was assessed subjectively in three ways by grading *(a)* Achilles tendon jerks (5-point scale with 0 denoting no reflex response and 4 denoting maximally hyperactive response), *(b)* resistance to slow passive stretch using a 5-point Ashworth scale), and *(c)* clonus (4-point scale with 1 denoting no clonus and 4 denoting sustained clonus). Spasticity was objectively assessed by monitoring the changes in the magnitude and latencies of plantar flexor H-reflexes and mechanically elicited stretch reflexes. Antagonist nerve stimulation did not consistently change any of the subjective clinical indices of spasticity in the subjects. However, the ES program did increase H-reflex and stretch reflex latencies in more than half of the subjects. In many subjects, these delays in onset of reflex responses were accompanied by reductions in reflex magnitude, and the effects persisted for up to 1 hour after cessation of the stimulation. These increases in reflex latencies and decreased amplitudes of responses are considered as evidence of the inhibitory effect of ES on spasticity by altering the excitability of alpha motoneurons to the spastic muscles.

The efficacy of sensory-level stimulation to the nerve-innervating antagonists of spastic muscle appears to depend on the frequency of treatment. Levin and Hui-Chan (34) reported in 1992 study results similar to those described in the previous paragraph. The stimulation program, however, was applied to the common peroneal nerve for 15 daily 60-minute sessions for a period of 3 weeks in 11 hemiplegic subjects. Spasticity was once again assessed using a battery of tests. The intensive electrotherapy program resulted in significant improvement of the clinical spasticity measures, significant increase in vibratory inhibition of soleus H-reflexes, and dramatic improvement in voluntary dorsiflexion forces after just 2 weeks. On completion of the third week, significant reductions in stretch reflexes of spastic triceps surae were noted. These workers suggested that the sensory-level stimulation may inhibit spasticity by activation of presynaptic inhibition of afferents to spastic muscle motoneurons. In addition, improvements in volitional dorsiflexion, which would serve to counteract foot-drop during gait, may have resulted from stimulation producing a disinhibition of descending volitional control pathways to tibialis anterior motoneurons.

ES remote from spastic muscle for hypertonicity control

In one of the more novel approaches to testing the efficacy of ES on spasticity reduction, Walker (35) reported on the effect that ES of peripheral nerve of the upper extremity had on lower limb spasticity. Subcutaneous electrodes were implanted in the median, radial, and saphenous nerves in patients with multiple sclerosis or postsurgical nerve "irritability." Each patient in the study had spasticity manifested as a persistent ankle clonus. Patients were stimulated using 20-pps spike waveforms for 1 hour, 2 times a day, for 1 week. Such stimulation at *sensory* levels suppressed the stretch-elicited clonus in all patients; however, the maximum reduction in clonus did not occur until approximately 1 hour after stimulation was terminated. Once reached, the maximum inhibition lasted for 3 hours in all patients. Control patients (those stimulated at points distal to the three stimulation sites) showed no suppression of clonus, although the precise manner of peripheral stimulation in the control subjects was not readily apparent. To determine whether the stimulation of the peripheral nerves in the upper extremities alone could inhibit the hypertonic signs in the lower extremities, radial and median nerves bilaterally were stimulated alone. Such stimulation was reported to have completely inhibited ankle clonus as well. Although clonus was reduced in these patients as a consequence of the experimental stimulation, other manifestations of spasticity such as the "extent of scissoring, hyperreflexia or Babinski sign" appeared to be unchanged.

About ten years later, Hui-Chan and Levin (33) reported on the effects of *upper extremity nerve stimulation* measures of spasticity in lower limb plantar flexor muscles. Surface electrodes were placed on the volar aspect of the wrist over the median nerve formed from nerves emanating from C6 through T1 nerve roots. Stimulation was applied using a continuous train of rectangular, asymmetric, biphasic pulses at 99 pps, pulse duration of 0.125 msec, and an amplitude set to twice the sensory threshold for 45 minutes. Spasticity was assessed subjectively in three ways and objectively in three ways as described above. In four of six subjects, remote sensory-level stimulation altered the latencies of both the mechanically and electrically elicited monosynaptic reflexes in a manner that would reflect a reduction in lower limb spasticity.

Paraspinal sensory-level stimulation has been applied in attempts to control lower extremity spasticity in individuals with multiple sclerosis. Fredriksen and coworkers (36) applied two electrodes in the midline at lower cervical and upper thoracic areas. ES was given using an asymmetric, biphasic pulsed current (50 pps, 250-μsec pulse duration, continuous mode) for 2 weeks. Spasticity was "clinically assessed" by the researchers and subjectively assessed by the patients. Only 4% of the subjects revealed reduced spasticity on clinical neurological assessment, whereas 40% of the participants reported subjective improvement in spasticity.

Long-term NMES and spasticity control

Conditioning programs using NMES in spinal cord–injured patients are currently the subject of research at numerous centers worldwide. The objective of NMES programs is initially to strengthen paralyzed muscle so that NMES can ultimately be used to produce functional activation of muscles in activities such as standing or walking. In some cases, these long-term NMES conditioning programs have been found to produce a reduction in muscle spasticity. In 1993 Granat (37) reported on the benefits of NMES conditioning and a gait-training program in six patients with incomplete spinal cord injury. The NMES muscle-strengthening programs varied from subject to subject but included NMES to the quadriceps, hamstrings, hip abductors, and erector spinae musculature as needed by the subject. All subjects completed a 6-month NMES progressive resistive exercise program to these muscles. Conditioning stimulation parameters were incompletely specified but did consist of 25-Hz stimuli with 300-μsec pulse duration, with 4-sec on time and 8-sec off time. While the conditioning program was continued, each subject was placed on an NMES gait program. Quadriceps, peroneals, abductors, and erector spinae were stimulated as necessary for synthesis of gait. Subjects were asked to perform the gait program for 30 minutes each day, 5 days per week. Spasticity was assessed using an Ashworth scale (opposition to slow, passive manual stretch ranging from "no increase in tone" to "rigid in extension or flexion") and the pendulum test as described previously. Four of the six subjects demonstrated increased relaxation indices from pendulum testing of the quadriceps bilaterally, performed 24 hours after cessation of the last stimulation session. Only one of the six subjects appeared to have increased spasticity following the 9 months of rehabilitation. However, analysis of Ashworth scale scores did not show significant reductions in passive resistance to stretch at the completion of the training program. In this program, stimulation was provided to the spastic quadriceps, to quadriceps antagonists, and paraspinally remote from the spastic muscles, and therefore one cannot draw any conclusion regarding which individual site may have been responsible for reducing the spasticity.

The beneficial effect of long-term NMES on spasticity has also been reported following peroneal gait-training stimulation in the hemiplegic patient population. Stefanovska and colleagues (38) measured passive resistance to sinusoidal stretch of the ankle dorsiflexor and plantar flexor groups in eight hemiplegic subjects before and after a 6-month gait-training program. Patients used an implanted peroneal nerve stimulator for up to 2 hours each day for 6 months. Spasticity as reflected in passive resistance to stretch was reduced at each of the speeds of stretch examined.

Not all studies examining the effects of long-term spastic agonist stimulation have shown reductions in spasticity. Robinson and coworkers (39) examined the effects of a 4-to-8-week NMES conditioning program on spasticity in 31 spinal cord–injured subjects. Spasticity was assessed by the relaxation index (R2n), calculated from the pendulum drop test. The program consisted of bilateral quadriceps stimulation using monophasic pulsed current (400-μsec pulse duration) at 20 pps, 2.5 sec on and 2.5 sec off. NMES exercise sessions lasted 20 minutes twice each day for 6 days each week. The majority of subjects who completed 4 weeks of training ($n = 22$) exhibited decreases in R2n values (increases in spasticity). For the eight subjects who completed 8 weeks of NMES training, about half of the quadriceps showed increases in spasticity and half revealed decreases in spasticity. The authors indicated that such results contrasted with those of other researchers (18); however, differences might be attributable to different subject populations (hemiplegic vs. quadriplegic or paraplegic groups).

Epidural spinal cord stimulation for spasticity control

Direct stimulation to the spinal cord with electrodes implanted in the epidural space was developed primarily as a treatment for intractable pain. In some CNS-injured patients, this procedure used for pain control (40–42) was also observed to have beneficial effects on hypertonic muscles (43–48). Barolat-Romana and coworkers (49) reported on the effects of this procedure in six spinal cord–injured patients. The cathode of the stimulation circuit was inserted in the dorsal epidural space with a Tuohy needle at a level that induced paresthesias in all four limbs with intraoperative stimulation. The anode of the circuit was placed subcutaneously in the flank. Rectangular pulses at frequencies of 75–100 pps, 100 to 250-μsec pulse durations, and amplitudes between sensory and motor thresholds were applied for periods lasting up to hours. Assessment was performed before and after stimulation using dynamometry, electromyography, and electrically elicited monosynaptic reflex (H-reflex) testing. In three of the patients, spasticity was "significantly reduced" immediately on initiation of stimulation and persisted until stimulation was terminated. "Good" reduction in spasticity was noted in the remaining three subjects.

Epidural stimulation for spasticity does not have as dramatic effects on all forms of spasticity. For example, Fredriksen and colleagues (36) applied epidural stimulation to patients with spasticity secondary to multiple sclerosis. Only 4% of the subjects (2 of 49) revealed any improvement in clinical measures of spasticity at both 1 month and 10 months following implantation. These results were comparable to those using transcutaneous paraspinal stimulation in the same subject population (described above). In addition to the poor results, a number of problems such as electrode breakage, electrode migration, hematomas, and infections were encountered using the epidural stimulation procedure.

Transrectal stimulation for spasticity control

As so often happens in clinical research, new discoveries are made when careful researchers observe unexpected responses to treatment. One of the most recent developments in electrotherapy for spasticity control emerged from the clinical use of ES for anejaculation in spinal cord–injured (SCI) men. Clinicians applying electroejaculation procedures noted that the majority of their SCI patients reported reductions in spasticity that lasted for up to several hours following treatment. Halstead and Seager (50) examined the efficacy of this approach to spasticity control in 14 SCI subjects. Treatment consisted of rectal probe electrical stimulation (RPES) using 60-Hz sinusoidal AC at 200–500 mA for 12–35 1-sec-duration shocks. Spasticity was assessed before and after treatment using both the Penn spasticity scale (spasm frequency) and the Ashworth scale (degree of muscle tone) (Table 5.1) (50). Approximately 40% of the subjects showed "excellent" relief of spasticity after about thirty RPES sessions, whereas about 30% experienced "good" reduction in spasticity after 14 treatments. Maximum effectiveness of treatment occurred within 1 hour of treatment. The mechanism of action of this new approach is as yet unknown. One adverse side effect of this procedure was dysreflexia in subjects with injuries at T6 or above. This side effect, however, was controlled when subjects were medicated with sublingual procardia and/or sublingual nitroglycerine before treatment.

Methodological concerns with ES for spasticity control studies

In reviewing the literature on ES for spasticity control, a number of research design and implementation problems are readily apparent in some of the studies. These common problems include the following.

1. Use of tests to characterize spasticity that were not reliable or for which reliability had not been examined.
2. Poor patient selection: inhomogeneous subject population with respect to etiology, onset time, severity of symptoms; inadequate numbers of subjects.
3. No control group used in the study.
4. Inadequate description and/or variation of stimulation parameters.
5. Inappropriate or no statistical anaylsis of data.

In spite of these shortcomings in some of the literature, electrotherapy for spasticity control appears to remain a viable treatment approach for many patients. Like all conservative management approaches, ES for spasticity will not benefit all patients or control all forms of spasticity. Additional carefully controlled and well-designed studies are needed to identify optimal stimulation approaches and specific patient selection criteria for successful treatment application.

Functional Neuromuscular Electrical Stimulation

Attainment of the goals of NMES applications such as the strengthening of muscle, increasing muscular endurance, improvement in joint range of motion, or reduction of spasticity does not ensure that patients will be able to produce voluntary muscle contraction sufficient to maintain posture or to produce purposeful movements. In many patients who have sustained CNS damage, the control exerted by higher nervous system centers over muscle contraction may be impaired. In such patients, a variety of advanced therapeutic exercise techniques have traditionally been employed to facilitate the return of controlled functional muscular activity or to maintain postural alignment until recovery from dysfunction occurs. In addition, orthotic devices have commonly been used to improve function or control posture in cases where a return of normal muscle function is either slow or unlikely to occur.

Over the past 30 years, a number of NMES applications have been developed for use in conjunction with classical management approaches to enhance the purposeful contraction of skeletal muscle. This area of laboratory and clinical interest was termed **functional electrotherapy** by early workers in the field (51). The use of NMES as an orthotic substitute (to take the place of a brace or support), to maintain posture, or to produce limb movements important for activities of daily living is currently referred to as **functional electrical stimulation (FES)** (52–54). A detailed review of the extensive research and wide variety of procedures in FES is beyond the scope of this text. The objective of this segment of the chapter is to organize the scope of FES applications and to review selected reports in order to illustrate how FES is being used to manage a number of commonly encountered clinical problems.

FES for dorsiflexion assist during gait

Among the most common clinical applications of NMES for functional muscle contraction is as a substitute for static or dynamic orthotic devices. Modern investigation into this application of NMES began in 1961 when Liberson and coworkers (51) used NMES on paralyzed ankle dorsiflexor and everter muscles of hemiplegic patients for the control of foot-drop during the swing phase of gait.

In their study, one conductive rubber electrode was applied over the peroneal nerve just below the knee and a second over either the thigh or tibialis anterior (Fig. 5.6A). Stimulation to the peroneal nerve was provided with one of several different types of transistorized stimulators. In all cases, the stimulators produced dorsiflexion and eversion of the involved ankle using stimuli with pulse durations ranging from 20 to 250 μsec, peak pulse currents below 90 mA, and stimulation frequencies between 30 and 300 pps. Stimulation timing was controlled by a switch in the sole of a subject's shoe on the involved side; the switch stopped stimulation in the stance phase of gait and triggered stimulation as the subject entered the swing phase. All seven subjects in this study were reported to have demonstrated "considerable" improvement in gait using the "electrophysiologic brace," and in some cases the prolonged use of the peroneal nerve stimulation improved their ability to volitionally dorsiflex the involved foot during walking. These findings, along with the rapid miniaturization of the components of electrical stimulators, fostered a significant resurgence of research interest in the use of NMES to improve the functional abilities of patients with disturbed motor control.

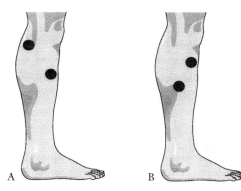

Figure 5.6 Electrode placement options for NMES for dorsiflexion during the swing phase of gait. A pressure-sensitive switch beneath the heel triggers stimulation as the heel is lifted and stops it when the heel strikes.

Functional ES of ankle dorsiflexion and eversion during the swing phase of gait in hemiplegics acts as a substitute for the commonly employed ankle-foot orthosis (AFO, dorsiflexion-assist brace). Surface electrode placement is generally bipolar, either directly over the peroneal nerve near the head of the fibula or over the motor points of the tibialis anterior and peroneal muscles (53, 55–57).

Since the original work of Liberson and colleagues, both surface electrodes and implanted percutaneous electrodes have been used. Stimulation has been provided with a variety of different pulsed waveforms with pulse durations generally in the 0.1-to-0.6-msec range at frequencies usually between 20 and 50 pps. The intensity of stimulation is adjusted to produce a smooth tetanic isotonic contraction sufficient to dorsiflex and evert the ankle. The timing of stimulation is controlled by either pressure-sensitive switches in the soles of shoes or a preprogrammed sequence of stimulation. When preprogrammed stimulation is used, modifications are made in the stimulation pattern to suit the gait pattern of each individual patient.

Surface stimulation to control foot-drop is often used in early clinical trials to determine if an individual is an appropriate candidate for long-term use to improve gait. A number of implantable stimulation systems have been developed; for example, in one study a small stimulation circuit was surgically implanted over the peroneal nerve trunk and sutured in place (57). Two platinum loops extending from the 17 × 8 mm electronic stimulator circuit acted as electrodes, and the unit was activated by an antenna placed on the skin connected to a pulse generator.

Gracacin (53) has stated that peroneal nerve stimulation improves the kinematics of gait in both children and adults with CNS disorders of motor control. In addition, electromyographic recordings revealed a restoration of the cyclic activity of muscles during walking.

Merletti and coworkers (58) found that 76% of 50 patients treated with peroneal nerve stimulation demonstrated good or excellent results. A "good" result was one characterized by (*a*) a marked temporary decrease in spasticity, (*b*) a limited but long-lasting drop in spasticity, (*c*) an increase in muscle-test grade of 1 to 2 points on a 5-point Kendall scale, or (*d*) a temporary improvement in nonassisted gait. An "excellent" result was characterized as (*a*) a marked and permanent reduction in spasticity, (*b*) a muscle-test grade improvement of more than 2 points, or (*c*) a marked, stable improvement in nonassisted gait. About one-third of the patients using the peroneal stimulator achieved "excellent" therapeutic results. When the selection criteria were strictly adhered to, only 3 of 50 patients failed to show any improvement with stimulation. The authors determined that, in general, approximately 20% of the ambulatory hemiparetic population would be good candidates for this form of therapeutic intervention.

The long-term use of implanted stimulation systems for foot-drop control is associated with a number of problems. Kljajic and coworkers (59) learned that in a population of 35 patients who had received peroneal nerve implants 4 subjects had implants removed as a result of "unpleasant" sensations linked to the stimulation. Three other patients did not achieve adequate control of their foot-drop and also discontinued their program. Nineteen members of this population underwent dynamic gait evaluation. Of this subgroup,

9 were found to have had electrode displacement after an average of 3.5 years, which resulted in the failure of stimulation to produce the desired correction. Reimplantations in these subjects resulted in an immediate improvement in the kinematics of gait to levels similar to those observed following the initial implantation. One subject required reimplantation because of fracture of the implanted electrodes, and others required reimplantation because of the formation of a fibrous capsule around the electrodes. Otherwise, no pathological changes were observed in or around the peroneal nerve at the stimulation site.

NMES has generally been used as an electronic peroneal nerve brace in patients who cannot or will not use a conventional AFO. Other important patient selection criteria include (58):

1. The patient's ability to cooperate and communicate.
2. Limited spasticity of the ankle plantar flexors.
3. NMES-induced improvement of gait pattern significantly greater than with the conventional AFO approach.
4. No significant limitation in ankle passive ROM.
5. No significant knee or hip volitional movement limitations.
6. No hypersensitivity to NMES.
7. A high level of subject motivation.

Peroneal nerve stimulation in patients who do not meet these selection criteria may not produce more functional improvement than conventional bracing management.

The application of NMES to control ankle position during gait is not limited to hemiplegic adults. Both Gracinin and coworkers (60) and Vodovnik's laboratory (61, 62) have reported successful use of functional electrical stimulation in children with cerebral palsy.

FES and muscle reeducation

One of the common observations made by investigators using NMES for orthotic substitution is that the volitional activation of involved muscles frequently improves following a period of FES. This observation was made as early as Liberson's work of the early 1960s on peroneal nerve stimulation (51). They reported that after training with the electronic peroneal brace, several patients acquired the transitory ability to voluntarily dorsiflex the involved foot.

Quantitative data regarding this phenomenon are available from a number of reports (20, 55, 61). In a study reported by Carnstam and coworkers (20), voluntary isometric dorsiflexion forces were monitored both before and after a 10-minute session of tetanic stimulation to the paretic tibialis anterior muscle. In one patient who demonstrated very weak dorsiflexion prior to stimulation, NMES of the dorsiflexors was followed by a fourfold increase in the volitional level of contraction. Other patients who were much stronger exhibited a "more or less evident" increase in strength also. The increases in voluntary force following stimulation were inversely related to the strength before stimulation. Similar results have been reported by Vodovnik and Rebersek (61) in five of six hemiparetic patients receiving NMES to dorsiflexors for 20 minutes of walking. In this report, however, the increase in volitional dorsiflexion following stimulation did not appear to be related to the prestimulation tensions produced.

An increase in the strength of voluntary muscle contraction following NMES is a reflection of an increase in either the number of motor units activated or the rate at which recruited units are activated. Such changes do not ensure that the use of these muscles in functional isotonic or isometric contractions is improved. That is, the skill or accuracy of isotonic movements requires not only adequate muscle force but also careful control of such force in order to improve motor performance.

The timing of contraction initiation is an important element in the production of skilled purposeful movement. The influence of NMES on the timing of voluntary activation has been examined by Fleury and Lagasse (63). In this study, 62 normal subjects were asked to produce a horizontal adduction of the suspended nondominant arm as rapidly as possible in response to a visual signal. The movement required that this ballistic contraction be controlled so that the arm moved to, but not beyond, the end point of a specified range of movement. The electromyographic activity of the anterior deltoid and pectoralis major muscles was monitored during each movement. Each participant performed 10 adduction movements in pretest sessions. Subjects were randomly divided into three groups. Control subjects performed 100 volitional adduction movements after pretest. Individuals in the stimulation-only group received NMES to the anterior deltoid and pectoralis to initiate adduction followed by posterior deltoid stimulation to stop the movement within the specified range. A third group of subjects received alternating stimulation with volitional activity for 100 trials. Ten volitional shoulder adduction movements were performed following the training sessions to determine the effect of the training programs on the timing of the electrical activity in the muscles associated with horizontal adduction. All three groups of subjects showed reduced reaction times (time from the presentation of start signal to initiation of movement). Reaction times for both groups receiving NMES were shorter than that for the control group.

On closer examination, the improvement in total reaction time could be attributed to a reduction in premotor time (the time interval between the presentation of the start signal and the initiation of electrical activity in the muscles). The greater reduction in premotor times of individuals in the two stimulated groups supports the contention that NMES is more effective than volitional training alone for improving motor performance. Similar studies in patients with motor control disturbances are not yet available, and consequently the effect of NMES training in such populations has yet to be established.

Improvements in the voluntary activation of skeletal muscle in hemiplegic patients by the use of NMES have been reported. As indicated earlier, NMES and positional feedback combined with other, more traditional forms of treatment have been shown to improve patients' abilities to produce voluntary movement as reflected by an increase in their active range of motion (15, 16).

From these studies one cannot attribute improvements in volitional contraction to the NMES alone. In fact, some workers in this area have suggested that such improvements may result more from the continuous performance feedback than the stimulation (64). Carefully controlled studies on larger numbers of individuals in different patient populations are needed before one can safely conclude that NMES in any form gives improved volitional muscle control that is functionally significant.

FES for scoliosis management

Idiopathic scoliosis is a lateral curvature of the spine of unknown etiology. This condition is commonly found in children and, if untreated, produces severe postural deformities (such as thoracic rib hump) that may lead to impaired cardiopulmonary function, joint disease of the spine, back pain, and limitation in daily functional activities.

The common approach for minimizing or correcting scoliosis has been the use of the Milwaukee brace. This orthotic device is applied in children with progressive spinal curves between 20° and 45°. It is designed to promote improved alignment in the developing spine and must be worn until spinal growth ceases.

Although this approach to scoliosis management is often successful, many problems have been associated with it. Patients fitted with the Milwaukee brace must wear it for 23 hours every day and consequently cannot participate in many forms of recreation. Pressure exerted by the brace may lead to skin irritation, rashes, or even breakdown. The position of the chin piece of the brace may make eating or other activities in the sitting position difficult.

Because of these and other problems, NMES has been investigated as an alternative treatment in the management of scoliosis. The first attempt to use ES to prevent or correct the progression of lateral spinal curvature was made by Bobechko and coworkers (65) using implanted intramuscular electrodes in the paraspinal musculature on the convex side of the curve. The technique was found to halt the progression of the curve in over 80% of the patients with curves measuring less than 45°. Since application and maintenance of the electrodes requires surgical intervention, other investigators have examined the efficacy of transcutaneous NMES, using surface electrodes.

In the late 1970s, Axelgaard (66) demonstrated that surface NMES could sharply reduce scoliosis curves. Electrode placement was either *(a)* over the paraspinal musculature, *(b)* laterally over the midaxillary line, or *(c)* in a position midway between them on the convex side. The most lateral electrode placement was associated with the greatest degree of curvature correction. As a result of this study, these workers coined the acronym *LESS* to specify this electrode placement, in lateral electrical surface stimulation.

Eckerson and Axelgaard (67) have summarized this NMES approach. Patients who are being considered candidates for this procedure must meet the following strict selection criteria.

1. Curves measuring 20–45° by the Cobb method.
2. At least 1 year of spinal growth remaining.
3. An idiopathic and progressive nature of the curve.
4. Cooperative and psychologically stable.
5. Compliant and tolerant of the stimulation.

Prestimulation assessment included measurements of sitting and standing height, leg length, trunk decompensation, height and location of rib hump, and spinal roentgenograms. NMES for scoliosis was provided by using commercially available conductive carbon-rubber electrodes placed over the midaxillary line on the convex side of the curve, symmetrically about the rib that is attached to the vertebra at the apex of the curve, and within the boundaries of the curvature (Fig. 5.7). The stimulator used produced rectangular constant-current pulses of 220-μsec duration and frequency of 25 pps. The on time/off time of stimulation was 6 sec on and 6 sec off. The amplitude is gradually increased until muscular contraction is strong enough to produce spinal movement in the "straightening" direction. The duration of stimulation is gradually increased until the patient can tolerate 8 continuous hours of stimulation. Subsequently the patient uses this stimulation each night until skeletal maturity is reached.

In long-term follow-up studies of this technique, Axelgaard and Brown (68) have reported that nearly three-fourths of the patients examined showed a reduction or halt in the progression of their scoliosis. Complete compliance with the treatment has been reported to be as high as 82%, and minor negative side effects such as skin irritation at the electrode sites may occur in more than half of the patients. A multicenter study reported by Brown and colleagues (69) described the effects of NMES for scoliosis management in 548 patients who had been treated for periods of up to 51 months. Seventy-two percent of these patients had either stabilized or reduced scoliosis curves, and another 13% had modest curve progression before stabilization occurred. Only 15% of the patients were discontinued from NMES treatment because of significant progression. Another multicenter study, by McCullough (70), reported success in curve management with NMES in over 90% of patients with curves of less than 30°. Other studies that have attempted to reproduce the findings of the Axelgaard and Brown study (68) have reported limited success with NMES for scoliosis. Using a somewhat different NMES protocol, Bradford (71) noted that curve stabilization or slight progression occurred for only 64% (16 of 25) of patients. This level of success is consistent with recent findings by Swank (72) and Fisher (73). Goldberg and coworkers (74), using a stimulation program similar to that of

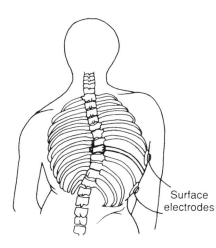

Surface electrodes

Figure 5.7 Electrode placement in midaxillary line for NMES correction of thoracic scoliosis. (Redrawn with permission from *Physical Therapy.* American Physical Therapy Assoc., 1984, Vol 64, p 486.)

Axelgaard (68) and of Brown (69), noted stabilization or slight progressions in only 32% of 41 patients. The low percentages of curve stabilization noted in this and other studies (75, 76) have led some researchers to conclude that NMES for scoliosis is not as effective as traditional orthotic management (75) or that it is actually ineffective in preventing curve progression (76).

The wide variation in the reported efficacy of NMES for scoliosis is difficult to explain. Some workers in this field have suggested that the essential ingredient for achieving success is strict patient compliance, a variable difficult to accurately measure. Compliance can prove to be a function of not only the individual patient but also of the health care professional who is responsible for the treatment program.

FES for shoulder subluxation

An early manifestation of stroke is a flaccid paralysis of limb musculature. If volitional muscular activity is lost in the shoulder, the force of gravity acting on the upper limb tends to stretch the ligamentous structures about the glenohumeral joint and may lead to shoulder subluxation. Such malalignment of the humerus and glenoid fossa may produce severe pain and may not be corrected even if voluntary control and shoulder muscle strength return during recovery. To combat this potential problem, hemiplegic patients are frequently fitted with a shoulder sling. Despite this effort, however, shoulder subluxation often results.

Baker (77) has reported on the use of NMES as an alternative to shoulder sling application for hemiplegic patients with flaccid shoulder musculature. In this technique, ES is applied to activate the posterior deltoid and supraspinatus (Fig. 5.8) muscles at amplitudes high enough to produce realignment of the humerus in the glenoid fossa. The stimulator used in this study produced an asymmetric, biphasic pulsed current at frequencies between 12 and 25 pps. The NMES was gradually increased from three 30-minute sessions until patients were able to tolerate up to 6 or 8 hours without significant fatigue of the shoulder musculature in a single session. On time/off time ratios were gradually changed from 1:3 to 12:1 over the course of 6 weeks. Mean shoulder subluxation measured from radiographs was reduced in the stimulated group of patients from 14.8 mm initially to 8.6 mm at the termination of the study. Control subjects did not show any change in shoulder subluxation.

Very recently, the effectiveness of NMES for shoulder subluxation was reexamined in a population of 26 stroke patients (78). Both control and experimental subjects received conventional physical therapy. Those in the experimental group received repetitive stimulation to the supraspinatus and posterior deltoid for up to 6 hours per day for 6 weeks. Subjects in the stimulated group showed significantly greater improvement in arm function, EMG activity of the posterior deltoid, range of motion, and reduction in subluxation than the control subjects. This controlled clinical study confirms the findings of Baker and colleagues (77). The apparent benefits demonstrated by the recent Faghri and coworkers (78) study should encourage the use of this simple procedure in the management of this common condition.

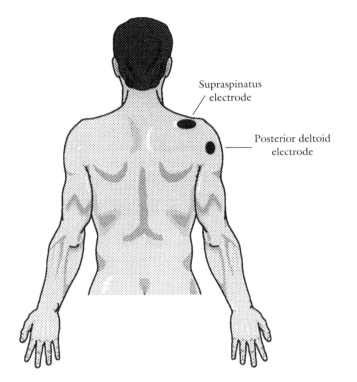

Supraspinatus
electrode

Posterior deltoid
electrode

Figure 5.8 Electrode placement over the supraspinatus and posterior deltoid muscles in FES for shoulder subluxation.

Electrical Stimulation to Minimize the Effects of Immobilzation

Over the past two decades, extensive research has been performed related to the restoration of movement in individuals who have sustained extensive damage to the central nervous system secondary to injury or disease. One of the long-term goals of this line of inquiry is the production of functional gait independent of assistive devices or orthoses. Despite the fact that independent walking has been produced in spinal cord patients in the laboratory setting, this technology is not yet in general clinical practice. However, the research effort to synthesize gait in individuals who have lost motor control has resulted in the development of a number of clinical procedures that are of immediate benefit. These procedures are directed toward preventing the detriment of prolonged disuse and immobilization. The correlaries of immobilization include pressure sores, circulatory disorders, diminished cardiovascular capacity, and muscle atrophy/fibrosis contributing to limitations in limb movement.

NMES for the prevention of pressure sores

Pressure sores are ulcerations of the skin and/or deeper tissues caused by unrelieved pressure, shear forces, and frictional forces (79). Pressure sores are also known as *decubitus ulcers,* or *bed sores,* and often arise in areas where soft tissues overlie bony prominences (e.g., ischial tuberosities, malleoli, sacrum). Prolonged pressures on soft tissues produce ischemia and hypoxia, which ultimately lead to tissue necrosis. Individuals who have sustained CNS damage are particulary prone to the development of pressure sores for a number of reasons. Reductions in the thickness of soft tissues (e.g., muscle atrophy) commonly associated with paralysis decrease the dispersion of compressive forces and increase the possibility of vascular collapse and subsequent necrosis. CNS-involved patients often have impaired sensation, such that they are unaware of the presence of sustained tissue pressure and consequently are not alerted of the need to move and relieve the pressure. Lastly, bladder and bowel incontinence in CNS-damaged individuals tends to produce maceration of the skin, which predisposes the skin to breakdown.

A variety of cushioning systems have been developed for the purpose of pressure sore prevention (79). Although these cushions prove to be adequate for most individuals, all too frequently pressure sores still occur. For that reason and because of the high cost of treatment procedures used to promote ulcer healing (80), NMES procedures are being investigated as an alternative approach to pressure sore prevention.

Levine and coworkers (81, 82) have systematically pursued this line of research. Their research has focused on the prevention of ulcers over ischial tuberosities by NMES to the gluteal musculature. A constant-current NMES device that produces an asymmetric, biphasic waveform (300-μsec pulse duration, 50 pps) has been used to evoke intermittent gluteal muscle contractions bilaterally. Early investigations have shown that NMES-induced gluteal contraction can *(a)* alter the pressure pattern at the seating interface and *(b)* vary the shape of the buttocks under load, and *(c)* measurably increase skin and gluteal muscle blood flow. Whether these effects of gluteal stimulation will significantly reduce the incidence of ischial pressure sores in various patient populations remains to be established in controlled clinical studies.

NMES for prevention of immobility-related circulatory disorders

Immobilization of the limbs for extended periods of time reduces the normal circulation of blood. In particular, loss of the routine muscular contraction in the extremities reduces venous return to the heart and fosters thrombus formation by the reduction in fibrinolytic activity. A thrombus is an aggregation of fibrin, clotting factors, platelets, and cellular elements of the blood that may further occlude venous flow. Thrombi may detach from peripheral veins, circulate through the right heart, block the circulation of blood to portions of the lung (pulmonary embolus), and create a life-threatening situation. Prevention of deep vein thrombosis (DVT) is therefore an important component in the care plan of those immobilized for long times.

Conservative preventive measures for DVT have included elevation of the limbs, intermittent pneumatic compression, or application of compression garments (e.g., elastic stockings) to prevent pooling of blood in the veins. Pharmacological preventive measures have included the use of anticoagulants such as heparin, aspirin, or warfarin to reduce the potential for clotting. Each of these approaches offers significant limitations, which have been addressed in detail in a recent review (83). The elderly and CNS-involved (e.g., spinal cord–injured) populations, who are particularly susceptible to DVT, have difficulty with donning and removing compression garments; as a result, compliance with their regular use is low. Long-term pharmacological therapy risks significant side effects, especially in the elderly, and a significant incidence of venous thromboembolism persists in spite of these approaches (83).

In the search for safe and effective conservative approaches to the prevention of DVT, NMES procedures have been studied. The primary focus of this research has revolved around the application of motor-level stimulation to the muscles of the lower legs (tibialis anterior, triceps surae) because the veins in the calf are thought to be a primary site of origin of thrombi.

NMES to the calf musculature in 10 spinal cord–injured patients was applied by Katz and colleagues (84) to examine the effects on venous blood flow and fibrinolytic activity. Tetanic contractions were evoked alternately in the dorsiflexors and plantar flexors for 60 minutes using a portable NMES device. Contractions were produced once per minute with on times of only 4 sec. The study found significantly increased fibrinolytic activity in 9 of 10 subjects within 45 minutes of cessation of stimulation. NMES was found to augment venous return moderately; however, this procedure was not as effective as manual compression of the calf. Merli and colleagues (85) examined the relative effectiveness of calf NMES, low-dose heparin, and combined heparin and NMES on the incidence of DVT in a prospective, randomized clinical study. Tibialis anterior and triceps surae muscle groups were stimulated bilaterally at a frequency of 10 pps to produce intermittent contractions (4 sec on, 8 sec off) for 23 hours daily. Stimulation was maintained for 28 days. Each day, each subject received radioactive iodine fibrinogen scanning to detect DVT. Positive findings with fibrinogen scanning were confirmed by venography. Approximately 50% (8 of 17) of SCI patients in the placebo and 50% (8 of 16) in the heparinized group developed DVT. Only 1 subject out of 17 (6%) in the NMES-plus-heparin group developed DVT.

The findings of these two studies provide strong support for the use of NMES in the SCI population to prevent immobilization-linked DVT.

The earliest reports on the effectiveness of calf NMES in the prevention of DVT appeared intermittently from the 1950s through the early 1970s on individuals and addressed treatment of patients who were immobilized due to medical conditions or after surgery (86) or under anesthesia during surgery (87, 88). Each of these early studies employed motor-level stimulation to the calf musculature at frequencies that produced repetitive twitchlike contractions. More recently, Lindstrom and coworkers (89–91) have reported that tetanic contractions of the calf musculature increases venous return from the legs and decreases venous pooling in the calves. They suggested that calf-muscle stimulation was a safe and cost-effective alternative to the use of mechanical approaches to prevent DVT (elastic stockings or pneumatic pumps) or the use of anticoagulant drugs.

NMES for improving cardiopulmonary capacity in the spinal cord injured

With the improved management of a number of medical complications that often accompany CNS injury, cardiopulmonary complications have emerged as a major cause of death in the spinal cord–injured population (92).

The paralysis of paraplegia and quadriplegia is accompanied by a reduction in cardiac mass (cardiac atrophy) and compromised pump function (decreased cardiac output) (93).

In healthy populations, deconditioning of the cardiopulmonary system is often managed by initiation of "aerobic" training regimens designed to improve heart and lung function while simultaneously enhancing the endurance characteristics of skeletal muscles. These programs are characterized by smooth tetanic muscular contractions in multiple muscle groups against low loads (<30% maximum voluntary levels) for a relatively high number of repetitions (30–100).

A number of different laboratories around the world have been interested in determining if analogous training programs performed by subjects in the SCI population can reverse the deconditioning effects associated with long-term paralysis.

One of the most commonly used machines in clinics to provide aerobic training to nonneurologically involved individuals is the bicycle ergometer. Several combined NMES/ergometer systems (Fig. 5.9) have been developed specifically for SCI patients to determine if low-level, repeated NMES-induced contractions against low resistances will produce cardiopulmonary responses and adaptations similar to those observed in healthy populations. These NMES/ergometer systems generally consist of a conventional ergometer linked with a microprocessor-based stimulation control unit and a seating arrangement adapted for the SCI subject. The systems are designed to allow microprocessor control of the stimulation amplitude and timing to maintain reciprocal LE cycling movements against a predetermined load. Electrodes are secured over the quadriceps, hamstrings, and gluteals bilaterally. Sensors on the ergometer and/or on the subject provide on-line feedback to allow the automatic adjustment of stimulation parameters. Safety features are often designed into the devices to limit the maximum current (approximately

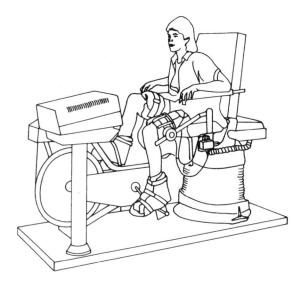

Figure 5.9 Computer-controlled NMES-ergometer for stimulation of lower extremity cycling in individuals with LE paralysis.

130 mA) and automatically shut off stimulation if driving voltage or impedance limits are reached (refer to the manufacturer's literature for details).

Investigations into this question are rather demanding on participants. Consequently, a number of researchers (93–95) have identified several criteria that must be met before an SCI person may participate in aerobic training. These criteria include the following:

1. Absence of medical complications
2. Upper MN lesion (complete or incomplete); no lower motor neuron lesion
3. No cardiac abnormalities
4. No autonomic dysreflexia
5. No muscle disease
6. No pressure sores
7. Low spasticity
8. No fractures
9. No heterotropic ossification of limbs or severe osteoporosis
10. No severe joint calcification, subluxation, or disarticulation

Potential participants who have long-standing paralysis may require an NMES strengthening program (addressed below) to have sufficient muscle endurance to participate.

The parameters of an optimal bicycle ergometer training program have yet to be established. The training programs usually begin with NMES-induced cycling sessions against no external flywheel resistance with rest periods of 5 to 10 minutes, as necessary. Sessions are continued until three consecutive 30-minute training sessions at a particular resistance are performed. The resistance is then adjusted upward in small increments. The program is carried out three times per week for a period of about three months.

Acute physiologic responses (during exercise) have included increased oxygen uptake, carbon dioxide production, pulmonary ventilation, heart rate, left ventricular stroke volume, and cardiac output (96, 97). Hooker and colleagues (96) believed that the observed increases in heart rate (33%–60%), stroke volume (45%–69%), and cardiac output (113%–142%) during prolonged FES-induced LE cycle ergometry were sufficient to promote cardiovascular conditioning in SCI participants. NMES endurance training with isolated knee extension alone in SCI-injured patients, however, does not produce cardiovascular response comparable to acute exercise bouts and hence will not likely produce long-term cardiovascular improvement (98). The recent work of Taylor and others (99) supports this contention. They examined the effect of isolated NMES training of quadriceps alone for three months on cardiac output and found no significant effects.

Long-term, beneficial cardiovascular adaptations have been shown to occur after NMES-induced bicycle ergometer training in SCI subjects (100). After three months of training both stroke volume and cardiac output were found to be increased at rest over pretraining levels. Posttraining heart rate and blood pressure appeared to be more stable after training, which may help to alleviate bouts of hypotension occasionally experienced by this group. Electrically induced lower-extremity training has not been shown to improve ventilatory capacity in the SCI population (101).

Other stimulation-linked aerobic training systems (NMES rowing machines) for SCI patients are currently being examined to determine if superior long-term cardiovascular benefits may be derived from lower-extremity, NMES-induced exercise combined with voluntary upper-extremity exercise (102, 103). Simultaneous upper and lower limb training offers the potential benefit of increasing exercise efficiency by reducing the time required by subjects to train. Increasing the training stress by combined training may in the future be found to produce significantly better cardiovascular adaptations than either upper or lower extremity training alone.

Further research on much larger numbers of subjects is required before the full scope of aerobic training effects on the cardiopulmonary systems in paralyzed people is well understood. Although many questions remain to be studied in this area, NMES ergometry is rapidly becoming a standard part of rehabilitation programs for the spinal cord injured.

NMES to prevent atrophy and improve contraction force in paralyzed muscle

Atrophy of skeletal muscle occurs whenever muscle is denervated, that is, when the motor axons to muscle degenerate. Atrophy will also occur when the CNS is no longer capable of activating motoneurons, as occurs following CNS injury. As opposed to the absolute denervation associated with peripheral nerve injury, the functional denervation of CNS disorders results in muscle atrophy and strength loss that may be managed using NMES procedures.

The general principles and procedures for prevention of atrophy or restoration of muscle strength after CNS injury are similar to those used for NMES strengthening programs applied to neurologically intact individuals (see Chapter 4). They require that NMES produces muscle contraction against relatively high resistances for a limited number of repetitions. Progressive–resistive NMES training of individual muscles performed over several weeks or months has led to significant hypertrophy in paralyzed quadriceps (93, 95, 99, 104, 105). Such NMES programs for strengthening of paralyzed muscles have also been shown to significantly increase the tension-generating capacity (strength) of muscle (105, 106). Rabischong (107) reported that NMES quadriceps training (30 minutes twice a day for two months) increased NMES-induced isometric torque production in quadriceps by 40% or more for all angles tested. These results are consistent with results from an earlier study (108) that indicated that progressive–resistive NMES training to paralyzed quadriceps for 4 weeks increased the mean load moved through 45 full knee extensions by 45% (5.56 lb before training; 8.38 lb after training). The increase in contractile capability of NMES-trained paralyzed muscle may be due to increases in contractile protein synthesis observed after stimulation training in subjects without neurologic dysfunction (109).

Increases in the strength of paralyzed muscle are also reported from studies that have employed NMES cycle ergometry training (97, 100, 108). Increases in power output following NMES ergometry programs have ranged from 45% when subjects had been pretrained with isolated NMES to the quadriceps (97) to over 300% when subjects had not undergone isolated quadriceps NMES for strengthening before the cycle ergometer training regimen (100).

Even though NMES strengthening programs may augment the torque-generating capacity of paralyzed muscle, the maximum NMES-evoked knee extensor torques are only about 20 to 50% of peak knee extension torques in normal subjects (107).

NMES to increase endurance of paralyzed muscle

NMES programs designed in part to increase strength of paralyzed muscles also simultaneously increase the ability of the muscles to sustain contraction forces during repeated contractions (95, 97, 100, 108). Increase in the endurance or fatigue resistance of NMES-trained paralyzed muscles has been shown to occur with both stimulation programs to individual muscles (110) and groups of muscles stimulated during cycle ergometry training (95, 97, 100).

Stein and coworkers (110) applied NMES to paralyzed tibialis anterior muscles for progressively longer periods (15 minutes per day to 8 hours per day) each session for six weeks. Training contractions were isotonic and elicited against no external load. Fatigue tests employing repetitive stimulation revealed increased paralyzed tibialis anterior (TA) endurance to levels similar to those measured in normal control subjects. Stimulation training sessions as short as two hours per day were sufficient to bring about optimal increases in muscle endurance. Failure to add external resistances analogous to those employed in NMES-based strength training programs may account for the finding that maximal force of stimulated contractions were not found to increase in this study.

In a companion study on the same subject population, Martin and coworkers (111) examined the morphological and histochemical characteristics of the NMES-trained TA muscles. They found an increase in the percentage of type I fatigue-resistant muscle fibers following NMES training as well as an increase in oxidative enzyme (succinate dehydrogenase) activity in both type I and type II fibers. In addition, some subjects experienced marked increases (mean 29%, range 0–71%) in capillary density within stimulated muscles. Although the percentage of type I fibers did not return to levels found in normal control muscles, the changes in oxidative and circulatory capacities account in part for the restoration of the endurance characteristics of the muscle. These increases in oxidative metabolic capacity and muscle blood flow confirm similar results from other laboratories (99, 106, 112, 113).

NMES for Standing and Gait in the CNS-injured Population

Traditional approaches to produce standing and gait in the SCI patient population have included the use of passive knee-ankle-foot orthoses (KAFOs) applied to the lower extremities and mechanical balance devices used in conjunction with the upper extremities to provide balance (see Jaeger and coworkers [114] for a detailed review). In such systems the muscles of the shoulders and upper limbs provide the power for standing, gait, and sitting activities.

Although these approaches have been successfully employed for some individuals, they are not without problems. Applying and removing the bilateral braces is time-consuming and the user rejection rate is high. Some orthotic appliances are also very heavy and at times increase the force required to move with these devices to levels that cannot be generated by users. For these reasons, laboratory investigations into the feasibility of NMES for standing and gait in paralyzed populations has been ongoing since the mid-1970s. Over the last 20 years both NMES-induced standing (115) and walking (116–121) have been achieved in SCI subjects in the laboratory setting. In spite of these developments, this technology is not yet in widespread use in the clinical setting. The following segment will briefly outline postulated benefits of NMES for standing and walking, patient selection criteria, basic principles and procedures for NMES for standing and walking in the SCI population, and a commercially produced system designed specifically for this application.

Postulated value of standing and walking in CNS-injured patients

A number of researchers in the field of functional electrical stimulation have hypothesized that NMES to produce standing and to synthesize gait would have a number of long-term benefits to SCI patients beyond the obvious increase in mobility (114, 122). The proposed benefits include the following:

1. Reduction/prevention contracture; maintenence of ROM in LE; reduction in spasticity
2. Prevention of osteoporosis
3. Improvement in bowel, renal, and bladder function
4. Stimulation of circulation
5. Reduction in seating pressure and hence pressure sore development
6. Increased ability to functionally reach
7. Psychological benefit

At this time, no large-scale clinical trials have been performed to examine whether any of these postulated benefits of NMES-induced standing and walking may be realized by regular use of FES standing and walking procedures. One smaller-scale investigation has suggested that spasticity may be reduced following FES gait training (119).

Limited evidence exists to suggest that some of the proposed benefits are not achieved through simply "passive" standing in patients who have been confined to wheelchairs for many years (>10 years). Kunkel and coworkers (123) used a mechanical frame for standing in six paralyzed patients for 45 minutes twice each day (average standing time = 144 hours over 135 days). They did not find any change in measures of spasticity, contracture, or osteoporosis following the stimulation program. Several subjects, however, reported psychological benefits from regular standing activity and continued to use the standing frame after termination of the study. The lack of beneficial side effects from passive (non–FES-induced) standing does not constitute evidence that such beneficial effects will not occur when NMES is used to produce standing in SCI individuals.

Selection criteria for subjects for NMES standing and walking

Not all individuals who have become paralyzed due to spinal cord injury or disease are appropriate candidates for NMES-induced standing or walking. Jaeger 1990 (124) has estimated that only 5 to 11% of all individuals with spinal cord injury would be potential users of electrical stimulation systems for standing and walking. One reason for the relatively small number of potential users is related to the strict subject selection criteria for potential users (122, 125). These criteria include mid- to low-thoracic level injury; no lower motoneuron (peripheral nerve) damage; no medical complications (such as infection, ulcers, open wounds, skin disease, cardiovascular or pulmonary insufficiency, marked visual or auditory impairment, scoliosis, morbid obesity, autonomic dysreflexia, pregnancy); and no severe ROM limitations (contractures) or spasticity.

Principles and procedures for NMES for standing and walking

Although standing in normal individuals is associated with volitional activation of hip, knee, and ankle antigravity musculature, a number of laboratories have demonstrated that standing can be achieved in the spinal cord injured by simultaneous bilateral quadriceps stimulation alone as long as assistive devices are available to maintain balance. Standing ability may be improved if bilateral quadriceps stimulation is coupled with bilateral gluteus maximus stimulation to provide hip extension along with knee extension. Restoration of gait with NMES in the paralyzed patient generally requires more complex stimulation approaches (126, 127). One approach to synthesize gait is to stimulate many of the lower extremity muscles alternately to produce the desired movements. A second approach involves direct stimulation of certain muscles (e.g., hip and knee extensors during stance phase) and reflex activation of other muscles (hip flexors, knee flexors, and dorsiflexors during swing phase) by activation of flexion reflexes. A third approach is to maintain fixed levels of stimulation to antigravity muscles and employ swing-to or swing-through gait with braces and forms of external support.

To produce contraction in paralyzed muscles, both surface and implanted elecrode systems have been used in NMES for standing or walking investigations. Surface electrodes are inexpensive and easy to apply and remove. Major disadvantages of surface electrodes include variation in recruitment characteristics, lack of selective muscle stimulation, greater superficial and less deep muscle activation, high cutaneous sensory stimulation, and long application and removal times (127). Implanted electrodes in either specific muscles or around peripheral nerves to muscles overcome some of the shortcomings of surface electrodes (e.g., donning/doffing times, variation in recruitment characteristics). On the other hand, implanted electrodes may break, move, or become infected, which would require either reinsertion or surgical replacement.

For safe NMES standing or walking, subjects are generally required to use some form of external support to maintain balance. These supports include conventional or adapted walkers, crutches, KAFOs, or reciprocating gait orthoses. NMES systems with assistive devices are referred to as *hybrid systems*. At present, such hybrid systems appear to be the most feasible for more widespread clinical application.

In normal human activity, muscular contractions and associated movements are accompanied by sensory feedback regarding force, speed, and direction of movements, which assists in modification of motor commands to produce smooth and coordinated activity. In the spinal cord–injured population, commands to produce muscle contraction are produced for standing and walking by electrical stimulators. In the simplest stimulators designed for standing and walking applications, a predetermined pattern of stimulation is provided to muscles regardless of the position of the subject's limbs. No feedback is provided to the stimulator regarding the evoked movements. Muscle contractions and subsequent movements in such an open loop control configuration depend on the performance characteristics of the stimulator (output amplitude, frequency, etc.), electrode characteristics (size, shape, location) and the contractile characteristics of the muscles (strength, endurance) stimulated. Such stimulation systems may provide too little contraction to achieve the desired movement or may provide too much contraction and result in excessive musculoskeletal stresses. For these and other reasons, investigators performing research in this area now use stimulation systems that allow movement information (e.g., force, position) to be fed back to the stimulator or user controls to adjust the stimulator output to meet the immediate functional requirements. Such systems are described as having closed loop control.

Commercial systems for synthesis of standing and gait

One hybrid system for providing standing and independent ambulation in the spinal cord–injured population is commercially marketed in Canada and parts of Europe and has recently received approval by the U.S. Food and Drug Administration for commercial marketing in the United States. The Parastep® System (Fig. 5.10; Sigmedics, Inc., Northfield, IL) (128) consists of a microprocessor-based NMES device, a belt-clipped battery pack, and a

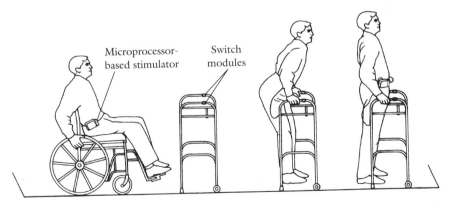

Figure 5.10 The Parastep system (Sigmedics, Inc., 1 Northfield Plaza, Northfield, IL) for standing and walking in spinal cord–injured individuals. Microprocessor-based stimulator is belt mounted. Switch modules are located on handles of walker. Battery pack (not shown) is also belt mounted.

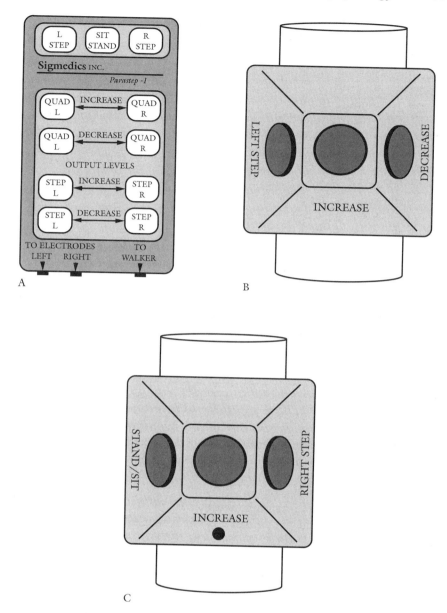

Figure 5.11 (**A**) Control panel of the Parastep system stimulator; (**B**) Parastep system left switch module mounted on the walker for manual control of the timing of stimulation for stepping on the left. (**C**) Parastep system right switch module mounted on the walker for manual control of the timing of stimulation for stepping on the right and controlling stimulation for sitting and standing.

specially adapted walker that contains control switch modules used to regulate the timing of muscular contractions. The stimulator (Fig. 5.11) has up to six output channels and produces rectangular monophasic pulsed current with maximum peak output to 300 mA and polarity reversal every other pulse. Initial stimulation parameters may be set using a keypad on the face of the stimulator/control unit. Both reciprocating and standard nonreciprocating walkers are available with switch modules (Fig. 5.11) that allow the user to initiate stimulation and control the timing for standing, sitting, or stepping and adjust the amplitude of stimulation associated with these activities. Electrodes used with the system are self-adhesive and reusable for several applications. Standing and sitting are achieved by simultaneous bilateral stimulation of the quadriceps and gluteus maximus muscles; stepping is evoked by peroneal stimulation of withdrawal reflexes. The manufacturer's clinical trials have indicated that some users of the system are capable of standing for periods of up to 45 minutes and may walk for distances of up to 500 feet. Successful use of the system requires preconditioning programs and extensive supervised training prior to actual attempts to stand and walk. At this time, this system appears to be a design suitable for a significant number of potential users in the SCI population.

Electrical Stimulation of Denervated Muscle

Denervation of skeletal muscle, the disruption of motor nerve supply of muscle, results in lost volitional control and produces profound structural and physiologic changes (Table 5.2). Researchers in this area have hypothesized that these morphological and functional changes in denervated muscle result from the marked reduction in the pattern of use (inactivity) of involved muscles, the loss of the influence of neurotrophic substances provided by the nerve fibers that normally innervate the muscle, or some combination of these two factors. Because many workers have believed that the degenerative changes in denervated muscle result from cessation of the normal patterns of muscle use, electrical stimulation of denervated muscle (**electrical muscle stimulation, EMS**) to increase muscle activity has been examined in laboratory and clinical studies to determine whether such activation can prevent or retard the degradative effects of denervation.

In 1841, Reid (129) first suggested that ES of denervated muscle could offset the structural and physiological changes that accompany loss of motor innervation. Since that time, interest in and support of this therapeutic procedure has waxed and waned. The bulk of the experimental evidence on the efficacy of EMS in combating denervation atrophy and other changes was accumulated during or shortly after World War II. For a critical review and synthesis of the experimental findings of the studies on denervated muscle through the 1950s, the interested reader is referred to Gutmann's definitive text on the subject (130). Hnik and colleagues (131, 132) indicated that through the late 1950s, the efficacy of electrotherapeutic stimulation of denervated muscle remained controversial. Even today, for every report that suggests that EMS prevents or retards the effects of denervation (133–140), another can be found that indicates that EMS is either ineffective (141, 142) or, in fact, detrimental (143–145) to the maintainence or recovery of denervated muscle.

Reviews of the literature accumulated over the last 30 years indicate that the controversy over the efficacy of EMS persists (146, 147). The main problem encountered in attempting to reconcile the contradictory findings is that researchers have employed EMS under markedly different conditions, making comparison of the results invalid. No workers in the field have performed a series of studies in which only one experimental variable has been systematically changed at a time.

Although the debate over the efficacy of EMS continues, stimulation protocols that have retarded the effects of denervation appear to have several features in common. Important conditions for obtaining positive clinical effects with EMS (132) include the following:

1. Use of **supramaximal stimulus intensities**: all of the denervated muscle fibers must be activated. EMS may be less successful in larger muscles because the stimulation intensities required to activate all muscle fibers may not be tolerated by the patient. Some authors (148) believe that stimulation of all muscle fibers can be attained only by use of implanted electrodes.

2. **Isometric contraction** in response to stimulation: isotonic contraction of denervated muscle during stimulation sessions has not been shown to be effective.

3. Initiation of **daily** sessions of stimulation as soon as possible after denervation.

4. Use of **bipolar electrode placement**: higher-magnitude evoked contractions have been obtained with this technique than with placement of one electrode of the stimulation circuit over the denervated muscle. A potential disadvantage of bipolar placement may be the spread of current to adjacent innervated muscle or sensory structures, causing pain or discomfort.

Optimal values for other important parameters of EMS protocols have not been established. Both monophasic and biphasic pulsed currents (often described in early literature as interrupted galvanic and sinusoidal currents) and sinusoidal AC at frequencies generally below 40 pps or Hz have been used to activate denervated muscle. In general, these and similar currents produce significant pain in subjects when applied at amplitudes sufficient to produce maximal contractions in EMS applications.

To overcome this limitation, some investigators have used a rather unusual form of monophasic pulsed current referred to as "exponential current." In exponential monophasic pulses, the current amplitude gradually rises in an

---

Table 5.2
Morphological and Physiological Changes in Denervated Skeletal Muscle

Muscle atrophy (red muscle atrophy > pale muscle atrophy)
   Decrease in muscle weight
   Decrease in muscle contractile protein (myofibril)
   Decrease in muscle sarcoplasm
   Decrease in number of muscle fibers

Replacement of muscle by fibrous and adipose tissue

Changes in muscle excitability
   Oscillations in resting membrane potential
   Fibrillation potentials
   Dispersion of acetylcholine receptors
   Increase in chronaxie and rheobase

exponential manner. Nervous tissue and normally innervated muscle in the region of such currents accommodate readily to this form of stimulation. Denervated muscle does not accommodate and is well activated by these pulsed waveforms. The main advantage of the exponentially increasing monophasic pulses is that such pulses reportedly do not produce pain during stimulation (149). As a result, EMS using this form of pulsatile current may produce maximal muscle contraction without undue patient discomfort. By improving subject tolerance, compliance with long-term stimulation programs may be improved. In addition, exponential pulses are thought to stimulate only the denervated muscle fibers while not activating any normally innervated fibers in a partially denervated muscle.

Research in EMS to date has not clearly established optimal values for numerous other treatment parameters, including

1. Number of contractions in a single stimulation session; **10 to 20 maximal contractions per session**
2. Number of stimulation sessions per day or week; **three or four sessions per day each day have been suggested in some references**
3. Frequency of stimulation (pps or Hz); **frequency should be sufficient to produce tetanic muscle contraction; some suggest frequencies in excess of 20–25 pps**
4. On time/off time of stimulation in denervation training; **10-sec on times and off times set to minimize fatigue (e.g., > 30 sec) have been suggested**

In spite of numerous studies reported since the turn of the century, electrical stimulation of denervated muscle is an area of electrotherapy in humans where fundamental questions remain unresolved. Can EMS *prevent* the structural and physiological changes of denervation? No. Can EMS *delay* denervation atrophy and associated changes? Yes, under certain conditions. Is the delay of denervation changes significant in terms of clinical recovery? Probably not. Is EMS worth the time, effort, and expense required to achieve positive results? Yes, but only in a very limited number of cases with denervation isolated to one or two muscles. Does any alternative form of management exist that has the potential of EMS for the maintainence of denervated muscle? Not currently.

Precautions and Contraindications for NMES

The precautions or contraindications for NMES (150–152) are similar for most applications described in this chapter. They include the following:

1. NMES over the thoracic region because current may interfere with the function of vital internal organs, including the heart.
2. NMES in the thoracic region of patients with demand cardiac pacemakers because the current may interfere with pacemaker activity and may lead to asystole or ventricular fibrillation.
3. NMES in regions of phrenic nerve or urinary bladder stimulators because current may interfere with the normal operation of these devices.
4. NMES over the carotid sinus because current may interfere with the normal regulation of blood pressure and cardiac contractility and may produce cardiac arrythmia or bradycardia.

5. NMES in hypertensive or hypotensive patients because autonomic responses may adversely affect control of blood pressure.

6. NMES in areas of peripheral vascular disorders such as venous thrombosis or thrombophlebitis because of the risk of releasing emboli.

7. NMES in regions of neoplasm or infection because muscular and circulatory effects may aggravate these conditions.

8. NMES on the trunk of pregnant females because of the risk of inducing uterine contractions that may influence the developing fetus.

9. NMES in close proximity to diathermy devices because of the potential for loss of control of stimulation parameters.

10. NMES in areas of excessive adipose tissue (as in obese patients) because levels of stimulation required to activate muscle in such patients may produce adverse autonomic reactions.

11. NMES in patients who are unable to provide clear feedback regarding the level of stimulation such as infants, senile subjects, or individuals with mental disorders.

Careful observation by the therapist is required in all applications of electrical stimulation. Detrimental responses to stimulation often occur rapidly and require quick reaction to avoid serious injury. Patients should not be allowed to use NMES independently until the therapist is confident that they have been trained adequately in the safe use of the device and are aware of the signs and symptoms of adverse reactions that may occur.

Case Studies

The following case studies serve as examples of the ways in which NMES can be used in clinical practice. Electrotherapeutic treatment plans are described in detail, but additional treatment is only outlined. These cases are not meant to provide an exhaustive description of the uses of NMES in patient care, but merely to illustrate clinical decision making in electrotherapy.

Case 1

A 67-year-old woman who is six weeks status post right cerebral vascular accident (CVA) presents with resolving left hemiplegia.

Objective: This patient has well-developed flexor synergy in her left upper extremity with no volitional movement out of synergy. Right upper extremity function, including fine motor skill, is excellent. Left lower extremity function is largely not synergistic, with demonstrable weakness of the ankle and toe extensors and resultant gait disturbances. Manual muscle testing reveals thigh, hip, and posterior leg musculature that is generally good to normal. Dorsiflexors and everters are poor to fair. Deep tendon reflexes are 4+ on the left and there is a positive left Babinski. Sporadic sensory loss is noted throughout the left upper extremity, but otherwise sensation is intact. Balance is good in standing. She ambulates independently with a steppage gait and uses a quad cane. She is able to follow multistage commands without difficulty.

Assessment: Resolving left hemiplegia with significant residual upper extremity involvement and lower extremity deficit that is largely limited to weakness in the anterolateral lower leg.

Plan:

1. Therapeutic exercise
2. ADL activities
3. Gait training
4. NMES as a dorsiflexion assist during gait

Detailed Electrotherapeutic Plan:

Mode of stimulation: Tetanic muscular contraction

Type of stimulator: Portable NMES with heel switch trigger

Electrode Placement: Two small electrodes placed to evoke at least a fair-grade muscle contraction of the dorsiflexors and peroneals (Fig. 5.6)

Duration of stimulation: Throughout the day, as needed

Rationale: The electrotherapeutic device is being used as an orthotic substitute and must be portable. The external trigger is needed to synchronize stimulation with the patient's gait cycle. Electrode placement must provide dorsiflexion of the ankle without excessive inversion or eversion. After training, the patient will wear the stimulator throughout the day to assist in ambulation.

Case 2

A 13-year-old girl presents with a radiographically diagnosed 40° right thoracic structural scoliosis.

Objective: Postural evaluation reveals a right thoracic scoliosis with a compensatory left lumbar curve. The thoracic curve does not reverse with lateral bending to the left. A rib hump is noted on the right on Adam's bend. Shoulder and pelvic levels are unequal and expected muscle imbalances are found.

Assessment: Findings consistent with idiopathic scoliosis requiring bracing or similar intervention.

Plan:

1. Postural exercise
2. Patient education
3. FES for scoliosis management

Detailed Electrotherapeutic Plan:

Mode of stimulation: Cycled NMES

Type of stimulator: Portable

Electrode placement: Laterally on the midaxillary line, just above and below the apical rib (Fig. 5.7)

Treatment duration: 8 hours per day, at night

Rationale: Research substantiation for this use of NMES describes duty cycles of 6 seconds on and 6 seconds off. The stimulator is worn at night while sleeping, necessitating a portable stimulator. Lateral electrode placement is demonstrably more effective than paraspinal placement.

The following case is an example of NMES used in place of a splint to help position a patient for the exercise of other musculature.

Case 3

A 46-year-old head-injured patient has fair-grade shoulder musculature, trace triceps, and spastic biceps on the right. To position him for gravity-minimized active horizontal abduction and adduction on a powder board, the elbow must be stabilized in extension to prevent the biceps from pulling the elbow into flexion during shoulder exercise. This could be accomplished using standard splinting devices or NMES.

Detailed Electrotherapeutic Plan:

Mode of stimulation: Externally triggered NMES

Type of stimulator: Clinical or portable

Electrode placement: Over the motor point of the triceps brachii muscle

Duration of treatment: Dependent on duration of exercise session

Rationale: External triggering by the therapist or patient to synchronize the NMES with the exercise. Either a clinical or portable stimulator can be used. Decision depends on the ability of the stimulator to be externally triggered and the size of the exercise area.

Summary

This chapter has reviewed the essential features of neuromuscular electrical stimulators, the spectrum of clinical applications of NMES, the stimulation parameters and techniques employed in NMES applications, precautions and contraindications for NMES applications, and the therapeutic application of ES for denervated muscle.

Advances in NMES applications and techniques are occurring at a faster pace than ever before. Clinicians using NMES and other electrical techniques should strive to stay abreast of new developments in the field. If possible, users of NMES procedures are encouraged to add to the knowledge base in NMES through the careful design and implementation of prospective, randomized, clinical research studies. Through the coordinated efforts of laboratory and clinical researchers, the breadth and depth of understanding and applications will increase. The patients we serve will benefit most from these efforts.

Study Questions

1. What are the major adaptations of skeletal muscle to long-term, low-amplitude, numerous repeated contractions?

2. What is the major adaptation of skeletal muscle to short-term, high-amplitude contractions repeated regularly for only a few repetitions?

3. What is the "best" current to employ in NMES applications?

4. List the types of stimulator controls required to perform most contemporary NMES applications. Identify the range of values over which adjustment may occur in a typical stimulator designed for NMES applications.

5. List several electrode placement sites that have been shown in preliminary studies to be successful for stimulation to control spasticity.

6. Identify several testing procedures used to quantify muscle spasticity.

7. Define the term "functional electrical stimulation."

8. Stimulation of dorsiflexion and eversion of the ankle during the swing phase of gait in hemiplegics acts as a substitute for the use of ___(a)___. Surface electrode placement is generally bipolar with electrodes located ___(b)___.

9. List seven selection criteria for patients considered for FES as a dorsiflexion assist during gait.

10. Outline the treatment parameters for an FES for a scoliosis management home program.

11. FES for shoulder subluxation consists of intermittent tetanic contraction of the ___(a)___ and the ___(b)___ muscles.

12. List four complications of long-term immobilization in spinal cord–injured individuals that may be managed with an electrotherapeutic program.

13. List seven possible benefits of NMES-induced standing and/or walking programs in the spinal cord–injured individual.

14. Describe the three approaches that have been used to produce standing and to synthesize gait using NMES in the spinal cord–injured population.

15. Electrical stimulation for denervated muscle should be performed under four explicit conditions in order to retard muscle atrophy. What are these four conditions?

References

1. Rose SJ, Rothstein JM. Muscle mutability. Part 1. General concepts and adaptations to altered patterns of use. Phys Ther 1982;62:1751-1830.

2. Salmons S, Henriksson J. The adaptive response of skeletal muscle to increased use. Muscle and Nerve 1981;4:94-105.

3. Delitto A, Rose SJ. Comparative comfort of three waveforms used in electrically eliciting quadriceps femoris contractions. Phys Ther 1986;66:1704-1707.

4. Bowman BR, Baker LL. Effects of waveform parameters on comfort during transcutaneous neuromuscular electrical stimulation. Ann Biomed Eng 1985;13:59-74.

5. Delitto A, Strube MJ, Shulman AD, Minor SD. A study of discomfort with electrical stimulation. Phys Ther 1992;72:410-421.

6. Association for the Advancement of Medical Instrumentation. American national standard for neuromuscular stimulators. Arlington, VA (in preparation, 1994).

7. Alon G, Allin J, Inbar GF. Optimization of pulse duration and pulse charge during transcutaneous electrical nerve stimulation. Aust J Physiother 1983;29:196-210.

8. Ferguson JP, Blackley MW, Knight RD, et al. Effects of varying electrode site placement on the torque output of an electrically stimulated involuntary quadriceps femoris muscle contraction. J Orthop Sports Phys Ther 1989;11:24-29.

9. Brooks ME, Smith EM, Currier DP. Effect of longitudinal versus transverse electrode placement on torque production by the quadriceps femoris muscle during neuromuscular electrical stimulation. J Orthop Sports Phys Ther 1990;11:530-534.

10. Nelson HE, Smith MB, Bowman BR, Waters RL. Electrode effectiveness during transcutaneous motor stimulation. Arch Phys Med Rehabil 1980;61:73-77.

11. Alon G. High voltage stimulation—effects of electrode size on basic excitatory responses. Phys Ther 1985;65:890-895.

12. Owens J, Malone T. Treatment parameters of high frequency electrical stimulation as established on the Electro-Stim 180. J Orthop Sports Phys Ther, 1983;4: 162-164.

13. Munsat TL, McNeal DR, Waters RL. Preliminary observations on prolonged stimulation of peripheral nerve in man. Arch Neurol 1976;33:608-617.

14. Baker LL, Yeh C, Wilson D, Waters RL. Electrical stimulation of wrist and fingers for hemiplegic patients. Phys Ther 1979;59:1495-1499.

15. Bowman BR, Baker LL, Waters RL. Positional feedback and electrical stimulation: an automated treatment for the hemiplegic wrist. Arch Phys Med Rehabil 1979;60:497-501.

16. Winchester P, Montgomery J, Bowman B, Hislop H. Effects of feedback stimulation training and cyclical electrical stimulation on knee extension in hemiparetic patients. Phys Ther 1983;63:1096-1103.

17. Landau WM. Spasticity: what is it? What is it not? In: Feldman RG, Young RR, Koella WP, eds. Spasticity: disordered motor control. Chicago: Year Book, 1980:17.

18. Levine MG, Knott M, Kabot H. Relaxation of spasticity by electrical stimulation of antagonist muscles. Arch Phys Med 1952;33:668-673.

19. Alfieri V. Electrical treatment of spasticity—reflex tonic activity in hemiplegic patients and selected specific electrostimulation. Scand J Rehab Med 1982;14: 177-182.

20. Carnstam B, Larsson LE, Prevec TS. Improvement of gait following functional electrical stimulation. 1. Investigations on changes in voluntary strength and proprioceptive reflexes. Scand J Rehab Med 1977;9:7-13.

21. Peterson T, Klemar KB. Electrical stimulation as a treatment of lower limb spasticity. J Neuro Rehab 1988;2:103-108.

22. Apkarian JA, Naumann S. Stretch reflex inhibition using electrical stimulation in normal subjects and subjects with spasticity. J Biomed Eng 1991;13:67-73.

23. Lee WJ, McGovern JP, Duvall EN. Continuous tetanizing (low voltage) currents for relief of spasm. Arch Phys Med 1950;31:766-771.

24. Vogel M, Weinstein L, Abramson AS. Use of tetanizing current for spasticity. Phys Ther Rev 1955;35:435-437.

25. Bowman BR, Bajd T. Influence of electrical stimulation on skeletal muscle spasticity. In: Proceedings of the International Symposium on External Control of Human Extremities, Belgrade, Yugoslav Committee for Electronics and Automation, 1981:561-576.

26. Robinson CJ, Kett NA, Bolam JM. Spasticity in spinal cord injured patients. 1. Short-term effects of surface electrical stimulation. Arch Phys Med Rehabil 1988;69:598-604.

27. Bajd T, Vodovnik L. Pendulum testing of spasticity. J Biomed Eng 1984; 6:9-16.

28. Baker LL. Clinical uses of neuromuscular electrical stimulation. In: Nelson RM, Currier DP, eds. Clinical electrotherapy. Norwalk, CT: Appleton and Lange, 1987.

29. Vodovnik L, Bowman BR, Hufford P. Effects of electrical stimulation on spinal spasticity. Scand J Rehab Med 1984;16:29-34.

30. Wartenburg R. Pendulousness of the legs as a diagnostic test. Neurology 1951;1:18-23.

31. Bajd T, Gregoric M, Vodovnik L, Benko H. Electrical stimulation in treating spasticity resulting from spinal cord injury. Arch Phys Med Rehabil 1985;66:515-517.

32. Andrews BJ, Bajd T, Baxendale RH. Cutaneous electrical stimulation and reductions in spinal spasticity in man. J Physiol 1984;367:86pp.

33. Hui-Chan CWY, Levin MF. Stretch reflex latencies in spastic hemiparetic subjects are prolonged after transcutaneous electrical nerve stimulation. Can J Neurol Sci 1993;20:97-106.

34. Levin MF, Hui-Chan CWY. Relief of hemiparetic spasticity by TENS is associated with improvement in reflex and voluntary motor functions. Electroenceph Clin Neurophys 1992;85:131-142.

35. Walker JB. Modulation of spasticity: prolonged suppression of spinal reflex by electrical stimulation. Science 1982;216:203-204.

36. Fredriksen TA, Bergmann S, Hesselberg JP, Stolt-Nielsen A, Ringkjob R, Sjaastad O. Electrical stimulation in multiple sclerosis—comparison of transcutaneous electrical stimulation and epidural spinal cord stimulation. Appl Neurophysiolo 1986;49:4-24.

37. Granat MH, Ferguson ACB, Andrews BJ, Delargy M. The role of functional electrical stimulation in the rehabilitation of patients with incomplete spinal cord injury—observed benefits during gait studies. Paraplegia 1993;331:207-215.

38. Stefanovska A, Gros N, Vodovnik L, Rebersek S, Acimovic-Janezic R. Chronic electrical stimulation for the modification of spasticity in hemiplegic patients. Scand J Rehabil Med Suppl 1988;17:115-121.

39. Robinson CJ, Kett NA, Bolam JM. Spasticity in spinal cord injured patients. 2. Initial measures and long-term effects of surface electrical stimulation. Arch Phys Med Rehabil 1988;69:862-868.

40. Richardson RR, Siqueira EB, Cerullo LJ. Spinal epidural stimulation for treatment of acute and chronic intractable pain: initial and long term results. Neurosurgery 1979;5:344-348.

41. Kumar K, Wyant GM, Ekong CEU. Epidural spinal cord stimulation for relief of chronic pain. Pain Clin 1986;1:91-99.

42. Koeze TH, Williams A, Reiman S. Spinal cord stimulation and the relief of chronic pain. J Neurol Neurosurg Psych 1987;50:1424-1429.

43. Richardson RR, McLone DG. Percutaneous epidural neurostimulation for spasticity. Surg Neurol 1978;9:153-155.

44. Reynolds AF, Oakley JC. High frequency cervical epidural stimulation for spasticity. Appl Neurophysiol 1982;45:93-97.

45. Siegfried J, Krainick JU, Haas H., et al. Electrical spinal cord stimulation for spastic movement disorders. Appl Neurophysiol 1978;41:134-141.

46. Siegfried J. Treatment of spasticity by dorsal column stimulation. Int Rehab Med 1980;2:31-44.

47. Campos RJ, Dimitrijevic MM, Faganel J, Sharkey PC. Clinical evaluation of the effect of chronic spinal cord stimulation on motor performance in patients with upper motor neuron lesions. Appl Neurophysiol 1981;44:141-151.

48. Dimitrijevic MR, Dimitrijevic MM, Sherwood AM, Faganel J. Neurophysiological evaluation of chronic spinal cord stimulation in patients with upper motor neuron disorders. Int Rehab Med 1980;2:82-85.

49. Barolat-Romana G, Myklebust JB, Hemmy DC, Myklebust B, Wenninger W. Immediate effects of spinal cord stimulation in spinal spasticity. J Neurosurg 1985;62:558-562.

50. Halstead LS, Seager SWJ. The effects of rectal probe electrostimulation on spinal cord injury spasticity. Paraplegia 1991;29:43-47.

51. Liberson WT, Holmquist HJ, Scot D, Dow M. Functional electrotherapy: stimulation of the peroneal nerve synchronized with the swing phase of gait of hemiplegic patients. Arch Phys Med Rehabil 1961;42:101-105.

52. Moe JM, Post MW. Functional electrical stimulation for ambulation in hemiplegia. Lancet 1962;82:285-288.

53. Gracinin F. Functional electrical stimulation in control of motor output and movements. Contemp Clin Neurophysiol 1978;34(EEG suppl):355-368.

54. Singer B. Functional electrical stimulation of the extremities in the neurological patient: a review. Aust J Physiotherapy 1987;33:33-42.

55. Merletti R, Zelaschi F, Latella D, Galli M, Angeli S, Bellucci Sessa L. A control study of muscle force recovery in hemiparetic patients during treatment with functional electrical stimulation. Scand J Rehab Med 1978;10:147-154.

56. Stanic U, Acimovic-Janezic R, Gros N, Trnkoczy A, Bajd T, Kljajic M. Multichannel electrical stimulation of correction of hemiplegic gait. Scand J Rehabil Med 1978;10:75-92.

57. Strojnik P, Acimovic R, Vavken E, Simic V, Stanic U. Treatment of drop foot using an implantable underknee stimulator. Scand J Rehabil Med 1987;19:37-43.

58. Merletti R, Andina A, Galante M, Furlan I. Clinical experience of electronic peroneal stimulators in 50 hemiparetic patients. Scand J Rehab Med 1979;11:111-121.

59. Kljajic M, Malezic M, Acimovic E, Vavken E, Stanic U, Pangrsic B, Rozman J. Gait evaluation in hemiparetic patients using subcutaneous peroneal electrical stimulation. Scand J Rehab Med 1992:24:121-126.

60. Gracanin F, Vrabic M, Vrabic G. Six years experience with FES methods applied to children. Europa Medicophysica 1976;12:61-68.

61. Vodovnik L, Rebersek S. Improvements in voluntary control of paretic muscles due to electrical stimulation. In: Fields WS, Leavitt LA, eds. Neural organization and its relevance to prosthetics. New York: International Medical Books, 1973:101-116.

62. Vodovnik L, Stanic U, Kralj A, Gracanin F, Strojnik P. Functional electrical stimulation for control of locomotor systems. CRC Critical Reviews in Bioengineering 1981;6:63-131.

63. Fleury M, Lagasse P. Influence of functional electrical stimulation training on premotor and motor reaction time. Percept Motor Skills 1979;48:387-393.

64. Teng EL, McNeal DR, Kralj A, Waters RL. Electrical stimulation and feedback training: effects on the voluntary control of paretic muscle. Arch Phys Med Rehabil 1976;57:228-233.

65. Bobechko WP, Herbert MA, Friedman HG. Electrospinal instrumentation for scoliosis. Orthop Clin North Am 1979;10:927-941.

66. Axelgaard J, Brown JC, Harada Y, McNeal DR, Nordwall A. Lateral surface stimulation for the correction of scoliosis. Proceedings of the 30th Annual Conference on Engineering in Medicine and Biology, Los Angeles, The Alliance of Engineering in Medicine and Biology, 1976.

67. Eckerson LF, Axelgaard J. Lateral electrical surface stimulation as an alternative to bracing in the treatment of idiopathic scoliosis. Phys Ther 1984;64:483-490.

68. Axelgaard J, Brown JC. Lateral electrical surface stimulation for the treatment of progressive idopathic scoliosis. Spine 1983;8:242-260.

69. Brown JC, Axelgaard J, Howson DC. Multicenter trial of a noninvasive stimulation method for idiopathic scoliosis. Spine 1984;9:382-387.

70. McCullough NC. Nonoperative treatment of idiopathic scoliosis using surface electrical stimulation. Spine 1986;11:802-804.

71. Bradford DS, Tanguy A, Vanselow J. Surface elctrical stimulation in the treatment of idiopathic scoliosis: preliminary results in 30 patients. Spine 1983;8:757-764.

72. Swank SM, Brown JC, Jennings MV, Conradi C. Lateral electrical surface stimulation in idiopathic scoliosis—experience in two private practices. Spine 1989;14:1293-1295.

73. Fisher DA, Rapp GF, Emkes M. Idiopathic scoliosis: transcutaneous muscle stimulation versus the Milwaukee brace. Spine 1987;12:987-991.

74. Goldberg C, Dowling FE, Fogarty EE, Regan BF, Blake NS. Electro-spinal stimulation in children with adolescent and juvenile scoliosis. Spine 1988;13:482-484.

75. O'Donnell CS, Bunnell WP, Betz RR, Bowen R, Tipping CR. Electrical stimulation in the treatment of idiopathic scoliosis. Clin Orthop Relat Res 1988;229: 107-113.

76. Durham JW, Moskowitz A, Whitney J. Surface electrical stimulation versus brace in treatment of idiopathic scoliosis. Spine 1990;15:888-891.

77. Baker LL. Neuromuscular electrical stimulation of the muscles surrounding the shoulder. Phys Ther 1986;66:1930-1934.

78. Faghri PD, Rodgers MM, Glaser RM, Bors JG, Akuthota P. The effects of functional electrical stimulation on shoulder subluxation, arm function recovery, and shoulder pain in hemiplegic stroke patients. Arch Phys Med Rehabil 1994;75: 73-79.

79. Crenshaw RP, Vistnes LM. A decade of pressure sore research: 1977-1987. J Rehabil Res Dev 1989;26:63-74.

80. Krouskop TA, Noble PC, Garber SL, Spencer WA. The effectiveness of preventive management in reducing the occurrence of pressure sores. J Rehabil Res Dev 1983;20:74-83.

81. Levine SP, Kett RL, Cedera PS, Bowers LD, Brooks SV. Electrical muscle stimulation for pressure variation at the seating interface. J Rehabil Res Dev 1989;26:1-8.

82. Levine SP, Kett RL, Cedera PS, Brooks SV. Electrical stimulation for pressure sore prevention: tissue shape variation. Arch Phys Med Rehabil 1990;71:210-215.

83. Clagett GP, Anderson FA, Levine MN, Salzman EW, Wheeler HB. Prevention of venous thromboembolism. Chest 1992;120(suppl):391S–407S.

84. Katz RT, Green D, Sullivan T, Yarkony G. Functional electrical stimulation to enhance fibrinolytic activity in spinal cord injured patients. Arch Phys Med Rehabil 1987;68:423-426.

85. Merli GJ, Herbison GJ, Ditunno JF, et al. Deep vein thrombosis: prophylaxis in acute spinal cord injured patients. Arch Phys Med Rehabil 1988;69:661-664.

86. Martella J, Cincotti JJ, Springer WP. Prevention of thromboembolic disease by electrical stimulation of the leg muscles. Arch Phys Med Rehabil 1954;35: 24-29.

87. Doran FSA, White HM. A demonstration that the risk of postoperative deep venous thrombosis is reduced by stimulating the calf muscles electrically during the operation. Br J Surg 1967;54:686-689.

88. Nicilaides AN, Kakkar VV, Field ES, Fish P. Optimal electrical stimulus for prevention of deep vein thrombosis. Br Med J 1972;3:756-758.

89. Lindstrom B. Prediction and prophylaxis of postoperative thromboembolism: a comparison between perioperative calf muscle stimulation with groups of impulses and dextran 40. Br J Surg 1982;69:633-637.

90. Lindstrom B, Korsan-Bengston K, Jonsson O, Petruson B, Pettersson S, Wikstrand J. Electrically induced short-lasting tetanus of calf muscle for prevention of deep vein thrombosis. Br J Surg 1982;69:203-206.

91. Jonsson O, Lindstrom B. Perioperative calf muscle stimulation for the prevention of postoperative thromboembolic complication. Geriatric Med Today 1983;29: 83-89.

92. DiVivo MJ, Fine PR, Stover SL. Cause of death following spinal cord injury [Abstract]. Arch Phys Med Rehabil 1984;65:622.

93. Nash MS. Computerized functional electrical stimulation: an emerging rehabilitation technology. Trends Rehabil 1986;Spring:5-13.

94. Cybulski GR, Penn RD, Jaeger RJ. Lower extremity functional neuromuscular stimulation in cases of spinal cord injury. Neurosurg 1984;15:132-146.

95. Bremner LA, Sloan KE, Day RE, Scull ER, Ackland T. A clinical exercise system for paraplegics using functional electrical stimulation. Paraplegia 1992;30:647-655.

96. Hooker SP, Figoni SF, Glaser RM, Ezenwa BN, Faghri PD. Physiologic responses to prolonged electrically stimulated leg-cycle exercise in the spinal cord injured. Arch Phys Med Rehabil 1990;71:863-869.

97. Hooker SP, Figoni SF, Rodgers MM, et al. Physiologic effects of electrical stimulation leg cycle exercise training in spinal cord injured persons. Arch Phys Med Rehabil 1992;73:470-476.

98. Figoni SF, Glaser RM, Rodgers MM, et al. Acute hemodynamic responses of spinal cord injured individuals to functional neuromuscular stimulation induced knee extension exercise. J Rehabil Res Dev 1991;28:9-18.

99. Taylor PN, Ewins DJ, Fox B, Grundy D, Swain ID. Limb blood flow, cardiac output and quadriceps muscle bulk following spinal cord injury and the effects of training for the Odstock functional electrical stimulation standing system. Paraplegia 1993; 31:303-310.

100. Faghri PD, Glaser RM, Figoni SF. Functional electrical stimulation leg cycle ergometer exercise: training effects on cardiorespiratory responses of spinal cord injured subjects at rest and during submaximal exercise. Arch Phys Med Rehabil 1992; 73:1085-1093.

101. Pollock SF, Axen K, Spielholtz N, Levin N, Haas F, Ragnarsson KT. Aerobic training effects of electrically induced lower extremity exercises in spinial cord injured people. Arch Phys Med Rehabil 1989;70:214-219.

102. Laskin JJ, Ashley EA, Olenik LM, et al. Electrical stimulation–assisted rowing exercise in spinal cord injured people. A pilot study. Paraplegia 1993;31:534-541.

103. Petrofsky JS, Smith J. Aerobic exercise trainer for both paralyzed and nonparalyzed muscles. J Clin Eng 1991;16:505-513.

104. Bajd TA, Kralj A, Turk R, Benko H, Sega J. Use of functional electrical stimulation in rehabilitation of patients with incomplete spinal cord injuries. J Biomed Eng 1989;11:96-102.

105. Peckham PH, Mortimer JT, Marsolais EB. Alterations in force and fatiguability of skeletal muscle in quadriplegic humans following exercises induced by chronic electrical stimulation. Clin Orthop Relat Res 1976;114:326-334.

106. Petrofsky JA, Phillips CA. Active physical therapy: a modern approach to rehabilitation. J Neurol Orthop Surg 1983;4:165-173.

107. Rabischong E, Ohanna F. Effects of functional electrical stimulation (FES) on evoked muscular output in paraplegic quadriceps muscle. Paraplegia 1992;30:467-473.

108. Ragnarsson KT, Pollack S, O'Daniel W, Edgar R, Petrofsky J, Nash MS. Clinical evaluation of computerized functional electrical stimulation after spinal cord injury: a multicenter pilot study. Arch Phys Med Rehabil 1988;69:672-677.

109. Gibson JNA, Jeffery R, Smith K, et al. The effect of therapeutic percutaneous electrical stimulation of atrophic human quadriceps on muscle composition, protein synthesis and contractile properties. Europ J Clin Invest 1989;19(2):206-12.

110. Stein RB, Gordon T, Jefferson J, et al. Optimal stimulation of paralyzed muscle after human spinal cord injury. J Appl Physiol 1992;72:1393-1400.

111. Martin TP, Stein RB, Hoeppner PH, Reid DC. Influence of electrical stimulation on the morphological and metabolic properties of paralyzed muscle. J Appl Physiol 1992;72:1401-1406.

112. Munsat TL, McNeal D, Waters R. Effects of nerve stimulation on human muscle. Arch Neurol 1976;33:608-617.

113. Riley DA, Allin EF. The effects of inactivity, programmed stimulation and denervation on the histochemistry of skeletal muscle fiber types. Exp Neurol 1973;40:391-413.

114. Jaeger RJ, Yarkony GM, Roth EJ. Rehabilitation technology for standing and walking after spinal cord injury. Am J Phys Med Rehabil 1989;68:128-133.

115. Yarkony GM, Jaeger RJ, Roth E, Krajl AR, Quintern J. Functional neuromuscular stimulation for standing after spinal cord injury. Arch Phys Med Rehabil 1990;71:201-206.

116. Phillips CA. Functional electrical stimulation and lower extremity bracing for ambulation exercise of the spinal cord injured individual: a medically prescribed system. Phys Ther 1989;69:842-849.

117. Isakov E, Douglas R, Berns P. Ambulation using the reciprocating gait orthosis and functional electrical stimulation. Paraplegia 1992;30:239-245.

118. Granat M, Keating JF, Smith ACB, Delargy M, Andrew BJ. The use of functional electrical stimulation to assist gait in patients with incomplete spinal cord injury. Disability and Rehabil 1992;14:93-97.

119. Granat MH, Ferguson ACB, Andrews BJ, Delargy M. The role of functional electrical stimulation in the rehabilitation of patients with incomplete spinal cord injury—observed benefits during gait studies. Paraplegia 1993;31:207-215.

120. Stein RB, Belanger M, Wheeler G, et al. Electrical systems for improving locomotion after incomplete spinal cord injury: an assessment. Arch Phys Med Rehabil 1993;74:954-959.

121. Kralj AR, Bajd T, Munith M, Turk R. FES gait restoration and balance control in spinal cord injured patients. Brain Res 1993;97:387-396.

122. Cybulski GR, Penn RD, Jaeger RJ. Lower extremity functional neuromuscular stimulation in cases of spinal cord injury. Neurosurg 1984;15:132-146.

123. Kunkel CF, Scremin AME, Eisenberg, B, Garcia JF, Roberts S, Martinez S. Effect of standing on spasticity, contracture, and osteoporosis in paralyzed males. Arch Phys Med Rehabil 1993;74:73-78.

124. Jaeger RJ, Yarkony GM, Roth EJ, Lovell L. Estimating the user population for a simple electrical stimulation system for standing. Paraplegia 1990;28:505-511.

125. Taylor PN, Ewins DJ, Swain ID. The Odstock closed loop FES standing system. Experience in clinical use. Proc Ljubljana FES conference, August 22–26, 1993.

126. Peckham PH, Creasey GH. Neural prostheses: clinical application of functional electrical stimulation in spinal cord injury. Paraplegia 1992;30:96-101.

127. Yarkony GM, Roth EJ, Cybulski G, Jaeger RJ. Neuromuscular stimulation in spinal cord injury. I. Restoration of functional movement of the extremities. Arch Phys Med Rehabil 1992;73:78-86.

128. The Parastep System User's Manual, Sigmetics, Inc., Northfield, Illinois, 1993.

129. Reid J. On the relation between muscular contractility and the nervous system. Lond Edinb Month J Med Sci 1841;1:320-329.

130. Gutmann E, ed. The denervated muscle. Prague: Publishing House of the Czechoslovak Academy of Sciences, 1962.

131. Hnik P. Rate of denervation muscle atrophy. In Gutmann E (ed): The Denervated Muscle. Prague, Publishing House of the Czechoslovak Academy of Sciences, 1962, pp 341-375.

132. Hnik P, Skorpil V, Vyklicky L: Diagnosis and therapy of denervation muscle atrophy. In: Gutmann E, ed. The denervated muscle. Prague: Publishing House of the Czechoslovak Academy of Sciences, 1962: pp 433-466.

133. Nemeth P. Electrical stimulation of denervated muscle prevents decreases in oxidative enzymes. Muscle and Nerve 1982;5:134-139.

134. Carraro U, Catani C, Sggin L, et al. Isomyosin changes after functional electrostimulation of denervated sheep muscle. Muscle and Nerve 1988;11:1016-1028.

135. Al-Amood WS, Lewis DM. The role of frequency in the effects of long-term intermittent stimulation of denervated slow-twitch muscle in the rat. J Physiol 1987;392:337-395.

136. Valencic V, Vodovnik L, Stefancic M, Jelnikar T. Improved motor response due to chronic electrical stimulation of denervated tibialis anterior muscle in humans. Muscle and Nerve 1986;9:612-617.

137. Nemoto K, Williams HB, Nemoto K, Lough J, Chiu RCJ. The effects of electrical stimulation on denervated muscle using implantable electrodes. J Resconstruct Microsurg 1988;4:251-255.

138. Yanai A, Harii K, Okabe K. Preventing denervation atrophy of a grafted muscle. J Reconstruct Microsurg 1991;7:85-92.

139. Pachnter BR, Ebersteint A, Goodgold J. Electrical stimulation effect on denervated skeletal myofibers in rats: a light and electron microscopic study. Arch Phys Med Rehabil 1982;63:427-430.

140. Cole BG, Gardiner PF. Does electrical stimulation of denervated muscle continued after reinnervation influence recovery of contractile function? Exp Neurol 1984;85:52-62.

141. Herbison GJ, Teng C, Gordon EE. Electrical stimulation of reinnervating rat muscle. Arch Phys Med Rehabil 1973;54:156-160.

142. Nix WA. Effects of intermittent high frequency electrical stimulation on denervated EDL muscle of rabbit. Muscle and Nerve 1990;13:580-585.

143. Girlanda P, Dattola R, Oteri VG, LoPresti F, Messina C. Effect of electrotherapy on denervated muscles in rabbits: an electrophysiological and morphological study. Exp Neurol 1982;77:483-491.

144. Schmirigk K, McLaughlin J, Gruninger W. The effect of electrical stimulation on the experimentally denervated rat muscle. Scand J Rehabil Med 1977;9:55-60.

145. Nix WA, Dahm M. The effect of isometric short-term electrical stimulation on denervated muscle. Muscle and Nerve 1987;10:136-143.

146. Davis HL. Is electrostimulation beneficial to denervated muscle? A review of results from basic research. Physiother Can 1983;35:306-310.

147. Hayes KW. Electrical stimulation and denervation: proposed program and equipment limitations. Top Acute Care Trauma Rehabil 1988;3:27-37.

148. Nix WA, Dahm M. The effect of isometric short-term electrical stimulation of denervated muscle. Muscle and Nerve 1987;10:136-143.

149. Cummings JP. Conservative management of peripheral nerve injuries utilizing selective electrical stimulation of denervated muscle with exponentially progressive current forms. J Orthop Sports Phys Ther 1985;7:11-15.

150. Barr JO. Transcutaneous electrical nerve stimulation. In: Nelson RM, Currier DP, eds. Clinical electrotherapy. Norwalk, CT: Appleton and Lange, 1991:274-277.

151. Kloth L. Interference current. In: Nelson RM, Currier DP, eds. Clinical electrotherapy. Norwalk, CT: Appleton and Lange, 1991:255-256.

152. Newton R. High-voltage pulsed galvanic stimulation: theoretical bases and clinical applications. In: Nelson RM, Currier DP, eds. Clinical electrotherapy. Norwalk, CT: Appleton and Lange, 1991:218.

Chapter 6

Neural Mechanisms of Pain

David J. Mayer
Donald D. Price

$\bigwedge$ $\mathbf{B}$onica (1) has estimated that over 80,000,000 persons in the United States suffer from some painful syndrome at any given time. Approximately 60 million of these suffer at least a partial disability. Thus, approximately 25% of the population is affected at any given time. The cost to the economy in the United States is estimated to be over 60 billion dollars. Reliable estimates of the magnitude of the problem in other countries are not available, but the problem is likely to be even worse in less-developed countries because overall health care is generally not as advanced as in the United States.

This chapter will review the neurobiology of pain. We will begin by reviewing the most important facts known about the afferent transmission of pain, from the periphery to the central nervous system to and including the cerebral cortex. This discussion will be followed by a review of exciting recent advances in the understanding of neuropathic pain conditions. Next, we will examine the complex literature on neural systems involved in the modulation of afferent pain transmission. This discussion will emphasize the role of endogenous opiates in pain modulation. Finally, we will examine the role that these pain modulatory systems may play in some treatment modalities applicable to humans.

Neurobiology of Afferent Pain Transmission

Introduction

This section is intended to provide a concise review of portions of the afferent pathways involved in pain transmission. Pain is generally analyzed as a sensory system, and this approach will be followed here although there are good reasons for alternative analytical approaches as discussed by Mayer and Price (2) and by Melzack and Wall (3). An analysis of pain as a sensory system involves a discussion of the receptors, primary afferent neurons, and central nervous system synapses involved in the processing of information about stimuli in the environment that produce or threaten to produce tissue damage.

The role of various types of primary nociceptive afferent neurons and central neurons in pain processing will be discussed in terms of an overall strategy used to identify pain-related (nociceptive) neurons. In brief, four distinct lines of evidence are used to identify these neurons: (*a*) selective stimulation of the candidate neuron(s) gives rise to a painful sensation, (*b*) responses of the candidate neuron to controlled nociceptive stimulation show parallels to human psychophysical responses to similar stimulation, (*c*) reduction or abolition of the neuron's responses to nociceptive stimulation results in a decrease in the behavioral response to nociceptive stimulation, and (*d*) the neurons have anatomical connections consistent with a role in the sensory-discriminative aspects of pain.

Primary afferent nociceptive neurons

Cutaneous Afferents

Low-Threshold Myelinated (A$_\beta$) Mechanoreceptive Afferents

There are several categories of this general type of afferent that function to detect either the position or velocity of gentle forms of mechanical stimulation applied to their receptive field (4), or that area of skin from which impulse discharge can be evoked. All such types are usually very sensitive to weak mechanical stimuli (100 mg) and do not respond with higher impulse frequencies to nociceptive mechanical stimuli. In fact, many are suppressed by noxious stimulation, particularly noxious heat (4). They have synaptic contacts with spinal interneurons and spinal projection neurons that are at the origin of pathways projecting to the brain (5, 6). It has been suggested that large myelinated afferents contribute to pain because they form part of the total afferent barrage that occurs in response to nociceptive stimulation (3). However, it is known that activation of A_β afferents alone cannot produce pain even when this activation occurs at extremely high frequencies (4, 7). In addition, pain sensations can occur in the absence of impulse conduction in these large afferents and in fact are more intense when they are blocked (8–10). Thus, A_β mechanosensitive afferents are neither necessary nor sufficient to evoke pain. However, some low-threshold A_β mechanosensitive afferents synapse on and excite spinothalamic and trigeminothalamic neurons that most likely participate in sensory aspects of pain (11, 12), and their activity may summate with that produced by nociceptive afferents terminating on these same central neurons. More likely, they have a role in the inhibition of nociceptive transmission since a major effect of the stimulation of A_β afferents is to reduce spontaneous activity and nociceptive responses in various types of spinal cord neurons (5, 6).

Sensitive Thermoreceptive Neurons

Distinct classes of thermoreceptive neurons respond maximally to warming (34 to 43°C) or cooling (34 to 20°C) of the skin (5, 13, 14). The pain threshold is approximately 45°C. "Warm" afferents respond maximally to temperatures below the painful range and generally do not increase their responses as temperatures become painful (15). Such neurons do not contribute to the ability to perceive different intensities of heat-induced pain. It is unlikely that they have synaptic contacts on dorsal horn nociceptive neurons because the vast majority of the latter respond very weakly if at all to mild warming temperatures (35 to 42°C), the range over which "warm" afferents are maximally sensitive (5, 6). At present there is little evidence that such afferents contribute to pain. However, it is possible that the transient pain that sometimes occurs in response to the immediate immersion of the skin in water maintained at 36–43°C may be related to high-frequency activation of warm afferents (15). For similar reasons, low-threshold "cold" thermoreceptive afferents are unlikely to play a major role in pain. Their maximum responses occur in the nonpainful cooling range, and they do not appear to synaptically activate central nociceptive neurons because nociceptive dorsal horn neurons do not respond to non-noxious cooling. However, some low-threshold "cold" afferents respond paradoxically to intense heat stimuli, which could account for the "paradoxical cold sensation" when "cold spots" on the skin are stimulated with temperatures above 45°C. High-threshold A_δ and C "cold" afferents respond maximally to noxious cold (<20°C), and they may activate central nociceptive neurons because some spinothalamic tract neurons are activated by noxious cold stimuli (5, 16). However, a fraction of C polymodal nociceptive afferents (see below) also respond to intense cold, so it is unclear whether high-threshold "cold" afferents are necessary or sufficient for cold-induced pain.

High-Threshold A_δ Mechanoreceptive Afferents

These afferents respond mainly to intense mechanical stimuli (4, 11, 15). Most nociceptive afferents in this class respond to stimulus intensities that produce overt tissue damage, though some respond to nondamaging intense stimuli.

Their response to mechanical nociceptive stimuli of varying intensity as well as their small receptive fields strongly implicate these afferents in pain. However, it would be difficult if not impossible to selectively stimulate only this class of neurons because there are other types of nociceptive afferents that also respond to intense mechanical stimuli. Thus, it is uncertain whether activation of this class of nociceptive afferents alone would evoke pain. A_δ mechanical-nociceptive afferents excite spinal cord dorsal horn and trigeminal nucleus caudalis neurons identified as projecting to the thalamus. The axon terminals of these afferents have been traced to the superficial laminae of the dorsal horn (laminae I–II) and to lamina V (6). Therefore, they have anatomical connections consistent with a role in sensory aspects of pain. In brief, the response characteristics and anatomical connections of this group strongly implicate them in pain, although it is not known whether their sole activation produces pain.

An interesting property of A_δ mechanical-nociceptive afferents is that they can be sensitized by noxious heat (17). Although they do not respond to an initial heat stimulus, repeated noxious heat stimuli or a burn may eventually produce vigorous impulse discharges. Therefore, sensitization of A_δ mechanical nociceptors is likely to have a crucial role in the hyperalgesia that results from damaging burns.

A_δ Heat-Nociceptive or Mechanothermal Afferents

This critical nociceptive afferent population has conduction velocities in the A_δ range (5, 15). Receptive fields are usually small (≤ 5 mm^2), though larger fields with multiple sensitive spots have been reported in the skin of limbs (5,15). As shown in Figure 6.1 and discussed previously, A_δ heat-

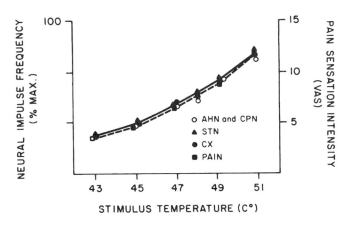

Figure 6.1 Nociceptive stimulus–pain rating *(top)* and nociceptive stimulus–neural response *(bottom)* functions generated from the same type and range of noxious contact thermal stimuli. *AHN*, A_δ heat-nociceptive afferent; *CPN*, C polymodal nociceptive afferent; *STN*, spinothalamic tract neuron; *CX*, cerebral cortical nociceptive neuron.

nociceptive afferents have positively accelerating stimulus–response functions with the steep portion of the curve in the 45–53°C range. Threshold temperatures are usually below noxious levels (40–46°C). They also respond to intense mechanical stimuli that, at threshold, are not painful to humans, and their responses to mechanical stimuli appear to increase monotonically with increases in stimulus intensity (15). Some also respond to intense cold (20°C) stimuli (5).

The role of A_δ heat-nociceptive afferents in pain discrimination is clearly established. They are likely to be critical for human judgments of the intensity of heat-induced pain because their 43–51°C stimulus–impulse response functions precisely parallel human stimulus–pain intensity rating functions (both are power functions with exponents of about 2.1), and their responses to sudden heat stimuli occur early enough to account for the brief latency escape responses of monkeys and the perceptions of first pain in humans to the same stimuli (10, 15). The close association between responses of A_δ heat-nociceptive afferents and first pain clearly indicates that their activation is sufficient to evoke pain and that a reduction of their responses is accompanied by a concomitant reduction in first pain (10). Finally, dorsal horn sensory projection neurons must receive direct or indirect synaptic connections with A_δ heat-nociceptive afferents, because such dorsal horn neurons respond to brief heat stimuli with latencies that can only be accounted for on the basis of impulse conduction in slow myelinated afferents (5). Therefore, the physiological characteristics of A_δ heat-nociceptive afferents fulfill all of the major criteria for implication in the sensory–discriminative dimension of pain.

It is important to distinguish A_δ heat-nociceptive afferents from A_δ mechanical-nociceptive afferents. The latter may contribute to heat-induced pain but only after repeated or very intense stimulation, that is, after they become sensitized (17). By contrast, A_δ heat-nociceptive afferents are excited the very first time a noxious heat stimulus is applied and undoubtedly account for the first pain felt as one touches a hot stove.

C Polymodal Nociceptive Afferents

This extremely important group of nociceptive afferents makes up the vast majority of unmyelinated cutaneous afferents in primates (80–90% in monkeys and more than 90% in humans) (13, 18–20), and unmyelinated afferents outnumber myelinated afferents 3–4 to 1 (4). They are characterized by the fact that they are optimally responsive to at least three distinct forms of nociceptive stimulation, including thermal, mechanical, and chemical stimuli (4, 5). Hence, the term *polymodal nociceptor* has been used to designate this class of afferent. Approximately one-third of these afferents also respond to noxious cold (10°C) stimuli (5, 18). Thus, threshold activation of these afferents may not signal the presence of tissue-threatening stimuli (4, 15), and there is a distinct possibility that they convey non-nociceptive as well as nociceptive information. Unfortunately, this possibility has been understated in the literature.

As shown in Figure 6.1, their responses to graded intensities of nociceptive temperatures, similar to pain ratings by human observers, are monotonic, positively accelerating functions, with maximum sensitivity in the 45–53°C range. Therefore, it is very possible that C polymodal nociceptive afferents contribute to perceived intensities of heat-induced pain. However, since A_δ

heat-nociceptive afferents' responses to these same temperatures also parallel psychophysical responses, the extent of C polymodal nociceptive afferents' contribution to judgments of heat-induced pain intensity is somewhat unclear. It is premature to conclude that input from C polymodal nociceptive afferents is either necessary or sufficient for scaling heat pain.

Similar to A_δ heat-nociceptive afferents and A_δ mechanical-nociceptive afferents, C polymodal nociceptive afferents can be sensitized (19, 20). Although C polymodal nociceptive afferents are not sensitized by the 53°C stimulus described earlier, a lower intensity, 50°C, 100-sec stimulus does produce both hyperalgesia in humans and sensitization of C polymodal nociceptive afferents. Combining these observations with those discussed earlier for A_δ mechanical nociceptive afferents, it is evident that both A_δ and C nociceptive afferents are involved in hyperalgesia, the former in the case of severe damage and the latter in the case of mild damage.

Muscle and Visceral Primary Afferents

Muscle Nociceptive Afferents

In addition to low-threshold sensitive mechanoreceptive afferents (21), muscle is innervated by other types of afferents including nociceptive afferents (21, 22). Similar to cutaneous nociceptive afferents, those innervating muscle include slow-conducting unmyelinated (C) and possibly A_δ myelinated afferents. Presumably, muscle nociceptive afferents supply many free nerve endings found in the connective tissue of the muscle, between muscle fibers, in blood vessel walls, or in tendons (22).

The evidence for A_δ (or group III) muscle nociceptive afferents is equivocal (23). The strongest evidence is that many A_δ muscle afferents can be excited by intraarterial injections of algesic chemicals, including bradykinin, serotonin, and solutions of potassium ions (21, 22). These injections would likely produce pain in humans. However, Mense and Stahnke (22) found that a high proportion of A_δ muscle afferents responded to muscle stretch and many also were activated by muscle contractions. The amount of response to contraction was graded according to the strength of contraction. Mense and Stahnke concluded that A_δ muscle afferents are more likely to be activated during exercise than are C muscle afferents, and that their role may be ergoreceptive rather than nociceptive.

On the other hand, there is more substantial support for the role of unmyelinated (C) muscle nociceptive afferents in pain (21–24). Similar to A_δ muscle afferents, they are excited by algesic chemicals (23, 24). In addition, many are optimally responsive to strong, presumably noxious, mechanical stimuli (21, 23, 25, 26). According to Kumazawa and Mizumura (23, 24), C muscle nociceptive afferents often are similar to C polymodal nociceptive afferents innervating the skin. They found that muscle C afferents that were excited by noxious heating were also excited by mechanical stimuli and by algesic chemicals. Thus, there is evidence for the existence of a general class of C nociceptive muscle afferents that is optimally excited by nociceptive stimuli. Moreover, the responses of these afferents to controlled noxious stimuli at least qualitatively parallel aspects of ischemic muscular pain in humans.

The evidence that slow-conducting muscle afferents have excitatory effects on sensory projection neurons of the dorsal horn is that excitation of dorsal horn neurons follows electrical stimulation of slow-conducting muscle afferents within peripheral nerves (6). The same central cells also respond to stimulation of slow-conducting visceral afferents and cutaneous afferents. Neurons identified as sensory projection neurons by means of antidromic activation techniques have been shown to receive excitatory effects from stimulation of slow-conducting (A_δ and/or C) muscle afferents. These include spinothalamic and spinocervical tract neurons (6). Thus, there is general support that slow-conducting A_δ and/or C muscle nociceptive afferents have excitatory synaptic connections with central spinal cord neurons that give rise to ascending nociceptive pathways, which fulfills criterion 4.

Visceral Nociceptive Afferents

Despite the clinical significance of visceral diseases that cause pain, little is known about visceral nociceptive afferents. However, there are several unique features of visceral pain that may serve as guidelines for the types of anatomical and physiological characteristics to look for in searching for visceral nociceptors. First of all, visceral pain is often poorly localized and poorly discriminated in terms of the type of stimulus evoking it (6). Second, visceral pain is commonly referred to a somatic region (6) and that area to which pain is referred can become tender. Finally, visceral pain is often accompanied by autonomic and somatic reflexes. In some ways, characteristics of visceral pain are like those of cutaneous pains evoked by C polymodal nociceptive afferents.

Perhaps one of the best experimental approaches to the characterization and identification of visceral nociceptive afferents comes from the work of Kumazawa and Mizumura (27), who studied the responses of testicular nociceptive afferents to precisely quantified mechanical and thermal stimuli as well as algesic chemicals. Both small myelinated afferents and unmyelinated afferents were found that responded to mechanical pressures of the testicles over a range of 30 to 2000 g. The upper end of this range would be distinctly painful when applied to dog or human testicles. These same afferents also responded in a graded fashion to increasing noxious temperatures and to algesic chemicals; they became sensitized with repeated nociceptive heat stimuli. Therefore, these nociceptive afferents have physiological characteristics that are very similar to those of C polymodal nociceptive afferents innervating the skin. However, these testicular polymodal nociceptive afferents include both A_δ and C afferents. Since monkey spinothalamic tract neurons in upper lumbar segments respond to very similar nociceptive stimuli as those that were applied to the testicles, it is likely that A_δ and/or C polymodal testicular afferents of the type just described have excitatory connections appropriate for visceral nociception. Thus, these types of primary nociceptive visceral afferents fulfill to some extent most of the criteria for a role in pain.

Evidence for specific classes of nociceptive afferents innervating other visceral structures is less clear and usually indirect. There is evidence that both A_δ and C nociceptive afferents innervate the heart, since afferent axons of both sizes are excited during coronary occlusion (6). Afferents presumed to innervate structures near capillaries of the lung, called type J receptors (6), are activated by lung deflation or large lung inflations, as well as by irritants or other stimuli that irritate lung tissue. Many of these are likely to have a nociceptive function and are unmyelinated. Nociceptive afferents innervating the bile duct also have been recently identified (6). Nociceptive afferents innervating the urinary bladder, gastrointestinal tract, and urinary tract have yet to be identified and characterized, though it is clear that pain can be evoked by intense stimulation of these structures (6).

Although much further research is needed to characterize nociceptive afferents innervating noncutaneous structures, it is becoming evident that similar principles of nociceptor functioning extend across different tissues. First, for most all innervated tissues, there exist classes of nociceptive afferents that can be physiologically distinguished from low-threshold mechanoreceptive, low-threshold thermoreceptive, or chemosensitive afferents. Second, C polymodal nociceptive afferents similar to those innervating skin appear to innervate muscle, the testes, and perhaps the lungs and cardiovascular system. Finally, the latter type of nociceptive afferent comprises a large proportion of the nociceptive afferents innervating different tissues. The physiological characteristics of polymodal nociceptors may well account for pains that are diffuse, poorly localized, and poorly discriminated in terms of modality. These pains are especially prevalent in inflammation and in certain diseases.

Summary and Conclusions

Experiments utilizing modern electrophysiological techniques have confirmed the proposal made throughout this century that A_δ and C primary afferents are critically involved in pain mechanisms. Distinct classes of cutaneous, muscular, and visceral primary A_δ and C nociceptive afferents have been identified and characterized in terms of responses to physiological stimuli and in terms of the effects they exert on spinal cord neurons. An interesting principle has begun to emerge regarding the possible differential functional role of A_δ and C nociceptive afferents. A_δ nociceptive afferents tend to be relatively modality-specific and often optimally respond to intense mechanical stimuli. They elicit brief-latency, intense responses critically necessary for rapid withdrawal and escape. C nociceptive afferents tend to be of the polymodal variety, responding to mechanical, thermal, and chemical stimuli. Their responses tend to be more delayed and their central actions are prolonged and slowly summate over time.

The role of neurotransmitters in primary afferent pain transmission

Several putative synaptic transmitters such as substance P (SP) and the excitatory amino acids have been suggested to serve as neurotransmitters for the conduction of nociceptive information in afferent pathways. In order for a substance to be considered as a neurotransmitter, it should meet a number of criteria. First, its presence must be demonstrated within presynaptic terminals of synapses in the afferent pathways. It must be released following painful stimulation of the receptive field it subserves. It must excite appropriate postsynaptic cells within the nociceptive pathways. On administration into the synapse it normally activates, it should produce a behavior similar to or consistent with perceived pain. Finally, depleting the putative transmitter must cause analgesia.

SP meets several of the criteria required to be considered as a candidate primary afferent transmitter for pain. SP is found in primary afferent C fibers and is released by painful stimulation; this release has excitatory effects on spinal cord pain transmission neurons and can be blocked by opiates. It is unclear, however, whether this peptide indeed serves as a traditional neurotransmitter in afferent pain transmission for the following reasons. First, its excitatory action upon iontophoretic application is conspicuously long-lasting, more characteristic of a neuromodulator than of a neurotransmitter. Second, SP depletion fails to block thermal and, in the cornea at least, mechanical pain. Third, the behavior elicited by intrathecal administration of SP seems not to be related to brief acute pain but rather constitutes a convulsive type of activation of predominantly spinal motor systems (28). Its role in some types of long-term pain without well-verified behavioral correlates remains open to question.

Taken together, it does not seem likely that SP fulfills the role of a critical synaptic transmitter in afferent pain processes. In view of its long-lasting excitation of neurons, it conceivably acts as a neuromodulator. Whether such a modulating role is specific to the pain modality or is limited to sensory systems and not motor systems is a question that remains open.

The excitatory amino acids (EAA) are a group of single amino acid neurotransmitters such as L-glutamate and L-aspartate that have been shown to account for a large percentage of excitatory transmission in the central nervous system. These transmitters have complex excitatory actions because they act at several postsynaptic receptor sites. Activation of some of these sites, such as the N-methyl-D-aspartate (NMDA) receptor, can result in long-lasting changes in postsynaptic neurons. EAA action on other non-NMDA receptors (e.g., the AMPA, kainate, or quisqualate receptors) can have more traditional rapid depolarizing effects. A considerable body of evidence has implicated the EAA transmitters and these various receptor sites in the transmission of pain from primary afferent neurons to postsynaptic sites in the spinal cord.

The EAAs meet several of the criteria for neurotransmitters described above. They can be localized in small (A_δ and C) primary afferent neurons (29), are released selectively by painful stimulation (30), and cause excitation when applied to postsynaptic neurons in the spinal cord (31). Importantly, however, they fulfill other necessary criteria less well. Application of EAA transmitters directly to the spinal cord does not always produce clear evidence of pain (32), and substances that block the action of EAA (antagonists) at their receptors do not produce unequivocal analgesia (28). Thus, although the excitatory amino acids meet some of the criteria described above, there is, as yet, insufficient evidence to either rule out or accept the possibility that excitatory amino acids play a role in afferent pain transmission. It is likely to be critical that EAA transmitters have actions at several receptor sites. Thus, whereas the role of these transmitters in pain caused by an acute painful stimulus remains equivocal, their role in chronic and pathophysiological pain seems better established as will be evident from the discussion of pathophysiological pain later in this chapter.

The involvement of neurotransmitters in pain is an important area of ongoing pain research. There is some evidence for a role of a number of neurotransmitters not discussed here. An excellent detailed review of this topic has recently been done by Wilcox (31). A description of the neurochemical events involved in the primary afferent transmission of pain now appears likely to provide valuable insights for the pharmacological treatment of many pain syndromes in the near future.

Spinal cord mechanisms subserving pain

The spinal cord dorsal horn and, to a lesser extent, the ventral horn are the origins of second-order neural pathways involved in pain. Impulses from thermoreceptive, mechanoreceptive, and nociceptive primary afferent neurons excite various types of sensory projection neurons and interneurons of the spinal cord gray matter. The responses of second-order neurons to somatosensory stimuli do not simply passively reflect the responses of primary afferents described above. As will be discussed, radical transformations of primary input have important implications for perception of pain. The dorsal horn also is a central focal point for mediating the autonomic and somatomotor reflexes initiated by nociceptive stimulation. Nociceptive processing at this level is influenced by several descending control mechanisms as well as local inhibitory and facilitatory interactions.

Functional classes of second-order nociceptive neurons will be described in terms of their major physiological characteristics. These characteristics will then be discussed in terms of factors that serve to encode nociceptive information; this discussion will emphasize the relationships of responses of neurons to the perceptual attributes that parallel them. Finally, the question of functional characterization of ascending nociceptive pathways will be addressed.

Spinal Cord Neurons That Subserve Pain: Excitatory Connections and Mechanisms

Functional Classes of Spinal Cord Sensory-Projection Neurons

There is now general agreement about the major functional classes of somatosensory projection neurons located within the spinal cord gray matter. Although slightly different classifications are made and different names are sometimes used for the same class, spinal cord sensory-projection neurons can be divided into the following five classes. These classes of neurons are differentially located in different layers of the spinal cord gray matter, as shown in Figure 6.2. The lines of evidence implicating each class of neurons in pain are summarized in Table 6.1.

Wide-Dynamic-Range Neurons

These neurons have been more extensively studied than any other class of dorsal horn neuron (5, 16, 33). Their potential involvement in pain is a critical aspect of the gate-control theory (3). This theory proposes that both large-diameter afferents and small-diameter afferents converge on central cells of this type and that the final output of these neurons (called *T cells*) is controlled by interactions between large- and small-fiber input and by descending pathways from the brain. Since the proposal of the gate-control theory, a large body of evidence has accumulated that indicates that wide-dynamic-range (WDR) neurons contribute to the perception of pain (5, 34).

As shown in Figure 6.2, wide-dynamic-range neurons exist in high concentrations in laminae V and VI of the dorsal horn and to a lesser extent in laminae I, II, and IV. Many WDR neurons in laminae I and V are spinothalamic or trigeminothalamic tract neurons (34, 35). Wide-dynamic-range neurons amount to approximately two-thirds of all neurons in the primate spinothalamic tract and are the main if not exclusive source of low-threshold

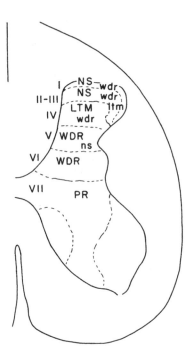

Figure 6.2 Schematic representation of relative distributions of different classes of somatosensory neurons within the spinal cord gray matter. Rexed laminae and boundaries are indicated by Roman numerals and dashed lines. *NS*, nociceptive-specific; *LTM*, low-threshold mechanoreceptive; *WDR*, wide-dynamic-range; *PR*, proprioceptive. Upper and lower case letters, respectively, represent high and low proportions of neurons.

Table 6.1
Lines of Evidence Supporting the Role of Spinal Cord and Medullary Dorsal Horn Neurons in Pain Discrimination

Neuron type	Lines of evidence*			
	1	2	3	4
Low-threshold mechanoreceptive (LTM)				
Thermoreceptive (warm)		+		+
Thermoreceptive (cold)	+		+	
Wide-dynamic-range (WDR)	++	++	++	++
Nociceptive-specific (NS)	++	++	++	

*Lines of evidence: 1, selective stimulation produces pain; 2, maximum response to nociceptive stimuli; 3, selective manipulations reduce neural responses and pain; 4, appropriate anatomical connections; ++, strong evidence; +, weak evidence (based on Price and Dunbar, 1977).

cutaneous mechanoreceptive input to this pathway (34, 35). Some WDR neurons in laminae III and IV are dorsal column postsynaptic neurons and/or spinocervical tract neurons (15). The WDR neurons receive excitatory effects from impulses in both large-diameter (A_β) sensitive mechanoreceptive cutaneous afferents and in small-diameter (A_δ and C) nociceptive cutaneous afferents, a pattern of convergence represented in Figure 6.3. Receptive fields of these neurons range considerably in size but are usually much larger than those of primary afferents (34, 35). WDR neuron receptive fields are often organized into central zones, in which the neuron responds with increasingly higher frequencies of impulse discharge to gentle touch, firm pressure, and pinch, and much larger zones wherein only nociceptive stimulation is effective

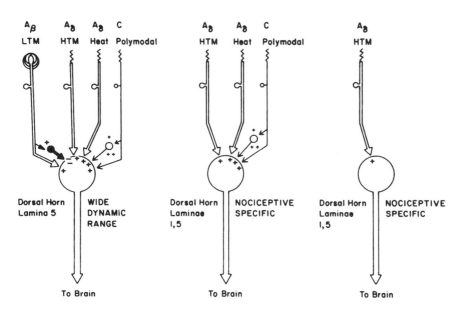

Figure 6.3 Two types of nociceptive sensory relay neurons of the dorsal horn and their patterns of primary afferent convergence. Facilitatory interneurons are indicated by small unfilled circles and plus signs. Inhibitory interneurons are indicated by filled circles and minus signs. *LTM*, low-threshold mechanoreceptive afferent; *HTM*, high-threshold mechanoreceptive afferent.

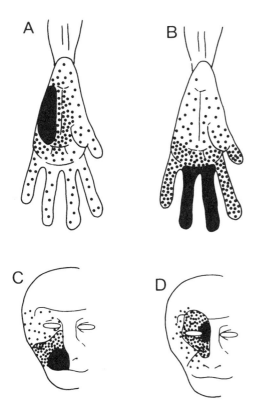

Figure 6.4 Illustration of the gradients of response sensitivity recorded from two different WDR neurons in anesthetized monkeys (**A** and **B**) and from two different WDR neurons in an awake, unanesthetized monkey (**C** and **D**). The central portions of their receptive fields *(darkly shaded)* are most sensitive and respond with increasing frequency as the intensity of mechanical stimuli is increased from touch to pressure to pinch. The more peripheral aspects of the receptive fields are systematically less sensitive, responding differentially only to pressure and pinch *(heavily stippled)* or to pinch only *(lightly stippled)*. **A** and **B** are from Price (unpublished data) and **C** and **D** are from Hayes et al. (55).

in evoking impulse discharges, as illustrated in Figure 6.4. The total receptive field sizes of WDR neurons are smallest in laminae I and II, larger in lamina V, and largest in lamina VI. For example, those in laminae I and II often include a portion of a monkey's foot, those in lamina V often include most of the foot, and those in lamina VI often include the foot and major portion of the rest of the limb (33–35).

The most distinguishing feature of WDR neurons is their capacity to respond with increasing impulse frequencies over a very wide range of non-nociceptive and nociceptive stimulus intensities. Although they respond to extremely gentle somatic stimuli, their maximum responses are to nociceptive stimuli, both in anesthetized as well as in unanesthetized preparations (5, 6, 35, 36).

Nociceptive-Specific Neurons

Spinal cord neurons that respond exclusively or nearly exclusively to nociceptive stimuli exist in high concentrations in the superficial layers of the dorsal horn (laminae I–II) and to a lesser extent in layer V (Fig. 6.2). The existence of nociceptive-specific (NS) neurons supports the specificity theory of pain, which states that pain is dependent on central neurons and pathways activated exclusively by nociceptive stimulation. Nociceptive-specific neurons make up 20–25% of spinothalamic neurons (33). The NS neurons receive exclusive excitatory effects from impulses in nociceptive afferents, most of which are slow-conducting. Major sources of input to these neurons are from high-threshold A_δ mechanosensitive afferents, A_δ heat-nociceptive afferents, and C polymodal nociceptive afferents (Fig. 6.3). The receptive fields of these cells are usually smaller than those of WDR neurons and sometimes show a gradient of sensitivity similar to that observed for WDR neurons. There is extensive somatotopic organization of these cells within the marginal layer. For example, monkey L-7 spinothalamic neurons whose receptive fields are located on the ventral surface of the foot exist in the medial part of the dorsal horn, whereas the spinothalamic cells with dorsal receptive fields are in the lateral part of the dorsal horn (6).

Low-Threshold Mechanoreceptive Neurons

A third general category of central neuron in the dorsal horn responds to light touch, pressure, or hair movement and appears to receive input exclusively from sensitive mechanoreceptive afferents (6). Most of these primary afferents are large myelinated A_β afferents. Low-threshold mechanoreceptive (LTM) neurons do not respond to noxious heat or exhibit higher discharge rates to noxious stimuli. The LTM neurons are concentrated in laminae III–IV (Fig. 6.2). Very few if any LTM neurons project in the spinothalamic tract (6).

Proprioceptive Spinothalamic Tract Neurons

A small population of spinothalamic neurons exists within the ventral laminae of the spinal gray matter (VI–VIII) that responds to deep tissue stimulation and/or to joint movement (6). Many of these lack cutaneous receptive fields.

Thermoreceptive Neurons

Second-order neurons responding exclusively to either warming (35–43°C) or cooling (35–20°C) stimuli exist in layer I and to a lesser extent in layers III–V (5, 16). They have small receptive fields and often respond as if they receive exclusive input from one type of primary thermoreceptive afferent, either cold afferents or warm afferents. These neurons are likely to form an important thermoreceptive afferent system involved in the perception of warmth or cooling.

Patterns of cutaneous, muscular, and visceral convergence onto WDR and NS neurons

Nociceptive afferent input from muscle and viscera also converge on WDR and NS neurons, thereby providing at least part of the basis of referred pain from these tissues (6, 37). Referred pain occurs when nociceptive stimulation of one type of tissue, such as a visceral structure, is perceived as originating in another type of tissue, such as skin. A fascinating pattern of convergence occurs such that the location of cutaneous, visceral, and muscular "receptive fields" can be predicted on the basis of known patterns of referred pain. For example, stimulation of cardiac sympathetic nerve axons excites WDR and NS spinothalamic tract neurons at upper thoracic segmental levels (6, 37). These same neurons often have a cutaneous receptive field that extends along the inner aspect of the forearm, the pattern of pain referral in angina pectoris (6). Similarly, spinothalamic WDR neurons of upper lumbar segments can be excited both by nociceptive stimulation of the testicle and nociceptive and non-nociceptive stimulation of upper flank and lower abdominal skin areas, areas of pain referral in testicular injury (6). Some of these same neurons also can be excited by overdistension of the urinary bladder. Finally, WDR and NS neurons of the sacral spinal segments can be excited by urinary bladder distension but not by testicular stimulation (6).

The patterns of pain referral and tenderness in humans are paralleled by patterns of nociceptive input to WDR and NS neurons from skin, muscle, and viscera (6, 37, 38). This parallel strongly supports the convergence theory of referred pain. Future detailed studies of these patterns of convergence are likely to be helpful in providing detailed knowledge about the peripheral origins of various pain symptoms, a problem of great clinical significance.

Inhibitory Mechanisms

Segmental and Intersegmental Inhibitory Mechanisms

The response characteristics of WDR and NS spinothalamic neurons reflect a variety of segmental inhibitory mechanisms related to nociception. A well-documented type is initiated by stimulation of low-threshold mechanoreceptive afferents (37–39). Most low-threshold mechanosensitive (touch) afferents have rapid conduction velocities and large-diameter axons. Their inhibitory role in pain mechanisms is one of the main features of the gate-control theory (3). High-frequency repetitive stimulation of these afferents can be accomplished by electrical stimulation of the dorsal columns or by low-intensity electrical stimulation of peripheral cutaneous nerves (40, 41). Both kinds of stimulation reduce the magnitude of spinothalamic neuron responses to nociceptive thermal and mechanical stimuli as well as their responses to electrical stimulation of A_δ and C afferents (40, 41). This form of inhibition, which does not outlast the stimulation period, is mediated at least partly by postsynaptic mechanisms. Dorsal column stimulation evokes brief inhibitory postsynaptic potentials (IPSPs) in spinothalamic nociceptive neurons (41). This inhibition is likely to be mediated by local interneuronal mechanisms because the latency to these inhibitory postsynaptic potentials follows within 1 to 2 msec after the onset of dorsal column stimulation (41). Presumably inhibitory mechanisms also may play a role in this form of inhibition, since dorsal column–evoked inhibitory periods are accompanied by increased excitability (i.e., lowered electrical thresholds) of primary afferent terminals (41). Presumably, then, these terminals are partially depolarized and release less transmitter substance during each action potential.

The inhibition of the nociceptive responses of WDR and NS spinothalamic neurons by dorsal column and peripheral nerve stimulation have their parallels in analgesia produced by dorsal column stimulation and by high-frequency low-intensity transcutaneous electrical stimulation of peripheral nerves in humans (42). Although the use of dorsal column stimulation has been reduced because of technical reasons, transcutaneous electrical nerve stimulation has been widely used as a conservative noninvasive form of chronic pain therapy.

A longer-lasting form of segmental inhibition of WDR and NS neurons can be produced by intense stimulation of peripheral nerves at a low frequency, around 2 Hz. For example, electrical stimulation of the common peroneal nerve at a strength sufficient to activate C afferents and at a frequency of 2 Hz produced a prolonged inhibition of responses of a monkey spinothalamic neuron to a volley in A and C afferents of the sural nerve (39). The C afferent–mediated response, which in primates can be considered mainly a nociceptive response, was more powerfully inhibited than the earlier A afferent–mediated response. This inhibition outlasted the period of conditioning stimulation by 20–30 min. This form of inhibition also has a parallel in human clinical studies that show analgesia from intense somatic stimulation (42).

Some Conclusions About the Involvement of Spinal WDR and NS Neurons in Pain

When the physiological characteristics of spinal cord somatosensory-projection neurons are considered in the context of the several coding factors just discussed, it is quite evident that NS and WDR neurons have differential roles in these factors. NS neurons are likely to code nociceptive stimulus location and, to some extent, the stimulus modality. WDR neurons code nociceptive stimulus intensity and differences in stimulus intensity, as well as provide a basis for spatial recruitment. Spatial recruitment, in turn, provides the basis for some spatial features of pain (i.e., radiation) and possibly part of the basis for pain sensation intensity. Both WDR and NS neurons have responses that indicate temporal transformations. Unlike responses of primary afferents, their responses to some types of abrupt repetitive nociceptive stimuli outlast the stimulus and summate slowly over time. In these respects, their responses more closely match the parameters of pain evoked by similar repeated noxious stimuli. Finally, it is likely that the functions of NS and WDR overlap, so that no one function, such as localization of pain, is subserved by only one class of spinal cord neurons.

Characteristics of Central Destinations of Spinal Cord Nociceptive Pathways

Nociceptive information is encoded by the unique spatiotemporal profiles of activated dorsal horn neurons. The number, type, and patterning of this output, and subsequently the patterning of the central neurons that receive this spinal outflow, are controlled by segmental and supraspinally organized modulation.

A question that remains is whether the different functional consequences of somatosensory events are represented in different outputs of the dorsal horn or whether the same coded outputs are transmitted simultaneously to different supraspinal centers. For example, it is now understood that pain is not just a sensory experience but one in which there are sensory–discriminative, arousal, and affective–motivational dimensions (3, 42). Are these different dimensions of the experience of pain differentially represented in subdivisions of the spinothalamic pathway or do neurons of this same ascending nociceptive pathway participate in multiple functions?

Although the question of different functional subdivisions of the spinothalamic tract is still open, quite different kinds of anatomic and physiological facts indicate that simple functional dichotomies represent an oversimplified and misleading characterization of ascending nociceptive systems. In fact, both the origins and central destinations of different spinothalamic nociceptive neurons appear to have overlapping distributions and probably overlapping functions.

This characteristic is represented in schematic form in Figure 6.5 for the spinothalamic pathway, the main nociceptive system of primates. The neurons of origin of this pathway can be subdivided into those projecting to the medial thalamus, those projecting to the caudal ventroposterolateral nucleus (VPL_c), and those projecting to both the VPL_c and medial thalamus (6). The first category, medial projecting neurons, tend to be located in more ventral layers VII–VIII of the spinal cord (6) and have extensive receptive fields encompassing a major portion of the body. Many such neurons respond only to nociceptive stimuli or to proprioceptive stimuli. However, some medial projecting spinothalamic neurons have restricted receptive fields whose response characteristics are indicative of WDR or NS neuron classes. Such neurons are located in layers I and V–VI. Neurons that project only to the lateral thalamus tend to have restricted receptive fields, WDR or NS response characteristics, and locations in layers I and IV–VI. Finally, those spinothalamic neurons projecting to both the medial and lateral thalamus are similar in anatomical locations and physiological characteristics to those projecting only to the lateral thalamus.

A similar principle of overlapping distributions applies to spinal nociceptive neurons that project to different levels of the brainstem. Many spinothalamic neurons have collateral axons to midbrain reticular formation nuclei or the central gray, others do not appear to have collaterals, and some spinal nociceptive neurons project only as far as the midbrain (6, 34). These three categories of neurons do not appear to be distinguished by major differences in physiological characteristics. Both WDR and NS neurons of layers I and IV–VII contribute to all three categories. The different patterns of spinothalamic and spinomesencephalic neuron termination indicate that individual spinal nociceptive projection neurons participate in multiple functions. The brain structures to which spinal nociceptive neurons project can be related to some extent to different aspects of pain experience and response.

Studies of human sensations evoked by anterolateral quadrant stimulation further support the concept that ascending nociceptive spinal neurons participate in multiple functions (43). First, stimulation of axons within the

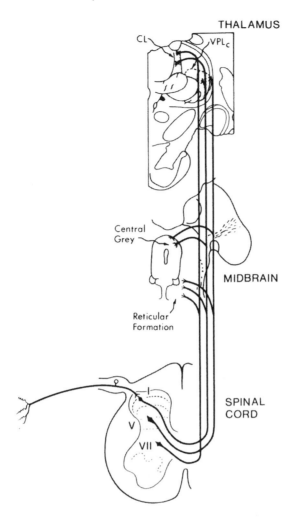

Figure 6.5 Schematic representation of patterns of central projection of spinothalamic tract neurons. Individual spinothalamic neurons originate mainly in *spinal cord* laminae I, V, and VII and project to the caudal portion of the ventroposterior lateral nucleus *(VPL$_c$)* and/or to the centralis lateralis *(CL)*. Individual spinothalamic neurons also send collateral axons to *midbrain* structures such as the *central gray* and *reticular formation*. This schematic is based on the work and ideas of several investigators.

cervical anterolateral quadrant gives rise to pains that have both sensory–discriminative and affective components. The pains evoked by such stimulation are clearly aversive, although certainly tolerable. Behavioral responses and reports indicating that the pain was unpleasant (i.e., "bad, hurts") were given without provocation or suggestion. Since the currents and frequencies were adjusted so as to activate the axons of WDR neurons (43), these indications of aversion are significant. They lead to the inference that wide-dynamic-range neurons activate central mechanisms related to the affective–motivational dimension of pain as well as the central mechanisms related to sensory discrimination.

The physiological characteristics of spinal nociceptive neurons do not clearly indicate functional subdivisions along the lines of sensation, affect, or arousal. For the most part, the responses of spinal nociceptive neurons to controlled nociceptive stimuli most closely parallel one or more aspects of the sensory–discriminative dimension of pain, yet many nociceptive WDR neurons have small tactile receptive fields and are exquisitely responsive to different intensities of non-nociceptive stimulation (5, 6, 35, 44). Stimulation of these neurons gives rise to non-nociceptive and nociceptive sensations and different types of affective responses depending on the number of axons stimulated and the frequency of stimulation (5, 34, 43). However, neurons that have high thresholds to cutaneous stimuli and very large receptive fields may be primarily involved in one function, such as arousal, orientation, or some generalized motoric response to pain. Their responses appear to encode the presence of a nociceptive stimulus but not its location or intensity. The projection of such neurons to medial thalamic nuclei further support this possibility because medial thalamic nuclei are involved in widespread cortical activation.

However, the responses of spinothalamic tract neurons are often of such a nature that they would be closely related to sensations that radiate, temporally summate, and outlast the stimulus. Such types of sensations are often experienced as painful or pleasurable. Thus, it is quite possible that sufficient activation of the same spinothalamic neurons results in both nociceptive sensations and unpleasant affective states. Temporal summation and extension of somatosensory signals beyond stimulus duration may be important for certain somatosensory events that have sensory, arousal, and emotional significance. Certainly, the types of after-responses and summation described above are most often experienced as unpleasant or pleasant. That spinothalamic tract neurons participate in both sensory and affective functions also is indirectly supported by the observation that interrupting the spinoreticular and spinothalamic pathways at cervical levels (i.e., anterolateral cordotomy) results in a loss not only of pain but also of other types of pleasant and unpleasant somatosensory sensations, including sexual sensations (45). The electrophysiological observation that spinothalamic neurons have collateral axons to reticular formation nuclei and the central gray (6) lends support to this idea, since these regions participate in arousal and autonomic nervous system functions that strongly influence affect and hence motivation.

If spinal neurons are not strictly subdivided along the lines of pain sensation, arousal, and affect, this may mean that separation of such functions occurs at supraspinal levels and that some of these functions may be in series rather than in parallel.

Brain processing of nociceptive information

Beyond the levels of primary afferents and spinal cord neurons, detailed information concerning the processing of nociceptive information, until recently, has been scant and controversial. Most of the information about pain mechanisms in the brainstem and cerebral cortex has come from human studies that evaluate the effects of destructive lesions or electrical stimulation. However, considerable neurophysiological and neuroanatomical information about brain mechanisms of pain has accumulated from animal experiments within the last decade. The following discussion will focus on these two divergent sources of knowledge about brain mechanisms of pain, and an attempt will be made to synthesize this information in a manner that should help explain underlying mechanisms. The types of nociceptive neurons and anatomical connections existing at each level of the neuraxis will be described, followed by a discussion of human studies and of the ways in which the different types of studies help explain brain processing of nociceptive information. Figure 6.6 serves as a schematic diagram of the relevant anatomical pathways and brain structures involved in pain.

Brain Processing: Medullary and Midbrain Mechanisms

Animal Studies

Nucleus Gigantocellularis (NGC)

With respect to their involvement in pain, the most extensively studied reticular formation neurons have been those near or within the nucleus gigantocellularis (NGC) of the medulla oblongata. Two types of neurons have been consistently found in this region (46–48). One type is excited exclusively by the stimulation of skin or underlying tissue at an intensity that would usually evoke pain in humans or reflex withdrawal in cats or rats. The other type is excited by both innocuous and intense cutaneous stimuli, but has a more prolonged response or higher frequency response to intense stimulation. In these respects, these two types of neurons are approximately similar to the wide-dynamic-range (WDR) and nociceptive-specific (NS) neurons of the dorsal horn. When tested for responses to electrical stimulation, both types of neurons usually require activation of both large A_β and small A_δ primary afferents for a maximum response (47), which consists of prolonged discharges that long outlast the arrival of primary afferent input (47). The possible contribution of C fiber impulse effects on these neurons remains unclear. Many NGC neurons are vigorously excited by arterial injections of bradykinin, a substance that produces painful sensations in humans (49). The receptive fields of NGC neurons are usually quite large, often including a major portion of the body surface.

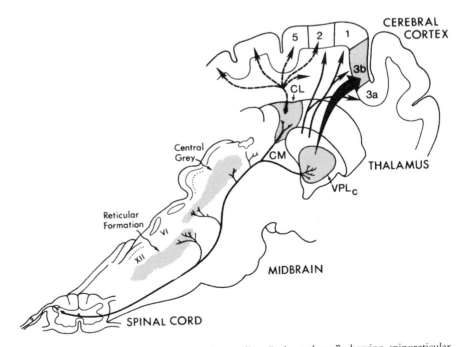

Figure 6.6 Schematic representation of ascending "pain pathway" showing spinoreticular, spinothalamic, and thalamocortical connections.

The evidence that such neurons are involved in sensory discriminations of noxious stimulation is equivocal. Many of these neurons appear to respond maximally to various types of stimulation presumed noxious. However, the relationship between graded natural noxious stimulus intensities and neuronal firing frequency has not been investigated; thus it is not known whether these neurons can signal intensity information in the noxious range. Their capacity to signal spatial information appears to be limited by their very large receptive fields. In awake, unrestrained cats, many NGC neurons respond after electrical stimulation of cutaneous nerves with high-frequency discharges as stimulus intensity is raised to levels eliciting escape behavior (46, 48). Other NGC neuronal responses are more closely associated with the imminence and execution of escape than with the intensity of noxious somatic stimulation (46, 48). Thus, in general, it is not clear whether NGC neuron responses are more closely related to the stimulus or behavioral response parameters of pain.

Casey (46, 48) has shown that electrical microstimulation elicits escape responses in cats at NGC sites where nociceptive units are recorded. This result could be interpreted as supporting NGC participation in pain sensory mechanisms, but there are problems with this interpretation. In this study, Casey found that stimulation of sites near the trigeminal nucleus elicited behavioral orientation toward facial areas as well as aversive responses (46). In contrast, when the same electrode was moved closer to the NGC, he noted that microstimulation elicited general arousal and escape but no indication of referral of sensation to a particular body location. Therefore, it is certainly possible that NGC stimulation elicits a general aversive feeling or reaction not associated with the usual sensory referents of pain. Consistent with results of NGC stimulation, lesions of NGC in cats result in long-lasting increases in escape latencies to presumably painful grid shock (46). However, neither the results of lesions nor stimulation have distinguished between the NGC's potential involvement in sensory–discriminative, arousal, aversive, or motoric dimensions of pain.

Studies of the anatomical connections of the NGC also do not allow discrimination between these four dimensions. Although the NGC receives input from anterolateral quadrant axons, anatomical studies in monkeys indicate that few of these originate from the contralateral dorsal horn (6). Ipsilateral spinal cord projections to the NGC are indicated by both anatomical results in monkeys and electrophysiological results in cats (6, 48). The origin of these projections in the cat appears to be deep within the spinal cord gray matter (laminae VII). Efferent projections of NGC have been traced to intralaminar thalamic nuclei, to cranial nerve motor nuclei, and to the spinal cord ventral horn (6). Thus the anatomical connections of NGC neurons, especially their efferent projections, do not clearly indicate a sensory–discriminative function in pain (5).

Deep Layers of the Superior Colliculus

The superior colliculus, particularly the deeper laminae, is likely to have a role in pain. Neurons of this region receive ascending afferent connections from the anterolateral quadrant and the lateral cervical nucleus (50). The neurons of deep layers of the hamster superior colliculus have been physiologically characterized as wide-dynamic-range (WDR) neurons or nociceptive-specific (NS) neurons (50). Their responses to peripheral somatosensory stimuli and their receptive field organizations were found to be strikingly similar to WDR and NS neurons of the dorsal horn.

Larson and colleagues (50) suggest the possibility that neurons of deep layers of the superior colliculus are involved in alerting and orienting an animal toward the source of nociceptive stimulation. This idea is supported by a wealth of data implicating the deep layers of the superior colliculus in orienting an animal toward the source of peripheral stimulation. Clearly, there is significant survival value in a neural mechanism that evokes orientation of an animal to the peripheral source of nociceptive stimulation.

Other Reticular Formation Areas

Reticular formation areas implicated in pain include the lateral tegmental fields of the caudal medulla, the midbrain tegmentum, and the central gray. The evidence consists mainly of neuronal responses to stimulation of A_δ and C fibers or to noxious mechanical or thermal stimuli (5, 6). Although there are indications that some neurons in these regions are affected by input from primary nociceptive afferents, it is not always clear that they respond maximally to nociceptive stimuli. There is no evidence that they are capable of transmitting sensory–discriminative information about noxious stimuli. The most positive statement that one can make is that some midbrain and medullary lateral reticular formation neurons appear to be accessed by nociceptive input. Such input to reticular formation structures could be related to arousal, aversive, motoric, or perhaps to sensory–discriminative functions. A role of these neurons in aversive responses to noxious stimuli is supported by the finding by Vertes and Miller (51) that rat reticular formation neurons increased their activity to a conditioned stimulus paired with footshock but not to one paired with positive reinforcement.

Human Studies

Information about the neural processing of pain at the medullary level of the neuraxis derives principally from attempts to interrupt spinothalamic fibers for the relief of shoulder and neck pain. A careful review of this work has been given by White and Sweet (45). Destruction of the spinothalamic tract at the medullary level produces results qualitatively similar to those resulting from anterolateral cordotomy. Results at the medullary level appear to be quantitatively similar to those at the spinal level. As with anterolateral cordotomy, medullary tractotomy typically results in at least temporary analgesia to acute pain on the contralateral body surface (45). No evidence emerges from these lesion results that there is any separation in the lateral medulla of the sensory–discriminative from other components of the pain experience.

Simulation of medullary structures in conscious humans has been restricted to attempts to verify electrode placement within the spinothalamic tract. Both sensory–discriminative and motivational–affective components of pain result from stimulation of the tract (43, 45), thus supporting the conclusion that the nature of pain pathways does not change significantly at this level. No attempt has been made to evaluate the effects of destruction in humans of the more medial pontobulbar structures in the vicinity of the nucleus reticularis gigantocellularis (NGC). Thus, the involvement of the NGC in pain mechanisms is exclusively based on the animal data described earlier. At the mesencephalic level, important information about neural processing of pain in humans derives from neurosurgical procedures attempting to alleviate pain by severing either spinothalamic pathways or spinoreticular connections. Studies at this level have begun to reveal a divergence of pain pathways with important theoretical and practical significance.

Spinothalamic tractotomy at the mesencephalic level is a procedure that has been for the most part abandoned because of the high incidence of deleterious side effects (45), but those cases that have been reported are highly instructive. Analysis of the review presented by White and Sweet (45) reveals that this procedure can produce at least temporary relief of clinical pain, although the probability of success, even when the lesion is correctly placed, is lower than at more caudal levels (12 out of 20 cases). The recurrence of pain within a short period after surgery appears likely.

An examination of the effects of more medial midbrain destruction and stimulation, although not without overlapping effects of spinothalamic tractotomy, is more revealing in the complementary effects observed. Destruction of the midbrain periaqueductal gray matter alone had been done only for the relief of pain of central origin. The procedure seems effective for the relief of this type of pain (52, 53), but its effect on chronic pain of peripheral origin remains unknown. The suffering aspect of central pain has been reported to be affected by this procedure (52). Of particular interest is the observation that lesions restricted to the mesencephalic periaqueductal gray matter do not appear to interfere with localization or detection of acute noxious stimuli, although the sensory sequelae of this lesion have not been analyzed in detail (52, 53).

The most striking effect of electrical stimulation of periaqueductal structures is the elicitation of an emotional complex of unpleasantness and fear that often results in the subject's not allowing further stimulation (45, 52, 53). At higher stimulation intensities, reports of frankly painful sensations occur. The pain is typically diffuse, deep, and localized to midline structures, particularly the face (45). This is in contrast to midbrain spinothalamic stimulation, which produces sharp, well-localized pain referred to the contralateral body surface (45). The emotional complex of unpleasantness, fear, and sometimes intense dread that results from stimulation of periaqueductal structures is consistent with animal data that implicate these structures in fear responses (54). For example, lesions of the periaqueductal central gray matter reduce fear responses in rats (54).

In summary, the available evidence from studies in humans suggests that a divergence of neural processing of nociceptive information occurs at least as caudally as the mesencephalic level. The lateral mesencephalic structures continue to carry information more concerned with the sensory–discriminative aspects of pain, whereas the most medial structures appear to be preferentially involved with the arousal, autonomic, and somatomotor responses associated with pain. It is important to point out that this apparent separation could be misleading since there is evidence that individual ascending nociceptive neurons project to multiple brainstem sites (6). In fact, it is this overlap of function that probably leads to the unpredictable results of surgical interventions at this level as well as at more rostral levels of the neuraxis. It should also be mentioned that there is clear evidence that medial mesencephalic structures are importantly involved not only in the afferent transmission of nociceptive information but also in the centrifugal control of this system, as will be discussed. Thus, studies of this area in humans are likely to produce complex results and should be interpreted with caution.

Brain Processing: Thalamus

Animal Studies

As pointed out above, individual spinothalamic neurons can be antidromically activated from the VPL nucleus and, in primates, the WDR and NS spinothalamic neurons projecting to this nucleus originate in laminae I and IV–VII of the contralateral spinal cord (6). The caudal VPL, termed the VPL_c, is a major locus of spinothalamic axon termination. The central lateral (CL) nucleus receives many spinothalamic tract terminals, and some spinothalamic axons bifurcate and project to both the VPL_c and CL (6).

Spinothalamic axonal projections are somatotopically organized within the VPL_c, with axons from the lower extremity ending laterally and those from the upper extremity ending medially. These electrophysiological findings are consistent with anatomical tracing techniques that, in addition, show that spinothalamic tract endings occur as patchlike clusters in transverse sections or as rodlike zones in three dimensions (6). In turn, the VPL_c nucleus projects to the S-1 and S-2 areas of the somatosensory cerebral cortex (55). Projections from VPL_c to the S-1 cortical area are likely to be restricted and somatotopically organized since both areas show somatotopy. In contrast, centralis lateralis and other medial thalamic nuclei involved in pain have diffuse projections to wide regions of the cerebral cortex (6). In sum, all of these anatomical facts point to the likelihood of a thalamic nucleus, VPL_c, whose anatomical input–output relationships strongly indicate a major role in transmission of sensory–discriminative aspects of pain to the cerebral cortex.

Until recently, physiological characterization of neurons within the mammalian ventroposterior lateral nucleus has emphasized the responses of neurons to mechanoreceptive inputs (6). Sporadic reports of nociceptive neurons within the VPL combined with well-established spinothalamic terminations in this nucleus have led to persistent efforts to identify nociceptive neurons in the VPL and to determine whether the VPL contains a subregion specialized to process nociceptive input.

Kenshalo and coworkers (56) systematically characterized nociceptive neurons in the VPL using methods similar to those used to characterize spinothalamic nociceptive neurons. The nociceptive neurons identified by these methods were greatly outnumbered by cells activated optimally by innocuous mechanical stimuli. However, nociceptive neurons appeared to be concentrated in the caudal portion of the VPL_c. The nociceptive neurons recorded within the VPL_c by Kenshalo and coworkers (56) were located in somatotopically appropriate regions of the nucleus, and these thalamic neurons could be antidromically activated from somatosensory S-1 cortical regions near the junction of areas 3b and 1. The response characteristics of the nociceptive neurons were in many ways similar to those of spinothalamic tract neurons. The cells could easily be divided into WDR (multireceptive) and NS (high-threshold) categories, similar to spinothalamic tract neurons.

One difference between responses of primary afferents, spinothalamic tract neurons, and VPL_c neurons to graded long-duration nociceptive temperatures is their rates of adaptation. VPL_c neurons have a distinctly slower rate of adaptation of response to long-duration nociceptive temperatures as compared to spinothalamic tract neurons. This slower adaptation is similar to human pain perception, which also adapts very slowly to maintained thermal nociceptive stimuli over 47°C (57). The slower adaptation rate of VPL_c neurons is suggestive of a temporal transformation mechanism similar to that occurring in the dorsal horn. This mechanism may serve to extend the nociceptive signal well beyond the duration of the stimulus and act to counter the effects of adaptation at the receptor level. Such a mechanism may have survival value in providing a warning signal that does not decline over time.

Studies of Thalamic Mechanisms of Pain in Humans

The specialization of thalamic neural systems mediating different aspects of the total pain experience in humans is perhaps more clear when preceded by analysis of neurophysiological studies of thalamic nociceptive neurons. The effects of lesions of specific thalamic nuclei and of electrical stimulation of these same nuclei in awake humans then become more comprehensible.

Thus, one kind of evidence is provided by clinical studies in which lesions of the thalamus have been made deliberately in order to alter some or all aspects of pain. Lesions that are directed toward the entire ventrobasal complex have been shown to give poor results in terms of pain relief in addition to the undesirable side effect of a large mechanoreceptive sensory deficit (45). This deficit does not appear to occur when the lesion is confined to the ventrocaudal nucleus, that portion of the VPL that corresponds approximately to the VPL_c in other mammalian species. Lesions of the ventrocaudal nucleus produce a distinct though temporary reduction in perception of acute pain. Better and longer-lasting relief has been observed when ventroposterior medial nucleus lesions are made for facial pain, possibly because such lesions may also interrupt projections to medial intralaminar thalamic nuclei (45). For similar reasons, medial thalamic lesions have been combined with ventrocaudal nucleus lesions in order to produce more permanent relief of pain.

Lesions made exclusively within the intralaminar thalamic complex also may produce temporary relief of pain. However, such temporary reduction appears to result from an interference with generalized thalamically mediated diffuse cortical activation, consistent with animal anatomical and physiological studies showing widespread projections from intralaminar nuclei to the cerebral cortex (45, 58, 59).

Similarly, lesions within the anterior and medial dorsal nuclei of the thalamus may produce pain relief by a reduction in widespread cortical activation in response to nociceptive stimulation. Such nuclei project heavily to prefrontal cortex, and, similar to prefrontal lobotomies, large medial dorsal thalamic lesions appear to reduce cognitive preoccupation with pain and hence affect, but not pain sensations (45, 58).

Human studies involving electrical stimulation of discrete thalamic sites further support functional differences between medial thalamic and ventrocaudal nuclear contributions to pain mechanisms. Two human studies support the hypothesis that the ventrocaudal nucleus (corresponding to the VPL_c in other species) is involved in the localization and discrimination of sensory aspects of pain. Hassler (60) found that stimulation caudal and inferior to the classical ventroposterolateral nucleus, that is, in the ventrocaudal nucleus, resulted in localized pain in conscious humans. These findings are important for several reasons. First, stimulation here is reported as similar in sensory quality to naturally occurring pains, despite the fact that artificial patterns of regularly spaced pulses were used. Second, stimulation of these sites was reported as unpleasant, supporting the hypothesis that sensory–discriminative neural systems at this level can access neural systems that produce the affective–motivational aspects of pain experience. Third, these thalamic sites appear to be very similar to those sites in which WDR and NS neurons are found in monkeys and from which spinothalamic tract neurons can be most effectively antidromically activated. Finally, stimulation with the same current intensities in the ventroposterior nucleus dorsal to this region produces tingling sensations or other innocuous sensations but not pain. These findings are in agreement with animal studies that demonstrate a highly specialized somatotopically organized nociceptive system within this region. The human stimulation studies further implicate neurons of this region in sensory–discriminative aspects of pain. It should be understood, however, that they do not implicate any specific class of ventrocaudal neurons in pain and that central nervous system electrical stimulation may activate axons "en passage" as well as cells located in the immediate vicinity.

In contradistinction to these specifically localized pain responses, stimulation of Cm-Pf nuclear sites in awake humans has been reported to produce diffuse sensations of discomfort but not discretely localized pain (45, 59). One study found, however, that stimulation of Cm-Pf and nucleus limitans evoked burning pain referred to large portions of the contralateral and even ipsilateral body surface (61). These results came from the same investigators who recorded nociceptive neural responses in these same nuclei.

Brain Processing: Cerebral Cortex

Animal Studies

As indicated above, thalamic nuclei receiving spinothalamic tract input project to cortical areas. The anatomical patterns of projections from VPL to somatosensory cortex are shown in Figure 6.6. Neurons of the VPL_c project to S-1 and S-2 somatosensory cortical areas (55), neurons of the central lateral nucleus project to wide areas of the cerebral cortex, including S-1 and S-2 (6), and the nucleus submedius projects to the orbitofrontal cortex (6). These thalamocortical projections appear to maintain the same principles of input–output organization that occur for different spinothalamic projections. Thus, the VPL_c, which itself is highly somatotopically organized, projects in a restricted fashion to somatosensory S-1 and S-2 cortical zones that are, in turn, highly somatotopically organized. By contrast, intralaminar medial thalamic nuclei that have highly convergent input and less somatotopic organization, in turn, have diffuse projections to wide regions of the cerebral cortex as well as to S-1 and S-2 cortical areas. Recently, several neurophysiological investigations have demonstrated the presence of different types of nociceptive neurons in the cerebral cortex that are consistent with the types of anatomical connections just described.

Kenshalo and Isensee (62) characterized different types of nociceptive neurons in monkey S-1 cortex, using methods similar to that used to characterize VPL_c and spinothalamic tract neurons. As might be expected at this point, nociceptive neurons could be characterized as WDR (multireceptive) cells or as NS (high-threshold) cells. Among 68 classified neurons, 37 were WDR and 31 were NS. However, unlike neurons of the VPL_c, S-1 cortical nociceptive neurons consisted of those that had restricted contralateral receptive fields (34 of 68) and those that had very large receptive fields that included most or even all of the body surface. Neurons with these two types of receptive fields were located side by side near the boundary between areas 3b and 1. Interestingly, this cortical region was found by Kenshalo and coworkers (56) to be the site from which most nociceptive VPL neurons could be most effectively antidromically activated.

The receptive field organization and other physiological response characteristics of S-1 neurons with restricted receptive fields were in many respects similar to those of VPL_c neurons and spinothalamic tract neurons. The WDR neurons showed increasing responses to gentle tactile stimuli, firm pressure, and pinching of the skin. They responded maximally to nociceptive stimuli and to both thermal and mechanical noxious stimulation. Their increased responses to a second series of nociceptive heat stimuli reflected the sensitization that occurs for primary mechanothermal nociceptive afferents as well as the hyperalgesia that can be measured psychophysically. NS cells responded only to intense pressure and nociceptive mechanical and thermal stimuli. The sizes of the receptive fields of the NS neurons could be as small as a fraction of a single digit. S-1 neurons with restricted contralateral receptive fields are likely to subserve the capacity to recognize the location and quality of a nociceptive stimulus.

The receptive fields' organization and other physiological response characteristics of S-1 neurons with large receptive fields had a striking resemblance to some spinothalamic tract neurons that project to the medial thalamus (6, 62). For this reason, Kenshalo and Isensee (62) proposed that cortical nociceptive neurons with large receptive fields may in part receive information from the projection of the spinothalamic tract via medial thalamic nuclei (e.g., central lateral nucleus). The latter are known to project diffusely to the somatosensory cortex. It also is possible that these types of cortical nociceptive neurons receive excitatory effects from other thalamic nuclei such as PO_m. The large-field cortical nociceptive neurons appear to represent the end stage of an ascending afferent system involved in widespread cortical activation.

The detailed analysis of responses of both restricted receptive field and large receptive field cortical nociceptive neurons to graded nociceptive temperatures sheds further light on the functional roles of these two neuronal populations. The nociceptive temperature–impulse response relationship of both types of neurons were found to be positively accelerating functions (Fig. 6.1). Thus, similar to primary afferent mechanothermal nociceptive afferents, spinothalamic tract neurons, and possibly VPL_c neurons, cortical nociceptive neuronal responses to graded nociceptive temperatures are consistent with nociceptive temperature–pain intensity rating responses obtained in several human psychophysical studies (Fig. 6.1). Therefore, both large-field and restricted-field nociceptive cortical neurons possess the capacity to provide precise information about nociceptive sensation intensity.

A major response characteristic differentiating restricted-field from wide-field cortical nociceptive neurons is their rate of adaptation to sustained nociceptive temperature stimuli. Those with restricted receptive fields, in general, exhibited almost no adaptation to a 30-sec temperature stimulus of intensity of 47 or greater. Those with large receptive fields adapted much more quickly. The former more closely parallel human pain perception.

During similar prolonged thermal stimulation, human subjects report almost no adaptation to continued stimulus intensities above 45°C (57). Thus, cortical nociceptive neurons with restricted receptive fields are most likely responsible for sustained nonadapting pain that results from long-duration nociceptive stimuli.

The slow-adapting responses of cortical nociceptive neurons with restricted receptive fields reflects, in part, temporal transformation mechanisms between input from thalamic nociceptive neurons and output from these cortical neurons. The rate of adaptation of cortical nociceptive neurons is slower than that of VPL_c nociceptive neurons (62). As pointed out earlier, VPL_c nociceptive neurons adapt more slowly than do spinothalamic neuron responses to sustained nociceptive temperatures. The latter, in turn, adapt more slowly than do primary nociceptive afferents. Thus, at each synaptic stage of central processing of nociceptive information exist mechanisms for prolonging or sustaining nociceptive responses. The slow rate of adaptation at cortical levels most closely parallels human pain perception.

Human Studies

Role of the Cerebral Cortex in Sensory Discriminative Aspects of Pain

Evidence for involvement of the human cerebral cortical areas in pain is limited by the fact that there have been few systematic studies of this problem, especially with regard to sensory–discriminative aspects of pain. However, two types of clinical and/or experimental observations implicate postcentral somatosensory cortex in pain perception. First, lesions of the postcentral gyrus produce a temporary reduction in ongoing clinical pain intensity (45, 63). The second type of observation implicating the postcentral gyrus in pain is that electrical stimulation of sites within this region have been shown to elicit pain. Although a systematic search was not undertaken to find cortical sites involved in pain, Penfield and Boldrey (64) found 11 out of 426 stimulation sites wherein electrical stimulation of the exposed postcentral gyrus in awake patients produced painful sensations. These reports were such rare occurrences that the investigators concluded that appreciation of pain was not represented in the cerebral cortex. However, the fact that stimulation of at least some postcentral gyrus sites produced pain is evidence in favor of a role of the postcentral gyrus in sensory–discriminative aspects of pain. As with electrical stimulation of other central nervous system structures, such as the anterolateral quadrant, evoking specific pain sensations by cortical stimulation may require higher current intensities to evoke sufficient spatial recruitment of neurons as compared to intensities required to evoke tactile sensation. The paucity of specific pain-related cortical sites therefore may partly reflect the technical limitations of Penfield and Boldrey's approach. Other attempts to evoke pain by stimulation of the postcentral gyrus have shown that intense and unpleasant pains can be consistently elicited by stimulation of these sites (65). The problem with this study is that it was carried out on patients suffering from phantom limb pain, a condition that may have potentiated pain or lowered the cortical threshold for pain.

Thus, the effects of lesions and electrical stimulation of the human postcentral gyrus, though producing somewhat equivocal results, tend to be consistent with animal neuroanatomical and neurophysiological evidence for a role of the primary somatosensory cortical area in sensory aspects of pain.

Role of Cerebral Cortex in Arousal, Affective, and Cognitive Aspects of Pain

The involvement of the human cerebral cortex in arousal, cognitive evaluative processes, and hence affective reactions associated with pain is unquestionable. It has long been known that pain is a potent means of producing arousal and widespread cortical activation. This effect has been observed as widespread electroencephalographic changes and increases in brain metabolism over wide areas of the cerebral cortex, especially the frontal cortex (6). Similarly, synchronous and repetitive nociceptive stimuli, produced by tooth pulp stimulation or laser beam heat stimulation, result in cortically evoked potentials that can be recorded over human somatosensory cortex but are maximal at the vertex (66, 67). Such a result indicates widespread cortical involvement in these responses. Generalized cortical activation by nociceptive stimulation may be mediated by mechanisms of the ascending reticular activating system (ARAS) or by medial thalamocortical pathways described earlier, or by both of these systems. Potent widespread cortical activation by noxious stimulation fits generally with the well-known observation that pain tends to dominate consciousness and has a prepotent influence over other ongoing sensory inputs.

The involvement of the prefrontal cortical lobes in complex aspects of cognitive-evaluative, arousal, and affective dimensions of pain is supported by detailed observations carried out on patients before and after prefrontal lobotomy (57). Hardy and colleagues (57) found that prefrontal lobotomy produced no overall change in heat-induced experimental pain thresholds in eight patients tested. Nevertheless, there was some reduction in the perceived intensity of clinical pain in four of five patients studied. By far the most striking changes were in patients' attitudes, emotional reactions, and cognitive processing of pain, and these changes have been corroborated by others (45). The lobotomized patients were emotionally indifferent to low-intensity pains, which, though perceived, evoked few affective reactions. This attitude was epitomized by such statements as "Yes, I feel the pain but it doesn't bother me." Moderate to high-intensity pains sometimes evoked overreactions manifested by a show of grimacing, fears, and agitation when direct questions forced them to focus on the pain. When left alone, however, spontaneous suffering or thoughts about pain were nearly absent. They showed little spontaneous concern about the negative implications of pain as regards damage to the body or threat to life. Evidently, lobotomy somehow interferes with the spontaneous ongoing cognitive evaluations that normally exert a somewhat prepotent influence during continuous moderate to severe pain. It is quite possible that lobotomy selectively reduces pain-related affect, a stage that is partly based on reflective processes related to the implications of pain.

Much of the human cerebral cortex appears to be involved in the different dimensions of pain and the cortical representations of these dimensions are still poorly understood. The problem, as pointed out by Willis (6), is that it is far more likely that the cerebral cortex has multiple representations of pain than that it has none. These multiple representations are likely to be at least partly related to the multiple dimensions of pain, including sensory–discrimination, arousal, cognitive evaluation, affect, and organized motor responses that often occur in response to pain.

Imaging Studies of Pain

Within recent years, neural imaging studies have confirmed increases in neural activity within all of the brain regions discussed above during both chronic and acute pain conditions. Regional increases in brain neural activity were examined in rats with painful peripheral mononeuropathy (produced by chronic constrictive injury of the sciatic nerve) by using the ^{14}C-2-deoxyglucose (2-DG) autoradiographic technique to measure local glucose utilization rate and hence neural activity (68). Although, as expected, brain regions contralateral to the injured nerve generally showed larger increases in neural activity, bilateral increases occurred in cortical somatosensory areas (S-1 and S-2), the cingulate cortex (within frontal lobes), amygdala, hypothalamic nuclei (VPM and arcuate nucleus), ventral posterior lateral nucleus, posterior thalamic nucleus, central gray matter, deep layers of the superior colliculus, pontine reticular formation, medullary gigantocellular nucleus, and paragigantocellular nucleus. These sciatic nerve injury–induced increases in neural activity within extensive brain regions previously implicated in sensory and affective–motivational dimensions of pain as well as in central modulation of pain are likely to reflect spontaneous pain. Similarly, neural imaging studies using positron emission tomography (PET) scanning in awake human subjects have shown increases in neural activity in S-1 and S-2 somatosensory cortex and cingulate cortex during an experimentally induced acute pain condition (69). Neural imaging studies such as these offer a unique advantage in understanding neural mechanisms of pain because they have the potential for characterizing simultaneous changes in neural activity within extensive regions of the brain that are produced by a single stimulus or chronic pain condition.

Summary and Conclusions

Among the many diverse studies of brain processing of nociceptive information, a number of unifying principles are beginning to emerge. First, the lines of evidence that have successfully implicated specific types of primary afferent and dorsal horn neurons in various aspects of pain can now begin to be applied to specific types of brainstem and cortical neurons. Second, many aspects of nociceptive stimulus–neural response relationships that occur at primary afferent and spinal cord levels appear to be preserved at midbrain, thalamic, and cortical levels. Finally, just as temporal and spatial transformations take place between the input from primary afferents and the output of spinal cord nociceptive neurons, so also are there similar transformations that take place at thalamic and cortical levels. One would expect the response patterns at these latter levels to even more closely parallel human pain perception than responses at earlier stages of nociceptive processing.

Pathophysiology of Chronic Pain

Although acute pain situations are not an insignificant clinical problem, the most difficult challenge for the health care professional is the problem of chronic pain. A comprehensive discussion of this topic is not possible here. Rather, we will focus on some significant recent advances that should have important implications for the treatment of chronic pain.

Chronic pain has been conceptualized as pain that persists past the normal time of healing (70). In practice, this will vary depending on the nature of the original injury or disease state. Thus, if the pain from a simple wrist fracture is still present 3 or 4 weeks after the injury, it is likely that pathophysiological processes have developed to maintain the pain even after healing of the damaged tissue. These processes include both peripheral sensitization or impulse-generating mechanisms and central neural plastic changes. The following discussion will first include a consideration of the varieties and different general mechanisms of chronic pain followed by considerations of specific peripheral and central neural mechanisms that are likely to subserve chronic pain.

The varieties and general mechanisms of chronic pain

Chronic pain can be caused by chronic pathologic processes in viscera or somatic structures, or by prolonged dysfunction of parts of the peripheral or central nervous system, or both. Moreover, it can be modulated by environmental or psychological factors (70). Although the neural and psychological mechanisms of chronic pain are more complex than those of acute pain, there is no reason to think that fundamentally different neural pathways, central nervous system regions, or even different types of central nociceptive neurons are involved in the two general types of pain. Just as there are not different chronic and acute primary nociceptive afferents, there also are not distinct spinal afferent or brain pathways differentially involved in acute and chronic pain. Rather, the distinctions between the two types of pain can be based on the presence or absence of neuroplastic changes that occur in neurons and neural pathways that subserve both types of pain.

From a therapeutic perspective, Bonica (70) has classified the suggested mechanisms of chronic pain into "peripheral," "peripheral–central," "central," and "psychologic." Peripheral mechanisms are said to be responsible for chronic pain associated with chronic musculoskeletal, visceral, and vascular disorders, such as myofascial syndromes, arthritis, chronic tendinitis, and peripheral vascular disease. This general category is thought to be due to persistent noxious stimulation of nociceptors or their sensitization or both, and some clinicians refer to such chronic pain syndromes as "nociceptive pain."

Peripheral–central mechanisms are likely to be operative in chronic pain syndromes wherein sustained input from primary nociceptive afferents produces significant long-term increases in the excitability of central neurons and/or decreases in central inhibitory mechanisms or both. These long-term changes may also be present to a lesser extent even in persistent inflammatory conditions, such as arthritis and chronic tendinitis. Peripheral–central mechanisms are distinguished by their greater resistance to conventional antinociceptive therapies and by the spread of painful symptoms to areas not directly innervated by nerves that contain the sustaining nociceptive afferent activity. For example, patients with reflex sympathetic dystrophy (RSD) or causalgia often have pains that extend well beyond the territory involved in the original injury or supplied by the nerve that is known to be injured. Thus, although tonic input from damaged nerves or other tissues may sustain the chronic pain condition, pathophysiological central mechanisms extend the area, duration, and intensity of the painful symptoms beyond that which would ordinarily occur as a result of peripheral nociceptive input.

Under conditions where permanent central changes have resulted from very long periods of tonic input from damaged nerves or with direct damage to central neural tissue itself, chronic pain conditions may exist that are not sustained by tonic input from peripheral nociceptive afferents. These conditions are generally referred to as "central pain," characterized by spontaneous burning or aching pain, hyperalgesia, dysesthesia, hyperpathia, and other sensory abnormalities. An example of a central pain condition is "thalamic pain," which often occurs after lesions of the lateral thalamus. Central pain also often accompanies accidental injury to the spinal cord, as occurs in paraplegics. In some patients with a very long history of RSD or causalgia, it is sometimes strongly suspected that the pain is no longer sustained by tonic input from peripheral nociceptive afferents (70–72).

There are rather concrete examples of how these three general categories can be identified in clinical settings. In particular, nerve blocks and sympathetic blocks can often be used to help discern which of the three general types of chronic pain is present. A simple case of nociceptive **peripheral pain** could be established if *(a)* the perceived area of a given pain is confined to within a territory supplied by a single nerve, for example the ulnar nerve; *(b)* the pain could be completely blocked by local anesthesia of that nerve; and *(c)* the clinical pain could be evoked or exaggerated by stimuli confined to the ulnar nerve territory. Such pain would not likely be peripheral–central pain, since the patient's ongoing pain generally does not extend beyond the territory of the nerve in question. Unfortunately, since many types of peripheral nociceptive pains may involve multiple nerves and since peripheral–central pain conditions are associated with the spread of painful symptoms well beyond the territory supplied by any given nerve, only multiple blocks of several different nerves at different times can clearly distinguish peripheral from peripheral–central mechanisms in many clinical instances.

On the other hand, the distinction between peripheral–central and **central** mechanisms can be made when one knows the original source of injury and/or knows the peripheral nerve(s) that are likely to be functioning in a pathophysiological manner. For example, Gracely and colleagues (73) have shown that focal local anesthetic injection of critical sites related to the original injury eliminates the allodynia in areas that are quite remote from these critical sites (i.e., within the territories of other nerves). These cases are likely to represent instances of peripheral–central pain.

Finally, there are patients in whom complete anesthetic block of the relevant peripheral nerves, even to the point of motor paralysis, does not reduce the patient's ongoing clinical pain. Although it is always possible that the relevant nerve is not blocked, these cases may well represent instances wherein the pathophysiological pain mechanisms have become entirely "centralized" and are not at all maintained by ongoing impulse activity in primary afferent neurons, as suggested by others (70, 74). However, central pains also can occur as a consequence of direct damage to the spinal cord or regions of the brain and may be partially identified by the observations that blocks of peripheral nerves and peripheral analgesics are completely ineffective in reducing the pain.

The recognition of three general categories of chronic pain and the capacity to identify them in individual patients carries some relatively straightforward therapeutic implications. As will be discussed in detail below, treatments can be directed toward the peripheral and central pathophysiological mechanisms involved in chronic pain. As pointed out below, a strategy of combining peripheral and central therapeutic treatments may well provide the optimal approach in many instances of chronic pain, particularly neuropathic pain.

Pathophysiological Pain Mechanisms

The possibilities for the use of both peripheral and central treatments for chronic pain are dependent on the extent of our present knowledge concerning pathophysiological mechanisms. It has been known for a long time that tissue injury results in spontaneous pain, hyperalgesia (increased pain sensitivity), and allodynia (pain evoked by a normally nonpainful stimulus), and it has been known for over 20 years that injury-induced sensitization of primary afferent nociceptive neurons accompanies and is the peripheral cause of these phenomena (74, 75). More recently, it has become evident that tissue injury is also normally followed by increased responsiveness of nociceptive neurons in the spinal cord (75) as well as in other regions of the central nervous system (72). The following discussion reviews even more recent evidence that at least some types of pathophysiological pains represent exaggerated or abnormally triggered expressions of the same central neural mechanisms that are evoked by tissue injury.

The role of tonic impulse discharge in afferents of peripheral nerves in persistent pain

Both ordinary tissue injury and damaged nerves can become a source of ongoing input from primary nociceptive afferents. Damaged tissue, including peripheral nerves, sometimes results in an abnormal sensitization of primary nociceptive afferents manifested as lowered thresholds, exaggerated responses to suprathreshold nociceptive stimuli, and spontaneous impulse discharge. However, this sensitization is sometimes abnormal in the sense that it lacks its usual association with injured non-neural tissues, such as skin, muscle, or joints. For example, primary nociceptive afferents of injured nerves can acquire an abnormal sensitivity to norepinephrine and hence to activity in sympathetic efferent neurons. It has been shown that, subsequent to a partial nerve injury, some cutaneous C nociceptive afferents that survive the injury acquire noradrenergic sensitivity and are subsequently more easily sensitized by tissue injury (76). A more recent report indicates that selective damage to sympathetic efferents somehow evokes the acquisition of sensitivity to norepinephrine in C nociceptive afferents (77). Regenerating sprouts of damaged nerves also acquire adrenergic sensitivity. Recent demonstrations of these specific mechanisms in animal models are consistent with the sympathetically maintained pains of reflex sympathetic dystrophy, causalgia, and possibly postherpetic neuralgia.

Besides the development of adrenergic sensitivity, other factors can contribute to sensitization or continuous impulse discharges in primary afferent axons of damaged nerves. Transected nerves form sprouts that end in neuromas that develop exquisite sensitivity to mechanical stimuli. Ectopic discharges often arise from regenerating sprouts and neuromas and from dorsal root ganglion cells related to the damaged nerves (74). Other possibilities include ephapses, wherein abnormal electrical connections and transfer of impulses occur between adjacent axons within damaged nerves, and extra "reflected" impulses in axons that are focally demyelinated (74). Clearly, there are multiple mechanisms possible whereby damaged nerves lead to tonic input over nociceptive as well as non-nociceptive afferents. Some of these are dependent on sympathetic efferent activity and some are not, relating to particular instances of sympathetically maintained and sympathetically independent pain, respectively.

Recent studies utilizing animal models of neuropathic pain as well as psychophysical studies of patients with neuropathic pain provide evidence for the crucial role of tonic input from nociceptive afferents in maintaining the persistent pain and sensory abnormalities present in such diseases. This tonic input occurs in an animal model of neuropathic pain produced by loosely constricting the common sciatic nerve of the rat (74, 78). A progression of anatomical and functional changes occurs in the nerve over the days following nerve ligation. First, the large A_β afferents cease to conduct impulses past the ligated region of the nerve and, second, spontaneous impulse discharges develop first in slowly conducting A_δ afferents and then in C polymodal nociceptive afferents (74). Many of these spontaneously active axons are likely to be nociceptive. That such spontaneous activity dynamically maintains the ongoing symptoms of hyperalgesia and spontaneous pain-related behaviors in these rats can be appreciated by the fact that local anesthetic block of the sciatic nerve reverses these symptoms for 24 hours after the block and hence even after the duration of the action of the local anesthetic itself (72). A similar interruption of pain symptoms in human neuropathic pain patients occurs with anesthetic block of the nerve related to the painful region and in some patients with anesthetic block of sympathetic efferents innervating the affected region (71, 79, 80).

Central pathophysiological pain mechanisms maintained by tonic primary nociceptive afferent input

Because there are multiple ways that damaged peripheral nerves can become a source of tonic input from nociceptive as well as other types of primary afferents and because similar tonic input occurs during inflammatory pain states, one might expect some general similarities between the symptoms of inflammatory pain and neuropathic pain. Indeed, neuropathic pain may reflect dysfunctional expressions of the same processes that occur during persistent inflammatory pain. This idea is supported by the general observation that both inflammatory pain and neuropathic pain are characterized by the presence of hyperalgesia, allodynia, and ongoing "spontaneous" pain. Moreover, recent psychophysical studies of neuropathic pain patients support this general observation and extend it by demonstrating the diversity of detailed sensory abnormalities of neuropathic pain patients (71, 73, 79). Some of these sensory characteristics extend beyond those usually found in persistent inflammatory pain.

Studies of neuropathic pain patients and studies using animal models of neuropathic pain have both shown that these neuropathic conditions are characterized by zones of skin in which heat hyperalgesia is present in some patients and larger zones in which mechanical hyperalgesia and/or allodynia is present in all or most patients. Similarly, the chronic constrictive nerve injury rat model of Bennett and Xie (78) shows heat hyperalgesia on the foot related to the injured nerve and mechanical hyperalgesia on both hind paws. Mechanical allodynia also occurs in skin territories outside those innervated by the injured sciatic nerve in this same model (74). Zones of hyperalgesia and allodynia that extend well beyond the cutaneous territory innervated by the injured nerve suggest that altered central processing is dynamically maintained by ongoing nociceptor input. Further evidence for this comes from experiments on patients who have one or more foci of unusually high sensitivity and areas of allodynic and hyperalgesic skin that are spatially remote from these small foci (73). Local anesthesia of these small foci was found to eliminate the patient's ongoing pain and eliminate the allodynia and hyperalgesia in areas of skin spatially remote from the local injections of anesthetics.

The sensory symptoms that are maintained by tonic nociceptor input are diverse among neuropathic pain patients, and, at the same time, their detailed analysis reveals important central mechanisms of these persistent pain states. Studies have characterized three sensory abnormalities in patients with a diagnosis of reflex sympathetic dystrophy (RSD), the extent to which the three abnormalities are associated with each other, and the extent to which these three abnormalities are associated with the intensity of spontaneous pain (71, 79). The three sensory abnormalities included heat-induced hyperalgesia, low-threshold A_β-mediated or high-threshold mechanical allodynia, and slow temporal summation of mechanical allodynia. These three sensory abnormalities occurred to a widely varying extent among the 31 patients tested, although all patients perceived normally innocuous mechanical stimuli as painful. For some of these patients, slow temporal summation of burning pain occurred when gentle mechanical stimuli or electrical stimulation of A_β afferents were applied at rates of once per 3 sec. For other patients, slow temporal summation occurred only with more intense but normally nonpainful mechanical stimuli. Still other patients did not exhibit slow temporal summation with repetitive stimuli. Both mechanical allodynia and slow temporal summation of allodynia were completely or nearly completely reversed by an anesthetic blockade of sympathetic ganglia, indicating that these sensory abnormalities were dynamically maintained by sympathetic efferent activity, presumably activity that induces continuous input over nociceptive afferents. Slow temporal summation of mechanical allodynia, particularly that induced by stimulation of A_β afferents, is abnormal since such types of stimuli do not evoke pain in pain-free subjects nor in these same pain patients when such stimuli are delivered to homologous contralateral pain-free zones. Abnormal slow temporal summation of mechanical allodynia may represent an exaggeration and/or abnormal triggering of physiological mechanisms that already exist in normal pain-free individuals. Such mechanisms can be demonstrated in the latter by the temporal summation of experimentally induced second pain as described earlier. Thus, under some pathological conditions after nerve injury, A_β input must somehow gain access to and either trigger or maintain the same NMDA receptor–slow temporal summation mechanism that is normally activated by C afferent stimulation. In other pathological conditions, sensitized nociceptors themselves are likely to be the direct proximal cause of the slow temporal summation of mechanical allodynia.

Mechanical allodynia and slow temporal summation of allodynia may well be integrally related to the patient ongoing "spontaneous" pain. This relationship could occur if continuous input from A_β low-threshold afferents (evoked in the normal course of mechanical stimulation from walking, sitting, or even contact with clothes) activated slow temporal summation of a type of burning, aching, or throbbing pain that built up slowly and dissipated slowly over time. This possibility was explicitly tested by comparing intensities of ongoing pain between patients who demonstrated slow temporal summation versus those who did not (71). The former had significantly higher intensities of ongoing pain than the latter. Therefore, exaggerated or abnormally triggered mechanisms of slow temporal summation are likely to form at least part of the basis of the persistent pain that sometimes follows nerve injury.

Specific neuronal and intracellular mechanisms of neuropathic pain

Two general spinal cord dorsal horn mechanisms that are not mutually exclusive may help explain the pathological pain states just described. The essential abnormality of the first mechanism lies in the inhibitory circuitry that controls the responses of WDR transmission neurons to A_β input, as proposed previously (71). The dorsal horn circuitry that generates A_β low-threshold-evoked inhibition (pre- or postsynaptic or both) is deficient or absent. This absence is evident in neuropathic pain patients in whom high-frequency low-intensity transcutaneous electrical nerve stimulation (HF-TENS) within the pathological zone evokes not the usual reduction of pain but pain itself. The absence of A_β-evoked inhibition in the spinal dorsal horn of rats who are likely to have neuropathic pain has been verified in several ways, including the demonstration of a loss of A_β-evoked presynaptic inhibition (74, 81) as well as the demonstration of morphological changes in small neurons in the substantia gelatinosa (81). The loss of inhibitory mechanisms could occur as a result of tonic input from nociceptive afferents and the excitotoxic release of glutamate which, in turn, results in the dysfunction of small inhibitory interneurons. The loss of inhibition then could result in an exaggeration of the excitatory effects of A_β afferents and hence A_β allodynia.

The essential abnormality of the second mechanism involves development of ongoing impulse discharge in nociceptive primary afferents, particularly C polymodal afferents, and an exaggeration of their central effects as a result of prolonged increases in the excitability of neurons in the spinal cord. If both of these general central mechanisms are not mutually exclusive, then either or both may coexist in the same patient and could partly explain the heterogeneity of pain symptoms in neuropathic pain patients. There is even the possibility that both mechanisms contribute to the same symptom. For example, loss of A_β-mediated inhibitory mechanisms and excitotoxic-induced sensitization of WDR neurons by tonic input from nociceptors could work in concert to produce A_β allodynia.

Electrophysiological recordings of spinothalamic neurons in animal models of neuropathic pain show important parallels to the sensory abnormalities described above for neuropathic pain patients. Recent studies by Palecek and coworkers (82) show that spinothalamic tract neurons of both rats and monkeys, particularly WDR neurons, increase their spontaneous activity and become hyperresponsive to innocuous brushing, noxious heating, and cooling of the skin as a result of an experimentally produced peripheral neuropathy. These changes are consistent with A_β allodynia, thermal hyperalgesia, and thermal allodynia in the case of cooling, all of which are observed to varying degrees in neuropathic pain patients. Consistent with these electrophysiological studies of single neurons, metabolic mapping of elevated neural activity in the rat chronic constrictive injury model (83) shows that the largest increase in activity occurs in spinal cord laminae V–VI, the region of highest concentration of WDR neurons. The spatial distribution of this elevated activity extended over considerable rostral–caudal (L_1–L_5), dorsal–ventral (laminae I–VII), and medial–lateral distances, consistent both with the idea that pathophysiological pain involves spatial recruitment mechanisms that extend even beyond that which occurs in normal pain and with the extensive spatial radiation of pain sensation that occurs in neuropathic pain patients (70, 84).

The idea that NMDA excitotoxic mechanisms are at least part of the basis for the slow temporal summation and central sensitization that occurs in neuropathic pain has led to behavioral and pharmacological studies of pain-related behaviors in rats with chronic constrictive injury of the sciatic nerve (72, 81). These studies indicate that thermal hyperalgesia and spontaneous pain behaviors are attenuated by pre- and postinjury spinal cord (intrathecal) treatment with NMDA and/or non-NMDA glutamate antagonists (72, 75). These results would be expected if the slow temporal summation mechanism described above is at least part of the basis of ongoing pain, hyperalgesia, and allodynia, and if such a mechanism is mediated by NMDA receptor activation.

Chronic activation of NMDA receptors, in turn, has been associated with intracellular processes characterized primarily by increased intracellular Ca^{++} concentrations (72), which induce long-term increases in the responsiveness of neurons. To take just one example, increased intracellular Ca^{++} activates protein kinase C (PKC) (72). Activation of PKC within neuron membranes results in phosphorylation of the proteins of the NMDA receptor channels and thereby increases their Ca^{++} conductance. This increased conductance is manifested functionally as an increased responsiveness of the neuron to its synaptic inputs. Since such altered neurons function in the afferent transmission of nociceptive information, their increased responsiveness is likely to contribute to hyperalgesia, allodynia, and spontaneous pain.

Central Nervous System Mechanisms of Pain Modulation

Introduction

It has long been recognized that a simple invariant relationship between stimulus intensity and the magnitude of pain perception is often not present. Two general classes of observations support the complexity of this relationship. The first is the clinical observation that pain is often present without any apparent precipitating pathology. This situation represents the clinical problem of pain treatment. More important for the topic of this section is the common observation that pain may not be experienced in the presence of factors that should produce it; that is, under a variety of circumstances, total or partial analgesia is seen. These observations were explicitly recognized in earlier models of pain perception in spite of the lack of direct evidence supporting the theoretical models (3, 85). Thus, the concept that the nervous system possesses intrinsic pain inhibitory mechanisms was recognized when only indirect evidence was available.

Endogenous opiate pain inhibitory systems

It has become clear that information about tissue damage is not passively received by the nervous system. Rather, it is filtered, even at the first synapse, by complex modulatory systems. The discovery of these systems has fostered, and has in turn been fostered by, the notion that the central nervous system contains endogenous substances, endorphins, that possess analgesic properties virtually identical to opiates of plant and synthetic origin. In this section, the development of these concepts is examined. We then discuss the existence of opiate and nonopiate central nervous system pain modulatory mechanisms activated by environmental stimuli.

The earliest work indicating that opiates produce analgesia, at least in part, by activation of endogenous pain inhibitory systems was done by Irwin and associates (86). They demonstrated that morphine was not effective in inhibiting the spinally mediated tail flick response in spinalized rats. They reasoned, based on this result, that morphine must activate supraspinal neural circuitry that has an output to the spinal cord and modulates the processing of nociceptive information at spinal level. This work was largely ignored until the early 1970s.

The first impetus for the detailed study of pain-modulatory circuitry resulted from the observation that electrical stimulation of the brain could powerfully suppress the perception of pain (87, 88). Further investigation of stimulation-produced analgesia (SPA) provided considerable detail about the neural circuitry involved, as reviewed by Mayer and Manning (89).

Significantly, at that time, several similarities were recognized between these observations and information emerging from a concomitant resurgence of interest in the mechanisms of opiate analgesia (88). The most important parallel facts revealed by these studies were the following: *(a)* effective loci for both opiate-microinjection analgesia (90) and stimulation-produced analgesia (88) lie within the periaqueductal and periventricular gray matter of the brain stem; *(b)* opiate analgesia and stimulation-produced analgesia are both mediated in part by the activation of a centrifugal control system that exits from the brain and modulates pain transmission at the level of the spinal cord (88); *(c)* and the ultimate inhibition of the transmission of nociceptive information occurs, at least in part, at the initial processing stages in the spinal cord dorsal horn and homologous trigeminal nucleus caudalis by selective inhibition of nociceptive neurons (91).

In addition to these correlative observations, studies of stimulation-produced analgesia provided direct evidence indicating that there are mechanisms extant in the central nervous system that depend upon endogenous opiates. For instance, subanalgesic doses of morphine were shown to synergize with subanalgesic levels of brain stimulation to produce behavioral analgesia (92). Tolerance, a phenomenon invariably associated with repeated administration of opiates, was observed to the analgesic effects of brain stimulation (93), and cross-tolerance between the analgesic effects of brain stimulation and opiates was demonstrated (93). Stimulation-produced analgesia (SPA) could be antagonized by naloxone, a specific narcotic antagonist (94, 95). This last observation, in particular, could be most parsimoniously explained if electrical stimulation resulted in the release of an endogenous opiatelike factor. Indeed, naloxone antagonism of stimulation-produced analgesia was a critical impetus leading to the eventual discovery of such a factor (96).

Another discovery of critical importance for our current concepts of endogenous analgesia systems coincided with work on SPA. Several laboratories, almost simultaneously, reported the existence of stereospecific binding sites for opiates in the central nervous system (97–99). These "receptor" sites were subsequently shown to be localized to neuronal synaptic regions (100) and to overlap anatomically with loci involved in the neural processing of pain (101). The existence of an opiate receptor again suggested the likelihood of an endogenous compound with opiate properties to occupy it.

In 1974, Hughes (96) and Kosterlitz reported the isolation of a factor (enkephalin) with such properties from neural tissue. An immense amount of subsequent work has characterized this and other neural and extraneural compounds with opiate properties. As with the opiate receptor, the anatomical distribution of endogenous opiate ligands shows overlap with sites involved in pain processing; Akil and colleagues (102) present a review of these studies.

Neural circuitry involved in analgesia resulting from the administration of exogenous opiates

This section will review the current data available on the sites and mechanisms involved in the modulation of pain by the administration of exogenous opiates. Primarily, two lines of experimentation will be examined: the locations in the central nervous system of sites at which administration of opiates results in analgesia and the administration of opiate antagonists block analgesia, and the locations in the nervous system where lesions block the action of exogenously administered opiates.

Following the work of Tsou and Jang (90), it was not until the early 1970s, with one exception (103), that opiate microinjection mapping studies began. Initially these studies concentrated on the periaqueductal–periventricular regions of the mesencephalon and diencephalon (104–106). This resurgence of interest in these particular sites of opiate action probably resulted from work showing that analgesia resulted from electrical stimulation of the periaqueductal gray matter (88), as well as the lead provided by the results of Tsou and Jang (90). Overall, these and other studies confirmed the importance of the periaqueductal–periventricular region in opiate analgesia and provided an impetus for the examination of other brain areas.

A second brain area that has proved to be of considerable importance for opiate action is the anatomically complex region of the ventromedial medulla. As shown in Figure 6.7, this region consists of at least three distinct nuclei: the medially located nucleus raphe magnus (NRM), the more laterally

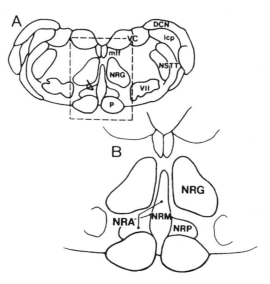

Figure 6.7 Detailed anatomy of the rat ventromedial medulla. **A,** frontal section through the rostral medulla of the rat. **B,** an enlargement of the box enclosed by the dashed line in **A.** The nucleus raphe alatus *(NRA)* is the combined cell groups of the nucleus raphe magnus *(NRM)* and nucleus reticularis paragigantocellularis *(NRP). MLF,* medial longitudinal faciculus; *NRG,* nucleus reticularis gigantocellularis; *VII,* nucleus of the facial nerve.

situated nucleus reticularis paragigantocellularis (NRP), and the dorsolaterally located nucleus reticularis gigantocellularis (NRG). Based on retrograde labeling criteria, Watkins and associates (107) have proposed the term nucleus raphe alatus (NRA) for the combined cell groups in NRM and NRP. Takagi and colleagues (108) were the first group to map this region for analgesia resulting from morphine microinjection. Overall, this group found the NRP to be approximately 20 times more sensitive to morphine than the NRG (109). They found microinjection of morphine into the NRM to be ineffective in the production of analgesia. This point is controversial since other groups (110, 111) have reported analgesia from microinjection into NRM. Azami and coworkers (110) did find, however, that the NRM was less sensitive than the NRPG.

A number of other brain areas including the amygdala (112, 113), the medial lemniscus (114), the nucleus medialis dorsalis of the thalamus (114), the mesencephalic reticular formation (105, 115), and the nucleus of the solitary tract (116) have been reported to produce analgesia when injected with opiates. However, the work on these areas is scant compared with those discussed above, and the relative potency of injections into these areas has not been explored.

A final but crucial point to be made in this section concerns the analgesic effects of microinjections of opiates directly into the intrathecal space of the spinal cord. Although Tsou and Jang (90) reported no analgesia from direct spinal application of morphine, subsequent work has consistently demonstrated the relatively potent effects of intrathecal morphine microinjection (117, 118). This observation has had important clinical application since direct intrathecal application of opiates has been shown to have analgesic effects without the concomitant psychoactive effects observed with systemic administration.

From this line of evidence it appears, then, that at least three general areas of the central nervous system are involved in opiate analgesia: the periaqueductal–periventricular gray matter, the ventromedial medulla, and the spinal cord. This observation indicates that the analgesic effects of a systemically administered opiate may produce analgesia by acting at any, all, or some combination of these distinct regions. The utilization of the microinjection of narcotic antagonists has provided at least a partial answer to this question.

A number of early studies concluded that supraspinal sites of opiate action are the effective ones since analgesia from systemically administered opiates was antagonized by either intracranioventricular or intracerebral microinjection of narcotic antagonists (119–122). Later work, however, demonstrated that naloxone administered intrathecally could antagonize the analgesia resulting from even relatively high doses of systemically administered opiates (117). Thus, these studies lead to the paradoxical conclusion that both supraspinal sites and spinal sites of opiate action are the critical ones involved in analgesia.

This seeming paradox was resolved in a series of complex but unusually important studies by Yeung and Rudy (123, 124). They demonstrated that by simultaneously administering various doses of morphine intrathecally (into the spinal cord) and intraventricularly (into the brain) a multiplicative dose–response function was observed. That is, simultaneous spinal and supraspinal morphine resulted in greater analgesia than the same total dose administered at either location alone. The effect is quite large, with the multiplicative factor being as much as 45 under certain circumstances (124).

Although the experiments just described elucidate the contribution of spinal versus supraspinal sites to opiate analgesia, the relative contribution of the various supraspinal sites at which opiates act to produce analgesia is less clear. The only work to examine this issue utilizing microinjection of narcotic antagonists was done by Azami and coworkers (110). They found, as did the work of Takagi's group, that the NRP was more sensitive to morphine than the NRM for the elicitation of analgesia. Surprisingly, however, they found that when analgesia was produced by systemically administered morphine, naloxone injection into the NRM-antagonized analgesia more effectively than injection into more lateral medullary regions, including the NRP. They concluded that the NRP does not make a significant contribution to the analgesia resulting from systemic administration of morphine. The relative contribution of medullary versus more rostral mesencephalic sites has not been examined.

An approach similar to the one just described for dissecting the neural circuitry participating in opiate analgesia utilizes the selective destruction of nuclei and pathways suspected of being involved in opiate analgesia. Opiates are administered systemically or at discrete sites in the nervous system and the effect of particular lesions are examined. Table 6.2 provides an overview of these experiments. An overview of this work supports the conclusion reached above utilizing injection of antagonists: several brain areas, including the periaqueductal gray matter, NRM, and NRP, need to be intact for the full expression of opiate analgesia.

Table 6.2
The Effect of Various CNS Lesions on Opiate Analgesia Resulting from Various Routes of Administration

Lesion site	Systemic	PAG MI	NRM MI	NRP MI
		Injection route		
DLF	< (125)	(179)		
	< (174)			
	< (110)	(183)		(110)
NRM	< (183)			
	< (181)			
	= (110)			
NRP	< (177)			
	< (178)			
NRA	% < (183)			
	% < (181)			
PAG	< (175)			
LC	< (176)			
POF	< (180)			

Symbols: "<", attenuates analgesia; "% <", partially attenuates analgesia; " = ", no effect on analgesia. Abbreviations: *DLF*, dorsolateral funiculus; *LC*, locus coeruleus; *MI*, microinjection; *NRA*, nucleus raphe alatus; *NRM*, nucleus raphe magnus; *NRP*, nucleus raphe paragigantocellularis; *PAG*, mesencephalic periaqueductal gray matter; *POF*, preoptic forebrain. The numbers in parentheses correspond to the references, given at the end of the chapter.

Environmental activation of endogenous analgesia systems

The demonstration that opiates activate well-defined neural systems capable of potently blocking pain transmission suggests, but by no means proves, that the function of this system is to dynamically modulate the perceived intensity of noxious stimuli. If, in fact, this system has such a physiological role, then one might expect that the level of activity within the system would be influenced by impinging environmental stimuli. If environmental situations could be identified that produce analgesia, it would give credibility to the idea that invasive procedures, such as brain stimulation or narcotic drugs, inhibit pain by mimicking the natural activity within these pathways.

A systematic search for environmental stimuli that activate pain inhibitory systems was begun by Hayes and associates (125, 126). They observed that potent analgesia could be produced by such diverse stimuli as brief footshock, centrifugal rotation, and injection of intraperitoneal saline. Another important, if unexpected, concept emerged from these experiments: the opiate antagonist, naloxone, did not block some forms of environmentally induced analgesia (126). Therefore, it appeared that nonopiate systems must exist in addition to the system activated by opiates described earlier.

Although the stimuli studied by Hayes and associates (125, 126) did not appear to activate an opiate system, subsequent investigations found clues that endogenous opioids might be involved in at least some types of environmentally induced analgesias. Akil and coworkers (127) studied the analgesic effects of prolonged footshock. In contrast to the results of Hayes and associates (125, 126), naloxone did partially antagonize the analgesia. The controversy over the involvement of opiates in footshock-induced analgesia was resolved, in part, by Lewis and colleagues (128) who noted that the duration of footshock used by the Hayes group (125, 126) and the Akil group (127) differed greatly and wondered whether this variable might explain the difference in their results. By comparing the effects of naloxone on analgesia produced by brief (3 min) versus prolonged (30 min) footshock, Lewis and colleagues (128) showed that only the latter could be blocked by naloxone. This suggested that different analgesia systems become active as the duration of footshock increases.

Concurrent with this work, Watkins and associates (129) made the observation that brief shock restricted to the front paws produced analgesia that was antagonized by low doses of naloxone. In contrast, even high doses of naloxone failed to reduce analgesia produced by hind-paw shock. In addition, they showed that animals made tolerant to morphine showed cross-tolerance to front-paw but not hind-paw footshock analgesia. Thus it appears that front-paw shock activates an endogenous opiate analgesia system whereas hind-paw shock activates an independent nonopiate analgesia system.

Additional work revealed the following facts about front-paw and hind-paw footshock induced analgesias (FSIA): *(a)*Front-paw FSIA is mediated by central nervous system (CNS) opioids since elimination of extraneural opiates by hypophysectomy, adrenalectomy, or sympathetic blockade does not block the effects (130). *(b)* Front-paw FSIA involves a neural circuit that ascends to the brain and then descends by way of the dorsolateral funiculus (DLF) to block pain transmission at the spinal level (131). This descending DLF pathway originates in the nucleus raphe alatus (NRA) (132). *(c)* The complete circuitry for the effect is caudal to the mesencephalon because decerebration does not affect the analgesia (25). *(d)* The critical opiate synapse for the system is situated in the spinal cord (see Fig. 6.8) at the segment of nociceptive input since intrathecal injection of naloxone in the lumbosacral but not thoracic cord blocks the effect (133). *(e)* The integrity of spinal cord serotonin is critical for the expression of front-paw FSIA (133). *(f)* Front-paw FSIA is blocked by small systemic or intrathecal doses of the peptide CCK-8 (134) and potentiated by the putative CCK antagonists, proglumide and benzotript (135).

Hind-paw FSIA is also a CNS-mediated phenomenon (130). However, this manipulation activates intraspinal as well as supraspinal pain inhibitory systems (131). The brain centers for hind-paw FSIA differ from those for front-paw FSIA; NRA lesions do not eliminate the analgesia (132). The neurochemical bases of hind-paw FSIA also differ from front paw FSIA: *(a)* CCK, serotonin, and norepinephrine do not appear to be involved in hind-paw FSIA (134, 136). *(b)* Brain, but not spinal cord, acetylcholine is necessary for the expression of hind-paw FSIA but does not appear to be involved in front-paw FSIA (137).

Of considerable interest is that both hind-paw and front-paw FSIA can be classically conditioned by repeated pairings of a conditioned stimulus with footshock (138). Regardless of whether hind-paw or front-paw shock is used as the unconditioned stimulus, an opiate analgesia system in many ways similar to the one involved in front-paw FSIA is activated because conditioned analgesia is eliminated by systemic and intrathecal naloxone, morphine tolerance, DLF lesions, and NRA lesions and is unaffected by hypophysectomy,

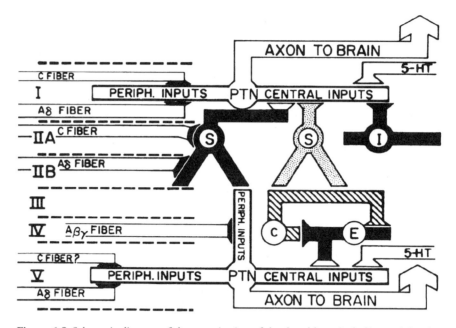

Figure 6.8 Schematic diagram of the organization of the dorsal horn including peripheral inputs, interneurons, and inputs originating from the central nervous system. Roman numerals at the left of the figure represent Rexed's laminae I–V. Filled terminals represent excitatory connections and unfilled terminals represent inhibitory connections. Filled cells indicate cells containing endogenous opioids. Note that an endorphinergic cell in laminae IV–V excites a cholecystokinin cell, which in turn inhibits the endorphinergic cell. *C*, cholecystokinin; *E*, endorphin; *I*, islet cell; *S*, stalked cell; *PTN*, pain transmission neuron.

adrenalectomy, and sympathetic blockade (130). In addition, as would be expected, higher structures are involved in the conditioned analgesia since it is eliminated by decerebration and reduced by periaqueductal gray (PAG) lesions (25).

A circuit diagram of these systems is shown in Figure 6.9. The details of the circuitry at the level of the spinal cord dorsal horn is given in Figure 6.8. A point that should be emphasized is that the involvement of other neurotransmitters or neuromodulators at the spinal cord level may be quite complex. For example, as shown in Figures 6.8 and 6.9, cholecystokinin (CCK) appears to modulate endogenous opioid systems. Intrathecal application of CCK antagonizes analgesia from application of exogenous opiates as well as analgesia elicited by activation of endogenous opiates (134). Also, CCK antagonists applied intrathecally potentiate these analgesias as well as reversing opiate tolerance (135). These findings suggest that other transmitters and/or modulators may interact with opiates to form complex circuits. Understanding of these circuits should offer important opportunities to pharmacologically manipulate clinical pain syndromes possibly without the drawbacks associated with opiate analgesia.

A number of other laboratories have now demonstrated that numerous environmental variables can be critical in determining the particular pain modulatory circuitry activated. Table 6.3 summarizes the voluminous literature on environmental events now known to influence the transmission of pain by utilizing endogenous opioid peptides. It is clear that numerous environmental manipulations result in modulation of pain transmission. From Table 6.3, it should be clear that the involvement of endogenous opioids in pain modulation is now beyond question. On the other hand, many questions remain unanswered. For example, it is generally not known where a particular endogenous opioid is released by a particular environmental manipulation nor is the endogenous ligand or the receptor type involved usually known. Nevertheless, the techniques and general strategies for answering these questions are available, and progress in this area should be forthcoming. It should also be pointed out that, in addition to systems that modulate nociceptive information utilizing endogenous opioids, there are nonopiate systems as well

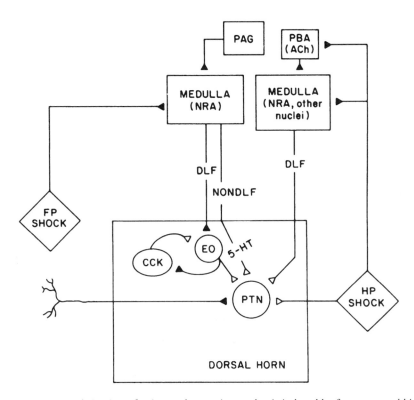

Figure 6.9 Neural circuitry of opiate and nonopiate analgesia induced by front-paw and hind-paw shock. Front-paw shock activates the nucleus raphe alatus *(NRA)* within the ventral medulla. This nucleus sends a descending projection through the *DLF* to the dorsal horn of the spinal cord. A serotonergic pathway lying outside of the DLF (non-DLF) is recruited as well. In turn, endogenous opioids are released, inhibiting pain transmission neurons *(PTN)*. Activation of endogenous opiates stimulates a negative feedback loop that utilizes CCK to reduce the activity of endogenous opioid systems. Hind-paw shock inhibits PTN via two nonopioid pathways: an intraspinal pathway and a descending DLF pathway. The latter originates from the NRA and from some other yet unidentified medullary area(s). Classically conditioned (opioid) analgesia seems to result from activation of the same DLF output pathway as front-paw (opioid) FSIA. After conditioning trials in which the conditioned stimulus is paired with either front-paw or hind-paw shock (the unconditioned stimulus), the conditioned stimulus becomes capable of activating rostral centers in the brain which, in turn, activate the periaqueductal gray *(PAG)* and subsequently the nucleus raphe alatus. This results, via a descending DLF pathway, in the release of endogenous opioids within the dorsal horn, producing analgesia.

Table 6.3
Summary of Representative Studies on Endogenous Opiate Analgesia Systems

	Systemic or IT Naloxone	Dynorphin antiserum	Enkephalin antiserum	Cross-tolerance	CNS endorphin level
Brief FPFS	(129)(184–189)			(129)	
CCA	(126, 138, 190)			(138)	(191)
ECS	(192–197)			(195–197)	
Defeat	(198–200)			(198)	(198)(201)
Acupuncture (Lo freq)	(157, 202, 203) (172, 205, 206)	(204)	(162)	(204)	(198)
Prolonged 4PFS	(188, 207, 208) (187, 197, 211, 212)				(209, 210) (213, 214)
Immobilization	(215, 216)			(215)	
15 min cold	(217)			(217)	
Exercise	(26, 163)				
Cond helpless	(201, 218, 219)			(201)	(218)
Prolonged TS	(220)				
SPA (PAG)	(94, 95, 221) (223–225) (228, 229)			(93)(201) (226)(230)	(153, 222) (95, 227) (150)
SPA (NRM)	(111, 225, 231)				
Kindled seizure	(232)				
TNS (Lo freq/ high int)	(233–235)				
Vaginal probing	(236)				

Table 6.3 (Continued)

	Systemic or IT Naloxone	Dynorphin antiserum	Enkephalin antiserum	Cross-tolerance	CNS endorphin level
DNIC	(237)				
Food deprivation	(238)				
Placebo	(169)				
Hypertension	(239)				
Anxiety (human)	(240, 241)				
Sex (male)					(242)
Pregnancy	(243)				
Spon hypertension	(244)				
Inter cold water	(245, 246)				
20°C swim	(247)			(245)	
Cold water	(246, 248–250)				
Warm water (32°C)	(250)				
Heat (40°C)	(251)				
Radiation	(126, 231)				
Hyperalgesia	(207, 252, 253)				
Mild footshock	(254)				

Note: The numbers in parentheses correspond to the references, listed at the end of the chapter. The studies used a particular opiate manipulation to implicate endogenous opiates in various forms of environmentally produced analgesia. The various opiate manipulations studied are indicated in the vertical columns and the environmental manipulations, in the horizontal rows. Abbreviations: *CCA*, classically conditioned analgesia; *Cond*, classically conditioned; *CNS*, central nervous system; *DNIC*, diffuse noxious inhibitory controls; *ECS*, electroconvulsive shock; *Freq*, frequency; *FPFS*, front-paw footshock; *Int*, intensity; *Inter*, intermittent; *IT*, intrathecal; *Lo*, low; *NRM*, region of the nucleus raphe magnus; *PAG*, periaqueductal gray matter; *SPA*, stimulation-produced analgesia; *Spon*, spontaneous; *TNS*, transcutaneous nerve stimulation; *TS*, tail shock; *4PFS*, four-paw footshock.

257

(see Fig. 6.9). In sum, a review of the animal data provides strong evidence for the existence of multiple pain modulatory systems. Work of this sort has already begun to yield information with important implications for the problem of human pain syndromes.

Evidence for endogenous opiate analgesia in humans

At this point, we would like to make some parallels between the work described above and experimental and clinical studies in humans. This will be done in order to highlight the potential relevance of this work to the very difficult problem of treating human pain syndromes. Throughout this discussion, it will be important to bear in mind that a number of distinct modulatory systems have been identified under controlled laboratory conditions. In the more naturalistic circumstances of clinical research, it is likely that more than one of these systems may be active at any given time, which may account for the variability and controversy in the clinical literature.

There are at least two situations available for study in which endogenous pain modulatory systems may be active in humans. The first involves the basal, tonic activity within these systems and allows the experimenter to assess whether pain inhibition occurs continuously, at least to some degree. The second involves clinical manipulations that attempt to activate pain inhibitory systems.

Attempts have been made to determine whether pain modulatory systems are tonically active. The assumption made by these studies has been that administration of opiate antagonists should alter the perception of pain if opiate systems are tonically active. This change in pain perception would be recorded either as a decreased pain threshold or an increased level of ongoing pain. In general, naloxone has failed to affect pain thresholds of normal human volunteers (139, 140). On the other hand, Buchsbaum and coworkers (141) found that naloxone lowered the thresholds of subjects with naturally high pain thresholds, yet had no effect in subjects with low pain thresholds. This observation is consistent with reports that the high pain thresholds seen in some cases of congenital insensitivity to pain can be lowered by naloxone (142–144). It should be pointed out that demonstrating analgesic or hyperalgesic effects of drugs in human subjects is difficult. These studies, taken together, suggest that endogenous opiate pain modulatory circuits may not always be tonically active in all people. Such a conclusion is supported by the report that naloxone decreases the higher pain thresholds seen in humans in the morning as compared to the afternoon (145).

Naloxone appears to be more consistently effective when delivered to experimental subjects who are experiencing some level of clinical pain. In this regard, these results are consistent with the animal studies described above in which pain was observed to be a powerful activator of endogenous analgesia systems. Thus, Lasagna (146), Levine and colleagues (147), and Gracely and associates (148) report that naloxone can increase the reported intensity of postoperative pain. In conclusion, it appears that, under normal circumstances, endogenous opiate pain inhibitory systems have little spontaneous activity. However, when some level of pain is present, these systems seem to be activated.

A second line of research on the involvement of endogenous opioids in pain modulation has examined a number of environmental manipulations known to have some degree of efficacy for the reduction of clinical and experimental pain. Most of these procedures were developed before the recent explosion of information about endogenous pain control systems. Indeed, many of them evolved from theoretical approaches that are now outdated or incorrect. Nevertheless, the procedures are efficacious and have inspired a considerable body of research aimed at determining the involvement of endogenous opioids in pain modulation.

This research has utilized two primary experimental strategies. The first strategy is based on the argument that if a particular environmental manipulation induces analgesia by utilizing endogenous opioids, that analgesia should be antagonized by a narcotic antagonist (usually naloxone). The second strategy depends on the argument that if endogenous opioids are involved in an environmentally elicited analgesia, then changes should be observed in the levels of these compounds in plasma or the central nervous system.

Stimulation-Produced Analgesia

Perhaps the most dramatic outcome of the basic science research on endogenous opioids has been the rapid and effective clinical application to the treatment of chronic pain syndromes in humans. As early as 1973, Richardson and Akil reported the use of periventricular gray matter stimulation to treat pain syndromes. Since then there have been over 20 reports in the literature describing various studies of this technique. In a review of this literature, Young and coworkers (149) conclude that "the method is reasonably effective in properly selected patients and, importantly, safe."

Several lines of evidence indicate a likely but not unequivocal role for endogenous opiates in stimulation-produced analgesia (SPA). Opiate antagonists are reported to reduce SPA (150, 151), tolerance develops to SPA (152), and dependence upon SPA has been reported (153). A more controversial literature exists with regard to the release of endogenous opioids, primarily β-endorphin, by electrical stimulation of the periaqueductal gray matter. Generally increased levels of endogenous opioids have been found to result from this stimulation. However, the possibility that these results are due to an artifact of using a contrast medium for electrode placement has been raised (154, 155). This criticism has been convincingly disputed for at least some circumstances (156). At this point, it seems likely that endogenous opioids mediate, at least in part, the analgesia elicited by periaqueductal gray stimulation in humans. The particular endogenous opioid and its site and mechanism of action have not been established.

Counterirritation Analgesia

The belief that an acute painful stimulus can be used to alleviate ongoing pain has been held since antiquity and is known as *counterirritation*. This procedure has a great deal in common with acupuncture and transcutaneous nerve stimulation (TNS); all use the application of somatic stimuli, either noxious or innocuous, to obtain relief from pain. Importantly, pain relief persists beyond the period of treatment in all cases. The site of treatment in relation to the painful area is highly variable, ranging from the painful dermatome itself to a theoretically unpredictable constellation of points in classical Chinese acupuncture. Lastly, the duration of treatment varies from less than a minute to hours. All of these factors are important determinants of the effects produced by footshock in animals. Thus, the highly variable effects observed in the clinic would be predicted from animal research. Nevertheless, human data suggest the involvement of the same systems described above.

Table 6.4 summarizes the studies that have examined the effect of naloxone on acupuncture analgesia as well as those studies that have measured acupuncture-induced changes in plasma or CSF β-endorphin or enkephalin levels. It is important to note that acupuncture is not a well-defined procedure—the only criterion for including a study in this table was that the authors call the procedure acupuncture. Many of the procedures discussed below under "TNS" are similar or identical to those defined as "acupuncture" here.

Thirteen studies have measured the effect of naloxone on clinical or experimental analgesia produced by acupuncture. Of these, 9 reported that naloxone reduced the analgesia while four found no effect. In two of the studies failing to find a naloxone effect (157,158) the negative interpretation of the results has been called into question (159). The third of the four nega-

Table 6.4
The Effect of Various Opiate Manipulations on Analgesia and Endorphin Levels Resulting from Transcutaneous Nerve Stimulation and Acupuncture

Acupuncture analgesia:

	Endorphin levels			
	β-endorphin		Enkephalin	
Naloxone	Plasma	CSF	Plasma	CSF
< (255)	= (261)	> (267)	> (264)	> (255)
= (160)	= (262)	> (268)	= (267)	
= (161)	= (263)			
= (158)	= (264)			
< (235)	= (265)			
= (157)	= (266)			
< (256)				
< (257)				
< (258)				
< (172)				
< (259)				
< (260)				

Transcutaneous nerve stimulation:

		Endorphin levels						
Naloxone		High-Freq				Lo-Freq		
		β-endorphin		Enkephalin		β-endorphin		Enkephalin
Hi-Freq	Low-Freq	Plasma	CSF	Plasma	CSF	Plasma	CSF	Plasma CSF
= (269)	< (269)	= (270)	= (271)			= (270)		
= (270)	= (270)	> (272)						
= (233)	< (274)	> (273)				> (273)		
= (274)	< (235)							
= (276)								

Symbols: "<", attenuates analgesia or endorphin level; "=", no effect on analgesia or endorphin level; ">", increases endorphin levels. (Numbers in the table refer to numbered references.)

tive studies examined long-term effects of acupuncture on migraine headache (160) and thus does not fit into the general paradigm of the other studies addressed here. The final negative study (161) utilized a dose of naloxone (0.4 mg) that is on the low end of the range of effective doses. Thus, it seems clear that naloxone, at least under most circumstances, appears to antagonize acupuncture analgesia.

The effects of acupuncture on CSF and plasma endorphin levels present a somewhat less consistent picture, but this is not surprising considering the complexities of these types of data. Considering that one could question the entire concept of plasma endorphin levels since they are indicative of CNS level in only very indirect ways, a nevertheless somewhat consistent picture emerges. As can be seen in Table 6.4, five studies have reported endorphin increases, and six have reported no effects. Such results should be interpreted with extreme caution since the meaning of increases in plasma endorphin levels is entirely unclear and even CSF endorphin levels are likely to be ambiguous because the site of endorphin release probably varies with the particular type of acupuncture stimulation. Nevertheless, an overview of these data is consistent with an involvement of endogenous opioids in at least some forms of acupuncture analgesia.

The literature concerning the involvement of endogenous opioids in TNS analgesia is considerably more complex than that associated with acupuncture analgesia, but this is likely to result from a greater variability in the intensity, frequency, duration, location, and other parameters of TNS. Despite this diversity in experimental paradigm, some general consistencies are apparent in the literature. While only 4 of 13 studies of TNS analgesia have reported naloxone antagonism, all 4 studies utilized low-frequency TNS. On the other hand, none of six studies utilizing high-frequency TNS found a naloxone antagonism (see Table 6.4 for references). The effects of TNS on endorphin levels have been less well studied. As seen in Table 6.4, three of the six reported studies have found an increase in endorphin levels, whereas the remainder have found no effects. Such results should be interpreted with the caveats discussed above in mind. Overall, these results are strikingly consistent with reports in the animal literature (162) and suggest the possibility that certain types of sensory stimulation either inactivate opiate systems or activate the opiate hyperalgesia systems as discussed above. Nevertheless, these results are consistent with an emerging picture that low-frequency, high-intensity acupuncturelike TNS invokes endogenous opioid mechanisms. Such studies, taken together, probably provide the most convincing evidence available that endogenous opioids can function to modulate human pain transmission.

In conclusion, acupuncture and transcutaneous nerve stimulation appear to be forms of counterirritation that activate both opiate and nonopiate systems. The variable clinical outcomes observed following these treatments probably result from differential recruitment of segmental, extrasegmental, opiate, and nonopiate pain inhibitory systems, all of which are now known to be activated by these types of stimulation in animals.

Stress Analgesia

A manipulation related to, but not identical with, counterirritation analgesia is a phenomenon most generally referred to as "stress analgesia." These studies have utilized environmental manipulations that are either severe physical or psychological stressors, including surgery, labor and childbirth, the application of overtly painful stimuli such as cold pressor pain or ischemic pain, chronic pain, anticipation of pain, and chronic stressful states such as life-threatening disease.

Examination of Table 6.5 reveals a strikingly consistent outcome for studies of this sort. It can be seen that in eight of eight studies in which a "stressful" manipulation increased nociceptive thresholds, naloxone at least partially reversed this increase. Another study (163) utilized vigorous exercise, at least a possible stressor, to increase the pain threshold and also showed naloxone reversibility. In addition, and not surprisingly, since β-endorphin is coreleased with ACTH, stress resulted in an increase in plasma β-endorphin levels in five of five studies,. Although such a result is not convincing alone, it is certainly consistent with the notion that endogenous opioids may underlie changes in the pain threshold produced by stress. In addition, such results are consistent with the results of counterirritation studies discussed above with respect to the observation that naloxone reversibility is more likely to occur with high-intensity peripheral stimulation (see Table 6.4). In conclusion, considering the diversity of procedures used in counterirritation studies and stress studies, a generally convincing picture of opioid involvement in pain modulation emerges. Many important questions about the nature of this involvement remain unanswered. The most important of these are the particular endogenous opiate involved and its site of action. Answers to such questions are unlikely to come from human studies since invasive procedures are probably necessary to acquire such information. The consistency of the animal and human studies, however, indicates the likelihood that such questions may be studied with animal models and verified in man.

Placebo Analgesia

Under conditions wherein patients have a strong need to be relieved of pain or wherein they have high expectations that pain relief will occur as a result of a medical procedure, a relatively inert substance, such as saline injection, will produce some relief of pain. This effect is the well-known placebo response.

Naloxone has also been used to examine whether endogenous opiates are involved in placebo analgesia. Levine and coworkers (164) reported that naloxone antagonized placebo analgesia in postsurgical patients. Although

Table 6.5
The Effect of Various Opiate Manipulations on Analgesia and Endorphin Levels Resulting from Stressful Manipulations

| | Endorphin levels | | | |
| | β-endorphin | | Enkephalin | |
Naloxone	Plasma	CSF	Plasma	CSF
< (277)	> (279)	> (283)	> (264)	> (255)
< (234)	> (280)			< (284)
< (278)	> (281)			
< (241)	> (282)			
< (240)				

Symbols: "<", attenuates analgesia or endorphin level; ">", increases endorphin levels. (Numbers in the table refer to numbered references.)

this conclusion has been questioned on technical grounds (165–167), little conflicting data has been published. Gracely and colleagures (148) have reported that naloxone results in an increase in the pain levels experienced by postsurgical patients that is independent of placebo effects. On the other hand, Levine and Gordon (168), in a study which rectified the technical problems of their earlier study (164), found convincing evidence for at least a partial reversal of placebo analgesia by naloxone. A difficulty with all of these studies is that surgical stress itself is a powerful activator of endogenous opioid systems (see above) and may influence the experimental results. A carefully controlled study utilizing experimentally induced ischemic pain (169) indicates that a partial antagonism of placebo analgesia can occur in the absence of any effects of naloxone on baseline pain responsivity. At this point, it seems most likely that endogenous opioids mediate at least some components of placebo analgesia. It should be kept in mind, however, that placebos, like other analgesic manipulations discussed above, are multidimensional manipulations likely to activate multiple pain inhibitory and possibly pain facilitatory systems. Thus, hypnosis, a manipulation that has at least a superficial resemblance to placebo procedures, consistently has failed to be antagonized by naloxone (170–173). The possibility that opiates are involved in some aspect of placebo analgesia appears particularly reasonable considering the fact that footshock analgesia can be classically conditioned in rats. Placebo analgesia can easily be conceived of as a classical conditioning paradigm wherein the placebo manipulation (e.g., injections, pills) serves as the conditioned stimulus and prior medication or treatment serves as the unconditioned stimulus. The observation by Grevert and associates (169) that placebo effects tend to extinguish with repeated trials supports such a conceptualization.

Although explanations of this sort are clearly speculative, they are indicative of the wealth of concepts from experimental pain research now available for clinical evaluation. Our increasing knowledge of pain modulatory systems has the potential not only of providing explanations of current therapies but of suggesting new approaches for the control of pain. The preponderance of current pain therapies involve either the surgical destruction of neural tissue or the use of addictive drugs. Such procedures offer great difficulties for the prolonged treatment of chronic pain. If multiple pain inhibitory systems could be activated pharmacologically or otherwise in an alternating sequence, the problems of tissue destruction and addiction could be circumvented.

Acknowledgment

Some of the experiments described here were supported by Public Health Service grants DA-00576 and NS-24009.

References

1. Bonica JJ. History of pain concepts and therapies. In: Bonica JJ, ed. The management of pain. Philadelphia: Lea & Febiger, 1990:2.

2. Mayer DJ, Price, DD. A physiological and psychological analysis of pain: a potential model of motivation. In: Pfaff DW ed. The physiological mechanisms of motivation. New York: Springer-Verlag, 1982: 433.

3. Melzack R, Wall PD. Pain mechanisms: a new theory. Science 1965;150: 971-979.

4. Burgess PR, Perl ER. Cutaneous mechanoreceptors and nociceptors. In: Iggo A ed. Handbook of sensory physiology, vol. 2. Heidelberg: Springer, 1973:29.

5. Price DD, Dubner R. Neurons that subserve the sensory-discriminative aspects of pain. Pain 1977;3:307-338.

6. Willis WD. The pain system. Basel: Karger, 1985.

7. Collins WF, Nulsen FE, Randt CT. Relation of peripheral nerve fiber size and sensation in man. Arch Neurol (Chic.) 1960;3:381-385.

8. Landau W, Bishop GH. Pain from dermal, periosteal, and fascial endings and from inflammation. Arch Neuro Psychiatry 1953;69:490-504.

9. Price DD. Characteristics of second pain and flexion reflexes indicative of prolonged central summation. Exp Neurol 1972;37:371-387.

10. Price DD, Hu JW, Dubner R, Gracely R. Peripheral suppression of first pain and central summation of second pain evoked by noxious heat pulses. Pain 1977;3: 57-68.

11. Perl ER. Myelinated afferent fibres innervating the primate skin and their response to noxious stimuli. J Physiol (Lond.) 1968;197:593-615.

12. Torebjork HE, Hallin RG. Identification of afferent C units in intact human skin nerves. Brain Res 1974;67:387-403.

13. Beitel RE, Dubner R. Fatigue and adaptation in unmyelinated (C) polymodal nociceptors to mechanical and thermal stimuli applied to the monkey's face. Brain Res 1976;112:402-406.

14. Dubner R, Bushnell MC, Duncan GH. Sensory-discriminative capacities of nociceptive pathways and their modulation by behavior. In: Yaksh TL, ed. Spinal afferent processing. New York: Plenum Press, 1986:331.

15. Dubner R, Beitel RE. Neural correlates of escape behavior in rhesus monkey to noxious heat applied to the face. In: Bonica JJ, Albe-Fessard D, eds. Advances in pain research and therapy, vol. 1. New York: Raven Press, 1976:155.

16. Dubner R, Bennett GJ. Spinal and trigeminal mechanisms of nociception. Annu Rev Neurosci 1983;6:381-418.

17. Meyer RA, Campbell JN. Myelinated nociceptive afferents account for the hyperalgesia that follows a burn to the hand. Science 1981;213:1527-1529.

18. Beitel RE, Dubner R. Sensitization and depression of C-polymodal nociceptors by noxious heat applied to the monkey's face. In: Advances in pain research and therapy. Bonica JJ, Albe-Fessard D, eds. New York: Raven Press, 1976:149.

19. LaMotte RH, Campbell JN. Comparison of responses of warm and nociceptive C-fiber afferents in monkey with human judgments of thermal pain. J Neurophysiol 1978;41:509-528.

20. LaMotte RH, Thalhammer JG, Torebjork HE, Robinson CJ. Peripheral neural mechanisms of cutaneous hyperalgesia following mild injury by heat. J Neurosci 1982;2:765-781.

21. Mense S, Schmidt RF. Muscle pain: which receptors are responsible for the transmission of noxious stimuli? In: Physiological aspects of clinical neurology. Rose JE, ed. Oxford: Blackwell, 1977:265.

22. Mense S, Stahnke M. Responses in muscle afferent fibres of slow conduction velocity to contractions and ischemia in the cat. J Physiol 1983;342:383-397.

23. Kumazawa T, Mizumura K. Thin-fibre receptors responding to mechanical, chemical, and thermal stimulation in the skeletal muscle of the dog. J Physiol 1977;273:179-194.

24. Kumazawa T, Mizumura K. The polymodal C-fiber receptor in the muscle of the dog. Brain Res 1976;101:489-493.

25. Watkins LR, Kinscheck IB, Mayer DJ. The neural basis of footshock analgesia: the effect of periaqueductal gray lesions and decerebration. Brain Res 1983;276: 317-324.

26. Shyu B-C, Andersson SA, Thoren P. Endorphin mediated increase in pain threshold induced by long-lasting exercise in rats. Life Sci 1982;30:833-841.

27. Kumazawa T, Mizumura K. The polymodal receptors in the testis of dog. Brain Res 1977;136:553-558.

28. Bossut D, Frenk H, Mayer DJ. Is substance P a primary afferent neurotransmitter for nociceptive input? IV. 2-Amino-5-phosphonovalerate (APV) and [D-Pro2,D-Trp7,9]-substance P exert different effects on behaviors induced by intrathecal substance P, strychnine and kainic acid. Brain Res 1988;455:247-253.

29. Battaglia G, Rustioni A. Coexistence of glutamate and substance P in dorsal root ganglion neurons of the rat and monkey. J Comp Neurol 1988;277:302-312.

30. Skilling SR, Smullin DH, Beitz AJ, Larson AA. Extracellular amino acid concentration in dorsal spinal cord of freely moving rats following veratridine and nociceptive stimulation. J Neurochem 1988;51:127-132.

31. Wilcox, GL. Excitatory neurotransmitters and pain. In: Bond MR, Charltron JE, Woolf CJ, eds. Proceedings of the VIth world congress on pain. Amsterdam: Elsevier, 1991:97.

32. Frenk H, Bossut D, Urca G, Mayer DJ. Is substance P a primary afferent neurotransmitter for nociceptive input? I. Analysis of pain-related behaviors resulting from intrathecal administration of substance P and 6 excitatory compounds. Brain Res 1988;455:223-231.

33. Price DD. Psychological and neural mechanisms of pain. New York: Raven Press, 1988.

34. Price DD, Hayes RL, Ruda MA, Dubner R. Spatial and temporal transformations of input to spinothalamic tract neurons and their relation to somatic sensation. J Neurophysiol 1978;41:933-947.

35. Price DD, Dubner R, Hu JW. Trigeminothalamic neurons in nucleus caudalis responsive to tactile, thermal, and nociceptive stimulation of monkey's face. J Neurophysiol 1976;39:936-953.

36. Wagman IH, Price DD. Responses of dorsal horn cells of *M. mulatta* to cutaneous and sural nerve A- and C-fiber stimulation. J Neurophysiol 1969;32:803-817.

37. Foreman RD. Viscerosomatic convergence onto spinal neurons responding to afferent fibers located in the inferior cardiac nerve. Brain Res 1977;137:164-168.

38. Bennett GJ, Hayashi H, Abdelmoumene M, Dubner R. Physiological properties of stalked cells of the substantia gelatinosa intracellularly stained with horseradish peroxidase. Brain Res 1979;164:285-289.

39. Chung JM, Fang ZR, Hori Y, Lee KH, Willis WD. Prolonged inhibition of primate spinothalamic tract cells by peripheral nerve stimulation. Pain 1984;19: 259-275.

40. Foreman RD, Applebaum AE, Beall JE, Trevino DL, Willis WD. Responses of primate spinothalamic tract neurons to electrical stimulation of hindlimb peripheral nerves. J Neurophysiol 1975;38:132-145.

41. Foreman RD, Beall JE, Applebaum AE, Coulter JD, Willis WD. Effects of dorsal column stimulation on primate spinothalamic tract neurons. J Neurophysiol 1976;39:534-546.

42. Melzack R. The puzzle of pain. New York: Basic Books, 1973.

43. Mayer DJ, Price DD, Becker DP. Neurophysiological characterization of the anterolateral spinal cord neurons contributing to pain perception in man. Pain 1975;1:51-58.

44. Price DD, Hayashi H, Dubner R, Ruda MA. Functional relationships between neurons of marginal and substantia gelatinosa layers of primate dorsal horn. J Neurophysiol 1979;42:1590-1608.

45. White JC, Sweet WH. Pain and the neurosurgeon. Springfield, Illinois: Thomas, 1969.

46. Casey KL. Escape elicited by bulboreticular stimulation in the cat. Int J Neurosci 1971;2:29-34.

47. Casey KL. Responses of bulboreticular units to somatic stimuli eliciting escape behavior in the cat. Int J Neurosci 1971;2:15-28.

48. Casey, KL, Keene, JJ, Morrow, T. Bulboreticular and medial thalamic unit activity in relation to aversive behavior and pain. In: Bonica JJ, ed. Advances in neurology, vol. 4. New York: Raven Press, 1974:197.

49. Lundeberg TCM. Vibratory stimulation for the alleviation of chronic pain. Acta Physiol Scand 1983;S523:1-51.

50. Larson MA, McHaffie JG, Stein BE. Response properties of nociceptive and low threshold mechanoreceptive neurons in the hamster superior colliculus. J Neurosci 1987;7:547-564.

51. Vertes RP, Miller NE. Brain stem neurons that fire selectively to a conditioned stimulus for shock. Arch Neuro Psychiatr 1976;63:739-748.

52. Nashold, BS,Jr., Wilson, WP, Slaughter, G. The midbrain and pain. In: Bonica JJ, ed. Advances in neurology. vol. 4. New York: Raven Press, 1974:191.

53. Nashold BS, Wilson WP, Slaughter DG. Stereotactic midbrain lesions for central dysesthesia and phantom pain: preliminary report. J Neurosurg 1969;30:116-126.

54. Mayer DJ, Price DD. Central nervous system mechanisms of analgesia. Pain 1976;2:379-404.

55. Jones EG, Friedman DP. Projection pattern of functional components of thalamic ventrobasal complex on monkey somatosensory cortex. J Neurophysiol 1982;48:521-544.

56. Kenshalo DR, Jr., Giesler GJ, Leonard RB, Willis WD. Responses of neurons in primate ventral posterior lateral nucleus to noxious stimuli. J Neurophysiol 1980:43:1594-1614.

57. Hardy JD, Wolff HG, Goodell H. Pain sensations and reactions, New York: Williams & Wilkins, 1952.

58. Mark VH, Ervin FR, Hackett TP. Clinical aspects of stereotactic thalamotomy in the human. I. The treatment of severe pain. Arch Neurol 1960;3:351-367.

59. Mundinger F, Becker P. Long-term results of central stereotactic interventions for pain. In: Sweet WH, Obrador S, Martin-Rodriguez JG, eds. Neurosurgical treatment in psychiatry, pain and epilepsy. Baltimore: University Park Press, 1977:685.

60. Hassler, R. Dichotomy of facial pain conduction in the diencephalon. In: Hassler R, Walker AE, eds. Trigeminal neuralgia. Philadelphia: W.B. Saunders, 1970:123.

61. Ishijima B, Yoshimasu N, Fukushima T, Hori T, Sekino H, Sano K. Nociceptive neurons in the human thalamus. Confin Neurol 1975;37:99-106.

62. Kenshalo DR, Jr., Isensee O. Responses of primate SI cortical neurons to noxious stimuli. J Neurophysiol 1983;50:1479-1496.

63. Lewin W, Phillips CG. Observations on partial removal of the postcentral gyrus for pain. J Neurol Neurosurg Psychiat 1952;15:143-147.

64. Penfield W, Boldrey E. Somatic motor and sensory representation in the cerebral cortex of man as studied by electrical stimulation. Brain 1937;60:389-443.

65. Echols DH, Cogclough JA. Abolition of painful phantom foot by resection of the sensory cortex. JAMA 1947;134:1476-1477.

66. Carmon A, Mor J, Goldberg J. Evoked cerebral responses to noxious thermal stimuli in humans. Brain Res 1976;25:103-107.

67. Chatrian GE, Canfield RC, Knauss RA, Lettich E. Cerebral responses to electrical tooth pulp stimulation in man: an objective correlate of acute experimental pain. Neurology (Minneap.) 1975;25:747-757.

68. Mao J, Mayer DJ, Price DD. Patterns of increased brain activity indicative of pain in a rat model of peripheral mononeuropathy. J Neurosci 1993;13:2689-2702.

69. Talbot JD, Marrett S, Evans AC, Meyer E, Bushnell MC, Duncan GH. Multiple representations of pain in human cerebral cortex. Science 1992;251:1355-1358.

70. Bonica, JJ. Causalgia and other reflex sympathetic dystrophies. In: Bonica JJ, ed. The management of pain. Philadelphia: Lea & Febiger, 1990:220.

71. Price DD, Long S, Huitt C. Sensory testing of pathophysiological mechanisms of pain in patients with reflex sympathetic dystrophy. Pain 1992;49:163-173.

72. Price, DD, Mao, J, Mayer, DJ. Central neural mechanisms of normal and abnormal pain states. In: Fields HL, Liebeskind JC, eds. Progress in pain research and management, vol 1. Seattle: IASP Press, 1994:61.

73. Gracely RH, Lynch SA, Bennett GJ. Painful neuropathy: altered central processing maintained dynamically by peripheral input. Published erratum appears in Pain 1993;52:251-3. Pain. 1992;51:175-194.

74. Bennett GJ. Evidence from animal models on the pathogenesis of painful neuropathy, and its relevance for pharmacotherapy. In: Basbaum AI, Besson J-M, eds. Towards a new pharmacotherapy. Chichester: John Wiley & Sons, 1991:365.

75. Dubner R. Neuronal plasticity and pain following peripheral tissue inflammation or nerve injury. In: Bond M, Charlton E, Woolf CJ, eds. Proceedings of Vth world congress on pain. Pain research and clinical management. vol. 5. Amsterdam: Elsevier, 1991:263.

76. Sato J, Perl ER. Adrenergic excitation of cutaneous pain receptors induced by peripheral nerve injury. Science 1991;251:1608-1610.

77. Bossut DF Perl ER. Sympathectomy induces novel adrenergic excitation of cutaneous nociceptors. Soc Neurosci Abstr 1992;18:x-x.

78. Bennett GJ, Xie YK. A peripheral mononeuropathy in rat that produces disorders of pain sensation like those seen in man. Pain 1988;33:87-107.

79. Price DD, Bennett GJ, Rafii A. Psychophysical observations on patients with neuropathic pain relieved by a sympathetic block. Pain 1989;36:273-288.

80. Thompson SWN, Woolf CJ. Primary afferent-evoked prolonged potentials in the spinal cord and their central summation: role of the NMDA receptor. In: Bond MR, Carlton JE, Woolf CJ, eds. Proceedings of the VIth world congress on pain. Amsterdam: Elsevier, 1991:291.

81. Bennett GJ, Kajander KC, Sahara Y, Iadarola MJ, Sugimoto T. Neurochemical and anatomical changes in the dorsal horn of rats with an experimental painful peripheral neuropathy. In: Cervero F, Bennett GJ, Headley PM, eds. Proceedings of sensory information in the superficial dorsal horn of the spinal cord. New York: Plenum Press, 1989:463.

82. Palecek J, Paleckova V, Dougherty PM, Carlton SM, Willis WD. Responses of spinothalamic tract cells to mechanical and thermal stimulation of skin in rats with experimental peripheral neuropathy. J Neurophysiol 1992;67:1562-1573.

83. Mao J, Price DD, Coghill RC, Mayer DJ, Hayes RL. Spatial patterns of spinal cord 2-deoxyglucose metabolic activity in a rat model of painful peripheral mononeuropathy. Pain 1992;50:89-100.

84. Thomas, PK. Clinical features and differential diagnosis of peripheral neuropathy. In: Dyck PJ, Thomas PK, Lambert EH, Bunge R, eds. Peripheral neuropathy. Philadelphia: W.B. Saunders, Co. 1984:1169.

85. Noordenbos W. Pain. Amsterdam: Elsevier, 1959:1-182.

86. Irwin S, Houde RW, Bennett DR, Hendershot LC, Seevers MH. The effects of morphine, methadone and meperidine on some reflex responses of spinal animals to nociceptive stimulation. J Pharmacol Exp Ther 1951;101:132-143.

87. Reynolds DV. Surgery in the rat during electrical analgesia induced by focal brain stimulation. Science 1969;164:444-445.

88. Mayer DJ, Wolfle TL, Akil H, Carder B, Liebeskind JC. Analgesia from electrical stimulation in the brainstem of the rat. Science 1971;174:1351-1354.

89. Mayer DJ, Manning BH. The role of opioid peptides in environmentally-induced analgesia. Harwood 1994. (in press)

90. Tsou K, Jang CS. Studies on the site of analgesic action of morphine by intracerebral micro-injection. Sci Sinica 1964;13:1099-1109.

91. Satoh M, Takagi H. Effect of morphine on the pre- and postsynaptic inhibitions in the spinal cord. Eur J Pharmacol 1971;14:150-154.

92. Samanin R, Valzelli L. Increase of morphine-induced analgesia by stimulation of the nucleus raphe dorsalis comment. Eur J Pharmacol 1971;16:298-302.

93. Mayer DJ, Hayes R. Stimulation-produced analgesia: development of tolerance and cross tolerance to morphine. Science 1975;188:941-943.

94. Akil H, Mayer D, Liebeskind J. Comparison chez le rat entre l'analgesie induite par stimulation de la substance grise periaqueducale et l'analgesie morphinique. CR Acad Sci 1972;274:3603-3605.

95. Akil H, Mayer DJ, Liebeskind JC. Antagonism of stimulation-produced analgesia by the narcotic antagonist, naloxone. Science 1976;191:961-962.

96. Hughes J. Search for the endogenous ligand of the opiate receptor. Neurosci Res Prog Bull 1975;13:55-58.

97. Hiller JM, Pearson J, Simon EJ. Distribution of stereospecific binding of the potent narcotic analgesic etorphine in the human brain: predominance in the limbic system. Res Commun Chem Pathol Pharm 1973;6:1052-1062.

98. Pert CB, Snyder SH. Opiate receptor: demonstration in nervous tissue. Science 1973;179:1011-1013.

99. Terenius L. Stereospecific interaction between narcotic analgesics and a synaptic plasma membrane fraction of rat cerebral cortex. Acta Pharmacol Toxicol 1973;32:317-320.

100. Pert CB, Snowman AM, Snyder SH. Localization of opiate receptor binding in synaptic membranes of rat brain. Brain Res 1974;70:184-188.

101. Pert CB, Kuhar MJ, Snyder SH. Autoradiographic localization of the opiate receptor in rat brain. Life Sci 1975;16:1849-1854.

102. Akil H, Watson SJ, Young E, Lewis ME, Khachaturian H, Walker JM. Endogenous opioids: Biology and function. Ann Rev Neurosci 1984;7:223-256.

103. Lotti VJ, Lomax P, George R. Temperature responses in the rat following intracerebral microinjection of morphine. J Pharmacol Exp Ther 1965;150:135-139.

104. Jacquet YF, Lajtha A. Morphine action at central nervous system sites in rat: analgesia or hyperalgesia depending on site and dose. Science 1973;182:490-491.

105. Pert A, Yaksh T. Sites of morphine induced analgesia in the primate brain: relation to pain pathways. Brain Res 1974;80:135-140.

106. Yaksh TL, Yeung JC, Rudy TA. Systematic examination in the rat of brain

sites sensitive to the direct application of morphine: observation of differential effects within the periaqueductal gray. Brain Res 1976;114:83-104.

107. Watkins LR, Griffin G, Leichnetz GR, Mayer DJ. The somatotopic organization of the nucleus raphe magnus and surrounding brainstem structures as revealed by HRP slow-release gels. Brain Res 1980;181:1-15.

108. Takagi, H, Doi, T, Akaike, A. Microinjection of morphine into the medial part of the bulbar reticular formation in rabbit and rat: inhibitory effects on lamina V cells of spinal dorsal horn and behavioral analgesia. In: Kosterlitz HW, ed. Opiates and endogenous opioid peptides. Amsterdam: N. Holland, 1976:191.

109. Takagi H. The nucleus reticularis paragigantocellularis as a site of analgesic action of morphine and enkephalin. Trends Pharmacol Sci 1980;1:182-184.

110. Azami J, Llewelyn MB, Roberts MHT. The contribution of nucleus reticularis paragigantocellularis and nucleus raphe magnus to the analgesia produced by systemically administered morphine, investigated with the microinjection technique. Pain 1982;12:229-246.

111. Zorman G, Belcher G, Adams JE, Fields HL. Lumbar intrathecal naloxone blocks analgesia produced by microstimulation of the ventromedial medulla in the rat. Brain Res 1982;236:77-89.

112. Rodgers RJ. Elevation of aversive threshold in rats by intra-amygdaloid injection of morphine sulphate. Pharmacol Biochem Behav 1977;6:385-390.

113. Rodgers RJ. Influence of intra-amygdaloid opiate injections on shock thresholds, tail-flick latencies and open field behaviour in rats. Brain Res 1978;153:211-216.

114. VanRee JM. Multiple brain sites involved in morphine antinociception. J Pharm Pharmacol 1977;29:765-766.

115. Haigler HJ, Spring DD. A comparison of the analgesic and behavioral effects of [D-Ala2] Met-enkephalinamide and morphine in the mesencephalic reticular formation of rats. Life Sci 1978;23:1229-1239.

116. Oley N, Cordova C, Kelly ML, Bronzino JD. Morphine administration to the region of the solitary tract nucleus produces analgesia in rats. Brain Res 1982;236:511-515.

117. Yaksh TL, Rudy TA. Studies on the direct spinal actin of narcotics in the production of analgesia in the rat. J Pharmacol Exp Ther 1977;202:411-428.

118. Yaksh TL, Rudy TA. Chronic catheterization of the spinal subarachnoid space. Physiol Behav 1976;17:1031-1036.

119. Albus K, Schott M, Herz A. Interaction between morphine and morphine antagonists after systemic and intraventricular application. Eur J Pharmacol 1970;12:53-64.

120. Jacquet YF, Lajtha A. Paradoxical effects after microinjection of morphine in the periaqueductal gray matter in the rat. Science 1974;185:1055-1057.

121. Tsou K. Antagonism of morphine analgesia by the intracerebral microinjection of nalorphine. Acta Physiol Sin 1963;26:332-337.

122. Toda K. Peripheral nerve stimulation for producing the suppressive effect on the tooth pulp-evoked jaw opening reflex in rat: relation between stimulus intensity and degree of suppression. Exp Neurol 1982;76:309-317.

123. Yeung JC, Rudy TA. Multiplicative interaction between narcotic agonisms expressed at spinal and supraspinal sites of antinociceptive action as revealed by concurrent intrathecal and intracerebroventricular injections of morphine. J Pharmacol Exp Ther 1980;215:633-642.

124. Yeung JC, Rudy TA. Sites of antinociceptive action of systemically injected morphine–involvement of supraspinal loci as revealed by intracerebroventricular injection of naloxone. J Pharmacol Exp Ther 1980;215:626-632.

125. Hayes RL, Price DD, Bennett GJ, Wilcox GL, Mayer DJ. Differential effects of spinal cord lesions on narcotic and non-narcotic suppression of nociceptive reflexes: further evidence for the physiologic multiplicity of pain modulation. Brain Res 1978;155:91-101.

126. Hayes RL, Bennett GJ, Newlon PG, Mayer DJ. Behavioral and physiological studies on non-narcotic analgesia in the rat elicited by certain environmental stimuli. Brain Res 1978;155:69-90.

127. Akil H, Madden J, Patrick RL, Barchas JD. Stress-induced increase in endogenous opiate peptides: Concurrent analgesia and its partial reversal by naloxone. In: Kosterlitz HW, ed. Opiates and endogenous opioid peptides. Amsterdam: Elsevier, 1976:63.

128. Lewis JW, Cannon JT, Liebeskind JC. Opioid and nonopioid mechanisms of stress analgesia. Science 1980;208:623-625.

129. Watkins LR, Cobelli DA, Faris P, Aceto MD, Mayer DJ. Opiate vs non-opiate footshock-induced analgesia: the body region shocked is a critical factor. Brain Res 1982;242:299-308.

130. Watkins LR, Cobelli DA, Newsome HH, Mayer DJ. Footshock induced analgesia is dependent neither on pituitary nor sympathetic activation. Brain Res 1982;245:81-96.

131. Watkins LR, Cobelli DA, Mayer DJ. Opiate vs non-opiate footshock induced analgesia (FSIA): descending and intraspinal components. Brain Res 1982;245:97-106.

132. Watkins LR, Young EG, Kinscheck IB, Mayer DJ. The neural basis of footshock analgesia: The role of specific ventral medullary nuclei. Brain Res 1983;276:305-315.

133. Watkins LR, Mayer DJ. Involvement of spinal opioid systems in footshock-induced analgesia: antagonism by naloxone is possible only before induction of analgesia. Brain Res 1982;242:309-316.

134. Faris P, Komisaruk B, Watkins L, Mayer DJ. Evidence for the neuropeptide cholecystokinin as an antagonist of opiate analgesia. Science 1983;219:310-312.

135. Watkins LR, Kinscheck IB, Mayer DJ. Potentiation of opiate analgesia and apparent reversal of morphine tolerance by proglumide. Science 1984;224:395-396.

136. Watkins LR, Johannessen JN, Kinscheck IB, Mayer DJ. The neurochemical basis of footshock analgesia: the role of spinal cord serotonin and norepinephrine. Brain Res 1984;290:107-117.

137. Watkins LR, Katayama Y, Kinscheck IB, Mayer DJ, Hayes RL. Muscarinic cholinergic mediation of opiate and nonopiate environmentally induced analgesias. Brain Res 1984;300:231-242.

138. Watkins LR, Cobelli DA, Mayer DJ. Classical conditioning of front paw and hind paw footshock induced analgesia (FSIA): naloxone reversibility and descending pathways. Brain Res 1982;243:119-132.

139. Grevert P, Baizman ER, Goldstein A. Naloxone effects on a nociceptive response of hypophysectomized and adrenalectomized mice. Life Sci 1978;23:723-728.

140. El-Sobky A, Dostrovsky JO, Wall PD. Lack of effect of naloxone on pain perception in humans. Nature 1976;263:783-784.

141. Buchsbaum MS, Davis GC, Bunney WE, Jr. Naloxone alters pain perception and somatosensory evoked potentials in normal subjects. Nature 1977;270:620-621.

142. Cesselin F, Bourgoin S, Hamon M, Artaud F, Testut MF, Rascol A. Normal CSF levels of met-enkephalin-like material in a case of naloxone-reversible congenital insensitivity to pain. Neuropeptides 1984;4:217-226.

143. Dehan H, Willer JC, Boureau F, Cambier J. Congenital insensitivity to pain, and endogenous morphine-like substances. Lancet 1977;2:293-294.

144. Dehen H, Willer JC, Prier S, Boureau F, Gonce M, Cambier J. Insensitivity to pain: electrophysiological study of the nociceptive reflex. Influence of naloxone. Rev Neurol 1978;134:255-262.

145. Davis GC, Buchsbaum MS, Bunney WE. Naloxone decreases diurnal variation in pain sensitivity and somatosensory evoked potentials. Life Sci 1978;23:1449-1460.

146. Lasagna L. Drug interaction in the field of analgesic drugs. Proc Roy Soc Med 1965;58:978-983.

147. Levine JD, Gordon NC, Bornstein JC, Fields HL. Role of pain in placebo analgesia. Proc Natl Acad Sci USA 1979;76:3528-3531.

148. Gracely RH, Dubner R, Wolskee PJ, Deeter WR. Placebo and naloxone can alter post-surgical pain by separate mechanisms. Nature 1983;306:264.

149. Young, RF, Feldman, RA, Kroening, R, Fulton, W, Morris, J. Electrical stimulation of the brain in the treatment of chronic pain in man. In: Liebeskind JC, Kruger L, eds. Advances in pain research and therapy: neural mechanisms of pain. New York: Raven Press, 1984:289.

150. Richardson DE, Akil H. Pain reduction by electrical brain stimulation in man. Part I: acute administration in periaqueductal and periventricular sites. J Neurosurg 1977;47:178-183.

151. Adams JE. Naloxone reversal of analgesia produced by brain stimulation in the human. Pain 1976;2:161-166.

152. Hosobuchi Y. Tryptophan reversal of tolerance to analgesia induced by central grey stimulation. Lancet 1978;2:47.

153. Hosobuchi Y, Adams JE, Linchitz R. Pain relief by electrical stimulation of the central gray matter in humans and its reversal by naloxone. Science 1977;196:183-186.

154. Fessler RG, Brown FD, Rachlin JR, Mullan S, Fang VS. Elevated Beta-endorphin in cerebrospinal fluid after electrical brain stimulation: artifact of contrast infusion. Science 1984;224:1017-1019.

155. Dionne RA, Mueller GP, Young RF, Greenberg RP, Hargreaves KM, Dubner R. Contrast medium causes the apparent increase in beta-endorphin levels in human cerebrospinal fluid following brain stimulation. Pain 1984;20:313-321.

156. Akil H, Richardson D, E. Contrast medium cause the apparent increase in beta-endorphin levels in human CSF following brain stimulation. Pain 1985;23:301-304.

157. Chapman CR, Colpitts YM, Benedetti C, Kitaeff R, Gehrig JD. Evoked potential assessment of acupunctural analgesia: attempted reversal with naloxone. Pain 1980;9:183-198.

158. Chapman CR, Benedetti C, Colpitts YH, Gerlach R. Naloxone fails to reverse pain thresholds elevated by acupuncture: acupuncture analgesia reconsidered. Pain 1983;16:13-31.

159. Miglecz E, Szekely JI. Intracerebroventricular saline treatment elevates the pain threshold. Is this phenomenon mediated by peripheral opiate receptors? Pharmacol Res Commun 1985;17:177-187.

160. Lenhard L, Waite PM. Acupuncture in the prophylactic treatment of migraine headaches: pilot study. N Z Med J 1983;96:663-666.

161. Kenyon JN, Knight CJ, Wells C. Centre for the study of alternative therapies. Acupunct Electrother Res 1983;8:17-24.

162. Han JS, Xie GX, Zhou ZF. Acupuncture mechanisms in rabbits studied with microinjection of antibodies against beta-endorphin, enkephalin and substance P. Neuropharmacol 1984;23:1-5.

163. Janal MN, Colt EWD, Clark WC, Glusman M. Pain sensitivity, mood and plasma endocrine levels in man following long-distance running: effects of naloxone. Pain 1984;19:13-26.

164. Levine JD, Gordon NC, Fields HL. The mechanism of placebo analgesia. Lancet 1978;2:654-657.

165. Goldstein A, Grevert P. Placebo analgesia, endorphins and naloxone. Lancet 1978;2:1385.

166. Korczyn A. Mechanism of placebo analgesia. Lancet 1978;2:1304-1305.

167. Skrabanek P. Naloxone and placebo. Lancet 1978;2:791.

168. Levine JD, Gordon NC. Influence of the method of drug administration on analgesic response. Nature 1984;312:755-756.

169. Grevert P, Albert LH, Goldstein A. Partial antagonism of placebo analgesia by naloxone. Pain 1983;16:129-144.

170. Barber J, Mayer D. Evaluation of the efficacy and neural mechanism of a hypnotic analgesia procedure in experimental and clinical dental pain. Pain 1977;4:41-48.

171. Goldstein A, Hilgard ER. Failure of the opiate antagonist naloxone to modify hypnotic analgesia. Proc Natl Acad Sci USA 1975;72:2041-2043.

172. Mayer DJ, Price DD, Rafii A, Barber J: Acupuncture hypalgesia: evidence for activation of a central control system as a mechanism of action. In: Bonica JJ, Albe-Fessard DG, eds.Advances in pain research and therapy. New York: Raven Press, 1976:751.

173. Spiegel D, Albert LH. Naloxone fails to reverse hypnotic alleviation of chronic pain. Psychopharmacol 1983;81:140-143.

174. Barton C, Basbaum AI, Fields HL. Dissociation of supraspinal and spinal actions of morphine: a quantitative evaluation. Brain Res 1980;188:487-498.

175. Dostrovsky JO, Deakin JFW. Periaqueductal grey lesions reduce morphine analgesia in the rat. Neurosci Lett 1977;4:99-103.

176. Hammond DL, Proudfit HK. Effects of locus coeruleus lesions on morphine-induced antinociception. Brain Res 1980;188:79-91.

177. Kishioka A, Iguchi Y, Ozaki M, Yamamoto H. Effect of electrical lesioning of nucleus reticularis gigantocellularis of rat medulla oblongata on morphine analgesia. Folia Pharmacol Jpn 1983;82:475-484.

178. Lai Y, Chan SHH. Antagonization of clonidine- and morphine-promoted antinociception by kainic acid lesion of nucleus reticularis gigantocellularis in the rat. Exp Neurol 1982;78:38-45.

179. Murfin R, Bennett GJ, Mayer DJ. The effects of dorsolateral spinal cord (DLF) lesions on analgesia from morphine microinjected into the periaqueductal gray matter (PAG) of the rat. Soc Neurosci Abstr 1976;2:946.

180. Pottoff P, Valentino D, Lal H. Attenuation of morphine analgesia by lesions of the preoptic forebrain region in the rat. Life Sci 1979;24:421-424.

181. Proudfit HK, Anderson EG. Morphine analgesia: blockade by raphe magnus lesions. Brain Res 1975;98:612-618.

182. Yeung JC, Yaksh TL, Rudy TA. Effects of brain lesions on the antinociceptive properties of morphine in rats. Clin Exp Pharm Physiol 1975;2:261-268.

183. Young EG, Watkins LR, Mayer DJ. Comparison of the effects of ventral medullary lesions on systemic and microinjection morphine analgesia. Brain Res 1984;290:119-129.

184. Cobelli DA, Watkins LR, Mayer DJ. Dissociation of opiate and non-opiate footshock produced analgesia. Soc Neurosci Abstr 1980;6:247.

185. Ross RT, Randich A. Unconditioned stress-induced analgesia following exposure to brief footshock. J Exp Psychol Anim Behav 1984;10:127-137.

186. Snow AE, Dewey WL. A comparison of antinociception induced by foot shock and morphine. J Pharmacol Exp Ther 1983;227:42-50.

187. Terman GW, Lewis JW, Liebeskind JC. The sensitivity of opioid mediated stress analgesia to narcotic antagonists. Proc W Pharmacol Soc 1983;26:49-52.

188. Cannon JT, Terman GW, Lewis JW, Liebeskind JC. Body region shocked need not critically define the neurochemical basis of stress analgesia. Brain Res 1984;323:316-319.

189. Chesher GB, Chan B. Footshock induced analgesia in mice-its reversal by naloxone and cross tolerance with morphine. Life Sci 1977;21:1569-1574.

190. Holaday JW, Tortella FC, Meyerhoff JL, Belenky GL, Hitzemann RJ. Electroconvulsive shock activates endogenous opioid systems: behavioral and biochemical correlates. Ann N Y Acad Sci 1986;467:249-255.

191. Chance WT, White AC, Krynock GM, Rosecrans JA. Conditional fear-induced antinociception and decreased binding of [3H]N-Leu-enkephalin to rat brain. Brain Res 1978;141:371-374.

192. Galligan JJ, Porreca F, Burks TF. Dissociation of analgesic and gastrointestinal effects of electroconvulsive shock-released opioids. Brain Res 1983;271:354-357.

193. Griffiths EC, Longson D, Whittam A. Biotransformation of dynorphin (1-17) by rat brain peptidases. J Physiol (London) 1983;342:39P.

194. Lewis JW, Cannon JT, Chudler EH, Liebeskind JC. Effects of naloxone and hypophysectomy on electroconvulsive shock-induced analgesia. Brain Res 1981;208:230-233.

195. Geller I. Effect of punishment on lever pressing maintained by food reward of brain stimulation. Physiol Behav 1970;5:203-206.

196. Urca G, Yitzhaky J, Frenk H. Different opioid systems may participate in post-electroconvulsive shocks (ECS) analgesia and catalepsy. Brain Res 1981;219:385-397.

197. Urca G, Harouni A, Sarne Y. Electroconvulsive shock (ECS) and h-endorphin-induced analgesia: unconventional interactions with naloxone. Eur J Pharmacol 1982;81:237-243.

198. Miczek KA, Thompson ML. Analgesia resulting from defeat in a social confrontation-the role of endogenous opioids in brain. In: Bandler R, ed. Modulation of sensorimotor activity during alterations in behavioral states. New York: Alan R. Liss, 1984:431.

199. Miczek KA, Thompson ML, Shuster L. Opioid-like analgesia in defeated mice. Science 1982;215:1520-1522.

200. Teskey GC, Kavaliers M. Ionizing radiation induces opioid-mediated analgesia in male mice. Life Sci 1984;35:1547-1552.

201. Drugan RC, Grau JW, Maier SF, Madden JIV, Barchas JD. Cross tolerance between morphine and the long-term analgesic reaction to inescapable shock. Pharmacol Biochem Behav 1981;14:677-682.

202. Chapman CR, Wilson ME, Gehrig JD. Comparative effects of acupuncture and transcutaneous stimulation on the perception of painful dental stimuli. Pain 1976;2:265-284.

203. Das S, Chatterjee TK, Ganguly A, Ghosh JJ. Role of adrenal steroids on electroacupuncture analgesia and on antagonizing potency of naloxone. Pain 1984;18: 135-144.

204. Han JS, Xie GX. Dynorphin: important mediator for electroacupuncture analgesia in the spinal cord of the rabbit. Pain 1984;18:367-376.

205. Han JS, Zhou ZF, Xuan YT. Acupuncture has an analgesic effect in rabbits. Pain 1983;15:83-92.

206. Willer JC, Roby A, Gerard A, Maulet C. Electrophysiological evidence for a possible serotonergic involvement in some endogenous opiate activity in humans. Eur J Pharmacol 1982;78:117-121.

207. Coderre TJ, Rollman GB. Stress analgesia: effects of PCPA, yohimbine, and naloxone. Pharmacol Biochem Behav 1984;21:681-686.

208. Fanselow MS. Naloxone attenuates rat's preference for signaled shock. Physiol Psychol 1979;7:70-74.

209. Rossier J, French ED, Rivier C, Ling N, Guillemin R, Bloom FE. Foot-shock stress increases beta-endorphin levels in blood but not brain. Nature 1977;270: 618-619.

210. Rossier J, Guillemin R, Bloom F. Foot shock induced stress decreases Leu5-enkephalin immunoreactivity in rat hypothalamus. Eur J Pharmacol 1978;48: 465-466.

211. Panerai AE, Martini A, Sacerdote P, Mantegazza P. Kappa-receptor antagonist reverse 'non-opioid' stress-induced analgesia. Brain Res 1984;304:153-156.

212. Tricklebank MD, Hutson PH, Curzon G. Analgesia induced by brief foot-shock is inhibited by 5-hydroxy-tryptamine but unaffected by antagonists of 5-hydroxy-tryptamine or by naloxone. Neuropharmacol 1982;21:51-57.

213. Madden JIV, Akil H, Patrick RL, Barchas JD. Stress-induced parallel changes in central opioid levels and pain responsiveness in the rat. Nature 1977;265:358-360.

214. Belenky GL, Gelinas-Sorell D, Kenner JR, Holaday JW. Evidence for delta-receptor involvement in the post-ictal antinociceptive responses to electroconvulsive shock in rats. Life Sci 1983;33 Suppl 1:583-585.

215. Hall ME, Stewart JM. Prevention of stress-induced analgesia by substance P. Behav Brain Res 1983;10:375-382.

216. Jorgensen HA, Fasmer OB, Berge OG, Tveiten L, Hole K. Immobilization-induced analgesia: Possible involvement of a non-opioid circulating substance. Pharmacol Biochem Behav 1984;20:289-292.

217. Schlen H, Bentley GA. The possibility that a component of morphine-induced analgesia is contributed indirectly via the release of endogenous opioids. Pain 1980;9:73-84.

218. Hyson RL, Ashcraft LJ, Drugan RC, Grau JW, Maier SF. Extent and control of shock affects naltrexone sensitivity of stress-induced analgesia and reactivity to morphine. Pharmacol Biochem Behav 1982;17:1019-1025.

219. Maier SF, Davies S, Grau JW, Jackson RL, Morrison DH, Moye T, Barchas JD. Opiate antagonists and long-term analgesic reaction induced by inescapable shock in rats. J Comp Physiol Psychol 1980;94:1172-1183.

220. Shimizu T, Koja T, Fujisaki T, Fukuda T. Effects of methysergide and naloxone on analgesia induced by the peripheral electric stimulation in mice. Brain Res 1981;208:463-467.

221. Akil H, Liebeskind JC. Monoaminergic mechanisms of stimulation-produced analgesia. Brain Res 1975;94:279-296.

222. Hosobuchi Y, Rossier J, Bloom FE, Guillemin R. Stimulation of human peri-aqueductal gray for pain relief increases immuno-reactive beta-endorphin in ventricular fluid. Science 1979;203:279-281.

223. Cannon JT, Prieto GJ, Lee A, Liebeskind JC. Evidence for opioid and non-opioid forms of stimulation-produced analgesia in the rat. Brain Res 1982;243:315-321.

224. Jurna I. Effect of stimulation in the periaqueductal grey matter on activity in ascending axons of the rat spinal cord: selective inhibition of activity evoked by afferent A-delta and C fibre stimulation and failure of naloxone to reduce inhibition comment. Brain Res 1980;196:33-42.

225. Kajander KC, Ebner TJ, Bloedel JR. Effects of periaqueductal gray and raphe magnus stimulation on the responses of spinocervical and other ascending projection neurons to non-noxious inputs. Brain Res 1984;291:29-37.

226. Mah C, Suissa A, Anisman H. Dissociation of antinociception and escape deficits induced by stress in mice. J Comp Physiol Psych 1980;94:1160-1171.

227. Akil H, Richardson DE, Hughes DJ, Barchas JD. Enkephalin-like material elevated in ventricular cerebrospinal fluid of pain patients after analgetic focal stimulation. Science 1978;201:463-465.

228. Meyerson BA, Boethius J, Carlsson AM. Percutaneous central gray stimulation for cancer pain. Appl Neurophysiol 1978;41:57-65.

229. Sessle BJ, Hu JW, Dubner R, Lucier GE. Functional properties of neurons in cat trigeminal subnucleus caudalis (medullary dorsal horn). II. Modulation of responses to noxious and nonnoxious stimuli by PAG, n. raphe magnus, cerebral cortex, and afferent influences and effect of naloxone comment. J Neurophysiol 1981;45:193-207.

230. Mayer DJ, Murfin R. Stimulation-produced analgesia (SPA) and morphine analgesia (MA): Cross tolerance from application at the same brain site. Fed Proc 1976;35:385.

231. Teskey GC, Kavaliers M, Hirst M. Social conflict activates opioid analgesic and ingestive behaviors in male mice. Life Sci 1984;35:303-316.

232. Frenk H, Yitzhaky J. Effects of amygdaloid kindling on the pain threshold of the rat. Exp Neurol 1981;71:487-496.

233. Freeman TB, Campbell JN, Long DM. Naloxone does not affect pain relief induced by electrical stimulation in man. Pain 1983;17:189-196.

234. Pertovaara A, Kemppainen P, Johansson G, Karonen SL. Ischemic pain non-segmentally produces a predominant reduction of pain and thermal sensitivity in man: a selective role for endogenous opioids. Brain Res 1982;251:83-92.

235. Willer J-C, Roby A, Boulu P, Boureau F. Comparative effects of electroacupuncture and transcutaneous nerve stimulation on the human blink reflex. Pain 1982;14:267-278.

236. Hill RG, Ayliffe SJ. The antinociceptive effect of vaginal stimulation in the rat is reduced by naloxone. Pharmacol Biochem Behav 1981;14:631-633.

237. LeBars D, Chitour D, Kraus E, Dickenson AH, Besson JM. Effect of naloxone upon diffuse noxious inhibitory controls (DNIC) in the rat. Brain Res 1981;204:387-402.

238. Mcgivern R, Berka C, Bernston GG, Walker JM, Sandman CA. Effect of naloxone on analgesia induced by food deprivation. Life Sci 1979;25:885-889.

239. Zamir N, Simantov R, Segal M. Pain sensitivity and opioid activity in genetically and experimentally hypertensive rats. Brain Res 1980;184:299-310.

240. Willer JC, Albe-Fessard D. Electrophysiological evidence for a release of endogenous opiates in stress-induced analgesia in man. Brain Res 1980;198:419-426.

241. Willer JC, Dehen H, Cambier J. Stress-induced analgesia in humans: endogenous opioids and naloxone-reversible depression of pain reflexes. Science 1981;212:689-690.

242. Szechtman H, Simantov R, Hershkowitz M. Sexual behavior decreases pain sensitivity and stimulates endogenous opioids in male rats. Eur J Pharmacol 1981;70:279-286.

243. Gintzler AR. Endorphin-mediated increases in pain threshold during pregnancy. Science 1980;210:193-195.

244. Saavedra JM. Naloxone reversible decrease in pain sensitivity in young and adult spontaneously hypertensive rats. Brain Res 1981;209:245-250.

245. Girardot MN, Holloway FA. Intermittent cold water stress-analgesia in rats: cross tolerance to morphine. Pharmacol Biochem Behav 1984;20:631-634.

246. Girardot MN, Holloway FA. Cold water stress analgesia in rats: differential effects of naltrexone. Physiol Behav 1984;32:547-555.

247. Hart SL, Slusarczyk H, Smith TW. The involvement of delta-receptors in stress induced antinociception in mice. Eur J Pharmacol 1983;95:283-286.

248. Bodnar RJ, Kelly DD, Spiaggia A, Ehrenberg C, Glusman M. Dose-dependent reductions by naloxone of analgesia induced by cold-water stress. Pharmacol Biochem Behav 1978;8:667-672.

249. Bodnar RJ, Kelly DD, Spiaggia A, Glusman M. Stress-induced analgesia: Adaptation following chronic cold water swims. Bull Psychonomic Soc 1978;11:337-340.

250. O'Conner P, Chipkin RE. Comparisons between warm and cold water swim stress in mice. Life Sci 1984;45:631-640.

251. Kulkarni SK. Heat and other physiological stress-induced analgesia; catecholamine mediated and naloxone reversible responses. Life Sci 1980;27:185-189.

252. Kayser V, Guilbaud G. Dose-dependent analgesic and hyperalgesic effects of systemic naloxone in arthritic rats. Brain Res 1981;226:344-348.

253. Augustinsson L-E, Bohlin P, Bundsen P, Carlsson C-A, Forssman L. Pain relief during delivery by transcutaneous electrical nerve stimulation. Pain 1977;4:59-66.

254. Grau JW, Hyson RL, Maier SF, Madden JIV, Barchas JD. Long-term stress-induced analgesia and activation of the opiate system. Science 1981;213:1409-1411.

255. He LF, Dong WQ. Activity of opioid peptidergic system in acupuncture analgesia. Acupunct Electrother Res 1983;8:257-266.

256. Tsunoda Y, Sakahira K, Nakano S, Matsumoto I, Yoshida T, Nagayama K, Ikezono E. Antagonism of acupuncture analgesia by naloxone in unconscious man. Bull Tokyo Med Dent Univ 1980;27:89-94.

257. Boureau F, Willer JC, Yamaguchi Y. Abolition by naloxone of the inhibitory effect of peripheral electrical stimulation on the blink reflex. Electroencephalogr Clin Neurophysiol 1979;47:322-328.

258. Chapman CR. Modulation of experimental dental pain in man with acupuncture and by transcutaneous electric stimulation. Ann Anesthesiol Fr 1978;19:427-433.

259. Mayer DJ, Price DD, Rafii A. Antagonism of acupuncture analgesia in man by the narcotic antagonist naloxone. Brain Res 1977;121:368-372.

260. Ernst M, Lee MH. Influence of naloxone on electro-acupuncture analgesia using an experimental dental pain test. Review of possible mechanisms of action. Acupunct Electrother Res 1987;12:5-22.

261. Szczudlik A, Kwasucki J. Beta endorphin-like immunoreactivity in the blood

of patients with chronic pain treated by pinpoint receptor stimulation (acupuncture). Neurol Neurochir Pol 1984;18:415-420.

262. Umimo M, Shimada M, Kubota Y. Effects of acupuncture anesthesia on the pituitary gland. Bull Tokyo Med Dent Univ 1984;31:93-98.

263. Szczudlik A, Lypka A. Plasma concentration of immunoreactive beta-endorphin in healthy persons due to pinpoint stimulation of receptors (acupuncture). Neurol Neurochir Pol 1983;17:535-540.

264. Kiser RS, Khatami MJ, Gatchel RJ, Huang XY, Bhatia K. Acupuncture relief of chronic pain syndrome correlates with increased plasma met-enkephalin concentrations. Lancet 1983;2:1394-1396.

265. Szczudlik A, Lypka A. Plasma immunoreactive beta-endorphin and enkephalin concentration in healthy subjects before and after electroacupuncture. Acupunct Electrother Res 1983;8:127-137.

266. Masala A, Satta G, Alagna S, Zolo TA, Rovasio PP, Rassu S. Suppression of electroacupuncture (EA)-induced beta-endorphin and ACTH release by hydrocortisone in man. Absence of effects on EA-induced anaesthesia. Acta Endocrinol 1983;103:469-472.

267. Clement-Jones V, McLoughlin L, Tomlin S, Besser GM, Rees LH, Wen HL. Increased beta-endorphin but not met-enkephalin levels in human cerebrospinal fluid after acupuncture for recurrent pain. Lancet 1980;2:946-949.

268. Sjolund B, Terenius L, Eriksson M. Increased cerebrospinal fluid levels of endorphins after electro-acupuncture. Acta Physiol Scand 1977;100:382-384.

269. Lundberg T, Bondesson L, Lundstrom V. Relief of primary dysmenorrhea by transcutaneous electrical nerve stimulation. Acta Obstet Gynecol Scand 1985;64: 491-497.

270. O'Brien WJ, Rutan FM, Sanborn C, Omer GE. Effect of transcutaneous electrical nerve stimulation on human blood beta-endorphin levels. Phys Ther 1984;64:1367-1374.

271. Johansson F, Almay BGL, Von Knorring L, Terenius L. Predictors for the outcome of treatment with high frequency transcutaneous electrical nerve stimulation in patients with chronic pain. Pain 1980;9:55-62.

272. Facchinetti F, Sandrini G, Petraglia F, Alfonsi E, Nappi G, Genazzani AR. Concomitant increase in nociceptive flexion reflex threshold and plasma opioids following transcutaneous nerve stimulation. Pain 1984;19:295-304.

273. Hughes GS, Lichstein PR, Whitlock D, Harker C. Response of plasma beta-endorphins to transcutaneous electrical nerve stimulation in healthy subjects. Phys Ther 1984;64:1062-1066.

274. Casale R, Zelaschi F, Guarnaschelli C, Bazzini G. Electroanalgesia by transcutaneous stimulation (TNS). Response to the naloxone test. Minerv Med 1983;74: 941-946.

275. Pertovaara A, Kemppainen P, Johansson G, Karonen S-L. Dental analgesia produced by non-painful, low-frequency stimulation is not influenced by stress or reversed by naloxone. Pain 1982;13:379-384.

276. Pertovaara A, Kemppainen P. The influence of naloxone on dental pain threshold elevation produced by peripheral conditioning stimulation at high frequency. Brain Res 1981;215:426-429.

277. Jungkinz G, Engel RR, King UG, Kuss HJ. Endogenous opiates increase pain tolerance after stress in humans. Psychiat Res 1983;8:13-18.

278. Frid M, Singer G, Oei T, Rana C. Reactions to ischemic pain interactions between individual, situational and naloxone effects. Psychopharmacol 1981;73: 116-120.

279. Delke I, Minkoff H, Grunebaum A. Effect of lamaze childbirth preparation on maternal plasma beta-endorphin immunoreactivity. Am J Perinatol 1985;2:317-319.

280. Smith R, Besser GM, Rees LH. The effect of surgery on plasma beta-endorphin and methionine-enkephalin. Neurosci Lett 1985;55:17-21.

281. Atkinson JH, Kremer EF, Risch SC, Morgan CD, Azad RF. Plasma measures of beta-endorphin/beta-lipotropin-like immunoreactivity in chronic pain syndrome and psychiatric subjects. Psychiat Res 1983;9:319-327.

282. Cohen MR, Pickar D, Dubois M, Bunney WE. Stress-induced plasma beta-endorphin immunoreactivity may predict postoperative morphine usage. Psychiat Res 1982;6:7-12.

283. Katz ER, Sharp B, Kellerman J, Marston AR, Hershman JM, Siegel SE. Beta-endorphin immunoreactivity and acute behavioral distress in children with leukemia. J Nerv Ment Dis 1982;170:72-77.

284. Puig MM, Laorden ML, Miralles FS, Olaso MJ. Endorphin levels in cerebrospinal fluid of patients with postoperative and chronic pain. Anesthesiol 1982; 57:1-4.

Chapter 7

Electrical Stimulation for Pain Modulation

Lynn Snyder-Mackler

In *The Puzzle of Pain*, Ronald Melzack describes a patient with a congenital inability to sense pain. By the time of her premature death at age 29, she had suffered third-degree burns, broken bones, arthritis, and other problems (1). She had what we as clinicians seek for our patients, freedom from pain, and that is in essence what she died from. Although most of this chapter addresses the amelioration of pain, remember that pain serves a protective function and that in all but a very small number of conditions, electrical stimulation for pain modulation should be used adjunctively with treatments of the underlying pathology.

Electroanalgesia can be traced to the presentation of the gate-control theory of pain transmission by Melzack and Wall in 1965 (2). Its application in clinical practice began with Shealy's and Long and colleagues' use of implanted dorsal column stimulators and their serendipitous discovery that transcutaneous stimulation appeared equally effective (3, 4). Subsequently the therapeutic use of electroanalgesia has increased exponentially. The objectives of this chapter are to present: *(a)* common methods used to assess and measure pain, *(b)* ways in which electrical stimulation is used to modulate pain, *(c)* a review of the literature with emphasis on prospective, randomized clinical trials, and *(d)* representative case studies as illustrative examples.

Clinical Assessment of Pain

Most forms of pain assessment attempt to quantify and objectify the patient's response to the query, "tell me about your pain." The study of pain assessment is a discipline unto itself; this section can only touch on its complexity. A brief discussion of pain assessment, however, is integral to the review of literature regarding pain control with electrical stimulation and to the evaluation of electrical stimulation as a therapeutic intervention for patients with pain.

Pain perception has sensory and affective components. That is, patients can describe pain intensity (sensory) and unpleasantness (affective) (5). Pain is multidimensional and pain assessment that focuses only on the sensory component, with the argument that it is the only physiologic or "real" pain, misses the boat clinically. The study of experimentally induced pain focuses on the sensory dimension of pain and does not capture the entirety of clinical pain. Pain perception is colored by many factors and accounting for those factors can be useful in the assessment of the effectiveness of therapeutic interventions.

Patient interview and history

Pain evaluation can be accomplished by the interview process, where the clinician elicits the pain history with a series of questions, avoiding the use of leading questions or those that are too broad. A question like "Does your back pain radiate down the back of your leg?" is suggestive. "Tell me about your pain" does not help direct the examiner, because it does not constrain the answer. "What activities worsen your pain?", "How do you relieve your pain?", "Where is your pain?", "When did the pain start?", and "Has the pain changed since it began?" are specific questions without being leading that help the examiner focus the problem. Although the reliability and validity of the pain interview, as it exists, are not clearly established, the interview provides a means for identifying confirming and disconfirming information about pain patterns.

Other means of assessing pain include pain drawings, questionnaires, verbal rating scales, visual analog scales, and indirect measures such as the assessment of analgesic use or relative levels of activity.

Pain drawings and questionnaires

Pain drawings are visual representations of the distribution and quality of a patient's pain (Fig. 7.1). Pain drawings can be completed by the clinician, the patient, or both. Evidence suggests that patient and clinician versions of pain drawings can be very different. The reliability and validity of these drawings have not been demonstrated (6).

Pain rating scales

Visual analog scales (VAS) are lines conventionally 10 cm long and numbered from 1–10 (most common), 1–100, or not numbered. These scales vary from rather detailed to very sparse. A patient is usually asked to describe the pain intensity at the moment. A variation of the numerical VAS is the **graphic rating scale (GRS)**, which has equally spaced descriptors on the line. These scales can be vertical or horizontal, and either the left (top) or right (bottom) margin may represent maximal pain intensity (Fig. 7.2). Reliability is highest for VAS and GRS on horizontal lines; validity is equivocal (7).

Verbal rating scales (VRS) use descriptive words, with patients picking the words that best categorize their pain. A verbal rating scale that is used often clinically is to ask the patient to verbally rate pain on a scale of 1 to 10. The reliability of the VAS is superior to that of the VRS (8, 9).

For both visual analog and verbal rating scales, the choice of descriptors or anchors for the scales is important. A scale that uses "no pain" and "worst pain imaginable" as anchors does not allow for discrimination of sensory and affective components of pain. The use of anchors like "no pain," "most intense pain," or "pleasant," "extremely unpleasant" allows for differentiation that has important clinical and experimental implications. (Fig. 7.2).

Pain questionnaires

A commonly used questionnaire is the McGill Pain Questionnaire, which exists in both a short and a long version. The McGill questionnaire is a global assessment tool that uses variations of the tools mentioned above. It uses grouped descriptors to attempt to quantify patients' pain characteristics. A pain interview, drawing, and VRS are also included in this questionnaire. The reliability of this questionnaire, as demonstrated by Melzack and others, is acceptable; validity is equivocal (10, 11).

Functional status questionnaires have been used to measure the pain and disability associated with chronic pain conditions. The Sickness Impact Profiles (SIP) and Oswestry Low Back Pain Questionnaire (Table 7.1) have been used both in research and in many clinical settings. These questionnaires cut to the heart of the question of whether pain-associated dysfunction is affected

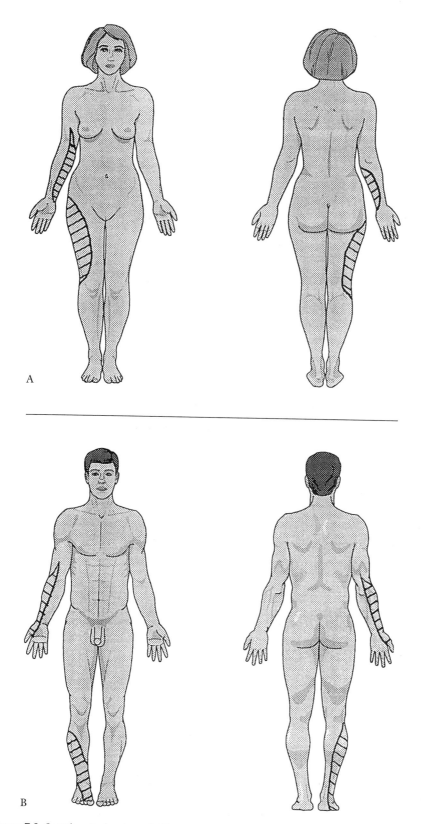

Figure 7.1 Sample pain drawings: **A**. Upper-extremity in the medial antebrachial cutaneous distribution. Lower-extremity pain in the lateral femoral cutaneous distribution. **B**. Upper-extremity pain in the C6 distribution. Lower-extremity pain in the L5 distribution.

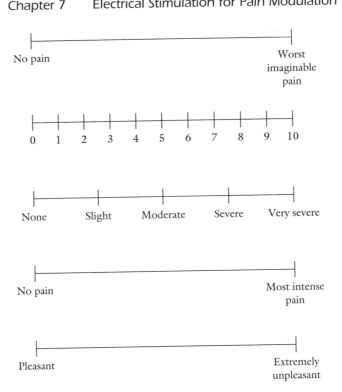

Figure 7.2 A sample of different types of visual analogue pain scales.

by treatment. These types of questionnaires are quite reliable and appear to also have high validity in the chronic pain population (12).

Alternative methods of pain assessment

Investigators have monitored analgesic intake to assess the relative intensity of pain (13). It is assumed that patients take fewer pills (lower dosage of analgesic) if pain intensity is lower, but this becomes problematic with different types of drugs. Some researchers have developed means of converting analgesics to aspirin equivalents, but the validity of this is questionable.

Evaluation of daily activity and levels of activity has also been used. The assumption is that patients are more active when pain is less intense, but evidence suggests that patients may not assess their activity level accurately (14).

Magnitude estimation is a technique in which pain is measured (either the affective component, the sensory component, or both) against a standard, representing the maximum value on a scale of intensity or unpleasantness. Delitto and colleagues used a 20-mA DC stimulus as the reference for maximum intensity/unpleasantness in a study of discomfort with electrical stimulation. Magnitude estimation yields ratio-scaled measurements with high reliability (16).

Methods of pain rating have been used to assess the effectiveness of agents for modulation of experimentally induced and clinical pain. Their use in clinical practice has been largely in the chronic pain population. The predictive value of these measures is unknown, primarily because of unsystematic and sporadic use in daily clinical practice. The use of one (or more) of the more reliable pain assessment measures combined with thorough, accurate documentation of treatment procedures could aid tremendously in the evaluation of pain control treatments.

Modes of Stimulation for Pain Control

Electrical stimulation has been used for pain modulation for over a century. Refinements in equipment design have made this approach to pain management significantly easier over the years. Even with the new delivery systems, attempts to modulate pain by electrotherapy are performed in one of four ways: subsensory-level stimulation, sensory-level stimulation, motor-level stimulation, and noxious-level stimulation (Table 7.2). Many of the commercially available stimulators can produce each of these forms of stimulation although the performance characteristics of some stimulators may be better suited to only one level of stimulation. Each mode of stimulation has unique responses and requirements, as described below.

Subsensory-level stimulation

Within the past few years, stimulators have been marketed that do not produce phase charges sufficient to excite peripheral nerve fibers and reach sensory threshold; such devices are termed **microcurrent electrical nerve stimulators (MENS)** by their purveyors. Because their peak current amplitudes are below 1 mA, these devices stimulate neither nerve nor muscle. Trade publications and clinical seminars have touted the effectiveness of MENS for a range of conditions, but, to date, no studies of the clinical effectiveness of this mode of stimulation demonstrate a better effect than sham stimulation (17,18,19). There is no evidence for the efficacy of electrical stimulation below sensory threshold for pain management.

Sensory-level stimulation

Sensory-level stimulation for pain modulation has been the best studied of any electrotherapeutic technique for pain control. Sensory-level stimulation is defined as stimulation at or above the sensory threshold and below the motor threshold. Sensory-level stimulation is most often accomplished by using a frequency in the 50–100 pps range, relatively short pulse or phase durations (2–50 μsec), and relatively low amplitudes. This form of stimulation (Figure 7.3*A*) is also called **conventional TENS**. Amplitudes are commonly determined by patient perception; in other words, current amplitude is increased until the patient perceives a comfortable paresthesia (tingling, or "pins and needles," sensation) beneath the electrodes. The clinician should not be able to see or palpate muscle contraction. At this level of stimulation, only the large-diameter, superficial cutaneous nerve fibers are activated. The likely mechanism of pain modulation is either a direct peripheral block of transmission or activation of central inhibition of pain transmission by large-diameter fiber stimulation (analogous to the original gate-control theory).

Table 7.1

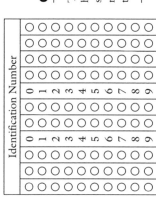

Identification Number								
○ 0	○ 0	○ 0	○ 0	○ 0	○ 0	○ 0	○ 0	○ 0
○ 1	○ 1	○ 1	○ 1	○ 1	○ 1	○ 1	○ 1	○ 1
○ 2	○ 2	○ 2	○ 2	○ 2	○ 2	○ 2	○ 2	○ 2
○ 3	○ 3	○ 3	○ 3	○ 3	○ 3	○ 3	○ 3	○ 3
○ 4	○ 4	○ 4	○ 4	○ 4	○ 4	○ 4	○ 4	○ 4
○ 5	○ 5	○ 5	○ 5	○ 5	○ 5	○ 5	○ 5	○ 5
○ 6	○ 6	○ 6	○ 6	○ 6	○ 6	○ 6	○ 6	○ 6
○ 7	○ 7	○ 7	○ 7	○ 7	○ 7	○ 7	○ 7	○ 7
○ 8	○ 8	○ 8	○ 8	○ 8	○ 8	○ 8	○ 8	○ 8
○ 9	○ 9	○ 9	○ 9	○ 9	○ 9	○ 9	○ 9	○ 9

OSWESTRY LOW BACK PAIN QUESTIONNAIRE

This questionnaire has been designed to give the physical therapist information as to how your pain has affected your ability to manage everyday life. Please answer all sections. Completely fill in only one circle in each section that applies to you. We realize that you may consider that two or more choices apply to you, but please fill the <u>one circle</u> that most clearly describes your problem.

Section 1 - PAIN INTENSITY

- ○ I have no pain.
- ○ I have no pain except when I move a certain way.
- ○ I have minimal pain most of the time.
- ○ I have moderate pain most of the time.
- ○ I have severe pain most of the time.
- ○ I have intense/intolerable pain most of the time.

Section 2 - PERSONAL CARE (washing, dressing, etc.)

- ○ I can take care of myself normally without causing extra pain.
- ○ I can take care of myself normally, but it causes extra pain.
- ○ It is painful to take care of myself and I am slow and careful.
- ○ I need some help, but manage most of my personal care.
- ○ I need help every day in most aspects of self care.
- ○ I do not get dressed, wash with difficulty, and stay in bed.

Section 6 - STANDING

- ○ I can stand as long as I want without pain.
- ○ I can stand as long as I want, but it gives me extra pain.
- ○ Pain prevents me from standing more than one hour.
- ○ Pain prevents me from standing more than 30 minutes.
- ○ Pain prevents me from standing more than 15 minutes.
- ○ Pain prevents me from standing at all.

Section 7 - SLEEPING

- ○ Pain does not prevent me from sleeping well.
- ○ I can sleep well only by taking medication.
- ○ I have less than 6 hours sleep because of pain.
- ○ I have less than 4 hours sleep because of pain.
- ○ I have less than 2 hours sleep because of pain.
- ○ Pain prevents me from sleeping at all.

Section 3 - LIFTING

○ I can lift heavy weights without pain.
○ I can lift heavy weights, but it causes extra pain.
○ Pain prevents me from lifting heavy weights off the floor but I can manage if they are conveniently positioned on a table.
○ Pain prevents me from lifting heavy weights, but I can manage light to medium weights if conveniently positioned.
○ I can lift only very light weights.
○ I cannot lift or carry anything at all.

Section 4 - WALKING

○ Pain does not prevent me from walking any distance.
○ Pain prevents me from walking more than one mile.
○ Pain prevents me from walking more than 1/2 mile.
○ Pain prevents me from walking more than 1/4 mile.
○ I can only walk using a cane or crutch.
○ I am in bed most of the time.

Section 5 - SITTING

○ I can sit in a chair as long as I like.
○ I can only sit in my favorite chair as long as I like.
○ Pain prevents me from sitting more than one hour.
○ Pain prevents me from sitting more than 30 minutes.
○ Pain prevents me from sitting more than 10 minutes.
○ Pain prevents me from sitting at all.

Section 8 - SEX LIFE

○ My sex life is normal and causes no extra pain.
○ My sex life is normal, but causes extra pain.
○ My sex life is nearly normal, but is very painful.
○ My sex life is severely restricted because of pain.
○ My sex life is nearly absent because of pain.
○ Pain prevents any sex life at all.

Section 9 - SOCIAL LIFE

○ My social life is normal and gives me no extra pain.
○ My social life is normal, but gives me extra pain.
○ Pain has no effect on my social life other than limiting some energetic interests like dancing.
○ Pain has restricted my social life and I do not go out as often.
○ Pain has restricted my social life to my home.
○ I have no social life because of pain.

Section 10 - TRAVELING

○ I can travel anywhere without extra pain.
○ I can travel anywhere, but it gives me extra pain.
○ Pain is bad, but I manage trips over 2 hours.
○ Pain restricts me to trips of less than one hour.
○ Pain restricts me to trips of less than 30 minutes.
○ Pain prevents me from traveling except to the doctor or hospital.

Table 7.2
Common Stimulation Characteristics for Electroanalgesia

Mode of stimulation	Phase duration (μsec)	Frequency (pps, bps, or beat/sec)	Amplitude	Duration of treatment	Duration of analgesia
Subsensory-level stimulation	Not specified	< 3 Hz	< 1 mA/cm²	Not known	Not known
Sensory-level stimulation	2–50	50–100	Perceptible tingling	20–30 min	Little residual post-treatment
Motor-level stimulation	> 150	2–4	Strong visible muscle contraction	30–45 min	Hours
Noxious-level stimulation	≤ 1 msec, up to 1 sec	1–5 or > 100	Noxious; below motor threshold	Seconds to minutes	Hours

Sensory-level stimulation is perceived by the patient as very comfortable and is often the first choice in electrotherapeutic intervention for pain control. A patient may be frightened or nervous about the use of electricity, and sensory-level stimulation introduces electrotherapy gently.

Patient response to sensory-level stimulation is usually immediate and does not persist after the stimulus is turned off (20). Patients use conventional TENS as needed throughout the day. The response depends on the chronicity of the problem and the activity level and pain level of the patient. Patients adapt readily to this type of stimulation, and perception of the stimulation can decline as the treatment progresses. Much as one becomes conditioned to external stimuli (e.g., night traffic noises in the city), one can adapt to a consistent stimulus. This is demonstrated in chronic problems where sensory-level stimulation is usually the first mode used, but is rarely sufficient for full remediation. For this reason, modulations of conventional TENS (usually amplitude or duration modulations, Fig. 7.3, B–D) are used to diminish the adaptation and enhance the effectiveness of the stimulation.

When sensory-level stimulation is chosen, electrodes are most commonly placed around or over the site of pain. There is some evidence that placement in remote areas (e.g., acupuncture points) may also result in the amelioration of pain, but the preponderance of evidence for effectiveness uses sensory-level stimulation directly over the pain site or very closely related anatomical structures, such as the peripheral nerve(s) that innervate(s) the site or the nerve root(s) to the area.

Many clinicians and investigators (most notably Mannheimer and Lampe (5) in their definitive text on TENS) consider electrode placement to be critical to the success of this type of stimulation. Clinicians should examine electrode placement sites carefully for each particular patient. Electrode placements must be individually tailored to each patient's unique set of

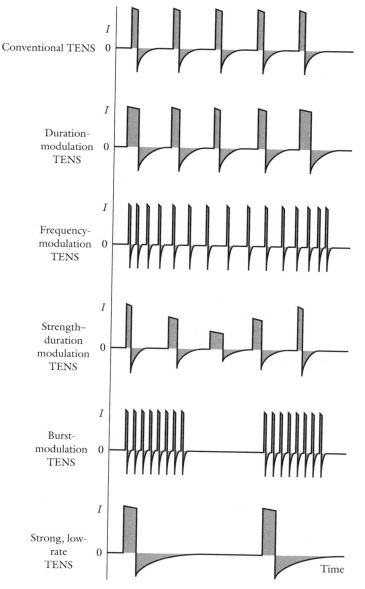

Figure 7.3 A sample of different types of pulse patterns utilized by clinical and portable TENS units.

circumstances. There is not, for example, an optimal low-back pain placement but rather a set of choices to be tried systematically and accepted or rejected based on patient response. No cookbook approach to electrode placement will replace a sound knowledge of anatomy, neuroanatomy, and pathology.

A wide range of current amplitudes cause electroanalgesia via sensory-level stimulation. Sensory-level stimulation at higher frequencies and longer pulse durations than those described for conventional TENS can be used to allow some specific procedures to be performed. These currents, with relatively close electrode placement, cause instantaneous interelectrode analgesia (17). This stimulation allows the clinician to perform short-duration therapeutic procedures that might prove too painful for the patient otherwise. Examples of such procedures include transverse friction massage, wound debridement, suture removal, and joint mobilization. The duration of the electrical stimulation treatment depends on the length of the procedure.

Motor-level stimulation

Motor-level stimulation for pain modulation is used primarily for control of nonacute pain. By definition, motor-level stimulation produces a visible muscle contraction. The efficacy of this stimulation is strongest when it is applied to a site anatomically or physiologically related to the site of pain. Motor-level stimulation for pain control most often involves continuous stimulation at a frequency in the 2–4 pps, bps, or beat/sec range, with relatively long phase durations (> 150 μsec), and amplitudes high enough to produce a strong, visible muscle contraction. This form of stimulation is also called **strong, low-rate TENS** (Fig. 7.3*F*). Higher frequencies and shorter pulse durations have been used; the common factor is the production of a rhythmic muscle contraction. In addition, brief bursts of pulses have been applied for pain control at levels that elicit muscle contraction. This form of stimulation (Fig. 7.3*E*) has been called **burst-modulated TENS**.

The magnitude of the induced muscle contraction varies from barely perceptable to extremely strong. Evidence supports the contention that stronger contractions produce better analgesia (17, 18). It is reasonable to assume that pain fibers are not activated at the lower levels of motor stimulation. A nonpainful contraction is induced and stimulation may simply effect pain relief in much the same way as sensory-level stimulation (peripheral block or activation of central inhibition). The induction of rhythmic contraction may also activate the endogenous opiate mechanisms of analgesia.

Motor-level stimulation has been shown to be effective in the modulation of experimentally induced and clinical pain. It is most often used clinically with patients who complain of a deep, throbbing, or chronic pain. This mode of stimulation has been called **noninvasive electroacupuncture** and **acupuncture-like TENS** as it is often applied to remote areas that may correspond to acupuncture points. Often, point stimulators are used to deliver this type of stimulation via small probe electrodes.

The response of patients to motor-level stimulation is usually not immediate but is often long-lasting. This supports an endogenous opiate mode of action. Patients do not readily adapt to this type of stimulation, and it is particularly well-suited for patients who have had pain relief from sensory-level stimulation in the past.

Motor-level stimulation may evoke a visible muscle contraction. Patients are often intolerant of this, especially with strong muscle contractions in the painful area. For this reason, electrodes are most commonly placed in an area remote from the pain site. Although acupuncture points have been used, it is advisable to apply electrodes to sites that are anatomically or physiologically related to the pain site, such as the related myotome. Reliance on motor point or acupuncture point charts for electrode placement is not advised because each patient has anatomical variations and individual problems. The charts are normative and as such fit everyone to a certain extent but no one well. Again, a sound anatomical and neuroanatomical base is necessary and sufficient for good electrode placement. Here, a knowledge of not just dermatomal and myotomal distributions, but also scleratomal (segmental innervation of bone) distributions is helpful.

Induction of analgesia is usually not immediate with motor-level stimulation. For this reason, treatment times are longer than with sensory-level stimulation (45 min or longer versus 20–30 min). This allows the induction of analgesia before treatment is terminated. When treatment is performed at home, it is usually done 2–3 times per day. This type of treatment is also performed on a several-times-per-week basis in clinical settings.

Noxious-level stimulation

Noxious-level stimulation for pain amelioration is the application of electrotherapeutic currents to produce a painful stimulus in or remote from the pain site. Noxious-level stimulation is produced using frequencies of either 1–5 pps, bps, or beat/sec or 80–100 pps, bps, or beat/sec, long pulse durations of up to 1 sec, and intensities that produce noxious sensory stimulation with or without muscle contraction. Although activation of motor nerves routinely occurs before activation of pain fibers, noxious-level stimulation is often applied to areas with no motor nerve fibers, minimizing the potential for producing muscle contraction. The pulse duration and frequency characteristics described above are those in common use, but the only requirement is that the patient perceive the stimulus as noxious.

Noxious-level stimulation is believed to effect pain amelioration by an endorphin-mediated mechanism. At higher stimulation levels all types of fibers in the peripheral nerve are activated, and this painful stimulus likely modulates pain by an endorphin-mediated mechanism. Pain production in a related or unrelated area is thought to cause a systemic release of endogenous opiates, which increases the patient's pain threshold. Noxious-level stimulation may also cause a quick-onset pain modulation called **hyperstimulation analgesia**, which has been hypothesized to interfere with the patterned-reverberation pain circuitry described by Nabe (21).

Noxious-level stimulation can be effective in the relief of clinical and experimentally induced pain. Because of its inherent discomfort, it is rarely used clinically as a first approach. It is most often used after adaptation to sensory-level stimulation or in cases where sensory-level or motor-level stimulation has not proven effective.

Electrode placement for noxious-level stimulation is the most varied and arbitrary of all the stimulation modes used for electroanalgesia. Electrodes are most often placed over the pain site, perhaps to reintroduce a noxious, regional stimulation into the homeostasis developed in the chronic pain situation. Associated acupuncture points, motor points, trigger points, and unrelated sites are also used and have similar efficacy. Noxious-level stimulation applied anywhere in the body may be as effective as stimulation applied to a site anatomically or physiologically related to the pain site in the amelioration of that pain (22, 23).

All means by which electroanalgesia can be produced fit into the above groups (Fig. 7.4) except iontophoresis (described in a later chapter).

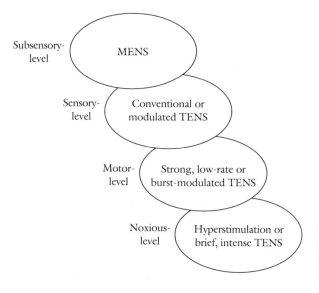

Figure 7.4 Methods of producing electroanalgesia and the corresponding intensity levels.

Literature on Effectiveness of Electrical Stimulation for Pain Control

This review of the literature regarding electrical stimulation and treatment of pain will attempt to correlate stimulation characteristics with the various theories of pain transmission and control in order to provide a rational basis for their use. The preponderance of literature reports on the use of portable TENS devices. Only one randomized, controlled clinical trial of the use of interferential current for pain control in the literature has demonstrated no effect of electrical stimulation over placebo (24). There are no similar studies of the effects of high-voltage pulsed current on pain control in the peer-reviewed literature.

Early reports, case studies, and retrospective reviews suggested that electrical stimulation was quite effective for the treatment of postoperative and acute pain (25–27). In problems as varied as low back pain (25), pancreatitis (26), and postherpetic neuralgia (27), sensory-level stimulation had been shown to be an effective modulator of acute pain. The findings in recent studies have been more equivocal. Reported benefits of postoperative electroanalgesia have included decreased length of stay, reduced pulmonary complications, and less narcotic usage. Schomburg and Carter-Baker (28) reported on a retrospective chart review of 150 patients who had undergone laparotomy, 75 of whom received sensory-level stimulation para-incisionally after surgery. The 75 patients in the comparison group were operated on prior to the institution of the postoperative electrical stimulation program. Length of stay and medication usage were the only comparisons made between groups, but pain intensity and unpleasantness were measured using VAS in the electrical stimulation group. The authors found a statistically lower medication usage in the electrical stimulation group than in the comparison group, but no difference in length of stay. Issenman and coworkers (29) compared 20 patients after spinal fusion with Harrington rods. Ten patients received sensory-level stimulation para-incisionally after surgery. Ten age-matched control patients were also studied. Length of stay and medication usage were lower in the electrical stimulation (ES) group. Results were presented descriptively without statistical analysis.

Some studies have demonstrated the effectiveness of ES for control of acute pain. Hargreaves and Lander randomly assigned 75 patients to electrical stimulation, sham electrical stimulation, and control groups after abdominal surgery (30). The intervention was applied 15 min prior to and during the dressing change that occurred on the second day after surgery. A visual analog scale was used. There was significantly greater pain relief in the electrical stimulation group and no difference between the sham electrical stimulation and control groups. Dawood and Ramos used a randomized crossover design to compare the effectiveness of electrical stimulation to that of sham electrical stimulation and ibuprofen in the treatment of primary dysmenorrhea. During the cycles of treatment with electrical stimulation, the women required less rescue medication than during the placebo electrical stimulation cycles and had pain relief comparable to that occurring during the ibuprofen cycles. (30, 31). These studies are in the minority in demonstrating effectiveness. In a recent review of the use of electrical stimulation for pain control, Long (32) stated, "There are no comparative studies of TENS with other modalities [in the treatment of acute pain]." Nevertheless, electrical stimulation is generally assumed to be the most effective treatment for acute and postoperative pain (28, 29).

The results of randomized prospective studies of the effectiveness of electrical stimulation on acute and postoperative pain, however, have not been as positive as one might have expected from earlier studies. Reuss and associates (33) randomly assigned 64 patients who were undergoing elective cholecystectomy to one of two groups, electrical stimulation and no electrical stimulation. A commercially available TENS device was set to deliver sensory-level stimulation and electrodes were placed within 2 cm of the surgical incision. There was no statistically significant difference in length of stay, narcotic usage, or pulmonary complications between groups (33). Carman and Roach studied children 11 to 21 years of age who were undergoing spinal surgery. Patients were randomly assigned to one of three groups (15 in each group): analgesic and electrical stimulation, analgesic and sham electrical stimulation, analgesics. Duration-modulated sensory-level stimulation at 60 pps was delivered. There was no difference in narcotic usage or length of stay between groups (34). Finson and coworkers randomly assigned 52 patients to one of three groups: electrical stimulation, sham electrical stimulation, sham electrical stimulation and analgesics. Again, there was no difference in narcotic usage or length of stay between groups (35). Smedley and colleagues randomly assigned 62 men scheduled for hernia repair into two groups: electrical stimulation or sham electrical stimulation. Sensory-level stimulation was used. There was no significant difference between groups in postoperative opiate usage, pulmonary function, or visual analog pain ratings (36). Walker and coworkers performed a sequential trial in the treatment of 30 patients after unilateral total knee arthroplasty. Sensory-level stimulation applied to the knee and constant passive motion was compared to CPM without electrical stimulation. There was no difference between treatments for length of stay, knee flexion range of motion, and analgesic use (37). These studies suggest that even in the cases where electrical stimulation is theoretically most likely to be effective, when subjected to the rigor of prospective randomized design, effectiveness is not readily demonstrated.

Studies of electrical stimulation for the relief of chronic pain had similar positive early results. Here, too, the preponderance of literature involved case reports, retrospective chart reviews, and, at best, crossover designs. Few prospective randomized clinical trials have been performed. Melzack and colleagues compared motor-level electrical stimulation (in the low back) at 4–8 pps to suction-cup massage in the treatment of patients with chronic low back pain. Forty-one patients were randomly assigned in this double-blind prospective study. The McGill Pain Questionnaire pain rating index and present pain index plus a straight leg raise measurement were used to assess outcome. The electrical stimulation group was statistically better than the massage group in all measures (4, 38). However, more recent examination suggests that sensory-level stimulation appears to have a minimal effect on chronic pain. Deyo and coworkers randomly assigned patients with low back pain for at least 3 months (average of 4 years) and who were not in active treatment to one of four groups: electrical stimulation, sham electrical stimulation, electrical stimulation and exercise, sham electrical stimulation and exercise (39). A modified Sickness Impact Profile, a visual analog scale, and an activity rating scale were used to determined effect of treatment. Exercise was more effective than electrical stimulation, and electrical stimulation was no more effective than placebo. This study had a profound effect on reimbursement for TENS for patients with low back pain and prompted a number of scholarly interchanges in the literature. Criticism ranged from the inclusion of patients who were not in active treatment and an apparent failure of the randomization with respect to previous surgery to disagreement regarding the current characteristics and timing of the stimulation. Many suggestions were made to "correct" the identified flaws with this study (40). Studies to refute Deyo's findings, however, have yet to be published. Gemignani and colleagues demonstrated no difference between sensory-level electrical stimulation and sham electrical stimulation in visual analog scores, a subjective score, mobility, or medication usage in patients with ankylosing spondylitis. They did, however, report a profound placebo effect (41). These studies must be viewed in the light of an interesting study by Johnson and colleagues, who studied long-term users of portable TENS devices (20). Approximately 50% of the patients reported that TENS relieved their pain by more than half. Electroanalgesia had a rapid onset and was not long-lasting after the device was turned off. One-third of the patients used TENS for more than 61 hours per week. Approximately half of the patients reported pain reduction from a burst modulation, and most used stimulation frequencies between 1 and 70 pps. Interestingly, two-thirds of those who reported no electroanalgesia nevertheless used the stimulators on a daily basis.

Dose–response issues in electrical stimulation for pain

One of the major criticisms of studies of electrical stimulation where effectiveness has not been demonstrated is that the investigators did not use the "right" stimulation characteristics (40). Some studies have been undertaken that compared the relative effectiveness of different stimulation characteristics. Recently, Johnson and colleagues have examined control of cold-induced pain by different frequencies of electrical stimulation. They found that the greatest analgesia occurred at frequencies between 20 and 80 pps for sensory-level stimulation. Frequencies above and below this range were less effective (42). Tulgar and coworkers examined patterns of stimulation in two studies. In the first study, three delivery modes were tested: constant-frequency sensory, level stimulation, burst-modulated sensory-level stimulation, and frequency-modulated sensory-level stimulation. The patients, all of whom had chronic pain, preferred the modulated modes to the constant-frequency mode (43). In the second study, constant-frequency, high-rate frequency-modulated (55–90 pps), low-rate frequency-modulated (20–60 pps), and burst-modulated sensory-level stimulation were tested for relief of chronic pain in 14 patients with a variety of pain conditions as measured by a visual analog scale of pain intensity (44). Six of the 14 patients did not report any effect of the stimulation on their pain. Among the remaining patients, there was a preference for the high-rate, frequency-modulated and burst-modulated modes. These studies are beginning to address the dose–response issues regarding frequency and frequency modulation, but other characteristics (current intensity, pulse duration) have yet to be systematically examined. This is a critical missing piece in the literature that serves both sides of the effectiveness debate in its absence.

Placebo effects of electrical stimulation for pain

Electrical stimulation has a profound placebo effect that has been underscored in many studies. Petrie and Hazleman demonstrated that sham electrical stimulation can be a strong placebo (45). Langley and coworkers, in a randomized, double-blind study of patients with chronic pain, demonstrated a clear placebo effect of sensory-level electrical stimulation (46). Although effectiveness has been reported in some studies, any clear benefit of TENS other than that which can be explained by the placebo effect has not been shown convincingly in randomized, prospective studies.

Precautions, Contraindications, and Adverse Effects

There are few additions to the general precautions and contraindications described in Chapter 2 for electroanalgesia devices. Some literature actually mitigates the listed precautions. Demand-inhibited (synchronous) pacemakers are sensitive to electromagnetic interference. Most pacemakers in use today are of this type. An electrical signal within the pacemaker's range can inhibit pacing by indicating a normal heart rhythm when none occurs. Performing an electrocardiogram (ECG) during a trial of TENS has been suggested to ensure normal pacemaker function. Chen and colleagues reported on two patients with synchronous pacemakers who were using TENS for control of chronic pain. ECG tracings showed no interference, but extended cardiac monitoring (Holter) showed inhibition of pacemaker function by the TENS devices. One patient used the stimulator with electrodes on the back and legs, the other with electrodes on the neck and shoulder. In both cases, the pacemakers were reprogrammed to change their sensitivities and no further inhibition of pacing occurred. This study's recommendation for extended cardiac monitoring for patients with synchronous pacemakers appears to be a prudent one (47).

The FDA suggests that electrical stimulation should not be used over the abdomen of women who are pregnant, but there are many reports of the use of electrical stimulation during labor and delivery and no adverse effects have been demonstrated. Interference with fetal monitoring has been reported (48). The safety for the fetus of prolonged or repetitive stimulation over the abdomen or back has not been determined.

Electrical stimulation for pain control should be used cautiously, if at all, when the patient's pain is serving a protective or useful function. For example, if pain is to be used as a limitation for progression of weight-bearing or range of motion, it would be inappropriate to use electrical stimulation to reduce the pain.

The most common complication from portable stimulator use is skin irritation, primarily near or under the electrodes (39). This complication is most likely a function of the electrode interface. The use of self-adhesive electrodes that are composed of synthetic materials should mitigate this response. The reader is referred to Chapter 5 for other contraindications and parameters that may be applicable in electrical stimulation for pain control applications.

Case Studies

The case studies that follow are examples of the types of patient pain problems that may be amenable to remediation by electrical stimulation. Regardless of the other treatment goals, pain modulation is common to all, and electrical stimulation for pain modulation is a component of each treatment plan. For brevity, only essential additional treatment is included (e.g., ultrasound in a combination treatment, or the therapeutic procedure being performed under electroanalgesia). Remember that electrotherapy is primarily adjunctive to other treatments crucial for a successful outcome.

Case 1

A 36-year-old longshoreman injured his low back 2 days ago while lifting a heavy object at work. Past medical history is noncontributory; x-rays of the spine are negative.

Objective Examination reveals marked limitation in all lumbar motions with pain rated as 10/10 on a verbal rating scale for all motions. Test maneuvers produce no peripheralization or centralization of symptoms but complete testing was precluded by pain. There is significant paralumbar muscle spasm and diffuse low back pain. Neurological signs are negative.

Assessment Significant resultant pain and muscle tenderness precludes complete examination. Muscle spasm and pain need to be diminished in order for examination to be completed.

Plan

Thermal agents

Electrical stimulation for pain control

ADL instruction and modification

Complete examination when pain is diminished sufficiently to allow for movement testing.

Detailed Electrotherapeutic Plan

Mode of stimulation: Sensory-level stimulation

Type of stimulator: Clinical; two available circuits

Electrodes and placement: four large electrodes placed to surround pain area (Fig. 7.5)

Duration of Treatment: 20–30 min

Rationale Sensory-level stimulation is an appropriate starting point for the management of acute pain because the patient should tolerate it well and it should be effective. A clinical stimulator is chosen because the problem should be amenable to in-clinic treatment. Since the painful area is rather large, larger electrodes are chosen and four electrodes from either one or two circuits are used. If the target area is too tender to allow placement over the site, placement over lower lumbar myotomes can be used. The duration of treatment should be sufficient to produce electroanalgesia with sensory-level stimulation.

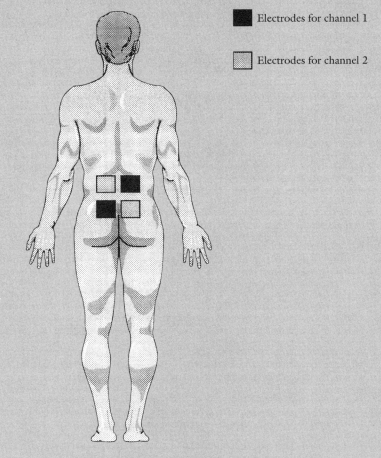

■ Electrodes for channel 1

□ Electrodes for channel 2

Figure 7.5 Electrode placement for Case Study 1.

Case 2

A 62-year-old woman sustained a Colles's fracture of the left wrist 8 weeks ago. X-rays showed excellent healing and her cast was removed this morning.

Objective Examination reveals marked, equal limitation in wrist extension and flexion with empty end feels, and slightly limited supination with a muscular end feel. The wrist and fingers are mildly edematous. She has at least "fair" grade distal upper-extremity musculature, but she is unable to accept any resistance secondary to pain. She is extremely tender to palpation and will not allow any passive motion to be performed.

Assessment Joint restriction at the midcarpal and radiocarpal joints and probable muscle weakness secondary to immobilization. Pain inhibits full evaluation and treatment.

Plan

Whirlpool

Mobilization to radiocarpal, midcarpal, and distal radioulnar joints

Remedial exercise

Electrical stimulation for the performance of mobilization and exercise

Detailed Electrotherapeutic Plan

Mode of stimulation: High-intensity sensory-level stimulation to allow passive and active ROM and joint mobilization techniques.

Type of stimulator: Clinical or portable

Electrodes and placement: Small, placed on volar and dorsal surface surrounding (but not over) the wrist joint (Fig. 7.6)

Duration of treatment: Dependent on duration of therapeutic procedures.

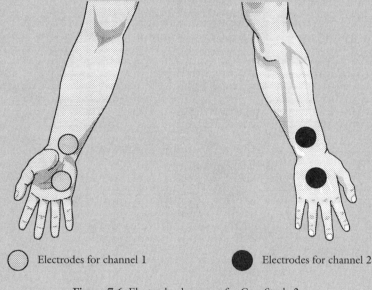

◯ Electrodes for channel 1   ⬤ Electrodes for channel 2

Figure 7.6 Electrode placement for Case Study 2.

Rationale Restricted joint motion after immobilization can cause significant pain. If the joint cannot be moved secondary to this pain, it will continue. This cycle can be broken by electrical stimulation. High-intensity sensory-level stimulation causes an instantaneous interelectrode analgesia that allows mobilization and exercise. Either clinical or portable stimulators of almost all types could be used. Small electrodes are necessary because the wrist area is small. Placement is around rather than over the joint so as not to interfere with movement. Stimulation should be applied throughout the procedures and may be left on for a short time afterward to minimize postmobilization soreness.

Case 3

A 45-year-old man, 2 years after left L4–5 laminectomy and discectomy, presents with complaints of continued, unrelenting left-sided low back pain with some radiation into the left buttock and posterior thigh.

Objective Examination reveals a slight loss of the lumbar lordosis and lateral shift to the right. There is a 50% loss of lumbosacral flexion, a complete loss of extension, and 50% loss of lateral bending to the left and rotation to the right with "stretching" pain at the end of these ranges. Movement tests reproduce these "stretching" symptoms but do not produce peripheralization or centralization of the patient's symptoms. Straight leg raise and sitting root tests are negative bilaterally. Deep tendon reflexes and sensation testing are normal. Motor function of the L2–S1 myotomes is normal. Sacroiliac torsion, compression, and distraction tests are negative. Oswestry score is 54/100.

Assessment Mechanical lumbar dysfunction accompanied by chronic low back pain

Plan

> Manipulation
>
> Remedial exercise
>
> Thermal agents
>
> Electrical stimulation for pain modulation

Detailed Electrotherapeutic Plan

> Mode of stimulation: Motor-level stimulation
>
> Type of stimulator: Portable and/or clinical
>
> Electrodes and placement: Two electrodes paraspinally at L4–5 and two at distal extent of pain or electrodes on the motor points of the hamstrings and gluteal muscles (Fig. 7.7, *A* and *B*)
>
> Duration of treatment: at least 30–45 min

Rationale This patient has a chronic problem and sensory-level stimulation will likely be insufficient to fully remediate his problem. Motor-level stimulation gives longer-lasting analgesia. A portable stimulator is preferable because the "unrelenting" nature of the patient's pain will probably require more than intermittent, clinic-based treatment. This treatment can be combined with the use of a clinical stimulator for in-clinic treatments. Electrodes can be placed directly over the target area or, if that is not well tolerated, over associated myotomes. The duration of treatment must be sufficient to allow induction of analgesia by endorphin-mediated mechanisms.

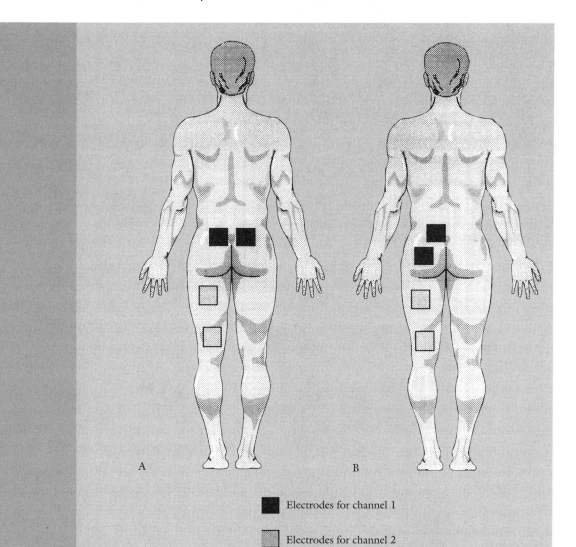

A   B

  ■ Electrodes for channel 1

  ▨ Electrodes for channel 2

Figure 7.7 Electrode placement for Case Study 3.

Case 4

A 35-year-old college professor presents with periodic complaints of severe headaches unaccompanied by any somatic complaints. The headaches are primarily concentrated in the suboccipital areas bilaterally but eventually involve the entire head. He states that they are accompanied by a "tight" feeling in his neck and shoulders and are brought on by "stress." Medical evaluation has been negative.

Objective Musculoskeletal evaluation is negative, with the exception of some muscle spasm and tenderness to palpation in both upper trapezei. He rates his pain as 10/10 when it occurs.

Assessment Muscle-contraction headache.

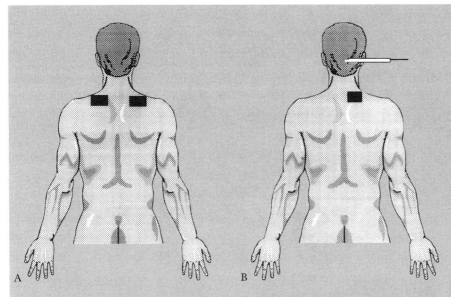

Figure 7.8 Electrode placement for Case Study 4.

Plan

Relaxation training

Electrical stimulation for pain modulation

Detailed Electrotherapeutic Plan

Mode of stimulation: Noxious- and motor-level stimulation

Type of stimulator: Clinical for noxious-level; portable for motor-level

Electrodes and placement: Probe electrode to stimulate bilaterally over occipital protuberances for noxious-level, bilaterally over motor points of the upper trapezei for motor-level (Fig. 7.8, *A* and *B*)

Duration of stimulation: 30–60 sec/stimulation site for noxious-level, at least 30–45 min for motor-level

Rationale Both motor-level and noxious-level stimulation provide long-lasting pain relief and are indicated for this type of problem. This patient may be amenable to clinic-based treatment, but if his complaints are only manifested when he has a headache, the electrotherapeutic treatment may need to be performed at home. A probe electrode is chosen for the noxious-level stimulation because the occipital protuberances are not accessible to traditional electrodes. The duration for noxious-level stimulation is brief as patient tolerance for this mode of stimulation is low.

Cases 5 and 6 demonstrate some special circumstances in which patients whose problems are not traditionally within the scope of physical therapy may be helped by electrotherapeutic intervention.

Case 5

A 31-year-old woman in the 8th month of an uneventful first pregnancy is referred for instruction in the use of electroanalgesia for labor and delivery. She is accompanied by her husband, who will be her labor "coach."

Plan Instruct patient and coach in the use of portable TENS for parturition pain

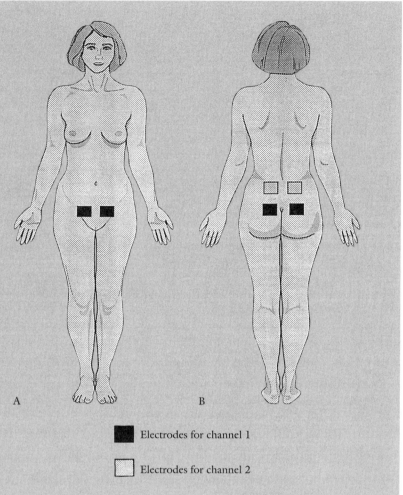

■ Electrodes for channel 1

☐ Electrodes for channel 2

Figure 7.9 Electrode placement for Case Study 5.

Detailed Electrotherapeutic Plan

 Mode of stimulation: Sensory-, motor-, and noxious-levels

 Type of stimulator: Portable

 Electrodes and placement: Stage I of labor, two electrodes paraspinally at L1 and two electrodes paraspinally at S1; Transition and Stage II, move electrodes from L1 to the suprapubic area (Fig. 7.9, *A* and *B*)

 Duration of treatment: Throughout labor

Rationale TENS has been demonstrated to be an effective means of ameliorating parturition pain. Patients and coaches are instructed in the administration of all modes of stimulation, to be used during labor as needed. Portable stimulators are well-suited for both home practice and the close confines of the labor room. Electrodes are placed for control of early and late labor pain with the lower and anterior placements used as the baby descends.

Case 6

A 44-year-old mildly obese woman with a 2-week history of right upper-quadrant pain is referred for instruction in the use of TENS for postoperative pain management. She is scheduled for a cholecystectomy tomorrow. She has a noncontributory past medical history and no history of narcotic use.

Plan

Electrical stimulation for postoperative pain

Detailed Electrotherapeutic Plan

Mode of stimulation: Sensory-level stimulation

Type of stimulator: Portable

Electrode placement: Para-incisional sterile postoperative electrodes placed in operating room (OR) by surgical team (Fig. 7.10)

Duration of treatment: As needed for incisional pain.

Rationale Sensory-level stimulation is used because only stimulation below the motor threshold will not interfere with the incision. Portable stimulators are used because the patient is moved from the OR to the recovery area to her room, all areas with little extra space. Electrodes are placed para-incisionally and are removed when the bandage is removed. Treatment is controlled by the patient but may be continuous throughout her hospital stay.

Case 7

A 21-year-old woman presents with chronic bilateral patellar tendinitis after participating in a cross country bicycle tour 6 months ago. Her patellar tendons are thickened and red. She cannot tolerate deep palpation of the tendons. She has pain with knee flexion past 90 degrees, resisted knee flexion, and inferior patellar glide. She rates her pain as 7/10 at rest and 10/10 with activity.

Plan Electrical stimulation to allow for increased weight-bearing activity and transverse friction massage.

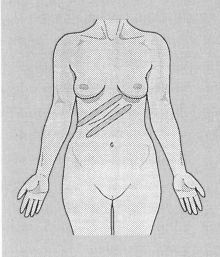

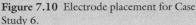

Figure 7.10 Electrode placement for Case Study 6.

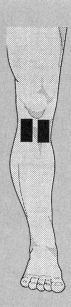

Figure 7.11 Electrode placement for Case Study 7.

Detailed Electrotherapeutic Plan

Mode of stimulation: Sensory-level stimulation, high-intensity

Type of stimulator: Clinical

Electrode placement: Small ($1'' \times 1''$), on either side of the patellar tendon (Fig. 7.11)

Duration of treatment: 10 min at the beginning of each treatment.

Rationale High-intensity sensory-level stimulation is used to cause analgesia of the patellar tendons in order to allow for them to be manipulated and to allow for weight-bearing activity. Clinical stimulator is used to provide very high-intensity stimulation and because short-duration, in-clinic treatment is planned. Electrodes are placed to focus the current on the tendons. The therapist will control the intensity to the highest tolerable level.

Case 8

A 12-year-old girl sustained a first-degree ankle sprain 1 month ago. Over the month she has developed severe reflex sympathetic dystrophy. She displays hyperesthesia and dystonia. She is unable to bear any weight at all on her left lower extremity. Her left foot and ankle are shiny, swollen, and red. She has had a sympathetic ganglion block, which resulted in a short duration decrease in her symptoms. Her pain level is rated as 90/100 using a visual analog pain rating scale.

Plan Sensory-level stimulation to facilitate progressive weight-bearing activity on the left ankle and foot

Detailed Electrotherapeutic Plan

Mode of stimulation: Sensory-level stimulation

Type of stimulator: Portable

Electrode placement: Over the peripheral sensory nerves that innervate the ankle and foot (Fig. 7.12)

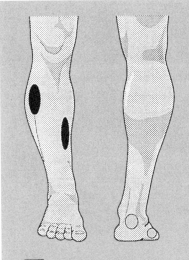

 Electrodes for channel 1

☐ Electrodes for channel 2

Figure 7.12 Electrode placement for Case Study 8.

Duration of treatment: In the clinic and at home during weight-bearing activities (Initially, begin with the patient seated, and have her attempt to bear a small amount of weight with her foot on a pillow)

Rationale

Sensory-level stimulation is used because of the patient's hypersensitivity and the fact that motor-level stimulation could interfere with the activities. A portable stimulator is used because the patient will be progressing to more vigorous activities that could not be performed if the patient were tethered. The patient will control the stimulation intensity.

The following case is used to detail a planned progression of electrotherapeutic treatment. The rationale for the first choice of device, mode, and electrode placement, and subsequent alterations in the plan of care will be explained.

Case 9

A 55-year-old businesswoman was driving 2 weeks ago and was hit from the rear while stopped at an intersection. She was wearing a seat belt. She was taken to the emergency room of General Hospital where x-rays were taken of the cervical spine and were reported as negative. She was given a soft cervical collar and a prescription for analgesics and was told to go home and rest. She has had continued neck pain and went to see her family physician yesterday, who referred her for physical therapy.

Examination The patient states that she has pain in the back of her neck and between her shoulder blades. It is exacerbated with weight bearing and activity and relieved with rest. There is no radiation into the upper extremities and no complaints of paresthesia or numbness.

Postural evaluation reveals a forward head, decreased cervical lordosis, and increased thoracic kyphosis. Passive range of motion (ROM) is full and pain-free

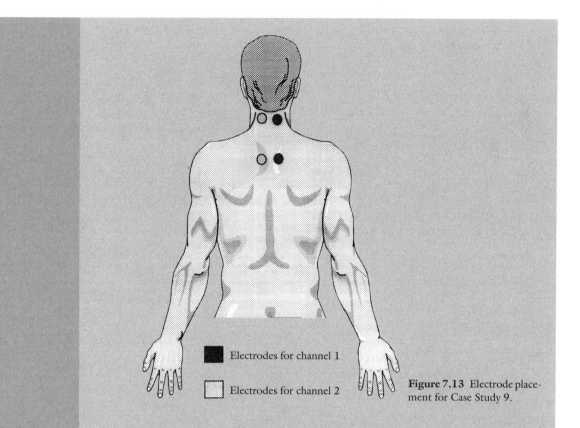

Electrodes for channel 1

Electrodes for channel 2

Figure 7.13 Electrode placement for Case Study 9.

with the exception of "stretching" pain with full flexion. Resisted isometrics and active ROM reveal pain with extension. Compression is negative; traction produces the same "stretching" pain. Deep tendon reflexes are 2+ and equal bilaterally. Sensory testing reveals no abnormality. Upper-extremity myotomal screening is negative. Patient is tender to palpation paraspinally from the occiput to approximately T5. There is no evidence of muscle spasm or trigger points.

Assessment Subacute strain of the cervical and thoracic extensors

Plan

    Remedial exercise

    Electrotherapy for pain modulation

Initial Electrotherapeutic Plan

    Mode of stimulation: Sensory-level

    Type of stimulator: Clinical

    Electrodes and placement: four electrodes placed paraspinally from C2–T5 (Fig. 7.13)

    Duration of stimulation: 20–30 min

Rationale See Case 1.

After approximately 10 min, the patient reports that there has been no change in her symptoms.

Action Change electrode placement so that the circuits cross.

Rationale Before drastically changing treatment modes or stimulator type, it is advisable to attempt electrode configuration changes as this often results in an improved outcome.

The patient responds well to this treatment and makes marked improvement over a 2-week period. She decides that she is ready to return to work, but you are concerned both that the increased activity may limit availability for treatment and necessitate stronger pain modulation. You determine that her clinic-based time should be spent on exercise.

Action Change to a portable stimulator and instruct her in its use to provide sensory-level and motor-level stimulation. Instruct her in appropriate electrode placement for motor-level stimulation.

Rationale The use of a portable stimulator at home and work will permit the clinic time to be used for exercise and may facilitate her return to work. The introduction of motor-level stimulation to the regimen will allow her to get longer-lasting analgesia from a brief treatment.

The patient gradually returns to her premorbid mobility level, is weaned from both treatment and stimulator over the next 2 months.

Summary

This chapter has addressed the ways and means of using electrical stimulation to modulate pain. There is much still to be investigated in this area. In the words of Lewis Thomas, "In real life, every field of science I can think of is incomplete, and most of them . . . are still in the earliest stage of their starting point. . . . The next week's issue of any scientific journal can turn a whole field upside down"(from his *Late Night Thoughts on Listening to Mahler's Ninth Symphony*). Regard this chapter as a beginning, a clinical guide to the ways in which various stimulators, modes of stimulation, and electrode placements can be used to effect electroanalgesia. Take this information and use it to expand the scope of your practice and enhance patient care.

Study Questions

1. What is the best way to ensure the deliverance of selective, sensory-level stimulation for pain control?

2. What are typical phase duration settings for sensory-level electrical stimulation for the relief of pain?

3. Differentiate between sensory- and motor-level stimulation based on stimulation characteristics and proposed mechanism of pain relief.

4. Differentiate between sensory- and noxious-level stimulation based on onset and duration of pain relief.

5. What are typical frequency settings for motor-level stimulation for pain control?

6. Differentiate between electrode placements for motor- versus sensory-

level stimulation for a patient with low back pain radiating down the posterior thigh.

7. List four contraindications or precautions to the use of electrical stimulation for pain control.

References

1. Melzack R. The puzzle of pain. New York: Basic Books, 1973.

2. Melzack R, Wall PD. Pain mechanisms: a new theory. Science 1965;150: 971-976.

3. Shealy CN. Six years experience with electrical stimulation for control of pain. Congr Neurol Surg 1974;4:775-782.

4. Long DM, Campbell JN, Gucer G. Transcutaneous electrical stimulation for relief of chronic pain. Adv Pain Res Ther 1979;3:593-597.

5. Gracely RH. Psychophysical assessment of human pain. In: Bonica J., ed. Advances in Pain Research and Therapy. New York: Raven Press, 1979:805-824.

6. Cummings GS, Routan JL. Validity of unassisted pain drawings by pains with chronic pain. (abstract) Phys Ther 1985;65:668.

7. Scott J, Huskisson EC. Graphic representation of pain. Pain 1976;2:175-185.

8. Agnew B, Mersky H. Words of chronic pain. Pain 1976;2:43-81.

9. Ohnhaus EE, Adler R. Methodological problems in the measurement of pain: a comparison between the VRS and the VAS. Pain 1975;1:379-384.

10. Melzack R. The McGill pain questionnaire: major properties and scoring methods. Pain 1975;1:277-299.

11. Graham C, Bond SS, Gerkovich MM, Cook ME. Use of the MPQ in the assessment of cancer pain: replicability and consistency. Pain 1982;8:377-387.

12. Deyo RA, Diehl AK. Measuring physical and psychosocial function in patients with low-back pain. Spine 1983;8:635-642.

13. Elton D, Burrows GD, Stanley GV. A multidimensional approach to the assessment of pain. Aust J Physiother 1979;25:33-37.

14. Kremer EF, Block A, Gaylor M. Behavioral approaches to treatment of chronic pain: the inaccuracy of patient self-report measure. Arch Phys Med Rehabil 1981; 62:188-191.

15. Delitto, A, Strube MJ, Shulman AD, Minor SD. A study of discomfort with electrical stimulation. Phys Ther 1992;72:410-412.

16. Lodge M. Magnitude scaling: quantitative measurement of opinion. In: Sullivan JL, ed. Quantitative applications in the social sciences. Beverly Hills, CA: Sage Publications, 1981:24-41.

17. Picker RI. Current trends: low-volt pulsed microamp stimulation, part I. Clinical Management in Physical Therapy 1988;9:10-14.

18. Picker RI. Current trends: low-volt pulsed microamp stimulation, part II. Clinical Management in Physical Therapy 1989;9:28-33.

19. Becker RO, Selden G. The body electric: electromagnetism and the foundation of life. New York: William Morrow & Co, 1985.

20. Johnson MI, Ashton CH, Thompson JW. An in-depth study of long-term users of transcutaneous electrical nerve stimulation: implications for clinical use or TENS. Pain 1991;44:221-229.

21. Nabe JB. A quantitative theory of feeling. J Gen Psychol 1929;2:199-204.

22. Chapman CR, Wilson ME, Gehrig JD. Comparative effects of acupuncture and TCS on the perception of painful dental stimuli. Pain 1976;2:265-291.

23. Chapman CR, Chen AC, Bonica JJ. Effects of intrasymental electrical acupuncture on central pain: evaluation of threshold estimation and sensory decision theory. Pain 1977;3:213-224.

24. Taylor K, Newton RA, Personius WJ, Bush FM. Effects of interferential current stimulation for treatments of subjects with recurrent jaw pain. Phys Ther 1987;67:346-349.

25. Ersek RA. Low back pain: prompt relief with transcutaneous neurostimulation. a report of 35 consecutive patients. Ortho Rev 1976;5:27-31.

26. Roberts HJ. TENS in the management of pancreatitis pain. South Med J 1978;71:396-399.

27. Nathan PW, Wall PD. Treatment of post-herpetic neuralgia by prolonged electrical stimulation. Br. Med J 1974;3:645-647.

28. Schomburg FL, Carter-Baker SA. Transcutaneous electrical nerve stimulation for postlaparotomy pain. Electrical stimulation: management of pain; Vol. 2. 191-196, 1983.

29. Issenman J, Nolan MF, Rowley J, and Hobby R. Transcutaneous electrical nerve stimulation for pain control after spinal fusion with Harrington rods. Phys Ther 1985;65(10):1517-1520.

30. Hargreaves A, Lander J. Use of transcutaneous electrical nerve stimulation for postoperative pain. Nursing Research 1989;38:159-161.

31. Dawood MY, Ramos J. Transcutaneous electrical nerve stimulation for the treatment of primary dysmenorrhea: a randomized crossover comparison with placebo TENS and Ibuprofen. Obstetrics and Gynecology 1990;75:656-660.

32. Long D. Fifteen years of transcutaneous electrical stimulation for pain control. Stereotact Funct Neurosurg 1991;56:2-19.

33. Reuss R, Cronen P, Abplanalp L. Transcutaneous electrical nerve stimulation for pain control after cholecystectomy: lack of expected benefits. Southern Medical Journal 1988;81:1361-1363.

34. Carman D, Roach JW. Transcutaneous electrical nerve stimulation for the relief of postoperative pain in children. Spine 1988;13:109-110.

35. Finsen V, Persen L, Lovlien M, Veslegaard EK, Simensen M, Gavsvann AK, Benum P. Transcutaneous electrical nerve stimulation after major amputation. The Journal of Bone and Joint Surgery 1988;70-B:109-112.

36. Smedley F, Taube M, Wastell C. Transcutaneous electrical nerve stimulation for pain relief following inguinal hernia repair: a controlled trial. Eur Surg Res 1988;20:233-237.

37. Walker RH, Morris BA, Angulo DL, Schneider J, Colwell, CW Jr. Postoperative use of continuous passive motion, transcutaneous electrical nerve stimulation, and continuous cooling pad following total knee arthroplasty. The Journal of Arthroplasty 1991;6:151-156.

38. Melzack R, Vetere P, Finch L. Transcutaneous electrical nerve stimulation for low back pain: a comparison of TENS and massage for pain and range of motion. Physical Therapy 1983;63:489-92.

39. Deyo R, Walsh NE, Martin DC, Schoenfield LS, Ramamurthy S. A controlled trial of transcutaneous electrical nerve stimulation and exercise for chronic low back pain. The New England Journal of Medicine 1990;322:1627-1634.

40. Barr JO, Winter A, Conwill DE, Cook BH, Merry A, Todd GA, Deyo RA. Letters to the editor. The New England Journal of Medicine 1990;323:1423-1425.

41. Gemignani G, Olivieri I, Ruju G, Pasero G. Letter to the editor. Lancet 788-789.

42. Johnson MI, Ashton CH, Bousfield DR, Thompson JW. Analgesic effects of different frequencies of transcutaneous electrical nerve stimulation on cold-induced pain in normal subjects. Pain 1989;39:231-236.

43. Tulgar M, McGlone F, Bowsher D, Miles JB. Comparative effectiveness of different stimulation modes in relieving pain. Part 1, a pilot study. Pain 1991;47:151-155.

44. Tulgar M, McGlone F, Bowsher D, Miles JB. Comparative effectiveness of different stimulation modes in relieving pain. Part 2, a double-blind controlled long-term clinical trial. Pain 1991;47:157-162.

45. Petrie I, Hazleman B. Credibility of placebo transcutaneous nerve stimulation and acupuncture. Clinical Experimental Rheumatology 1985;6:936-9.

46. Langley BG, Sheppard H, Johnson M, Wigley RD. The analgesic effects of transcutaneous electrical nerve stimulation and placebo in chronic pain patients: a double-blind non-crossover comparison. Rheumatol International 1984;4:119-23.

47. Chen D, Philip M, Philip PA, Monga TN. Cardiac pacemaker inhibition by transcutaneous electrical nerve stimulation. Arch Phys Med Rehabilitation 1990;7:27-30.

48. Grim LC, Morey SH. Transcutaneous electrical nerve stimulation for relief of parturition pain. Physical Therapy 1985;65:337-340.

Suggested Readings

Electrotherapeutic terminology in physical therapy. Alexandria, VA: Section on Clinical Electrophysiology of the American Physical Therapy Assoc., 1990.

Hymes AC, et al. Acute pain control by electrostimulation: a preliminary report. Adv Neurol 1974;4:761.

Mannheimer C, Carllsson CA. The analgesic effect of transcutaneous electrical nerve stimulation in patients with rheumatoid arthritis: a comparative study of different pulse patterns. Pain 1979;6:329–334.

Mannheimer J, Lampe G. Clinical transcutaneous electrical nerve stimulation. Philadelphia: FA Davis, 1983.

Melzack R. The McGill pain questionnaire: major properties and scoring methods. Pain 1975;1:277–299.

Melzack R, Wall PD. Pain mechanisms: a new theory. Science 1965;150:971–976.

Nolan, MF. A chronological indexing of the clinical and basic science literature concerning TENS, 1967–1987. Alexandria, VA: American Physical Therapy Assoc., 1987.

Wolf SL, Gersh MR, Rao VR. Examination of electrode placement and stimulating parameters in treating chronic pain with conventional electrical nerve stimulation (TENS). Pain 1981;11:37–47.

Chapter 8

Electrical
Stimulation
for Tissue Repair

Lynn Snyder-Mackler

"**A**nd that which is done is that which shall be done: and there is no new thing under the sun." This biblical quotation from Ecclesiastes is a fitting introduction to the concept of electrically induced tissue repair. As early as the 17th century, electricity was used to improve tissue healing (1). The renaissance of electrotherapy since the late 1960s has included an increased interest in the effects of electrotherapy on the healing rates of various tissues. In the 1980s and 1990s, the literature has continued to expand as electrical stimulation has been used in a variety of ways to promote tissue repair. Not all of the studies have supported the use of electrical currents for tissue repair; some applications that were described in the first edition of this text are no longer supportable based on new evidence. As JM Rothstein stated in a recent editorial, "over the course of our profession's evolution, we have developed many new techniques and have come to use a wide variety of devices and techniques...Eliminating devices and techniques [that are not effective] from our armamentarium does not weaken practice, rather it fortifies scientifically based physical therapy" (2).

Electrical stimulation for tissue repair (ESTR) is the use of electrical stimulation to directly or indirectly enhance tissue healing. ESTR may effect a reduction in inflammation by increased vascular supply, iontophoretic application of pharmacologic agents, edema control, or other unknown mechanisms. This chapter addresses the physiological effects of electrical currents on nonexcitable tissue and electrical stimulation for tissue repair in its various forms: electrical stimulation for improvement of vascular status, edema control, direct enhancement of wound healing, and osteogenesis. Representative case studies are presented as illustrative examples.

Electrical Stimulation and Circulation

Electrical stimulation can be used to improve peripheral, vascular perfusion by one of two possible mechanisms: reflexly, by activation of autonomic nerves, or by muscle contraction acting either as a pump or as a stimulus to group III and IV afferent fibers via a purported somatosympathetic reflex (3). Evidence for the effectiveness of the techniques used to produce these responses varies and is detailed below.

Effects of electrical stimulation on sympathetic tone and microvascular perfusion

Increased capillary perfusion has been accomplished using both **sensory-level** and **motor-level stimulation** (see Chapter 6). Stimulation has been applied either directly over the target area or in remote areas to achieve this effect. Although some experimenters report increased sympathetic responses or no change in sympathetic tone after electrical stimulation, most find a sympatholytic effect (e.g., cessation of sympathetic activity) (4–7). Changes in microvascular perfusion may occur in the absence of an increase in regional blood flow and may reflect changes other than an increase in local blood flow.

The effect of **transcutaneous electrical nerve stimulation (TENS)** on peripheral vasodilation was investigated by Dooley and Kasprak. A portable

TENS device was used to deliver sensory-level stimulation (SLS) to electrode placement sites at C-5, T-10, the ulnar nerve, and the proximal sciatic nerve. TENS over the spinal cord (but not over peripheral nerve) was shown to cause peripheral vasodilation, measured by electrical plethysmography (5). Owens and associates found transient increases in the temperature of the palms as measured by thermography after SLS to the ulnar nerve at 75 pps (8).

Kaada and colleagues have demonstrated increased vasodilation in patients with Raynaud's disease, diabetic polyneuropathy, and chronic lower-extremity wounds. Motor-level stimulation (MLS) to the dorsal web space of the hand and the ulnar forearm was delivered by a portable TENS device. A 100-pps balanced, asymmetrical, biphasic pulse was burst-modulated at 5 bps and applied for 30 min (7). The vasodilation was blocked by serotonin inhibitors, but not changed by either naloxone (an opiate antagonist) or antagonists of endogenous vasodilators (9–12). Proposed mechanisms for this response included nonperipheral release of vasoactive peptide, a serotonin-mediated effect, or blocking of sympathetic vasoconstriction.

Ernst and Lee found a transient increase in sympathetic tone after MLS using an 800-μsec monophasic pulse delivered to the dorsal web space of the hand at 1 pps at amplitudes just below the pain threshold (13). This effect persisted for up to 50 min and was followed by vasodilation. Ebersold and colleagues used SLS and failed to effect any changes in sympathetic tone in patients with chronic pain (14).

Wong and Jette found the SLS and MLS of acupuncture points in the upper extremity caused a decrease in fingertip temperature in healthy subjects. Asymmetrical, biphasic, pulsed current was delivered at a pulse duration of 100 μsec and frequency of 85 pps (SLS), a pulse duration of 250 μsec and frequency of 2 pps (MLS), and a pulse duration of 250 μsec, frequency of 85 pps in 2 bps (MLS) for 25 min. All modes of stimulation caused a sympathomimetic vasoconstriction response in the fingertip (4).

There is evidence that increased sympathetic activity is associated with pain. Wong and Jette suggest that relaxation due to pain relief may explain the vasodilation demonstrated by others. A more likely explanation of the contrary findings in their study may be that others studied patients whereas Wong and Jette used healthy subjects. The preponderance of evidence in patient studies shows that electrical stimulation can be used to decrease sympathetic activity by one or more of the mechanisms cited above.

Clemente and colleagues stimulated the tibialis anterior and extensor digitorum longus muscles of rats with motor-level stimulation. A 2500-Hz sine wave with a 50% duty cycle (10 msec) was delivered at three times the amplitude required for a minimal visible contraction with a 12 sec : 10 sec on/off ratio for 30 min. Muscle circulation was monitored at intervals from 0–30 min after cessation of the NMES. Microvascular perfusion was increased in both muscles immediately after stimulation and gradually returned to normal levels over the next 10–30 min. The authors attributed the increase in circulation to both (*a*) a direct effect of muscle contraction on group III and IV via a somatosympathetic reflex arc resulting in a depressor response and (*b*) the effects of metabolic by-products of muscle contraction (lactate and CO_2) on local circulation (15). Reed, in his commentary on this paper (16), points out that although the precise mechanism cannot be definitively ascertained from this study, the authors' use of a stimulator capable of producing strong muscle contractions provided a "best shot" means of trying to detect changes in microvascular perfusion because strong contractions both purportedly activate the somatosympathetic reflex and also effectively produce metabolites.

Therapeutic use of these stimulation techniques has been largely confined to patients with vasospastic disorders, surgical or disease-related sympathectomy, Raynaud's disease, and reflex sympathetic dystrophy.

Effects of electrical stimulation on regional blood flow

The preponderance of evidence suggests that rhythmic muscle contraction is required to increase arterial blood flow to the stimulated area. Strong evidence indicates that electrical stimulation below the motor threshold does not affect arterial flow to the target area.

In studies of animals and human subjects since the 1950s, stimuli that produce even weak contractions (10% of maximum voluntary contraction) and allow for relaxation cause poststimulation increases in blood flow to the target area (17, 18). Wakim examined the effect of monophasic pulsed current at frequencies ranging from 4–256 pps on blood flow in dogs. He found that blood flow increases were greatest when the continuous series of pulses was delivered at 16 pps. Frequencies of 64 pps and greater resulted in a continuous tetanic muscle contraction and relaxation never occurred. Frequencies of 4 and 2 pps resulted in unfused contractions and he hypothesized that not enough "work" was being done to effect a change in circulation (19). Randall, Imig, and Hines (20) and Petrofsky and colleagues (21) found blood flow to be elevated after electrical stimulation. Currier, Petrilli, and Threlkeld tested the effect of electrically elicited isometric contractions of 10 and 30% of MVC of the gastrocnemius on blood flow to the leg during and following contractions. They used a 2500-Hz alternating current packaged in 10-msec bursts. A Doppler device was used to measure the pulsatility index of the popliteal artery, and a significant increase in blood flow to the ipsilateral leg occurred during the first minute of stimulation and remained poststimulation (22).

In contrast, Barcroft and others in the 1930s studied sustained and rhythmic volitional contractions at 10 and 30% of MVC and found decreased blood flow during contraction and increased arterial blood flow after stimulation ceased (23, 24). Carlsson, Currier, and Threlkeld recently duplicated this study using one 5-min continuous electrically elicited contraction using HVPC (high voltage pulsed current) at 30 pps and 30 volitional contractions per minute. They concluded that HVPC-evoking contractions at 10 and 30% of MVC did not increase circulation over controls, whereas volitional exercise did (17). The effect of the continuous contraction on blood flow was likely constrictive and cannot be compared with the rhythmic contractions evoked in previous studies. Hecker and coworkers also found no significant changes in arterial blood flow in the legs of healthy human subjects using HVPC at a variety of frequencies, although they noted a trend toward increasing blood flow with increasing frequencies (25). There is no reason to assume that electrically elicited contractions evoked by HVPC should behave differently than other stimulators.

In recent work, Liu and colleagues, using the same protocol as the Currier group, demonstrated that the increase in blood flow to the stimulated area is accompanied by a decrease in blood flow to another area of the body (the contralateral long finger) (26). This suggests that electrically elicited contractions produce a systemic effect much like that seen with voluntary contractions, where blood is shunted toward the working muscle at the expense of circulation to "nonworking" areas.

The effect of varying stimulation frequency during motor-level stimulation on circulation has not been investigated thoroughly. Wakim's work demonstrated that, to a point, blood flow increased with increasing frequency. Once frequencies exceeded those required for tetany, blood flow decreased. These results must be interpreted with caution, however, because the stimulation was not cycled; there were no rest periods. Uninterrupted tetanic contractions produced by an uninterrupted series of pulses or AC impedes blood flow by mechanical compression of the vasculature similar to what occurs during continuous, voluntary, isometric contraction. One cannot generalize to protocols where cycled stimulation is used. Likewise, Hecker and associates' finding that blood flow tended to increase with increasing frequency could be due to the increasing force generation that accompanies an increase in stimulus frequency.

Tracy and associates tested the effect of increasing frequency on blood flow using the 200-μsec, symmetrical, biphasic pulsed current and the 2500-Hz alternating current burst-modulated every 2 to 10 msec (27). Blood flow increased as stimulus frequency increased from 1 to 50 pps. Stimulation at 1 pps was not significantly different from control values. Stimulation at 50 pps resulted in blood flow increases that were greater than 10 and 20 pps but not significantly. All three frequencies resulted in more blood flow than controls. There was no difference in the effect of the two stimulators. Mohr and colleagues studied the effect of high-voltage pulsed current on blood flow in the rat hind limb as measured by a Doppler flowmeter. They studied the effect of HVPC at various frequencies (2, 20, 80, 120 pps), intensities (20–200 V), and polarities on blood flow. They found that frequencies of 20 pps, more intense muscle contractions, and negative polarity stimulation resulted in greater increases in blood flow (28). Mohr and associates concluded that blood flow increases with increasing contraction strength, with intermittent muscle contraction producing a better "muscle pumping" action (29). Mohr's conclusion has more experimental support than any differing effect of frequency alone.

Twist used NMES to the quadriceps of a patient with an incomplete C5–C6 spinal cord injury to treat clinical acrocyanosis (30). Twenty years after injury, the patient developed this circulatory problem in both feet as evidenced by diminished pedal pulses and blackened, ulcerated toes. The patient was treated with NMES (350 μsec duration, 30 pps, 8 sec : 12 sec on/off ratio for 30–40 min 1 to 3 times per week) to the quadriceps femoris muscles bilaterally for 6 weeks. After 5 weeks, the discoloration in his feet was eliminated and the ulcers had nearly healed. This case report provides some interesting preliminary information about the possible effect of muscle contraction on circulation. These studies as a group provide some direction and renewed evidence for the use of strong muscle contraction to improve circulation (Table 8.1). The mechanisms are not clearly established, but there appears to be a strong correlation between strong, rhythmic muscle contraction and

Table 8.1
Common Stimulation Characteristics for Various ESTR Applications

Application	Type of current	Phase duration (μsec)	Amplitude	Treatment duration
Edema control	1) Pulsed or burst-mod AC	>100	Rhythmic muscle contraction	*
	2) High-voltage, pulsed monophasic, negative polarity	2–50	90% of motor threshold	30 min every 4 hours
Improve vascular status	1) Pulsed or burst-mod AC	>2	Sensory level	*
	2) Pulsed or burst-mod AC	>100	>10% MVC	10 min
Wound healing	1) DC	NA	<1 mA	1–2 hrs BID
	2) pulsed	2–150	<Motor threshold	45 min

* = Undetermined

improved arterial blood flow. There are strong parallels between the use of electrically elicited muscle contraction to increase circulation and its use to control edema, which is detailed in the next section.

Electrical Stimulation and Edema Control

Electrically elicited muscle contraction, used to mimic rhythmic, voluntary, muscle contraction, has been implicated in the resolution of posttraumatic and chronic edema. The role of voluntary muscle contraction in aiding lymphatic and venous drainage is well-documented (31, 32). Electrical stimulation may provide an effective increase in venous or lymphatic drainage when voluntary exercise cannot be performed or is insufficient.

Apperty and Cary used electrically stimulated muscle contractions of the lower extremities to treat persons with impaired venous blood pressure and decreased systolic blood pressure and demonstrated increased venous return (33). Doran and colleagues have reported the use of intraoperative, motor-level, lower-extremity electrical stimulation to prevent thromboembolic problems resulting from venous stasis (32). Stimulation characteristics are detailed in Table 8.1. Other reports of direct effects on edema are anecdotal. The above studies of increased venous return suggest that MLS may be effective for the control of posttraumatic or chronic edema.

Since 1974 when the first commercially available "high-voltage" stimulator was marketed, claims such as those found in a commercial monograph written in 1984 (34) have been made about its efficacy in reducing posttraumatic edema using sensory-level stimulation. This type of stimulator continues to be used with presumed effectiveness by many clinicians. Until recently, only case reports and testimonials have supported the use of sensory-level HVPC for edema control (35–39).

The protocols detailed in the monograph were tested by Michlovitz and associates in the treatment of acute ankle sprains (40). No differences in edema control were noted when HVPC was added to the standard ice, compression, and elevation treatment. Griffin and coworkers compared motor-level HVPC (8 pps) with intermittent pneumatic compression (IPC) and sham HVPC in 30 patients with chronic hand edema who were randomly assigned (41). Volumetric measurements were made before and after a single 30-min treatment session. IPC treatment significantly reduced the edema as compared to placebo HVPC; there was no significant difference between either the IPC and HVPC groups ($p = .446$) or the HVPC and placebo HVPC groups ($p = .04$). The authors stated that they judged the difference between the HVPC and placebo HVPC to be clinically significant.

Mohr and coworkers, in a recent study of the effect of HVPC on edema reduction in the traumatized rat hind limb, also found that sensory-level stimulation with this device produced no greater decrease in edema than that observed in untreated controls (28). Cosgrove and colleagues compared HVPC, symmetrical, biphasic pulsed current and sham stimulation in a study of post-traumatic edema in the rat hind limb (42). Stimulation was delivered just below the motor threshold for 1 hour at 24, 48, and 72 hours after injury. There were no differences among the groups at 96 hours after injury.

Reed studied the effect of HVPC above and below motor threshold on microvessel leakiness in hamsters (43). HVPC at 120 pps was delivered at either 10 V, 30 V (both subthreshold for visible muscle contraction), or 50 V coincident with application of histamine. Posthistamine microvessel leakiness to plasma proteins was significantly reduced in the 30 and 50 V groups over the 10 V group and control conditions. These results suggest that HVPC may reduce the formation of edema, which has interesting implications for the use of HVPC for edema control.

A recent series of studies from the laboratory of Fish has provided some interesting information about the effects of electrical stimulation on edema formation in frogs and rats. Edema was produced by traumatizing limbs with a fixed weight dropped at a fixed distance. In all but one study, HVPC at 120 pps (5–8 μsec in duration) was used (44). In all but one of the studies using HVPC, the dose was 90% of motor threshold (10% less than motor-level). In all but one of the remaining studies using HVPC, the cathode was placed over site of injury. They demonstrated that four sessions of HVPC, administered via the cathode immediately after injury for 30 min with either a 30-min or 60-min rest between sessions, reduced edema formation in frogs and rats (45–47). In two additional studies using a frog model, one 30-min session of cathodal HVPC retarded edema formation for 4 hours and one 6-hour session of cathodal HVPC applied 4.5 hours after injury reduced edema formation. Neither four 30-min sessions of anodal HVPC, nor motor level HVPC at 1 pps, nor monophasic "low-voltage" pulsed current at 100 pps curbed edema formation in frogs (44, 48).

These studies provide some insight into the physiology by which edema formation be reduced by electrical stimulation early after injury. First, muscle pumping does not help. Second, the effect is likely not neural because neither anodal HVPC nor low-voltage electrical stimulation at 90% of motor threshold retarded edema formation. If sensory-fiber excitation mediated the response, both of these treatments should have been indistinguishable from that of the cathodal HVPC at a similar intensity. Perhaps, as suggested by Reed (43), the cathodal HVPC decreased microvascular leakiness, thereby retarding atrophy.

These studies did not demonstrate a reduction in edema that had already formed, but rather a retardation in edema formation. In all studies, the volumes of the treated and untreated limbs converged within 24 hours after treatment ceased. In all but one case, treatment commenced within minutes of the injury. The relevance of these studies to clinical practice is debatable. Arguably, it is unlikely that treatment would begin within minutes of traumatic injury in many instances. It is possible, however, that treatment with HVPC as used in these studies could retard edema formation if used immediately after a treatment that might cause edema formation. The finding that edema formation approaches (and in some cases exceeds) that of the untreated extremity within 24 hours of cessation of stimulation is problematic, as is the every-4-hour treatment regimen. These investigators have just scratched the surface of the dose–response question and may clarify effective current characteristics further with future studies.

The studies cited above suggest that the effects of muscle pumping may resolve edema once it is formed and that sensory-level cathodal stimulation with HVPC may retard edema formation. Comparison with other treatments (ice, compression, IPC) has not demonstrated a differentially better or even equivalent effect of MLS in resolving edema (49, 50). The reduction in edema formation by cathodal HVPC has yet to be tested in a patient model. Mendel and Fish have recommended a tentative protocol for clinical practice that includes application of cathodal HVPC at 120 pps at 90% of motor threshold for 30 min every 4 hours as long as edema is likely to be forming (51).

Wound Healing

The problem of chronically open wounds is a severe one. Care of these lesions is protracted, costly, and often unsuccessful. The cost-per-wound for treatment of a pressure ulcer in a spinal cord–injured individual has been estimated to approach $30,000. Acceleration of the healing process, or, in some instances, any healing at all would be welcome. Many therapies, some with scientific merit and some that resemble the medicine-show "tonics" of the 1800s, have been tried.

The first reported use of electrical current to aid tissue healing was the use of electrically charged gold leaf to decrease scarring from smallpox (1). This and all future successful uses of electrically charged gold leaf may not be an example of ESTR, but rather an example of gold iontophoresis (52–54). Treatment rationales are described below. For a description of electrotherapy techniques for wound management in common use, see Table 8.1.

The polarity of nonexcitable tissue such as epidermis, dermis, and subcutaneous tissue has been documented. Becker and Murray report that normally the skin in the spinal cord region is positively charged with respect to the periphery (55). They hypothesize that tissue polarity reverses after injury. This "current of injury" theory proposes that wounds are initially positive with respect to the surrounding tissue, that this positive polarity triggers the onset of the repair processes, and that maintaining this positive polarity would potentiate healing. The "current of injury" theory would seem to be supported by the fact that the positive polarity is transient; chronic open wounds no longer posses this triggering error signal. Alvarez and colleagues studied the effects of low-amperage direct current on induced wounds in pigs and found increased rates of collagen synthesis (56). Cheng and colleagues studied the effect of low-amperage direct current on skin slice preparations and demonstrated increased amino acid uptake, ATP resynthesis, and protein synthesis (57). Bourguignon and Bourguignon showed that HVPC increased protein synthesis in human fibroblasts in vitro (58).

Dunn and coworkers recently examined the effect of direct current on collagen matrix implanted into full-thickness dermal wounds in a guinea pig model. They found that low-amperage direct current (cathode over wound) enhanced fibroblast migration and collagen alignment, whereas the anode placed over the wound attracted inflammatory cells (59).

Several clinical studies, however, suggest that placement of the anode in a direct-current circuit directly over the wound enhances tissue healing. Evidence for the use of the anode over the wound with direct-current amplitudes below 1 mA for enhancement of wound healing is strong. However, the absence of controls inherent in wound healing studies and the reporting methods used make interstudy comparison difficult.

Wolcott and colleagues treated 75 chronic wound patients with direct current and anode over wounds (60). Eight of these patients had bilateral, symmetrical lesions; only one of each pair of wounds was treated with ESTR. Current of 200 μA to 400 μA were used for sentient patients and currents of 400 μA to 800 μA were used for patients with chronic wounds, six of whom had bilateral lesions (61). In both studies, healing rates for the electrically stimulated wounds were significantly faster than the control lesions.

Carley and Wainapel randomly assigned 30 patients, by pairs, to ESTR or conventional treatment groups (62). After being matched for etiology, diagnosis, wound size, and age, positive direct current was found to produce healing rates, 1.5–2.5 times faster than in matched controls. All of the above studies used treatment durations of 2 hours, two to three times per day until closure, with anodal placement over the wounds. Treatment intensities were similar to those used in the Wolcott study.

Wound healing is also impeded by infection. Electrical stimulation using the negative lead of a direct-current generator has been shown in culture and in vivo either to be bacteriostatic or to retard the growth of common gram-negative and gram-positive microorganisms (63, 65). Rowley demonstrated a 38.8% reduction in the growth rate of *Escherichia coli* in vitro with the application of low-amplitude cathodal current (65). Rowley and colleagues have demonstrated a bacteriostatic effect of low-intensity, negative, direct current in rabbit wounds, contaminated by *Pseudomonas aeruginosa*, a gram-negative rod that commonly contaminates human wounds (65). Kincaid and Lavoie studied the effect of HVPC on bacterial growth in vitro (66). Three bacterial species, *Pseudomonas aeruginosa*, *Staphylococcus aureus*, and *Escherichia coli*, were grown and exposed to HVPC at 120 pps at voltages of 150, 200, 250, and 300 V for 1, 2, 3, and 4 hours. There was a significant, direct, linear relationship between duration of exposure and voltage at the cathode, and width of the zone of inhibition of bacterial growth of the organisms. All three organisms were equally affected. The dosages (voltages and exposure times) were higher than those used in most clinical studies, and results cannot be extrapolated to clinical protocols. It is encouraging, however, that short-duration pulsed current such as HVPC may have a bacteriocidal effect.

In 1966 a report was published on the effect of pulsed current on experimentally induced ischemic wounds in dogs (67). An HVPC was used to produce muscle contraction. Brown and Gogia (68) and Brown and coworkers (69) studied the effect of HVPC on the healing of surgically induced wounds in rabbits. HVPC at 80 pps at an amplitude sufficient to induce a palpable muscle contraction was delivered for 2 hours per day for 4 or 7 days. In the first set of experiments, HVPC was used with the cathode over the wound and control animals healed better after 7 days than the HVPC-treated animals. In the second set of experiments, HVPC with the anode over the wound was used and wound closure was accelerated between days 4 and 7 for the treated group as compared to the control animals. Tensile strength was not different between groups, but reepithelization appeared to be positively affected by the anodal HVPC treatment.

Im and associates compared a monophasic pulsed current (duration 132 μsec) delivered at 128 pps at an amplitude of 35 mA to sham and no stimulation in the healing of skin flaps in pigs (70). The experimental period was 9 days of BID, 30-min treatments with polarity alternated every 3 days beginning with negative current. The survival of the treated skin flaps was significantly better than those that were not treated. Mertz and colleagues studied the effect of polarity of an almost identical current on wound epithelization in surgically induced wounds in pigs (71). Four different regimens were used for 7 days: alternating polarity each day, all positive, all negative, and negative for 1 day followed by positive for the remaining days. All were compared to sham-treated wounds. Alternating polarity significantly inhibited epithelization. Negative polarity also had an inhibitory effect. Positive polarity and negative followed by positive enhanced wound epithelization over sham-treated animals.

Subsequent investigation of pulsed current for wound management has included case studies and reports and has, for the most part, been positive (72–74). In a recent study, Kloth and Feedar have demonstrated the effectiveness of HVPC in the acceleration of wound healing in humans, using a frequency of 105 pps with stimulus intensity below the motor threshold and treatment durations of 4–5 min/day, 5 day/week. Nine patients received HVPC and seven received sham treatment. Over an 8-week period, the treatment group had 100% healing while the sham group's wounds increased in size an average of 28.9% (75). Griffin and coworkers randomly assigned 17 patients with spinal cord injury and pelvic region pressure ulcers to a treatment or control group (76). Treatment consisted of cathodal HVPC at 100 pps and 200 V for 20 days. Despite some initial differences between the groups (randomization is not likely to account for all differences in a sample this small), there was significantly better healing in the group treated with HVPC.

Mulder randomly assigned 47 patients with stage II, II and IV ulcers to a treatment or sham treatment group (77). A monophasic pulsed-current stimulator (128 pps, 132 μsec, 29.3 mA, 30 min, BID) was used to treat the wounds with negative polarity until 3 days after they were debrided or exhibited serosanguineous drainage, and then polarity was changed every 3 days until the ulcers reached stage II. The pulse frequency was reduced to 64 pps and polarity was changed daily. After 4 weeks, treated wounds were significantly smaller than the sham-treated wounds. Gentzkow and colleagues used an almost identical stimulator and treatment paradigm in a study of 61 stage II and IV pressure ulcers in a baseline controlled study (78). After 4 weeks of "standard care," wounds that had not responded or had deteriorated were entered into the treatment paradigm. Improvement, defined as a change of one wound grade or two wound characters, was demonstrated in 82% of the wounds, and complete closure in 23% (Table 8.2).

Other studies that have used modulated alternating or symmetrical biphasic pulsed current also appear to be encouraging (79). The use of these currents cannot be based upon "currents of injury" or any other polar phenomenon. Even HVPC is of such short duration as to render its polar effects negligible; however, the evidence suggests that cathodal and anodal monophasic pulsed currents, including HVPC, have different effects of healing.

Sub-Sensory-level stimulation

Sub-sensory-level stimulation (also called *microcurrent stimulation*, or *MENS*) has been reported to aid tissue healing (80, 81). Most recently, two studies of surgically induced wounds in animal models by Byl and colleagues and Leffman have challenged this assertion. Neither group found any effect of treatment over sham stimulation using commercially available stimulators, recommended current characteristics, and ideal conditions for wound healing. Robinson, in his commentary on these papers, points out that although the two studies used current densities at the low and high end of those recommended by the purveyors of these devices, neither had an effect (82). There is no other evidence for the effectiveness of sub-sensory-level stimulation for the healing of open wounds, and these two studies should close the door on the use of this modality in this manner to aid wound healing.

Table 8.2
Definition of Wound Character

1. Covered by eschar
2. Necrotic purulent drainage
3. Exudate/seropurulent drainage
4. Creamy exudate/pale yellow drainage
5. Granulating base/serosanguineous drainage
6. Granulating tissue/serous drainage
7. Full granulation with epithelial progress
8. Total epithelization

Electrical Stimulation for Bone Healing

Nonunion and malunion fractures are as problematic for physicians and therapists as chronic open wounds. The response of bone growth to stress and its electrical correlates has been described. The "current of injury" theory for bone involves a relative negativity of the injured tissue with respect to the uninjured. Yasuda first reported in 1953 that cathodal current augments osteogenesis (83). The first reported clinical use of electrical stimulation to induce fracture healing was by Friedenberg and associates in the early 1970s (84). Since that time, low-amperage direct currents and implanted electrodes have been used to induce new bone growth in patients with nonunion fractures. This technique is often used immediately in conjunction with traditional fixation for fractures that are prone to nonunion (e.g., the scaphoid and the distal tibia). This technique had also been used to enhance bone graft fixation in patients (85) and animal models (86) after spinal fusion.

The current used originally was a low-amperage direct current. This type of current appears to rigidly adhere to the Arndt-Schultz biological law. Currents below 5 μA are below the threshold for osteogenesis; those from 5 to 20 μA appear to produce significant osteogenesis; currents above 20 μA are destructive. The three best-studied and most commonly used techniques are (*a*) cathodal placement in the fracture site and anodal placement on the skin at some distance from the cathode (semiinvasive DC) (87), (*b*) implantation of the entire system (totally invasive DC) (88), (*c*) and the use of pulsed electromagnetic fields (PEMFs). PEMF is the use of inductive coils similar to those used in diathermy external to the skin or cast to deliver an asymmetrical, biphasic pulse at a frequency of about 15 pps. Some recent application of PEMFs in animal models suggest that disuse osteoporosis can be prevented by the use of this noninvasive technique (89).

Semiinvasive DC, totally invasive DC, and PEMF were the only FDA-approved (and physician administered) osteogenic means until recently. Now, 60-Hz sinusoidal AC, pulsed currents, and interference modulations of higher-frequency alternating currents are also being used (90). Transcutaneous bone-growth stimulators have been developed for all of these types of currents. Both indwelling and transcutaneous electrodes are now being used and their efficacies appear comparable.

Case Studies

The following case studies are meant to serve as illustrative examples of the clinical decisions involved in the use of electrotherapy for tissue repair.

Case 1

A 40-year-old woman with a history of systemic lupus erythematosus, including Raynaud's syndrome, is referred with complaints of painful vasoconstriction of both hands. She is currently without any complaints of arthralgias and is ambulatory and self-sufficient.

Examination Examination reveals cold, painful hands, restricted active motion of the fingers and thumbs, and full passive motion. Other joints of the upper extremity display no abnormalities on musculoskeletal testing.

Assessment Active Raynaud's syndrome

Plan

Electrotherapy to increase circulation

Temperature biofeedback

Detailed Electrotherapeutic Plan

Mode of stimulation: Sensory-level stimulation

Type of stimulator: Portable and/or clinical

Electrodes and placement: four electrodes, two over the spinal nerve roots at C6–C8 bilaterally and two over the trunks of the brachial plexus bilaterally (Fig. 8.1)

Duration of treatment: 10–30 min, up to several times/day

Rationale Sensory-level stimulation over nerve roots and trunks has been shown to increase peripheral vasodilation. Motor-level stimulation of proximal muscles could also be used. This patient has a problem that interferes with function and is painful and debilitating. It is a chronic condition so a portable stimulator is chosen. If the patient's motor skills were insufficient for manipulation of the device, clinic-based treatment would be initiated. Ten minutes has been demonstrated as sufficient time for the induction of the vasodilatory response.

Case 2

A 67-year-old man with peripheral vascular disease is referred for management of impaired circulation and progressive lower-extremity dysfunction.

Examination Musculoskeletal evaluation reveals multiple problems with posture, gait, range of motion, and strength of the lower extremities. Examination of the skin shows marked trophic changes below the knee and stocking-type hypesthesia to the mid-calf bilaterally.

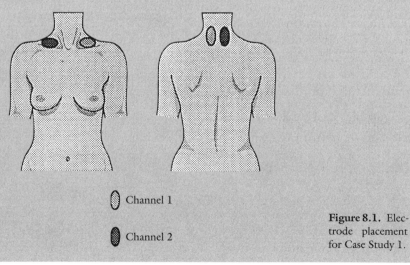

◯ Channel 1

⬤ Channel 2

Figure 8.1. Electrode placement for Case Study 1.

Plan

1. Tepid whirlpool
2. Electrotherapy to increase circulation
3. Remedial exercise
4. Gait training

Detailed Electrotherapeutic Plan

Mode of stimulation: Rhythmic low-level muscle contraction (see Table 8.1)

Type of stimulator: Clinical (perhaps portable later) phase-duration

Electrodes and placement: Two electrodes placed on the motor points of the gastrocnemius muscle bilaterally and set to cause intermittent isometric or isotonic contractions (Fig. 8.2)

Duration of treatment: At least 10 minutes

Rationale This type of neuromuscular electrical stimulation has been shown to produce increases in the circulation to the lower extremities. Ten minutes has been demonstrated to be sufficient for induction of this response. A clinical stimulator would be used if clinic-based treatment were sufficient for remediation of the problem. If treatment at home were necessary, a portable stimulator could be used.

Case 3

A 36-year-old recreational volleyball player states he "turned his ankle" during a game 1 hour ago.

Objective The patient is unable to bear weight on his left lower extremity. His left ankle is beginning to swell. He is tender to palpation over the anterior talofibular ligament and the insertion of the peroneus brevis. Contractile testing is negative. Passive range of motion testing reveals an increased inversion range. There is demonstrable ligamentous laxity of the anterior talofibular ligament and calcaneofibular ligament with firm endpoints for both.

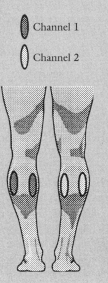

Channel 1

Channel 2

Figure 8.2. Electrode placement for Case Study 2.

Assessment Acute, second-degree, inversion sprain of left ankle with developing edema

Plan

1. Ice bath
2. Electrotherapy in water
3. Gait training
4. Remedial exercise
5. Compression wrap

Detailed Electrotherapeutic Plan

Mode of stimulation: Sensory-level (90% of motor threshold amplitude)

Type of stimulator: Portable HVPC, negative polarity

Electrodes and placement: Two, in cold water bath (Fig. 8.3)

Duration of treatment: 30 minutes every 4 hours while edema is likely to be forming

Rationale The parameters are suggested by the work of Fish and colleagues. The cold-water bath is an efficient and effective way to deliver cryotherapy and electrical stimulation. Timing has also been supported by the work of the Fish group; dose–response for edema control, however, has not been established.

Case 4

A 65-year-old diabetic woman presents with a persistent (4 years) ulcer over the right medial malleolus.

Objective Musculoskeletal evaluation reveals a moderately obese woman with a wide-based gait. Sensory testing reveals a stocking-type hypesthesia and analgesia to the knee, bilaterally. The patient has a well-circumscribed, erythematous, draining wound, measuring 4 cm by 7 cm over her right medial malleolus. Wound culture grows *Staphylococcus aureus.*

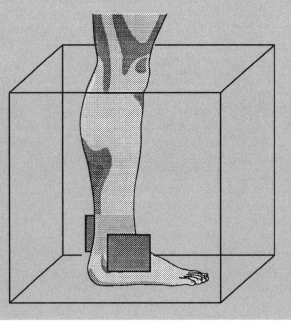

Figure 8.3. Electrode placement for Case Study 3.

Assessment Infected wound on right lower leg, diabetic polyneuropathy, and decreased circulation

Plan

1. Cleaning with surgical Water Pik™
2. Debridement
3. ESTR
4. Occlusive dressing

Detailed Electrotherapeutic Plan

Mode of stimulation: Direct current of less than 1 mA or pulsed monophasic

Type of stimulator: Clinical or portable direct-current generator or pulsed monophasic

Electrodes and placement: One sterile metal electrode wrapped in saline soaked gauze

Initial placement is cathode in the wound, anode (regular electrode) placed distal to the wound (Fig. 8.4). After wound is clean, switch orientation of electrodes

Duration of treatment: 2 hours, twice per day for DC; 30 min, twice per day for monophasic.

Rationale Direct current or monophasic pulsed current have been chosen to promote wound healing. The negative electrode is placed over the wound initially for bacteriocidal effects. When the wound looks clear, the positive electrode is used to induce healing. The treatment times and current characteristics are representative of those described in the literature.

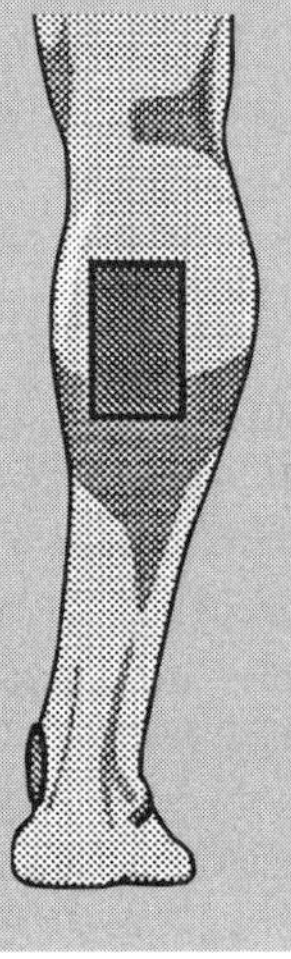

Figure 8.4. Electrode placement for Case Study 4.

Summary

Electrical stimulation can be used in a variety of ways to promote tissue healing. This chapter has described those techniques and the research substantiation for their use. ESTR applications represent an area of electrotherapy that is ripe for clinically research. The new generation of stimulators is more portable and user-friendly than the old. Examination of the effects that the use of these devices has on patient outcome is critical to the development of electrotherapy. Their efficacy in tissue repair has only begun to be investigated.

Study Questions

1. List the critical stimulation characteristics for edema control using a high-voltage pulsed-current generator.

2. List the critical stimulation characteristics for wound healing using a high-voltage pulsed-current generator.

3. Compare the stimulation characteristics for the use of electrical stimulation to increase large-vessel circulation with those used to increase microcirculation.

4. What would be the best choice for treatment of an infected trochanteric pressure sore in a T12 paraplegic and why?

5. List two mechanisms by which electrical stimulation is thought to limit formation of or reduce edema.

6. What mechanisms would the therapist try to invoke when using electrical stimulation to increase peripheral blood flow in the upper extremity of a patient with reflex sympathetic dystrophy?

7. By what mechanism does electrode polarity purportedly affect the healing of wounds?

References

1. Robertson WAS. Digby's receipts. Ann Med Hist 1925;7:216-219.

2. Rothstein JM. Microamperage; a lesson to be learned. Phys Ther 1994;74:194.

3. Clement DL, Shepherd JT. Regulation of peripheral circulation during muscular exercise. Prog Cardiovasc Dis 1976;19:23-31.

4. Wong RA, Jette PV. Changes in sympathetic tone associated with different forms of transcutaneous electrical nerve stimulation in healthy subject. Phys Ther 1984;64:478-482.

5. Dooley DM, Kasprak M. Modification of blood flow to the extremities by electrical stimulation of the nervous system. South Med J 1976;69:1309-1311.

6. Owens S, Atkinson ER, Lees DE. Thermographic evidence of reduced sympathetic tone with transcutaneous nerve stimulation. Anesthesiology 1979;50:62-65.

7. Kaada B. Vasodilation induced by transcutaneous nerve stimulation in peripheral ischemia (Raynaud's phenomenon and diabetic polyneuropathy). Eur Heart J 1982;3:303-314.

8. Owens S, Atkinson ER, Lees DE. Thermographic evidence of reduced sympathetic tone with transcutaneous nerve stimulation. Anesthesiology 1979;50:62-65.

9. Kaada B, Ensen O. In search of the mediators of skin vasodilation induced by transcutaneous nerve stimulation: II. serotonin implicated. Gen Pharmac 1983;14:635-641.

10. Kaada B, Eielson O. In search of the mediators of skin vasodilation induced by transcutaneous nerve stimulation: I. failure to block the response by antagonists of endogenous vasodilators. Gen Pharmac 1983;14:623-633.

11. Kaada B, Helle KB. In search of the mediators of skin vasodilation induced by transcutaneous nerve stimulation: IV. in vitro bioassay of the vasoinhibitory activity

of sera from patients suffering from peripheral ischemia. Gen Pharmacol 1984;15: 115-122.

12. Kaada B, Hegland O, Okledalan O, Opstad PK. Failure to influence the VIP level in the cerebrospinal fluid by transcutaneous nerve stimulation in humans. Gen Pharmacol 1984;15:563-565.

13. Ernst M, Lee MHW. Sympathetic vasomotor changes induced by manual and electrical acupuncture of the hoku point visualized by thermography. Pain 1985;21:25-33.

14. Ebersold MJ, Laws ER, Albers JW. Measurements of autonomic function before, during, and after transcutaneous stimulation inpatients with chronic pain and in control subjects. May Clin Proc 1977;52:228-232.

15. Clemente FR, et al. Effect of motor neuromuscular electrical stimulation on microvascular perfusion of stimulated rat skeletal muscle. Phys Ther 1991;71: 397-406.

16. Reed. Commentary on. Clemente FR, et al: Effect of motor neuromusclar electrical stimulation on microvascular perfusion of stimulated rat skeletal muscle. Phys Ther 1991;71:407-408.

17. Carlson Walker D, Currier DR, Thelkeld AJ. Effect of high voltage pulsed electrical stimulation on blood flow. Phys Ther 1988;68:481-485.

18. Randall BF, Imig CJ, Hines HM. Effect of electrical stimulation upon blood flow and temperature of skeletal muscle. Am J Phys Med 1953;32:22.

19. Wakim KC. Influence of frequency of muscle stimulation on circulation in the stimulated extremity. Arch Phys Med Rehabil 1953;34:291-295.

20. Randall BF, Imig CJ, Hines HM. Effect of electrical stimulation upon blood flow and temperature of skeletal muscle. Am J Phys Med 1953;32:22.

21. Petrofsky JS, Phillips CA, Sawka MN, et al. Blood flow and metabolism during isometric constrictions in cat skeletal muscle. J Appl Physiol 1981;50:493.

22. Currier DP, Petrilli CR, Threlkeld AJ. Effect of medium frequency electrical stimulation on local blood circulation to healthy muscle. Phys Ther 1986;66:937.

23. Barcroft H, Millen JLE. The blood flow through muscle during sustained contractions. J Physiol 1939;97:17-31.

24. Barcroft H, Dornhorst AC. The blood flow through the human calf during rhythmic exercise. J Physiol 1949;109:402-411.

25. Hecker B, Carron H, Schwartz DP. Pulsed galvanic stimulation: effects of current frequency and polarity on blood flow in healthy subjects. Arch Phys Med Rehabil 1985;66:369-371.

26. Liu HI, Currier DP, Threlkeld AJ. Circulatory response of unexercised body part to electrical stimulation of calf muscle in healthy subjects. Phys Ther 1987;67:340-345.

27. Tracy JE, Currier DP, and Threlkeld AJ. Comparison of selected pulse frequencies from two different electrical stimulators on blood flow in healthy subjects. Phys Ther 1988;68:1526-1532.

28. Mohr T, Akers TK, Landry RG. Effect of high voltage stimulation on edema reduction in the rat hindlimb. Phys Ther 1987;67:1703-1707.

29. Mohr R, Akers TK, Wessman HC. Effect of high voltage stimulation on blood flow in the rat hindlimb. Phys Ther 1987;67:526-533.

30. Twist DJ. Acrocyanosis in a spinal cord injured patient—effects of computer-controlled neuromuscular electrical stimulation: a case report. Phys Ther 1990;70: 45-49.

31. Apperty FL, Cary MK. The control of circulatory stasis by the electrical stimulation of large muscle groups. Am J Med Sci 1948;216:403-406.

32. Doran FSA, White M, Frury M. A clinical trial designed to test the relative value of two simple methods of reducing the risk of venous stasis in the lower limbs during surgical operations, the danger of thrombosis and a subsequent pulmonary embolism, with a survey of the problem. Br J Surg 1970;57:20-30.

33. Apperty FL, Cary MK. The control of circulatory stasis by the electrical stimulation of large muscle groups. Am J Med Sci 1948;216:403-406.

34. Alon G. High voltage galvanic stimulation. Chattanooga, Tenn: Chattanooga Corp, 1981:1-5.

35. Hobler CK. Reduction of chronic posttraumatic knee edema using interferential stimulation: a case study. Athletic Training, JNATA 1991;26:364-367.

36. Crisler GR. Sprains and strains treated with the ultrafaradic M-4 impulse generator. J Fla Med Assoc 1953;11:32-34.

37. Ross CR, Sega D. High voltage galvanic stimulation—an aid to post-operative healing. Curr Podiatr 1981;30:19-25.

38. Lamboni P, Harris B. The use of ice, airsplints and high voltage galvanic stimulation in effusion reduction. Athletic Training 1983;18:23.

39. Voight ML. Reduction of post-traumatic ankle edema with high voltage pulsed galvanic stimulation. Athletic Training 1984;19:278-279, 311.

40. Michlovitz S, Smith W, Watkins M. Ice and high voltage pulsed stimulation in treatment of lateral ankle sprains. Orthop Sports Phys Tgher 1988;9:301-304.

41. Griffin JW, Newsome LS, Stralka SW, et al. Reduction of chronic posttraumatic hand edema: a comparison of high voltage pulsed current, intermittent pneumatic compression and placebo treatments. Phys Ther 1990;70:279-286.

42. Cosgrove KA, Alon G, Bell SF, et al. The electrical effect of two commonly used clinical stimulators on traumatic edema in rats. Phys Ther 1992;72:227-233.

43. Reed BV. Effect of high voltage pulsed electrical stimulation on microvascular permeability to plasma proteins. Research 1988;68:491-495.

44. Karnes JL, Mendel FC, Fish DR. Effects of low voltage pulsed current on edema formation in frog hindlimbs following impact injury. Phys Ther 1992;72:273-278.

45. Bettany JA, Fish DR, Mendel FC. High voltage pulsed direct current: effect on edema formation following hyperflexion injury. Arch Phys Med Rehabil 1990;71:677-681.

46. Bettany JA, Fish DR, Mendel FC. Influence of high voltage pulsed direct current on edema formation following impact injury. Phys Ther 1990;70:219-224.

47. Mendel FC, Sylegala J, Fish DR. Influence of high voltage pulsed current on edema formation following impact injury in rats. Phys Ther 1992;72:668-673.

48. Taylor K, Fish DR, Mendel FC, Burton HW. Effects of electrically induced muscle contraction on posttraumatic edema formation in frog hind limbs. Phys Ther 1992;72:127-132.

49. Michlovitz S, Smith W, Watkins M. Ice and high voltage pulsed stimulation in treatment of lateral ankle sprains. Orthop Sports Phys Ther 1988;9:301-304.

50. Griffin JW, Newsome LS, Stralka SW, Wright PE. Reduction of chronic posttraumatic hand edema: a comparison of high voltage pulsed current, intermittent pneumatic compression and placebo treatments. Phys Ther 1990;70:279-286.

51. Mendel FC, Fish DR. New perspectives in edema control via electrical stimulation. Journal of Athletic Training 1993;28:63-74.

52. Wolf M, Wheeler P, Wolcott L. Gold-leaf treatment of ischemic skin ulcers. JAMA 1964;189:923-933.

53. Kanof N. Gold leaf in the treatment of cutaneous ulcers. J Invest Derm 1964;43:441-442.

54. Gallagher J, Geschickter C. The use of charged gold leaf in surgery. JAMA 1964;189:923-933.

55. Becker RO, Murray DG. Method for producing cellular dedifferentiation by means of very small electric currents. Ann NY Acad Sci 1967;29:606-615.

56. Alvarez OM, Mertz PM, Smerbeck RV et al. The healing of superficial skin wounds in stimulated by external electrical current. J Invest Dermatol 1983;81: 144-148.

57. Cheng. Effect of low amperage DC on rat skin slice preparations and demonstrated an increased amino acid uptake.

58. Bourguignon GJ, Bourguignon LY. Electric stimulation of protein and DNA synthesis in human fibroplasts in vitro. FASEB J 1987;1:398-402.

59. Dunn MG, et al. Wound healing using a collagen matrix: effect of DC electrical stimulation. J Biomed Mater Res: Applied Biomaterials 1988;22:191-206.

60. Wolcott L, Wheeler P, Hardwicke H, et al. Accelerated, healing of skin ulcers by electrotherapy. So Med J 1969;62:795-801.

61. Gault W, Gatens P. Use of low density direct current in management of ischemic skin ulcers. Phys Ther 1976;56:265-269.

62. Carley P, Wainapel S. Electrotherapy for acceleration of wound healing: low intensity direct current. Arch Phys Med Rehab 1985;66:443-446.

63. Rowley B. Electrical current effects on *E. coli* growth rates. Proc Soc Exp Biol Med 1972;139:929-934.

64. Barranco S, Berger T. In vitro effect of weak direct current on *Staphylococcus aureus*. Clin Ortho 1974;100:250.

65. Rowley B, McKenna J, Chase G, et al. The influence of electrical current on an infecting micro organism in wounds. Ann NY Acad Sci 1974;238:543-551.

66. Kincaid CB, Lavoie KH. Inhibition of bacterial growth in vitro following stimulation with high voltage, monophasic, pulsed current. Phys Ther 1989;69: 651-655.

67. Young H. Electric impulse therapy aids wound healing. Mod Vet Prac 1966;476:60-62.

68. Brown M, Gogia PP. Effects of high voltage stimulation on cutaneous wound healing in rabbits. Phys Ther 1987;67:662-667.

69. Brown M, McDonnell MK, Menton DN. Electrical stimulation effects on cutaneous wound healing in rabbits. Research 1988;68:955-960.

70. Im MJ, Lee WPA, Hoopes JE. Effect of electrical stimulation on survival of skin flaps in pigs. Phys Ther 1990;70:37-40.

71. Mertz PM, Davis SC, Cazzaniga AL, Cheng K, Reich JD, Eaglstein WH. Electrical stimulation: acceleration of soft tissue repair by varying the polarity. Wounds 1993;5:153-159.

72. Thurman B, Christian E. Response of a serous circulatory lesion to electrical stimulation. Phys Ther 1971;51:1107-1110.

73. Akers T, Gabrielson A. The effect of high voltage galvanic stimulation on the rate of healing of decubitus ulcers. Bioelect Biomech 1984;20:99-100.

74. Ross C, Segal D. High voltage stimulation: an aid to post-operative healing. Current Podiatry May, 1981.

75. Kloth LC, Feedar JA. Acceleration of wound healing with high voltage monophasic pulsed current. Phys Ther 1988;68:503-508.

76. Griffin JW et al. Efficacy of high voltage pulsed current for healing of pressure ulcers in patients with spinal cord injury. Phys Ther 1991;71:433-444.

77. Mulder GD. Treatment of open-skin wounds with electric stimulation. Arch Phys Med Rehabil 1991;72:375-377.

78. Feedar JA, Kloth LC, Gentzkow GD. Chronic dermal ulcer healing enhanced with monophasic pulsed electrical stimulation. Phys Ther 1991;71:639-648.

79. Barron JJ, Jacobson WE, Tidd G. Treatment of decubitus ulcers, a new approach. Minn Med 1985;68:103-106.

80. Byl NN, et al. Pulsed microamperage stimulation: a controlled study of healing of surgically induced wounds in Yucatan pigs. Phys Ther 1994;74:201-213.

81. Leffman D. Author Response. Phys Ther 1994;74:216.

82. Robinson AJ. Invited Commentary. Phys Ther 1994;74:213-215.

83. Yasuda I, Noguchi K, Iida H. Application of electrical callus. J Ipn Orthop Assoc 1955;29:351-355.

84. Freidenberg ZB, Harlow MC, Brighton CT. Healing of nonunion of the medial malleolus by means of direct current: a case study. J Trauma 1971;II:883-885.

85. Cane V, et al. Electromagnetic stimulation of bone repair: a histomorphometric study. J Orthop Res 1991;9:908-917.

86. Kahanovitz N, Arnoczky SP. The efficacy of direct current electrical stimulation to enhance canine spinal fusion. Clinical Orthopaedics and Related Research 1990;251:295-299.

87. Brighton CT. Treatment of nonunion of the tibia with constant direct current (1980 Fitts Lecture, AAST). J Trauma 1981;21:189-195.

88. Basset CAL, et al. Treatment of therapeutically resistant nonunions with bone grafts and pulsing electromagnetic fields. J Bone Joint Surg (Am) 1982;64:1214-1220.

89. Skerry TM, Pead MJ, and Lanyon LE. Modulation of bone loss during disuse by pulsed electromagnetic fields. J Orthop Res 1991;9:600-608.

90. Ganne J, Spechland B, Mayne L. Interferential therapy to promote union of mandibular fractures. Aust NZ J Surg 1979;49:81.

Chapter 9

Iontophoresis

Charles D. Ciccone

Iontophoresis is the use of direct current to enhance the transcutaneous administration of ionizable substances. These substances typically consist of medications such as antiinflammatory steroids and local anesthetics, as well as a variety of other prescription and non-prescription agents. In theory, iontophoresis uses direct current to drive the medication through the skin and into the underlying tissues. This procedure may provide a relatively safe and painless way to deliver clinically significant amounts of medication to cutaneous and subcutaneous tissues.

The basic principles underlying iontophoresis techniques were first described almost a century ago (1). Iontophoresis has been used as a clinical intervention for the past 40 years. As with many traditional therapeutic interventions, few experimental research studies exist that document the clinical effects of iontophoresis. The literature consists largely of case reports and similar clinical commentaries. A number of these case reports have been very helpful in describing the potential uses and effects of various iontophoresis techniques. Although it is regrettable that more studies have not been forthcoming, preliminary evidence does suggest that iontophoresis can be considered an appropriate intervention in the treatment of certain conditions.

The purpose of this chapter is to provide an overview of how iontophoresis can be used to help achieve specific therapeutic goals. The chapter will first present some of the basic principles that govern the use of direct current to enhance drug administration. A discussion of some of the practical issues involved in iontophoresis, including the selection of appropriate instrumentation and the clinical methods typically used during iontophoresis application, will follow. Finally, medications that have been applied iontophoretically to achieve specific outcomes will be discussed.

Basic Principles of Iontophoresis Application

Ion transfer

Many compounds are composed of positively and negatively charged structural units called *ions*. When these compounds are placed in an appropriate solution they dissociate into these polar (electrically charged) components and assume a positive charge or a negative charge depending on whether the atom loses or gains an electron. While in a charged state, each ion can be influenced by an electrical field created within the solution. Positively charged ions (cations) will be attracted to the negative pole (cathode) and repelled from the positive pole (anode). Negatively charged ions (anions) will be attracted to the anode and repelled from the cathode. The electrostatic repulsion of like charges is the driving force for iontophoresis (Fig. 9.1).

Many drugs placed in an aqueous solution are ionic compounds that dissociate into positive and negative components. When drugs ionize in solution, the drug portion of the molecule will assume either a negative or a positive charge while some ionic side group will assume the opposite charge. For instance, the antiinflammatory preparation dexamethasone sodium phosphate will ionize in an aqueous solution to form the negatively charged dexamethasone phosphate ion and the positively charged sodium ion (see Fig. 9.1). Other drugs such as lidocaine hydrochloride typically assume a positive ionic

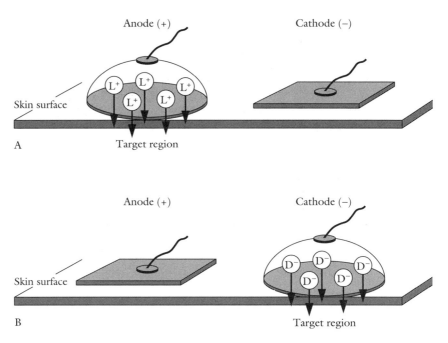

Figure 9.1 Schematic diagram of ion transfer during iontophoresis. **A**, positive ions such as lidocaine ($L+$) are driven away from the anode. **B**, negative ions such as dexamethasone ($D-$) are driven away from the cathode.

charge in aqueous solution. Identification of the drug's polarity dictates the polarity of the electrode used to drive the ion toward underlying tissues. Positive drugs are placed under the positive electrode (anode) and negative drugs are placed under the negative electrode (cathode). The electrode containing or overlying the drug is typically referred to as the "active" or "delivery" electrode, with the opposite electrode often called the "return" or "dispersive" electrode.

Table 9.1 lists some of the drugs commonly used during iontophoresis, the polarity of the ionized form of each drug, and indications for the use of each drug. These drug indications are addressed in more detail later in this chapter.

Electricity as a drug administration vehicle

The primary role of electric current during iontophoresis is to help transmit medication to a target site. In this sense, electricity can be viewed as a vehicle for transdermal drug administration much in the same way that a hypodermic syringe serves as the vehicle to administer a drug via a subcutaneous or intravenous injection. Specific parameters of electricity should be selected to optimize iontophoretic effects much in the same way that a certain size and caliber syringe will be chosen during different types of injection. These optimal electric parameters for iontophoresis are summarized in Table 9.2, and are presented in more detail here.

Type of Current

Exclusively direct (galvanic) current (DC) is used during iontophoresis. Direct current creates a constant, unidirectional electrostatic field between the electrodes to allow continuous transmission of the medication. General aspects of DC were discussed in Chapter 1, and the types of DC generators that can supply the appropriate current for iontophoresis are discussed in Chapter 2 and later in this chapter. One aspect of DC that is particularly important during iontophoresis is the tendency for specific clinical and electrophysiological effects to occur beneath each electrode (Table 9.3). Even in the absence of any medication, an alkaline reaction occurs at the cathode due to the formation of sodium hydroxide while an acidic reaction occurs at the anode due to the formation of hydrochloric acid. These reactions are especially important when the cathode is used as the delivery electrode because alkaline reactions are usually more caustic to human skin during application of DC. Likewise, other effects on protein density and nerve excitability can influence the patient's ability to tolerate repeated applications of iontophoresis. Finally, changes in pH at the site of the delivery electrode may affect drug stability and ionization, thus having the potential to dramatically affect the amount of drug delivered. This issue of pH and drug stability is discussed in more detail later in this chapter.

Current Amplitude

Most clinical studies using iontophoresis have reported using current amplitudes ranging from 1.0 to 5.0 mA (2–8). Contemporary iontophoresis devices typically provide a maximum current of 4.0 mA, with the minimum amplitude ranging anywhere from 0.01 to 1.0 mA. Manufacturers of these devices advocate amplitudes in the range of 1.0 to 4.0 mA. The exact current amplitude used during an iontophoresis treatment, however, is dictated by several factors including patient tolerance, polarity of the active electrode, size of the active electrode, and duration of treatment.

Duration of Current Application

The length of time (duration) that the current is applied during iontophoresis varies greatly from study to study. Durations as short as 5 min or as long as several hours have been reported (9, 10). Generally, the length of time the current is applied is inversely proportional to the magnitude of the current. Larger amplitude currents (approaching or exceeding 4.0 mA) are typically applied for shorter periods of time, whereas smaller currents are applied for longer durations.

Concept of current dosage during iontophoresis

Both current amplitude and duration are factors that will influence how much electricity will be provided to facilitate drug administration during iontophoresis. Rather than referring to these two parameters independently, it is much more practical to consider the product of these factors as the total electrical "dosage" that is used to facilitate drug application. That is, the current amplitude (mA) is multiplied by the duration (minutes) to provide the current dosage in units of mA·min. For instance, a current applied at an amplitude of 2 mA for a duration of 20 min would yield a dosage of 40 mA·min.

Thus, many contemporary iontophoresis protocols describe treatment parameters in terms of such a mA·min dosage. Dosages ranging from of 40 to 80 mA·min have often been suggested for treating conditions such as pain

Table 9.1
Primary Medications Administered by Iontophoresis *

Drug	Principal indication(s)	Treatment rationale	Iontophoresis
Acetic acid	Calcific tendonitis; myositis ossificans	Acetate is believed to increase solubility of calcium deposits in tendons and other soft tissues	2–5% aqueous solution from negative pole
Calcium chloride	Skeletal muscle spasms	Calcium stabilizes excitable membranes; appears to decrease excitability threshold in peripheral nerves and skeletal muscle	2% aqueous solution from positive pole
Dexamethasone	Inflammation	Synthetic steroidal antiinflammatory agent	4 mg/ml in aqueous solution from negative pole
Iodine	Adhesive capsulitis and other soft-tissue adhesions; microbial infections	Iodine is a broad-spectrum antibiotic, hence its use in infections, and so on; the sclerolytic actions of iodine are not fully understood	5–10% solution or ointment from negative pole
Lidocaine	Soft-tissue pain and inflammation (e.g., bursitis, tendosynovitis)	Local anesthetic effects produce transient analgesia	4–5% solution or ointment from positive pole

and inflammation (2, 3, 11–13). Many clinical studies have reported a current–duration combination that resulted in dosages that approach the upper end of this dosage range (see Table 9.4). As indicated in Table 9.4, it may be worthwhile to use dosages of 70 or 80 mA·min to achieve optimal effects. Clearly, however, an ideal current dosage for iontophoresis treatments remains to be determined and clinicians should adjust current dosage as needed for each individual patient.

Table 9.1 Continued

Drug	Principal indication(s)	Treatment rationale	Iontophoresis
Magnesium sulfate	Skeletal muscle spasms; myositis	Muscle relaxant effect may be due to decreased excitability of the skeletal muscle membrane and decreased transmission at the neuromuscular junction	2% aqueous solution or ointment from positive pole
Hyaluronidase	Local edema (subacute and chronic stage)	Appears to increase permeability in connective tissue by hydrolizing hyaluronic acid, thus decreasing encapsulation and allowing dispersement local edema	Reconstitute with 0.9% sodium chloride to provide a 150 μg/ml solution from positive pole
Salicylates	Muscle and joint pain in acute and chronic conditions (e.g., overuse injuries, rheumatoid arthritis)	Aspirinlike drugs with analgesic and antiinflammatory effects	10% trolamine salicylate ointment or 2–3% sodium salicylate solution from negative pole
Zinc oxide	Skin ulcers, other dermatologic disorders	Zinc acts as a general antiseptic; may increase tissue healing	20% ointment from positive pole

*Adapted from Ciccone CD (24).

Table 9.2
Summary of Optimal Current Parameters During Iontophoresis

Type:	Direct current (DC)
Amplitude:	1.0–4.0 mA
Duration:	20–40 min
Total Current Dosage:	40–80 mA · min

Table 9.3
Summary of Reactions at the Anode and Cathode

Cathode	Anode
Attraction of (+) ions	Attraction of (−) ions
Alkaline reaction by formation of NaOH	Acid reaction by formation of HCl
Increased density of protein (sclerotic)	Decreased density of proteins (sclerolytic)
Increased nerve excitability via depolarization	Decreased nerve excitability via hyperpolarization (anode blockade)

Table 9.4
Current Dosages Used in "Successful" Iontophoresis Studies*

Author, date	Number of subjects	Current dosage
Bertolucci, 1982 (11)	$n = 30$	65 mA · min
Braun, 1987 (2)	$n = 1$	76 mA · min
Delacerda, 1982 (3)	$n = 8$	85 mA · min
Harris, 1982 (12)	$n = 50$	100 mA · min
Hasson, 1992 (13)	$n = 1$	65 mA · min
		$X = 78$ mA · min

*All studies treated conditions of pain and inflammation using a combination of dexamethasone and lidocaine.

It should also be noted that any given current dosage can be achieved by an infinite number of amplitude-duration combinations. For instance, a dosage of 40 mA·min could be achieved by applying 1 mA for 40 min, 2 mA for 20 min, 3 mA for 13.3 min, or 4 mA for 10 min. This illustrates how the concept of current dosage can be very helpful in standardizing the amount of electrical current that is used as a drug vehicle during iontophoresis. Patients who are not able to tolerate higher current amplitudes can still be given a designated current dosage via an extension of the treatment duration. Clinicians should be aware, however, that there is no evidence that different combinations of amplitude and duration provide equivalent amounts of ion transfer even if the current dosages are all the same. For instance, it is unclear if 1 mA applied for 40 min will actually deliver the same amount of medication as 4 mA applied for 10 min. Likewise, we do not know if there is a minimum amplitude that will produce a meaningful amount of ion transfer. Additional laboratory and clinical studies are needed to definitively address issues regarding the amplitude and duration combinations that will ultimately achieve optimal current dosages.

Instrumentation for Iontophoresis

DC generators

Any DC generator capable of delivering direct current with fine control in the 0 to 5 mA range could be used to administer iontophoresis. Several commercial devices are available, however, that are designed to be used exclusively for iontophoresis. These iontophoresis-dedicated devices are typically small, battery-powered units that provide one or two channels for current application. Manufacturers of these devices have included additional control features to assist in providing iontophoresis treatments. For instance, many devices provide an automatic current ramp-up at the beginning of the treatment and current ramp-down at the end of the treatment. Some devices also allow the therapist to preset the parameters of current amplitude and total desired dosage (in mA·min), and the treatment duration is then adjusted automatically within the device. Other features such as a built-in timer, automatic shut-off at the end of the treatment, and an audible warning signal that indicates an interruption in current delivery are also available on some devices. Individuals who are interested in using these devices should compare these units based on cost, ease of use, and various other features.

Electrodes

Types of Electrodes for Iontophoresis

Electrodes used for iontophoresis can either be constructed by the clinician or obtained commercially (Figure 9.2). Noncommercial electrodes are typically made from some type of malleable metal such as tin, aluminum, or copper. Sheets of tin or copper foil can be purchased and cut to fit over the treatment site. Alternatively, several folds of heavy-duty aluminum foil can be cut and pressed into a treatment electrode. The metal delivery electrode is placed over gauze pads or paper towels that have been soaked in the drug that is being applied. Towels or gauze soaked in tap water are placed beneath the return electrode. The electrodes are then held in place by weights, rubber straps, or elastic bandages, and alligator clips are used to connect the electrodes to wires coming from the DC generator.

Several types of commercial electrodes have been introduced for use during iontophoresis (Fig. 9.2; see also Fig. 2.17). These self-adhesive electrodes usually come in a package containing a delivery and a return pad. The delivery electrode typically consists of a fiber pad or gel matrix that can be impregnated with the desired medication. The return electrode may be either a fiber pad that must be filled with tap water or a gumlike polymer that does not need filling. These electrodes have some specific connector to allow the attachment of wires leading to the DC generator (usually a commercial iontophoresis device). Some commercial electrodes are designed for a single application, whereas others may be used for two to three applications of the same drug.

A popular commercial electrode used in the past consisted of a reservoir that could be filled with the drug solution. These reservoir or "bubble" electrodes were difficult to apply properly and tended to cause skin irritation in some patients. Hence, they have essentially been replaced by the gel or fiber electrodes described above.

There are advantages and disadvantages to using noncommercial and commercial electrodes. Commercial electrodes come in fixed sizes, and the maximum area of drug delivery will be limited by the size and shape of the delivery electrode. Self-constructed electrodes of tin or aluminum are very inexpensive and can be custom made for each patient. This may be beneficial for treating rather large areas—areas that cannot be covered adequately by the commercial electrodes. Some disadvantages of self-constructed electrodes include the fact that it may be difficult to conform the electrode to irregularly shaped areas

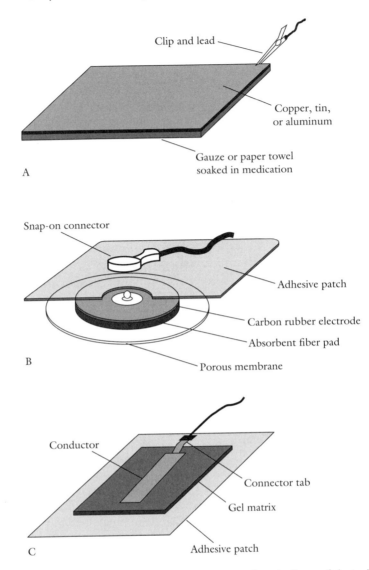

Figure 9.2 Electrodes used for drug delivery during iontophoresis. Types of electrodes shown here are self-constructed electrodes (**A**), and commercially available fiber pads (**B**) and gel pads (**C**).

of the body, and burning can occur because uniform contact of the electrode to the skin may be difficult to maintain.

Commercial electrodes offer the advantages of convenience and ease of application, but these electrodes are often quite expensive (one set of electrodes may cost upwards of $10.00 U.S.). Commercial electrodes may, however, have other features that could make these more attractive for use in some patients. For instance, certain commercial electrodes contain a chemical feature such as a buffering agent or silver chloride that supposedly helps regulate pH at the electrode site during the treatment. This may help moderate the acidic and alkaline reactions that typically occur at the anode and cathode, respectively, thus helping decrease the chance of skin irritation and burns. No real evidence exists, however, that these features actually help prevent burns in a clinical setting. Hence, clinicians must continue to make a decision between commercial electrodes or some type of self-constructed electrode based on their own experience and practical issues such as time and cost.

Electrode Size and Current Density

Current density is the magnitude of the applied current (in milliamperes) divided by the conductive surface area of the electrode (in cm^2). It has been suggested that current density should not exceed 0.5 mA/cm^2 if the cathode is used as the delivery electrode, and current density should not exceed 1.0 mA/cm^2 if the anode is used to apply the drug (14). These parameters must be considered during iontophoresis in order to minimize skin irritations and burns.

Self-constructed electrodes can be made large enough to provide an adequate current density even when larger current amplitudes are used (3.0–4.0 mA). The commercially available electrodes have fixed areas ranging from 6.5 cm^2 for the smaller electrodes to over 30 cm^2 for the largest electrodes. It is critical, therefore, to select a current that does not exceed the maximal current density when using each type of commercial electrode. This is usually not a problem when medium or large commercial electrodes are used. That is, current density will fall below 0.5 mA/cm^2 even when a current of 4.0 mA is applied at medium and large commercial electrodes. Current densities exceeding 0.5 mA/cm^2 may occur, however, when higher amplitudes (4.0 mA) are applied through some of the smaller commercial electrodes (see Table 9.5). Again, this could precipitate skin irritation and burns, especially if the drug is delivered from the cathode. Thus, clinicians should consider the issue of current density when using iontophoresis techniques, and be especially aware of maintaining safe density levels when applying drugs with cathodal current through smaller electrodes.

Application Principles in Iontophoresis

Drug concentrations

As indicated in Table 9.1, specific dosages and concentrations have been reported for the various medications administered via iontophoresis. These dosages tend to be fairly low or contain relatively low concentrations of the

Table 9.5
Current Density Using Small Commercial Electrodes

	EMPI[a] Buffered	IOMED[b] (Trans-O)	Life-Tech[c] Meditrode
Active area (cm^2)	11.0	7.2	6.5
Density at 4.0 mA (mA/cm^2)	0.36	0.56	0.62
Current (in mA) that will generate current density of 0.5 mA/cm^2	5.5	3.6	3.25

[a]Empi, Inc. 1275 Grey Fox Road, St. Paul, Minnesota 55112
[b]IOMED, Inc. 1290 West 2320 South, Suite A, Salt Lake City, Utah 84119
[c]Life-Tech, Inc. 10920 Kinghurst, Houston, TX 77099

drug. Increasing the concentration of drugs administered by iontophoresis does not appear to increase the amount of the drug delivered (14). This is probably due to the fact that increasing the ionic strength of the drug at the delivery site will increase the interionic attraction of the ions in solution— that is, the drug ions will tend to be attracted more strongly to one another at higher concentrations and retard the ability of the active electrode to drive the drug into the underlying tissues. Likewise, iontophoresis is often described as being "current limited," meaning that the amplitude and duration of the applied current is the primary factor that influences how much drug will be applied through the skin (14). Hence, the dosages and concentrations listed in Table 9.1 should probably remain as the standard convention for clinical practice.

Simultaneous use of two drugs

Several studies have reported using two drugs simultaneously at the same delivery electrode (2, 3, 11, 12, 15). Typically this is done by combining lidocaine (a local anesthetic) with dexamethasone (an antiinflammatory steroid). It was originally believed that both of these drugs are positively charged, and that both could be delivered from the anode to achieve a concomitant reduction in pain and inflammation. It has since been realized that dexamethasone is negatively charged, and this drug should be administered from the cathode for optimal penetration (16). Studies that applied these two medications from the anode may have produced some movement of dexamethasone away from the delivery electrode by the process of ionohydrokinesis (14), where the iontophoresed ion (lidocaine) uses electrostatic attraction to the counterion (dexamethasone) to carry or pull the opposite ion into the underlying tissue. It has been estimated, however, that this process of ionohydrokinesis is only about one-tenth as effective as delivering the ion from the proper electrode (14).

Using two drugs of opposite polarity has therefore prompted some clinicians to reverse the polarity of the delivery electrode at some point during the iontophoresis treatment. For instance, Gangarosa and associates advocates beginning the treatment with 2 mA for 10 min from the anode (to facilitate lidocaine delivery) and then switching to the cathode for 20 min at 2 mA (to facilitate dexamethasone delivery) (15). There is some discussion, however, that providing two drugs at the same time may not be beneficial even if both drugs are similarly charged. This relates to the issue discussed above about drug concentration. Including two drugs in the delivery solution increases the concentration of the ions at the delivery site. In addition, the fact that iontophoresis is "current limited" means that the two drugs must compete for the available current when both drugs are delivered simultaneously from the same electrode (14). This competition could potentially limit the amount of each medication that is phoresed. It may therefore be advisable for clinicians to select only one medication during iontophoresis in order to achieve optimal transfer of that particular drug.

Use of other physical agents to accentuate iontophoretic effects

Some studies and clinical reports have advocated other physical agents as useful adjuncts to iontophoresis. For instance, ultrasound has been used directly over the site following iontophoresis to facilitate further drug entry by phonophoresis (4). Heating agents (continuous ultrasound, diathermy) have also been applied following iontophoretic drug delivery to increase the dispersal and penetration of the drug. It must be realized, however, that one of the benefits of using iontophoresis is that the drug is administered to a fairly localized and specific site. Use of heating agents or other interventions (exercise, massage) that might hasten drug dispersal by increasing blood flow may be counterproductive if the intention is to keep the drug localized at the target site.

Effects of pH at the delivery electrode

The degree of alkalinity or acidity (pH) at the delivery electrode can directly influence the amount of ion transfer during iontophoresis. Many drugs are weak acids or weak bases and the ionization potential of these drugs is directly influenced by the pH of the drug solution. As the pH of the solution decreases, acidic compounds will begin to lose their ability to ionize. Basic compounds will lose their ionization potential as pH increases. Since DC produces an acidic reaction at the anode and an alkaline reaction at the cathode (14) (see Table 9.3), these changes in pH at the delivery electrode may facilitate or impair the amount of drug available for ion transfer. For instance, applying an acidic drug from the anode (acidic reaction) could reduce the amount of drug available for ion transfer because the drug molecule begins to become less ionized and more neutral. Conversely, a basic drug applied from the cathode could result in decreased ionization and less drug available for transfer.

Very little information is available about the extent to which these changes in pH at the delivery electrode can affect ionization of specific medications. It is simply not known if these alterations in pH are large enough to influence the effectiveness of specific iontophoresis treatments. Likewise, it is unclear what effect other interventions such as providing a buffer at the delivery electrode will have on pH and the amount of the drug delivered. Future laboratory and clinical studies will hopefully shed some light on whether or not changes in pH can influence iontophoresis treatments.

Clinical Procedures of Iontophoresis Application

Patient screening, indications, and contraindications

A thorough examination should verify that iontophoresis of a specific medication has the potential to help alleviate the patient's condition without producing any untoward effects. Patients should be queried about drug allergies or prior incidence of adverse drug reactions. The patient's physician should be contacted before proceeding with iontophoresis if the patient has experienced a problem with the type of drug selected for iontophoresis. Conditions amenable to iontophoresis should be fairly localized so that the active electrode can adequately cover the affected area. For example, conditions involving a single tendon (bicipital, supraspinatus) have been reported to respond more favorably to iontophoresis of dexamethasone than those involving larger areas such as adhesive capsulitis of the shoulder (11).

The target tissues of iontophoresis generally include the skin and relatively superficial tissues including muscle, tendon, and bursae. The maximum depth of penetration of iontophoresed drugs has not been established because studies of drug penetration in humans are very difficult to perform. In addition, the depth of penetration in individual patients will be influenced by a number of factors including the type of drug used, current dosage, and current density. Various anatomical factors such as skin thickness, subcutaneous adipose tissue, and the size of other structures including skeletal muscle will also influence the depth of penetration from patient to patient. It may, for example, be much more difficult to reach the supraspinatus tendon in an athletic patient who has a hypertrophied deltoid muscle than in a patient where the supraspinatus tendon is relatively more accessible. Hence, each patient must be evaluated carefully to determine if iontophoresis may be a viable intervention.

Iontophoresis is contraindicated if the skin at the treatment site is damaged or broken (17, 18). A decrease in skin sensation at the treatment site may also be a contraindication to iontophoresis, although several studies have reported using iontophoresis with caution to help resolve scar tissue and other conditions where sensation may be impaired (19–21). Sensitivity to the drug being administered or sensitivity to direct current (excessive skin irritation) may also contraindicate iontophoresis treatments. Iontophoresis should not be applied if the direct current will affect cardiac pacemakers or some other implanted electrical device (18).

Skin preparation and electrode placement

Most sources advocate preparing the skin by vigorous rubbing with isopropyl alcohol (17, 22). This is believed to lessen the chance of skin irritation by removing any oils and impurities. The skin should be shaved if excessive hair is present. The solution or preparation containing the drug is usually impregnated into gauze or paper towels, or it is placed directly onto the delivery electrode if an absorbable commercial electrode is used. Medications contained in thicker creams (zinc oxide, salicylate ointments) can be placed directly on the skin and then covered by a few thicknesses of gauze or paper towels soaked in warm tap water. The amount of drug applied varies according to each preparation and according to the size of the delivery electrode and area being treated.

The delivery electrode is placed over the desired site of application with care to maintain uniform contact between the skin and electrode throughout the treatment. The return electrode is placed at a nearby site on the same limb or an adjacent body segment. Some sources advocate a minimum distance of 18 inches between the delivery and return electrodes (17), but there does not appear to be any evidence that this will enhance drug application. Commercial electrodes typically have a fixed distance between the delivery and return electrodes, and this distance usually dictates how far the return electrode can be placed away from the delivery electrode. Efforts should be made, however, to separate the delivery and return electrodes as much as possible to minimize the chance of irritation and burns. Placing these electrodes too close to one another increases the risk of irritation and burns because the direct current may tend to bridge across the skin's surface rather than penetrating the underlying tissues.

Connection to the DC generator and application of current

After verifying that the current source is turned off, wire leads should be attached to the treatment electrodes. The leads are then connected to the DC generator, with care taken to match the polarity of the delivery electrode to the polarity of the drug being administered. Current is then slowly increased (ramped up) to the initial desired amplitude. Most protocols advocate fairly low amplitudes (1–2 mA) for the initial part of the treatment (5–10 min). If tolerated well, current can then be increased to higher levels (3–4 mA) for the remainder of the treatment. As discussed earlier, total treatment duration will depend on the desired current dosage. Hence, increasing the current will shorten the overall treatment time, whereas a longer treatment duration is needed if current is maintained at lower levels.

Completing the treatment

Current should be slowly decreased (ramped down) at the end of the treatment. Electrodes are then removed and the skin at each electrode site should be carefully inspected. Some redness may be present at the skin of the treatment site, especially if the drug is delivered from the cathode. Likewise, small blisterlike vesicles may be present on the skin where the drug is delivered. These signs of redness and irritation typically disappear within several minutes to several hours. Failure of these signs to resolve, however, may indicate a more serious reaction, and subsequent treatments may be contraindicated because the patient is not able to tolerate the direct current, the drug, or a combination of current and drug.

Many clinicians conclude iontophoresis treatments by applying a skin lotion containing lanolin, aloe vera, or a similar ingredient. This may help alleviate any redness and accelerate recovery of the skin.

Clinical Indications for Iontophoresis

Treatment of inflammation

Evidence of Clinical Efficacy Using Antiinflammatory Agents

Iontophoresis has often been reported to be helpful in treating inflammation of the skin and subcutaneous tissues. Antiinflammatory steroids known as glucocorticoids are commonly used in this situation. Glucocorticoids consist of a group of chemically related compounds that includes hydrocortisone, methylprednisolone, dexamethasone, and several similar medications. The most common glucocorticoid currently used for iontophoresis is dexamethasone sodium phosphate (Decadron). It was originally believed that this drug formed a positive ion and should be administered from the anode. Contemporary thought is that this drug forms a negative ion when administered as dexamethasone sodium phosphate, and this preparation should be delivered from the negative pole (16, 18).

Several studies have noted that glucocorticoid iontophoresis may be beneficial in treating musculoskeletal inflammation. Bertolucci used dexamethasone combined with a local anesthetic (lidocaine) to treat patients having various inflammatory conditions including lateral epicondylitis, sacroiliitis, and tendinitis in the shoulder area (11). Results suggested that patients who actually received the drug responded more favorably than those receiving a placebo (sodium chloride), and that patients were more apt to respond favorably if they were younger (less than 45 years old) and if they did not also have degenerative conditions involving the shoulder or neck. Similar results were noted by Harris, who used dexamethasone and lidocaine to treat conditions of tendinitis and bursitis at various anatomical sites (12). Delacerda reported that iontophoresis of dexamethasone with lidocaine was superior to systemic medications (analgesics, antispasmodics) or physical agents (hydrocollator packs, ultrasound) in resolving soft tissue inflammation around the shoulder girdle (3).

Several investigators have reported that various glucocorticoids can be introduced iontophoretically into the temporomandibular joint (TMJ), and that this intervention combined with other drugs (lidocaine) and exercise may be especially helpful in alleviating TMJ pain and inflammation (2, 4, 15, 18). Some preliminary evidence has indicated that chronic inflammatory conditions such as rheumatoid arthritis may also respond favorably to iontophoresis. Case reports are available that document the benefits of dexamethasone iontophoresis either used alone or in conjunction with exercise to help manage inflammation in the knees of patients with rheumatoid arthritis (13, 23).

Mechanism of Glucocorticoid Effects

Glucocorticoids such as dexamethasone exert a number of effects that decrease inflammation. These drugs inhibit the synthesis of proinflammatory substances such as prostaglandins and leukotrienes (24, 25). These agents also inhibit the migration of scavenger white blood cells to the site of inflammation, and they have the ability to stabilize the lysosomal membrane within the affected cell, thus attenuating autologous cell destruction by preventing rupture and release of destructive enzymes within the cell (26). These effects working in concert with one another help account for the rather powerful antiinflammatory properties of these agents.

Potential Side Effects

Although glucocorticoids are very effective in decreasing the symptoms of inflammation, these drugs are associated with a number of side effects. Primary among these is the potential for causing breakdown (catabolism) of muscle, tendon, bone, and other collagenous tissues in the body (27). Likewise, prolonged administration of glucocorticoids can inhibit the body's endogenous production of these hormones, thus producing a potentially serious problem known as adrenocortical suppression (26, 28). These side effects are very common when high doses of glucocorticoids are given systemically for prolonged periods of time. The use of iontophoresis to administer these drugs to specific local tissues at relatively lower doses should diminish the risk of these serious side effects. Still, indiscriminate and repeated use of glucocorticoid iontophoresis could expose the patient to some potentially serious side effects.

To avoid these potential adverse effects, it has been suggested that treatments should not be given every day and that at least one recovery day should be provided between each iontophoresis session (12, 13). Lark and Gangarosa recommend a 3-to-7-day interval between each treatment (18). They feel that this allows the drug to exert antiinflammatory effects without producing a sustained inhibition of collagen synthesis, thus minimizing the catabolic effects. In addition, the total number of glucocorticoid treatments should probably be limited during any given treatment sequence. Several sources suggest that beneficial effects should begin to occur within three or four treatments (2, 12, 18). The lack of any positive effects after three or four treatments may indicate that glucocorticoid iontophoresis is not going to be effective in that patient and an alternative intervention should be considered.

Treatment of pain

As discussed above, iontophoresis can be used to reduce inflammation and thus reduce the pain that is typically associated with the inflammatory process. It may also be desirable, however, to iontophorese drugs that have direct analgesic effects. Iontophoresis has been used to administer analgesic agents such as local anesthetics, salicylates, opioids, and other drugs. Examples are provided here describing how iontophoresis of these agents can be used to decrease pain.

Iontophoresis of Local Anesthetics

Local anesthetics such as lidocaine have been used frequently during iontophoresis. These drugs produce an anesthetic effect by blocking the transmission of impulses along peripheral nerve axons. Lidocaine and similar drugs bind to sodium channels in the nerve membrane and prevent sodium from entering the axon (29, 30). The affected portion of the axon is not able to initiate an action potential, and anesthesia occurs in the tissues innervated by that neuron because afferent impulses cannot reach the central nervous system.

As indicated earlier, several studies reported using the simultaneous administration of a local anesthetic (lidocaine) and an antiinflammatory steroid (dexamethasone) to decrease pain and inflammation in various musculoskeletal disorders (2, 3, 11, 12). Although this protocol has been used frequently by clinicians, there are some rather obvious problems with this technique. For instance, these drugs are now regarded as having opposite polarity, with lidocaine being positively charged and dexamethasone being negatively charged. Some sources feel that both drugs can be administered from the positive pole because lidocaine will pull the negatively charged dexamethasone by the process of ionohydrokinesis. Other clinicians report that these drugs can be given simultaneously, but the polarity of the delivery electrode should be reversed at some point during the treatment to allow optimal delivery of both drugs (15). This has practical implications because it prolongs the treatment time and may increase the chance for skin irritation.

Moreover, local anesthetics may cause a decrease in sensation by anesthetizing the skin at the delivery site. An interesting paradox is created by this approach to pain control because decreased skin sensation is usually considered a contraindication to iontophoresis treatment. Lidocaine iontophoresis could put the patient at a greater risk for an electrical burn if sensation is decreased at the delivery site. Hence, clinicians who elect to administer local anesthetics either alone or in combination with other drugs should probably avoid using higher current amplitudes to diminish the chances of electrical burns.

Lastly, clinicians should consider what constitutes appropriate clinical use of iontophoresis for the delivery of local anesthetics. Ample evidence exists to indicate that lidocaine iontophoresis can be used to produce local anesthesia to allow some relatively minor surgical or diagnostic procedure. For instance, iontophoresis has been used to administer lidocaine or a similar drug to the eardrum before myringotomy, to the teeth prior to tooth extraction, or to other cutaneous tissues to allow some other minor surgical procedures to be performed (31–35).

The benefits of iontophoresis of local anesthetics in the management of musculoskeletal disorders, however, are questionable. These drugs create an anesthetic effect by inhibiting the excitation of sensory nerve membranes, but they do not influence the actual source of pain. This could have beneficial effects if this transient decrease in pain allows some intervention to be performed (e.g., a vigorous manual technique like transverse friction massage) or if the intent is to try to interrupt some type of cycle like a pain-spasm syndrome. Therapists should realize, however, that local anesthetics are typically used only to produce a temporary decrease in sensation so that some other procedure can be performed; use of these agents may not have any long-term effects.

Rather than using lidocaine iontophoresis as a primary intervention, it has been suggested that this technique could be used as a tool to determine the efficacy of subsequent iontophoresis treatments. Clinicians considering iontophoretic administration of other drugs to treat pain and inflammation (e.g., dexamethasone) could administer an initial treatment of lidocaine to see if iontophoresis is successful in delivering the drug to the appropriate tissues. If pain is temporarily diminished following lidocaine iontophoresis, this is a fairly good indication that iontophoresis is successful in driving the drug to the affected tissues (18). Subsequent treatments with drugs that may have a more lasting effect should then be attempted. This assumes, of course, that these other drugs will penetrate to the same depth as lidocaine, a fact that may or may not be true. Still, a failure of lidocaine iontophoresis to decrease pain suggests that iontophoresis may not be successful in reaching the target tissue and that alternative treatments may be more appropriate.

Salicylate Iontophoresis

Non-narcotic analgesics such as the salicylates are weak acids that produce several therapeutic benefits including analgesic and antiinflammatory effects because of their ability to inhibit the biosynthesis of prostaglandins (36, 37). Prostaglandins are hormonelike substances that are produced locally in injured tissues, and excessive production of these substances potentiates pain and inflammation in the affected tissues (38). Salicylates and similar nonopioid analgesics are potent inhibitors of the enzyme that produces prostaglandins, and these drugs are effective in controlling pain and inflammation in a variety of conditions. Likewise, these drugs may be ideally suited for iontophoretic administration in conditions such as bursitis, tendinitis, and similar musculoskeletal problems. Regrettably, there is little information about the clinical efficacy of salicylate iontophoresis. One case study documented the successful use of salicylate iontophoresis in treating residual thigh pain following hip arthroplasty (39). Additional research is needed to determine if iontophoresis of these drugs is effective in treating a variety of patients who have other types of problems.

Opioid Iontophoresis

Pain following extensive surgery or severe trauma is often treated with opioid analgesics such as meperidine or morphine. These drugs are powerful analgesics that have traditionally been used to act on neurons in the central nervous system (brain and spinal cord) to impair the transmission and perception of painful stimuli. Recent evidence also suggests that peripheral opioid receptors exist, and that administration of these drugs directly into peripheral tissues (e.g., an inflamed joint) may be helpful in decreasing pain (40). Opioids are usually administered orally or by some type of injection (intravenous, intramuscular). A recent study suggested that opioids may also be administered iontophoretically to provide adequate postoperative analgesia (9). In this study, patients who had undergone hip or knee arthroplasty were able to decrease their need for other analgesic drugs if they received morphine iontophoresis. Morphine was delivered postoperatively for an extended period of time (6 hours) by an electrode placed on the patient's anterior forearm. Morphine iontophoresis was used in this situation to provide a slow, continuous supply of the drug into the patient's blood stream rather than to apply the drug to a specific site or tissue.

Although the use of opioid iontophoresis is still experimental at this time, perhaps this technique will gain acceptance and be used in more clinical situations in the future. Therapists may want to consult with physicians about the possibility of including opioid iontophoresis as a means of providing postoperative pain relief in certain patients.

Vinca Alkaloids Iontophoresis

Postherpetic neuralgia (PHN) is the severe pain that typically occurs in specific dermatomes following a herpes zoster infection. Several studies have documented that PHN may be resolved in many patients by iontophoretic administration of drugs known as vinca alkaloids (41–43). Vinca alkaloids are a group of nitrogen-based compounds including vincristine and vinblastine. These drugs are usually classified as cancer chemotherapy agents, and they are administered systemically to inhibit cell replication in certain forms of cancer (44). These drugs may also be administered by iontophoresis over the affected dermatomes in patients with PHN. Vinca alkaloids work by inhibiting cellular microtubule function. Hence, their ability to resolve pain in PHN seems to be due to inhibition of retrograde axoplasmic transport in the sensory neurons of the affected dermatome (41). This inhibition supposedly causes degenerative changes in afferent pain pathways so that painful stimuli are not able to reach central areas of pain perception (45).

Most of the reports documenting the benefits of vinca alkaloid iontophoresis originated over a decade ago, and this technique has not gained overwhelming acceptance in contemporary practice. As a result, it does not appear that health care professionals have attempted to use this technique to any great extent in treating patients with PHN as evidenced by the apparent lack of any mention of this procedure in the clinical and scientific literature. Clinicians who routinely treat patients with PHN should consider the possibility of using vinca alkaloid iontophoresis to help manage this condition.

Resolution of soft-tissue mineralization

It has been suggested that iontophoresis may be helpful in dissolving and dispersing painful mineral deposits such as calcium deposits and urate crystal deposits (46, 47). Acetic acid appears to be the drug of choice for treating calcium deposits. Iontophoresis of acetic acid from the cathode induces the negatively charged acetate ion to combine with the relatively insoluble calcium carbonate to form a more soluble compound, calcium acetate. The use of acetic acid to resolve calcium deposits is summarized by the following chemical reaction:

$$\underset{\text{calcium carbonate (insoluble)}}{CaCO_3} + \underset{\text{acetic acid}}{2H(C_2H_3O_2)} \rightarrow$$

$$\underset{\text{calcium acetate (soluble)}}{Ca(C_2H_3O_2)} + H_2O + CO_2$$

Since it is more soluble, calcium acetate can then be more easily dispersed from the affected tissues by local blood flow.

As with many iontophoresis techniques, there is little documentation of the clinical effectiveness of acetic acid iontophoresis. One rather convincing case report does exist that describes how this technique can be used to successfully resolve the soft-tissue mineralization that occurs in myositis ossificans (22). In this study, a series of nine iontophoresis treatments using 2% acetic acid was applied over a 3-week period. This treatment was helpful in resolving a large ossified mass in the patient's thigh, suggesting that acetic acid iontophoresis may be useful in treating some types of soft-tissue calcification. Clinicians should consider using this technique in patients with myositis ossificans as well as other types of calcium deposits.

In an analogous manner, lithium iontophoresis has been suggested as being useful in treating urate deposits that typically occur in gouty arthritis (46). Gouty arthritis is characterized by the precipitation of sodium urate crystals in specific areas such as the first metatarsalphalangeal joint. Administration of lithium from the anode supposedly replaces sodium in these deposits, thus forming lithium urate, which is more soluble in the blood (46). Once again, there is no experimental evidence documenting the effectiveness of this procedure. Still, therapists should be aware that this may be a plausible course of action in patients with localized, painful urate crystal deposits.

Treatment of wounds and infection

Iontophoresis has been reported to be useful in administering several types of drugs to treat infections and facilitate wound healing. Zinc oxide administered from the positive pole has been reported to have several beneficial properties including bactericidal effects and an ability to accelerate tissue growth and repair. Like many heavy metals, zinc may exert an antibacterial effect by interfering with critical metabolic activities in microbial cells, and tissue healing may be facilitated by zinc's ability to precipitate proteins. The effectiveness of zinc oxide iontophoresis in treating recalcitrant skin ulcers has been documented in several case reports appearing in the literature (19, 48).

Other antibacterial drugs (penicillin, gentamicin) have been administered by iontophoresis to treat certain types of infection. In particular, these agents have been described as being beneficial in treating chondritis following burns of the ear (6, 49). Iontophoresis appears to offer the advantage of administering fairly high doses of antibacterial drug through eschar into the underlying avascular tissue (10, 49). Clinicians who treat patients with burns may want to pursue the possibility of using antibacterial iontophoresis in managing conditions such as ear chondritis.

Finally, it has been suggested that iontophoresis of certain antiviral drugs could be a possible means of controlling cutaneous herpes virus infections. Iontophoresis has been used to administer vidarabine from the negative pole to treat herpes simplex virus in experimental animals (50). More recently, cathodal application of idoxuridine has been shown to be beneficial in treating herpetic infection of the finger in two patients (51). Perhaps future reports will continue to document iontophoresis as a means of controlling bacterial, viral, and other types of infection. This potential treatment could be significant because of the critical need for adequate management of infection, especially in patients with a compromised immune system.

Treatment of edema

Iontophoresis of a drug known as hyaluronidase has been reported to be helpful in reducing certain types of edema. Hyaluronidase is an enzyme that appears to increase permeability in connective tissue by hydrolyzing hyaluronic acid. Edema that is encapsulated by the connective tissue can then disperse into surrounding tissues and be carried away by the vascular and lymphatic systems. One study documented the ability of this technique to reduce edema in a large sample of patients with either acute injury (sprains) or chronic edema (postsurgical lymphedema) (7). No control or placebo group was used in this study, however, thus making it difficult to differentiate the drug effects from other factors (electrical current, spontaneous resolution). Furthermore, edema reduction was assessed using circumferential measurements rather than using a more exact and reliable measurement such as water displacement. In a different study, patients with soft-tissue and intraarticular hemorrhage secondary to hemophilia were treated with either hyaluronidase or a placebo (acetate buffer) (52). Circumferential measurements failed to reveal any difference in edema reduction between the two groups, suggesting that hyaluronidase iontophoresis may have limited value in treating this type of edema in patients with hemophilia.

Although hyaluronidase iontophoresis is often described as being helpful in reducing edema, no experimental evidence documents the clinical efficacy of this technique. It would be interesting to reexamine this technique using a valid and reliable measure of edema reduction (water displacement). Clinicians may then be able to make sound judgements about whether this technique has any real merit in treating certain types of edema.

Treatment of hyperhidrosis

Iontophoresis has been shown to be successful in reducing sweating in the hands, feet, and armpits in many patients with hyperhidrosis (53–55). Electrodes containing tap water are typically placed over the affected areas, and cathodal current is applied for the first half of the treatment followed by anodal current. This simple technique appears to induce the formation of keratinous plugs in the sweat glands, thus blocking the flow of sweat to the skin's surface (55). A series of daily treatments lasting anywhere from 8 to 20 days is usually needed before an appreciable decrease in sweating is noted. Treatments must either be continued on a weekly or semiweekly basis, or the daily treatments must be repeated after about 6 weeks to maintain a relatively euhidrotic state.

Treatment of scar tissue and adhesions

Iontophoresis may help reduce adhesions and scar tissue (20, 21). Although iodine is usually applied for antimicrobial purposes, this substance is also reported to have sclerolytic effects. The exact cellular basis for the sclerolytic action of iodine has not been clearly defined. Iodine is typically administered from the negative pole in the form of 5 to 10% ointment. A limited number of case reports have described the use of iodine iontophoresis in helping decrease joint restriction and tendon adhesions following surgical procedures (20, 21). It would be interesting to see if these preliminary results are corroborated in the future by well-designed, placebo-controlled studies. Clinicians should be aware, however, that application of iontophoresis over scar tissue is sometimes contraindicated because of the decreased cutaneous sensation at the scar site. Hence, this technique should be used cautiously in any future clinical applications or research studies.

Nontraditional and newer uses of iontophoresis

Several other potential applications of iontophoresis techniques have been described in the literature. Calcium iontophoresis has been reported to be helpful in reducing symptoms in a patient with suspected myopathy in the laryngeal musculature (5). Salicylate iontophoresis was used to successfully treat plantar warts in a small sample of patients (56). Iontophoresis of vinca alkaloids, a technique discussed earlier in the treatment of PHN, has also been used to treat intractable pain in patients with advanced cancer (45), and this technique has been used to diminish the skin lesions associated with Kaposi's sarcoma in people who have acquired immunodeficiency syndrome (57). Experimental studies on animals have even investigated the idea of using iontophoresis to administer other medications such as cardiovascular agents (58). Clearly, the potential exists for using iontophoresis as a noninvasive method to deliver a drug either locally or into the systemic circulation. It remains to be seen, however, which iontophoresis techniques will be incorporated into widespread clinical use in the future.

Summary

Iontophoresis offers a means of introducing medications through the surface of the skin in a relatively safe, easy, and painless manner. This chapter presented some of the fundamental principles of using electric current as a means of enhancing transdermal drug delivery. Basic principles and clinical procedures for applying iontophoresis treatments were also described. Finally, this chapter described how iontophoresis can be used to administer specific medications to achieve specific clinical outcomes including decreased inflammation, decreased pain, and several other beneficial effects.

Although iontophoresis techniques have been available to clinicians for several decades, iontophoresis appears to have rather limited acceptance as a primary therapeutic intervention. This is due at least in part to the relative lack of well-designed experimental research documenting the clinical efficacy of these techniques. As with many electrotherapeutic interventions, there is a need for more studies that document the use and effects of iontophoresis in various clinical situations. Sufficient evidence does exist that suggests that clinicians should consider using iontophoresis, especially in certain conditions such as localized soft-tissue inflammation. Iontophoresis has the potential to provide substantial benefits when this intervention is applied in the appropriate manner to an appropriate patient population. More widespread clinical use combined with additional research may allow iontophoresis to assume a more prominent role in the armamentaria of contemporary clinical practice.

Study Questions

1. The basic principle of iontophoresis is electrostatic ___(a)___ of charged particles, with positively charged ions being delivered from the ___(b)___ and negatively charged ions being delivered from the ___(c)___ .

2. Iontophoresis uses ___(a)___ current at amplitudes that typically range from ___(b)___ to ___(c)___ mA.

3. Current dosage used during iontophoresis is determined by the product of current ___(a)___ and current ___(b)___ , and current dosages are expressed in units of ___(c)___ .

4. When iontophoresis is used to treat pain and inflammation, current dosages typically range between ___(a)___ and ___(b)___ .

5. An iontophoresis treatment is initiated at an amplitude of 2 mA, and this amplitude is maintained for 5 min. Amplitude is then increased to 4 mA. How much longer should the treatment be applied to achieve a total current dosage of 50 mA·min?

6. With regard to polarity of the delivery electrode, administration of medications from the ___(a)___ typically produces more skin irritation because this electrode causes the formation of ___(b)___ which induces an ___(c)___ reaction on the skin's surface.

7. Although commercial electrodes may be quite expensive, these electrodes offer advantages of [list three].

8. It has been recommended that current density should not exceed ___(a)___ mA/cm^2 when the cathode is used for drug delivery, and ___(b)___ mA/cm^2 when the anode is used for drug delivery.

9. True or false: doubling the recommended concentration of a medication will double the amount of drug administered during iontophoresis.

10. Current evidence indicates that dexamethasone carries a ___(a)___ charge and lidocaine carries a ___(b)___ charge when these drugs are used during iontophoresis.

11. The cathode is being used to deliver a drug that is a weak base. If pH increases at the cathode, this could ___(a)___ the amount of drug delivered because ___(b)___ the drug exists in an ionized state.

12. List three contraindications or precautions to iontophoresis.

13. Glucocorticoids such as dexamethasone are used during iontophoresis for their powerful _____(a)_____ effects, but prolonged or extensive use of these drugs can produce a _____(b)_____ effect on collagenous tissues.

14. List three different types of medications that have been administered iontophoretically to specifically produce an analgesic effect

15. Conditions involving soft-tissue calcification have been treated with iontophoretic application of _____(a)_____, and the polarity of this medication necessitates delivery from the _____(b)_____.

References

1. LeDuc S. Electric ions and their use in medicine. Liverpool: Redman Ltd, 1908.

2. Braun BL. Treatment of acute anterior disk displacement in the temporomandibular joint: a case report. Phys Ther 1987;67:1234-1236.

3. Delacerda FG. A comparative study of three methods of treatment for shoulder girdle myofascial syndrome. J Orthop Sports Phys Ther 1982;4:51-54.

4. Kahn J. Iontophoresis and ultrasound for postsurgical temporomandibular trismus and paresthesia: case report. Phys Ther 1980;60:307-308.

5. Kahn J. Calcium iontophoresis in suspected myopathy. Phys Ther 1975;55: 376-377.

6. LaForest NT, Cofrancesco C. Antibiotic iontophoresis in the treatment of ear chondritis: clinical report. Phys Ther 1978;58:32-34.

7. Magistro CM. Hyaluronidase by iontophoresis. Phys Ther 1964;44:169-175.

8. Murray W, Lavine LS, Seifter E. The iontophoresis of C_{21} esterified glucocorticoids: preliminary report. Phys Ther 1963;43:579-581.

9. Ashburn MA, Stephen RL, Ackerman E, et al. Iontophoretic delivery of morphine for postoperative analgesia. J Pain Symptom Management 1992;7:27-33.

10. Rapperport AS, Larson DL, Henges DF, et al. Iontophoresis: a method of antibiotic administration in the burn patient. Plastic and Reconstructive Surgery 1965;36:547-552.

11. Bertolucci LE. Introduction of antiinflammatory drugs by iontophoresis: double blind study. J Orthop Sports Phys Ther 1982;4:103-108.

12. Harris PR. Iontophoresis: clinical research in musculoskeletal inflammatory conditions. J Orthop Sports Phys Ther 1982;4:109-112.

13. Hasson SH, Henderson GH, Daniels JC, Schieb DA. Exercise training and dexamethasone iontophoresis in rheumatoid arthritis: a case study. Physiotherapy Can 1991;43:11-29.

14. Henley EJ. Transcutaneous drug delivery: iontophoresis and phonophoresis. Critical Reviews in Physical and Rehabilitation Medicine 1991;2:139-151.

15. Gangarosa LP, Mahan PE, Ciarlone AE. Pharmacologic management of temporomandibular joint disorders and chronic head and neck pain. Cranio 1991;9: 328-338.

16. Petelenz TJ, Buttke JA, Bonds C, et al. Iontophoresis of dexamethasone: laboratory studies. J Controlled Release 1992;20:55-66.

17. Cummings J. Iontophoresis. In: Nelson RM, Currier DP, eds. Clinical electrotherapy. 2nd ed. Norwalk, CT: Appleton and Lange, 1991:317-329.

18. Lark MR, Gangarosa LP. Iontophoresis: an effective modality for the treatment of inflammatory disorders of the temporomandibular joint and myofascial pain. Cranio 1990;8:108-119.

19. Cornwall MW. Zinc iontophoresis to treat skin ulcers. Phys Ther 1981;61: 359-366.

20. Langley PL. Iontophoresis to aid in releasing tendon adhesions: suggestions from the field. Phys Ther 1984;64:1395.

21. Tannenbaum M. Iodine iontophoresis in reduction of scar tissue. Phys Ther 1980;60:792.

22. Weider DL. Treatment of traumatic myositis ossificans with acetic acid iontophoresis. Phys Ther 1992;72:133-137.

23. Hasson SM, English SE, Daniels JC, Reich M. Effect of iontophoretically delivered dexamethasone on muscle performance in a rheumatoid arthritic joint. Arthritis Care Res 1988;1:177-182.

24. Ciccone CD. Pharmacology in rehabilitation. Philadelphia: FA Davis, 1990: 336-353.

25. Lewis GD, Campbell WB, Johnson AR. Inhibition of prostaglandin synthesis of glucocorticoids in human endothelial cells. Endocrinology 1986;119:62-69.

26. Haynes RC. Adrenocorticotropic hormone; adrenocortical steroids and their synthetic analogs; inhibitors of the synthesis and actions of adrenocortical hormones. In: Gilman AG, Rall TW, Nies AS, Taylor P, eds. The pharmacological basis of therapeutics. 8th ed. New York: Pergammon Press, 1990:1431-1462.

27. Oikarinen AI, Vuorio EI, Zaragoza EJ, et al. Modulation of collagen metabolism by glucocorticoids. Receptor-mediated effects of dexamethasone on collagen biosynthesis in chick embryo fibroblasts and chondrocytes. Biochem Pharmacol 1988;37: 1451-1462.

28. Zora JA, Zimmerman D, Carey TL, et al. Hypothalamic-pituitary axis suppression after short-term, high-dose glucocorticoid therapy in children with asthma. J Allergy Clin Immunol 1986;77:9-13.

29. Hille B. Local anesthetics: hydrophilic and hydrophobic pathways for the drug-receptor reaction. J Gen Physiol 1977;69:497-515.

30. Savarese JJ, Covino BG. Basic and clinical pharmacology of local anesthetic drugs. In: Miller RD, ed. Anesthesia. vol. 2., 2nd ed. New York: Churchill Livingston, 1986:985-1013.

31. Bezzant JL, Stephen RL, Petelenz TJ, Jacobsen SC. Painless cauterization of spider veins with the use of iontophoretic local anesthesia. J Am Acad Dermatol 1988;19:869-875.

32. Comeau M, et al. Anesthesia of the human tympanic membrane by iontophoresis of a local anesthetic. Laryngoscope 1978; 88:277-285.

33. Gangarosa LP. Iontophoresis for surface local anesthesia. JADA 1974;88: 125-128.

34. Petelenz T, Axenti I, Petelenz TJ, et al. Mini set for iontophoresis for topical analgesia before injection. Int J Clin Pharmacol Ther Toxicol 1984;22:152-155.

35. Sirimanna KS, Madden GJ, Miles S. Anesthesia of the tympanic membrane: comparison of EMLA cream and iontophoresis. J Laryngol Otol 1990;104:195-196.

36. Born GVR, Gorog P, Begent NA. The biologic background to some therapeutic uses of aspirin. Am J Med 1983;74(suppl 6a):2-9.

37. Insel PA. Analgesic-antipyretics and antiinflammatory agents; drugs employed in the treatment of rheumatoid arthritis and gout. In: Gilman AG, Rall TW, Nies AS,

Taylor P, eds. The pharmacological basis of therapeutics. 8th ed. New York: Pergammon Press, 1990:638-681.

38. Robinson DR. Prostaglandins and the mechanism of action of antiinflammatory drugs. Am J Med 1983;75(suppl 4b):26-31.

39. Garzione JE. Salycilate iontophoresis as an alternative treatment for persistent thigh pain following hip surgery. Phys Ther 1978;58:570-571.

40. Stein C. Peripheral mechanisms of opioid analgesia. Anesth Analg 1993;76: 182-191.

41. Csillik B, Knyihar-Csillik E, Szucs A. Treatment of chronic pain syndromes with iontophoresis of vinca alkaloids to the skin of patients. Neurosci Lett 1982;31: 87-90.

42. Layman PR, Argyras E, Glynn CJ. Iontophoresis of vincristine versus saline in post-herpetic neuralgia. A controlled trial. Pain 1986;25:165-170.

43. Tajti J, Somogyi I, Szilard J. Treatment of chronic pain syndromes with transcutaneous iontophoresis of vinca alkaloids, with special regard to post-herpetic neuralgia. Acta Medica Hugarica 1989;46:3-12.

44. Calabresi P, Chabner BA. Antineoplastic agents. In: Gilman AG, Rall TW, Nies AS, Taylor P, eds. The pharmacological basis of therapeutics. 8th ed. New York: Pergammon Press, 1990:1209-1263.

45. Szucs A, Csillik B, Knyihar-Csillik E: Treatment of terminal pain in cancer patients by means of iontophoresis of vinca alkaloids. Recent Results in Cancer Research 1984;89:185-189.

46. Kahn J. A case report: lithium iontophoresis for gouty arthritis. J Orthop Sports Phys Ther 1982;4:113-114.

47. Kahn J. Acetic acid iontophoresis for calcium deposits. Phys Ther 1977;57: 658-659.

48. Balogun J, Abidoye AB, Akala EO. Zinc iontophoresis in the management of bacterial colonized wounds: a case report. Physiotherapy Can 1990;42:147-151.

49. Greminger RF, Elliott RA, Rapperport A. Antibiotic iontophoresis for the management of burned ear chondritis. Plastic Reconstructive Surg 1980;66:356-360.

50. Kwon BS, Hill JM, Wiggins C, et al. Iontophoretic application of adenine arabinoside monophosphate for the treatment of herpes simplex virus type 2 skin infections in hairless mice. J Infectious Diseases 1979;140:1014.

51. Gangarosa LP, Payne LJ, Hayakawa K, et al. Iontophoretic treatment of herpetic whitlow. Arch Phys Med Rehabil 1989;70:336-340.

52. Boone DC. Hyaluronidase iontophoresis. Phys Ther 1969;49:139-145.

53. Akins DL, Meisenheimer JL, Dobson RL. Efficacy of the Drionic unit in the treatment of hyperhidrosis. J Am Acad Dermatol 1987;16:828-833.

54. Levit R. Simple device for treatment of hyperhidrosis by iontophoresis. Arch Dermatol 1968;98:505-507.

55. Stolman LP. Treatment of excess sweating of the palms by iontophoresis. Arch Dermatol 1987;123:893-896.

56. Gordon AH, Weinstein MV. Sodium salicylate iontophoresis in the treatment of plantar warts: case report. Phys Ther 1969;49:869-870.

57. Smith KJ, Konzelman JL, Lombardo FA, et al. Iontophoresis of vinblastine into normal skin and for treatment of Kaposi's sarcoma in human immunodeficiency virus-positive patients. Arch Dermatol 1992;128:1365-1370.

58. Sanderson JE, Caldwell RW, Hsiao, et al. Noninvasive delivery of a novel inotropic catecholamine: iontophoretic versus intravenous infusion in dogs. J Pharmaceutical Sciences 1987;76:215-218.

Chapter 10

Clinical Electrophysiologic Assessment

Andrew J. Robinson
Robert Kellogg

$\sqrt[V]{V}$ **C**linical electrophysiologic assessment consists of the observation, analysis, and interpretation of the bioelectrical activity of muscle and nerve in response to volitional activation or electrical stimulation. Clinical electrophysiologic assessment is also called **electroneuromyography (ENMG)**. The process of electrophysiologic assessment includes the performance of a wide array of test procedures. The results of ENMG tests are integrated with clinical measurements, laboratory findings, and symptomatology in order to assist in the establishment of a diagnosis and subsequent plans of care. Electrophysiologic test results viewed in isolation are neither characteristic nor indicative of specific diseases or disorders. That is, ENMG findings alone will not establish the diagnosis of the underlying pathology but rather supplement the findings from numerous other clinical and laboratory tests used in differential diagnosis.

The main purpose of ENMG is to determine the integrity of specific components of the neuromuscular system. Special testing procedures are available for examining α motoneurons and their axons, the neuromuscular junction, skeletal muscle, peripheral sensory nerve fibers, selected reflexes, and certain central nervous system pathways. Electroneuromyographic examination of these structures will assist the practitioner in the determination of the location, magnitude, and chronicity of neuromuscular impairment.

The objectives of this chapter are to discuss the instrumentation and recording procedures used in electrophysiologic assessment procedures, describe specific modern and traditional approaches of ENMG testing, briefly discuss supplemental clinical tests of neuromuscular function, and present clinical case studies to illustrate the usefulness of ENMG testing procedures.

Extracellular and intracellular recording techniques

The bioelectrical responses of excitable tissue to volitional or electrically elicited activation are nerve and muscle action potentials occurring across the membrane of all sufficiently depolarized fibers. When muscle or nerve fibers are depolarized to threshold, action potentials are propagated along the membranes of these tissues. In short, sodium ions rush into these cells followed by potassium ions rushing out of each cell, and this process sweeps along the membrane. As a result of these transmembrane ionic fluxes, changes occur in the concentration of ions in the fluids surrounding these tissues. The ionic movements that occur in single nerve cells are monitored in research laboratory experiments on very large axons or cell bodies by actually penetrating the excitable cell membrane with a glass microelectrode and placing a second recording electrode outside the cell. Such a system of one intracellular and one extracellular recording electrode (Fig. 10.1*A*) monitors the transmembrane (across the membrane) electrical potential differences produced when ions cross the membrane. This **intracellular recording technique** cannot be used on the nerve and muscle cells of humans in clinical electrophysiologic testing because these cells are too small and cannot be stabilized sufficiently to perform this technique.

Disclaimer: The opinions, views, and comments contained within this chapter are those of the authors and do not reflect the official position of the United States government, the Department of Defense, or the Navy.

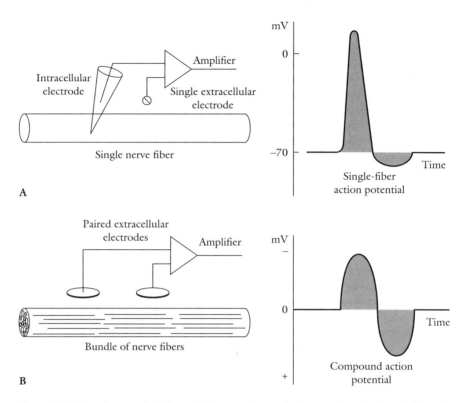

Figure 10.1 A, Diagram of the intracellular recording technique used on single excitable cells and the resultant trace of the intracellular action potential. **B**, Diagram of the extracellular recording technique used on multiple excitable cells and the resultant trace of the extracellular compound action potential.

Clinical ENMG requires the use of extracellular as opposed to intracellular recording techniques. In the **extracellular recording technique**, one or two small conductors (called *recording electrodes*) are placed near but outside nerve or muscle cells (Fig. 10.1*B*). The electrodes are then connected by wires to an electronic device that amplifies and displays the relative voltage (electrical potential) difference between the two electrodes. When no action potentials are transmitted along excitable cells, the concentration of ions beneath each of the electrodes is approximately equal and the voltage difference between the electrodes is zero. However, when action potential currents are propagated along nerve or muscle membranes, the relative concentration of ions beneath each of the electrodes changes over time and an electrical potential difference is recorded.

To understand the origin of extracellular voltage changes recorded during nerve or muscle excitation, consider a situation where action potentials are propagated along a bundle of nerve fibers. As sodium rushes into the nerve fibers beneath the first of the recording electrodes, the region below the electrode becomes more negative (less positive) than that below the second recording electrode (Fig. 10.2*A*). An instant later, the potassium leaves the fibers beneath the first electrode while sodium is rushing into the fibers

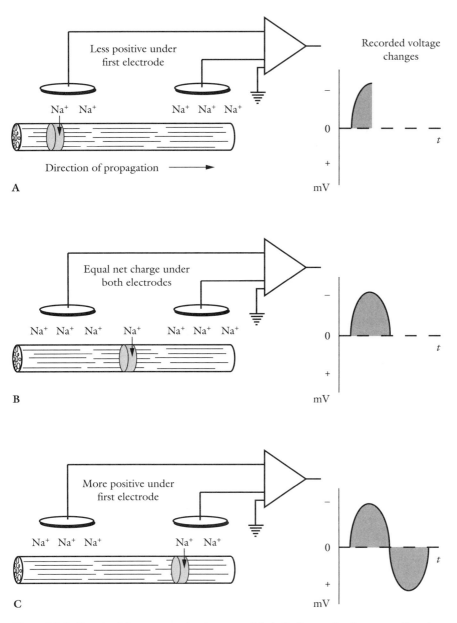

Figure 10.2 Genesis of the compound action potential. **A,** Sodium rushes into nerve fibers beneath the first recording electrode and the area beneath the electrode becomes more negative. **B,** Sodium moves into excitable cells between the two electrodes and ionic concentrations beneath each electrode equal. **C,** Sodium rushes into excitable tissues beneath the second electrode; hence, the first electrode appears more positive.

between the two electrodes. At this instant, the concentration of ions beneath each electrode is about the same and the electrical potential difference between the electrodes falls to zero (Fig. 10.2*B*). An instant later, as the inward sodium currents sweep beneath the region of the second electrode, the relative difference in voltage between the two electrodes becomes reversed. That is, the first electrode detects a higher concentration of positive ions than the second electrode. Hence the relative voltage difference between electrodes is positive (Fig. 10.2 C). This set of voltage changes recorded near nerve fibers is called a **compound nerve action potential (CNAP)**. This type of response recorded from a group of muscle fibers is called a **compound muscle action potential (CMAP)**.

The extracellular recording technique differs from the intracellular approach in that extracellular electrodes are used to monitor the electrical activity in many nerve or muscle fibers, whereas the intracellular technique only monitors the electrical activity in a single fiber. Hence, the electrical signals recorded are referred to as compound nerve or muscle action potentials.

Compound nerve and muscle action potentials can be characterized by their shape, amplitude, and duration. The time delay between the electrical stimulus and the appearance of the evoked potential can also be measured. Measurements of these parameters obtained during electrophysiologic assessment procedures can be compared with established normal values to identify conditions such as denervation, demyelination, conduction block, and axonal loss.

With appropriate selection of stimulation and recording sites, information can be obtained about the function of different segments of peripheral nerves, neuromuscular junctions, skeletal muscles, the reflex pathways, anterior horn cells (motoneurons), spinal nerve roots, and even some central nervous system pathways. The application of the extracellular recording technique and selected patterns of stimulation for testing the integrity of each of these structures will be detailed in subsequent segments of this chapter.

In clinical electrophysiologic assessment, needle electrodes are frequently used in the examination of skeletal muscle activity. In contrast, routine clinical testing of peripheral nerves is performed by placement of surface electrodes placed on the skin over the nerve. At times, needle electrodes are used to record the electrical activity in peripheral nerve. These electrodes do not penetrate the nerve but are placed near peripheral nerve bundles.

Monopolar and Bipolar Extracellular Recording

When using the extracellular recording technique, at times both recording electrodes are placed close to the excitable tissue monitored. This positioning of electrodes is referred to as a **bipolar recording electrode placement**. At other times in electrophysiologic assessment only one recording electrode will be placed near the excitable tissue examined, and the second electrode will be placed at some distance from the excitable tissue. This electrode pattern is referred to as **monopolar recording electrode placement**.

The closer the two recording electrodes are to each other and to the nerve or muscle examined, the smaller will be the area from which electrical potential changed will be detected. Consequently, the bipolar electrode arrangement will sense electrical activity in excitable tissues from a smaller volume of tissue than will the monopolar technique.

Electromyography and electroneurography

Compound muscle action potentials can be elicited by voluntary activation of the muscle or by electrically stimulating a motor nerve to a particular muscle.

Recording muscle action potentials that occur spontaneously or in response to volitional or stimulated activation using an extracellular recording technique is called **electromyography** and an individual record of compound muscle action potentials is called an **electromyogram (EMG)**.

Compound nerve action potentials can be elicited by either voluntary effort or by electrical stimulation. For example, in electrophysiological assessment, nerve action potentials are evoked by electrical stimulation in the cutaneous distribution of the sensory nerve and the evoked peripheral sensory potentials are monitored using the extracellular surface recording technique over peripheral nerves at some point remote to the site of stimulation. In more sophisticated procedures, extracellular recordings of compound nerve action potentials are made by actually placing electrodes over the surface of brain or spinal cord tissue. Recording nerve action potentials in this manner is called **electroneurography**, and the record of the nerve action potentials is called an **electroneurogram**.

Instrumentation for Electrophysiologic Assessment

Recording electrodes in electrophysiological assessment

In monitoring the responses of excitable tissues to voluntary or evoked activation, compound nerve and muscle action potentials are recorded through recording electrodes. The physiologic response to stimulation is recorded as a voltage change by the recording electrodes. The placement of the recording electrodes, their orientation, contact area, shape, and composition influence the size and shape of the recorded response.

Surface Electrodes

Surface recording electrodes are those placed on the skin. They are generally made of bare and polished metals, such as silver, gold, or platinum, that conduct electricity well (Fig. 10.3, *A* and *B*). These metals are more effective conductors if the skin is properly prepared by abrasion and an electrolyte coupling medium is placed between the electrode and the skin. Coupling media such as electrolytic pastes or creams decrease the impedance at the interface between the recording electrode and the skin. The use of coupling media allows surface electrodes to record smaller signals more accurately, improving the quality of the recorded signal. Concave recording discs retain the electrode paste or gel, thereby increasing the time the electrode can be used. Surface electrodes must be secured firmly in place with adhesive pads or straps in order to obtain consistent recordings of nerve or muscle activity.

Surface recording electrodes are used when the excitable tissues encompass a fairly large area. For example, surface electrodes are used to record EMG signals from skeletal muscle and ENG signals from superficial nerve bundles.

Needle Electrodes

Needle or other types of subcutaneous recording electrodes are those placed beneath the skin either near or directly in the excitable tissue of interest (Fig. 10.3, *C–E*). Indwelling electrodes used for recordings are often made of platinum, silver, or stainless steel. Platinum tolerates sterilization well, and tissue reactance to platinum is small. Stiff wire electrodes of platinum or silver

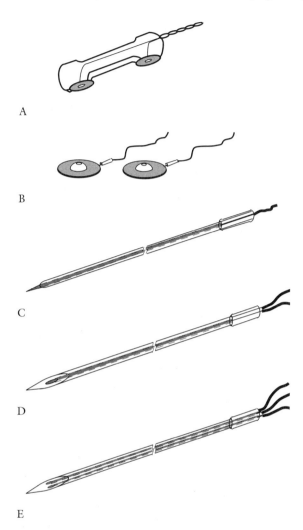

A

B

C

D

E

Figure 10.3 **A** and **B**, Surface recording electrodes, **C**, Monopolar needle electrode. **D**, Concentric needle electrode. **E**, Bipolar needle electrode.

can be used for recording from deep muscles. These electrodes are insulated with Teflon or nylon except at the tip. Subcutaneous electrodes do not require an electrolyte because they come in direct contact with body fluids which have a low resistance to the flow of electricity. Typical applications include the use of needle electrodes in electromyography (EMG) and electroneurography (ENG), and fine needle electrodes for recording potentials from the scalp in the somatosensory evoked potential. Fine wire electrodes are used to record EMG signals from deep small muscles whose potentials would be masked by overlying muscles when recording with surface electrodes. Fine wire electrodes are also used to record nerve signals (ENG) from small and deep nerves.

Needle electrodes are preferred for recording from small areas. For example, needles are chosen to monitor motor unit recruitment patterns. Because the electrodes are in direct contact with body fluids, the impedance at the recording site is low. Unfortunately, however, insertion of needle electrodes can cause some discomfort and possibly tissue damage if movement of the electrode is not minimized.

Not all needle electrodes have the same construction. The simplest of the needle electrodes is the **monopolar electrode** (Fig. 10.3*C*). Monopolar electrodes consist of a central stainless steel core conductor surrounded by a thin coating of Teflon. Only the pointed tip of the central core is exposed for approximately 0.4 mm. These electrodes are available in sizes ranging from 24 gauge to 30 gauge, depending on the overall length of the electrode.

Monopolar electrodes placed into or near the excitable tissue of interest must be used in conjunction with either additional needle or surface electrodes that act as reference and ground electrodes. Monopolar needle electrodes are commonly used in EMG studies in the United States.

Another common type of electrode is the **concentric needle electrode** (Fig. 10.3*D*). One of the commercially available types of concentric needle electrodes consists of a central platinum or stainless steel core surrounded by a polyimide insulator and contained within a stainless steel cannula. These electrodes are ground to a very fine point at an angle of 15°. The central conductive core is exposed only at the tip and acts as one of the recording electrodes, whereas the stainless steel housing acts as the reference electrode of the recording circuit. A separate ground electrode must also be used.

A third type of electrode available for use in electrophysiologic assessment procedures is the **bipolar needle electrode** (Fig. 10.3*E*). This electrode consists of two very fine platinum conductors insulated from each other by polyimide and from the stainless steel cannula in which they are housed. Like the concentric needle electrodes, these bipolar electrodes are ground to a point so that both central conductors are exposed to the tissues only at the tip of the electrode. Electrical potential differences are recorded between the two core electrodes. The cannula acts as the ground.

Components of signal processors in ENMG testing

When electrical potential differences (electromyogram or electroneurogram) are recorded in response to the activation of excitable tissue, the signal is generally too small to be immediately visualized on common display devices (e.g., oscilloscope). Biological electrical signals recorded from nerve and muscle are in the range of several microvolts to several millivolts. Furthermore, the frequency components of signals may range from DC to several thousand hertz. Adequate display of the response(s) requires some signal processing by other electronic devices such as amplifiers, filters, signal averagers, integrators, and analog-to-digital convertors.

Amplifiers

An amplifier is an electronic device used to increase the amplitude of electrical voltages monitored in electrophysiological assessment. Such devices are used to make small electrical signals larger on signal display instruments such as oscilloscopes, computer screens, or strip chart recorders. In some cases when low-voltage signals are being monitored, the amplifier built into the display device may increase the signal size enough for accurate measurements. In other cases, amplifiers with special features may be required to create an adequate display of the electrical potentials.

A feature characteristic of all amplifiers is the device's **gain**. The gain of an amplifier describes the relationship between the input-voltage amplitude of

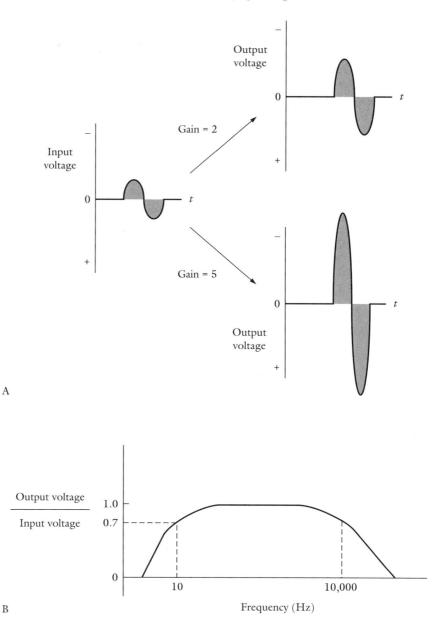

Figure 10.4 A, Diagrammatic representation of amplifier gain; **B,** frequency response characteristics of an amplifier.

the signal and the output-voltage amplitude from the amplifier (Fig. 10.4*A*). If the peak voltage of a monophasic input signal is doubled, from 1 mV to 2 mV, the gain of the amplifier is 2. If the same input signal's peak amplitude is increased by the amplifier to 5 mV, the gain of the amplifier is 5. Most amplifiers are constructed so that the user can set the amplifier gain to an appropriate level. The amplifier control for such an adjustment is often labeled "Sensitivity."

Amplifiers used in electrophysiological assessment should uniformly increase signal size for all signals within a specified frequency range (or bandwidth) of 2 Hz to 10,000 Hz (10 kHz). The **frequency response** of an amplifier (Fig. 10.4*B*) is determined by applying sinusoidal alternating voltages at various frequencies and known amplitude to the inputs of the amplifier. The ratio of the peak-to-peak voltages (V out/V in) should remain constant over the normal frequency spectrum of the recorded signals (2 Hz–10 kHz) to ensure that recorded potentials are faithfully reproduced by the amplification process. Any signals composed of frequencies falling above or below the bandwidth of the particular amplifier will be attenuated (decreased) in relative amplitude or not displayed at all.

Electrophysiological testing amplifiers should have a **high input impedance** (> 1 megohm) relative to the impedance of the recording electrodes used. A high amplifier input impedance will not allow the amplifier to draw much current from the biological tissue so the voltage changes recorded will more accurately reflect the ionic movements actually occurring in the tissue.

Simple and Differential Amplification

A **simple** (or **single-ended**) **amplifier** is a device that enhances all voltage differences between recording electrodes. At times, the ionic movements beneath each of the recording electrodes will be similar. When 60-Hz line current is sensed by each electrode or wire leading to the amplifier, this electrical "interference" is amplified along with any biological signal. Since biological potentials are generally several orders of magnitude lower in amplitude than 60-Hz interference, the use of simple amplification in electrophysiological testing may not allow the examiner to adequately visualize the bioelectric potentials of interest.

To overcome this problem, the signals recorded from test electrodes are generally fed to a **differential amplifier**. The differential amplifier increases the difference between the two signals. In the case of bipolar electrodes, the differential amplifier enhances the differences of the signals from the two closely spaced recording electrodes. In monopolar electrode placement, the differential amplifier increases the differences of the signals from a single recording electrode and a distant reference electrode. Differential amplifiers are usually used with bipolar or concentric needle electrodes. The differential amplification technique rejects signals that are common to both electrodes and enhances signals that are different. This is called **common-mode rejection**. Two recording surfaces tend to see different potentials for a given localized electrochemical event (e.g., a nerve or muscle potential). These signals are "passed through" the amplifier and recorded. Any signals that are common to the two recording electrodes, such as 60-Hz noise from the power supply and DC noise, are blocked by the amplifier. In this way, the amplifier takes the signals recorded from electrodes, filters out noise, and improves the clarity of the recorded electrical event.

The differential amplifier uses a ground electrode to compare the signals of the two recording electrodes. The amplifier actually subtracts the difference between one recording electrode and the reference electrode, and compares that with the difference between the second recording electrode and the reference. Whether a particular source of noise is "common mode" depends on its location and orientation with respect to the recording electrodes. Thus, for recording electrodes that are close together (e.g., bipolar), "noise" sources are far away from the recording site. This noise appears the same at both recording electrodes, and will be rejected as common mode. Conversely, with electrodes that are far apart relative to the noise, the noise may appear to be different at the two recording sites, and it will be passed through and amplified by the differential amplifier.

Filters

The physiological responses of nerves and muscles have characteristic frequency ranges. The "noise" of the external environment can be electronically removed from the excitable tissue signal (as long as the frequency ranges do not overlap) by an electronic device called a **filter**. Several types of bandpass filters are commonly found on modern electrophysiological testing instruments. For example, DC and 60-Hz AC line noise can be eliminated from higher frequency signals by using a device called a **notch filter**. This type of filter rejects signals at 60 Hz and "passes" (or "keeps") a band of frequencies less than and greater than 60 Hz. A second type of bandpass filter, called a **high-pass filter**, passes (keeps) the high-frequency signals and filters out the low-frequency signals. A third type of bandpass filter, called a **low-pass filter**, passes the low-frequency signal and rejects the higher frequencies.

Signal Averagers

If a single response to an electrical stimulus is very small, and the electrical "noise" in the environment is large with respect to it, visualization of the electrical signal of interest may not be immediately possible. To overcome this problem of an important electrical signal being "buried" in noise, the technique of signal averaging was developed. In signal averaging, an electrical stimulus is applied to evoke electrical activity in an excitable tissue. For a short period of time following the stimulus, all electrical activity (voltage changes) from the tissue is monitored and stored. When a second stimulus is applied, the electrical response is again stored and electronically added to the record from the first stimulus. This process is repeated with each successive record being added to the sum of all previous records. Since the elicited potentials are time-locked to the stimulus and occur at a fixed time following the stimulus, signal averaging will enhance these signals. Electrical interference (noise) is not time-locked to the stimulus, and the averaged noise voltage transients will be eliminated by the averager. In other words, the effect of signal averaging is that all voltage changes that occur at the same time following the stimulus are enhanced, whereas all randomly occurring voltage fluctuations are reduced.

The signal averaging technique is frequently used to record sensory-nerve action potentials and somatosensory evoked potentials. Sensory-nerve compound action potentials are generally on the order of a few microvolts in peak amplitude while electrical noise for 60 Hz interference is 10 to 100 times greater in amplitude. Signal averaging allows the clinician to separate the sensory signal from the background noise as shown in Figure 10.5. Signal averaging is also used in monitoring brain and spinal cord electrical activity in response to peripheral sensory stimulation. The somatosensory evoked potentials recorded in electrophysiological techniques are normally about 1–10

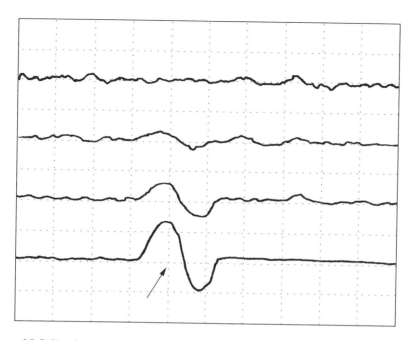

Figure 10.5 Signal averaging technique: successive traces of evoked potentials illustrating how the evoked potential *(arrow)* is processed from the electrical "noise" during testing.

μV (10^{-6} V) in amplitude, but the background EEG size is about 0.1–1 mV (10^{-3} V). The somatosensory evoked potential is distinguished from the noise of the EEG by averaging the response to 100 to 2000 stimuli. Once again, the signal of interest is separated from the background or interference signals through the use of the signal averaging procedure.

Integrators

Integration of an electrical signal is the calculation of the area under a signal waveform or curve. The units of the output of an electronic integrator are volt-seconds. An observed signal with average value of zero has a total integrated value of zero. Therefore, integration only applies to a full-wave rectified signal; the rectified value is always positive. The integrated value of a rectified signal increases as a function of time.

Analog-to-Digital Convertors

For many clinical applications, the physiologic electrical responses from nerve or muscle are recorded, processed (e.g., amplified and filtered), and displayed as **analog output**. This is the form of the output signal displayed on strip charts and many oscilloscopes, and stored on FM tape recorders. However, for computer applications, the analog signal must be converted into a digital signal for precise data acquisition, storage, and later retrieval for analysis. The data are "digitized" by an electronic device called an **analog-to-digital (A/D) converter**, whose sampling rates and signal processing times are specified. The characteristics of the A/D converter are dictated by the frequency of the recorded signal and the number of channels of data being simultaneously sampled. A/D convertors used in electrophysiological testing should

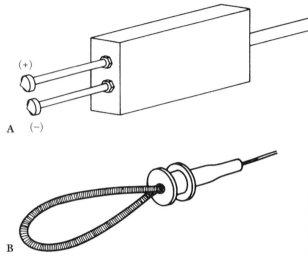

Figure 10.6 Surface stimulating electrodes in electrophysiologic assessment: **A,** bipolar; **B,** ring or "loop" electrodes.

have sampling rates high enough to faithfully digitize even the very highest frequency components of the EMG signal.

Display and Storage Devices

An electrical response of excitable tissue can be displayed in analog (continuous) or digital form (discrete or digitized measurements). An entire waveform may be displayed on an oscilloscope, computer screen, or strip chart recorder (polygraph). While a photograph may be taken of the waveform on an oscilloscope display and measurements may be made from this or from the waveform displayed on a strip chart recorder, neither of these methods allows further automated data processing. That is to say, the data cannot be reprocessed (filtered, amplified) or redisplayed in any other manner. In addition, some strip chart recorders using electromechanically controlled pens cannot accurately reproduce the highest frequency EMG signals.

An entire event or series of events may be preserved for future analysis on an FM tape recorder (analog data) or in computer memory (digitized data). Both of these data storage methods allow further analysis.

Neuromuscular stimulators in ENMG

The electrical stimulators used in activation of peripheral nerve or muscle in ENMG are similar to yet simpler in many respects than neuromuscular electrical stimulators used in therapeutic applications. Both constant-current and constant-voltage stimulators have been used in electrophysiologic assessment procedures and either appear to be satisfactory for clinical testing, although the constant-current type has been recommended for some applications (1). In general, these stimulators produce a rectangular monophasic pulse. Control over the stimulation parameters is usually limited to pulse amplitude, pulse duration, and pulse frequency. Amplitude and timing modulations are not required for standard assessment application and hence are not included. Pulse durations may generally be selected from 0.05 msec to 1.0 msec. Maximum amplitudes of stimulation may reach 600 V for constant-voltage devices or about 100 mA for constant-current stimulators. Frequencies of stimulation may be selected through a range of 1 pps to 50 pps.

Stimulating electrodes in electrophysiological assessment

A handheld "bipolar electrode" (Fig. 10.6*A*) is one type of instrument used to activate peripheral nerve and/or muscle in the many commonly applied electrophysiological assessment techniques. In this unit, the two electrodes (anode and cathode) of the stimulating circuit are fixed at a distance of 2 to 3 cm apart. The cathode and anode are commonly color-coded black and red, respectively. In the application of stimuli to evoke nerve or muscle action currents, this pair of electrodes is placed longitudinally along the particular peripheral nerve of interest with the cathode–anode orientation determined by the particular ENMG procedure performed.

Another type of stimulating electrode used in electrophysiological testing is the flexible "ring" or "loop" electrode (Fig. 10.6*B*). This type of electrode can be quickly applied to the digits and is used most often to stimulate the sensory nerves in the fingers to determine sensory nerve conduction velocity. For both of these stimulating electrodes, an electrolytic paste or gel must be applied to the electrodes to insure sufficient electrical conduction.

In some cases, needle electrodes like those described above may be connected to a stimulator and used as stimulating electrodes to obtain very localized activation of electrically excitable tissues.

Modern Procedures of Electrophysiologic Assessment

Nerve conduction studies

Nerve conduction studies assess peripheral motor and sensory nerve function by recording the evoked potential generated by electrical stimulation of a peripheral nerve. Studies of action potential conduction in peripheral nerve fibers are used clinically to answer the following questions:

1. Are peripheral nerve fibers involved?
2. Are sensory fibers, motor fibers, or both involved?
3. What is (are) the location(s) of the peripheral lesion? How many peripheral nerves are involved?
4. What is the magnitude of peripheral nerve involvement? Is the lesion partial or complete?
5. Is the peripheral nerve impairment increasing or decreasing over time? Is there evidence for recovery or further degeneration?
6. Does evidence for localized nerve block, axonal degeneration, or segmental demyelination exist?
7. Does the pattern of nerve involvement suggest a localized or a systemic disorder?

In theory all peripheral nerves can be tested electrophysiologically, but some nerves are much easier to test and are more commonly examined. Traditionally, these include the median and ulnar nerves in the upper extremity and the peroneal, posterior tibial, and sural nerves in the lower extremity. Other peripheral nerves can be tested if the need for further information dictates.

Motor-nerve conduction studies

To examine the conduction properties of motor nerve fibers, a classical electrophysiological approach is used. That is, a stimulus is applied in one region and the response to that stimulation is monitored using an extracellular technique in another area.

In performing motor-nerve conduction studies, a mixed peripheral nerve is electrically stimulated with single, short-duration (0.1–0.2 msec) stimuli and action potentials are generated in α motoneuron axons (Fig. 10.7). The evoked nerve action potentials are propagated along the motor fibers and activate the neuromuscular junctions (NMJ) of the stimulated motor axons. NMJ transmission results in muscle action potentials on innervated muscle fibers and triggers a twitch contraction in the muscle innervated by these fibers. In motor-nerve conduction studies, the actual force of twitch contraction is not measured. Rather, electrodes are placed either over the muscle or in the muscle to record the biphasic (or triphasic) compound muscle action potential that accompanies the twitch contraction. The amplitude of the evoked CMAP (also called the *M wave*) is proportional to the number of motor units depolarized and hence is a reflection of the extent of activation of muscle fibers produced as a result of motor nerve stimulation.

As the current amplitude is increased in motor nerve stimulation, motor nerve fibers are recruited progressively until a maximum twitch contraction is produced. With this level of stimulation, a further increase in stimulus amplitude will not increase the amplitude of the evoked response. At this point, all of the motor axons directly beneath the stimulator have been excited, the maximum number of neuromuscular junctions are triggered, and action potentials are generated on the membranes of the innervated muscle fibers. Stimulation of peripheral motor fibers at amplitudes greater than those required to recruit all motor axons near the electrodes and subsequently

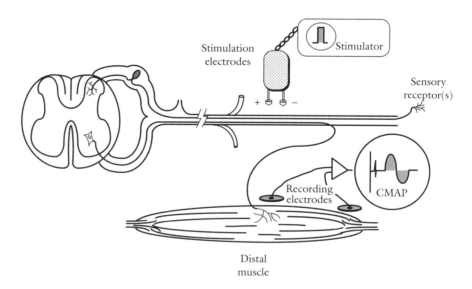

Figure 10.7 Diagrammatic representation of the stimulation and recording sites for performing motor-nerve conduction studies.

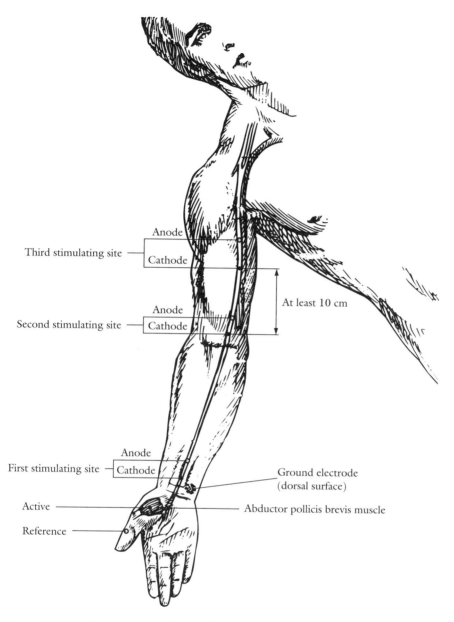

Anode

Third stimulating site —

Cathode

At least 10 cm

Anode

Second stimulating site —

Cathode

Anode

First stimulating site —

Cathode

Ground electrode
(dorsal surface)

Active —

Abductor pollicis brevis muscle

Reference —

Figure 10.8 Stimulation and recording sites for performing motor-nerve conduction studies in the median nerve.

all innervated muscle fibers is called **supramaximal stimulation**. By ensuring maximal activation of all innervated muscle fibers, supramaximal stimulation results in the recording of maximal compound muscle action potentials.

Example application: Median nerve motor conduction

To assess motor-nerve conduction in the median nerve (Fig. 10.8) (2), surface recording electrodes are placed on one of the most distal muscles innervated by the nerve such as the abductor pollicis brevis (APB). One recording electrode (called *active*) is placed over the APB motor point and the second electrode (called *reference*) is placed distally over the tendinous insertion area of the muscle. A third electrode, the ground electrode, is placed between the stimulating electrodes and the recording electrodes on a bony area such as the dorsum of the wrist. Electrical stimulation is provided by use of a handheld bipolar electrode connected to a stimulator that produces a rectangular, monophasic pulsed current with 0.1–0.5 msec pulse duration and an adjustable amplitude. Frequency of stimulation may be controlled manually or preset at 1 pps. The cathode and anode of the handheld electrode are separated by 3 cm. The cathode is placed distal to the anode over the median nerve, closest to the most proximal recording electrode over the muscle. The response (CMAP of abductor pollicis brevis) is amplified, filtered (10 Hz–10 kHz), and recorded on an oscilloscope or similar display device.

Testing is begun by increasing the current amplitude until the maximal amplitude of the recorded compound muscle action potential is reached. This signal is the evoked compound-muscle action potential (CMAP) and represents the summated electrical activity of all of the muscle fibers in the region of the recording electrodes that are innervated by the nerve being stimulated (Fig. 10.9). To ensure that the maximal response is recorded, a supramaximal stimulus must be delivered to the nerve. Supramaximal stimulation is technically defined as stimulus amplitude 20% greater than that required to obtain the maximal CMAP. The characteristics of the CMAP can then be measured to give a quantitative measure of functioning nerve and muscle fibers.

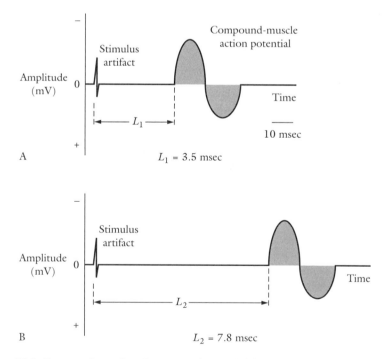

Figure 10.9 Compound-muscle action potentials recorded from the APB in response to supramaximal stimulation of the median nerve at the wrist (**A**), and the elbow (**B**).

The CMAPs shown in Figure 10.9 represent the summated electrical signals from muscle fibers innervated by large, intermediate, and small myelinated motor fibers as the amplitude of stimulation is gradually increased. The amplitude of the CMAP increases as greater numbers of motor units are recruited. Because the AP conduction speed of these nerve fibers depends on their size, the initial deflection from the baseline represents the beginning of activation of the largest diameter muscle fibers innervated by the largest motor axons. The beginning of activation of intermediate fibers contributes to the middle of the response, and the slower-conducting fibers contribute to the end of the negative portion of the waveform.

The display of the CMAP appears in a consistent position on the face of the display device because the display sweep is triggered at the beginning of each applied stimulus. The time elapsed between the onset of the stimulus and the beginning of the CMAP represents the time taken for motor-nerve action potentials to propagate from the point of stimulation to the neuromuscular junction plus the neuromuscular junction transmission time plus the time required for muscle action potentials to sweep by the first recording electrode. The time between the onset of the stimulus and the beginning (initial negative deflection) of the CMAP is called the **latency** of the response. Such latency values (measured in milliseconds) are one of the primary measurements taken in motor-nerve conduction studies. The recorded latency is proportional to the distance between the stimulating and recording electrodes. The farther the stimulating electrodes are from the recording electrodes, the longer the latency of the CMAP. Latency measurements to the initial takeoff from the baseline are a reflection of conduction for the largest-diameter and fastest-conducting fibers.

By applying bipolar stimulation to the median nerve at several sites, several latency values may be taken and used to calculate motor-nerve conduction velocity for various segments of the median nerve. Let's say for example that we are interested in determining the motor-nerve conduction velocity in the segment of the median nerve lying between the elbow and the wrist. First, stimulation as described above would be applied to the median nerve at the wrist and the latency (L_1) between stimulation and recording the CMAP from the abductor policis brevis would be recorded (Fig. 10.9A). Since median nerve stimulation at the wrist is the most distal point of stimulation of this nerve, the latency (L_1) is referred to as the **distal latency**. The point of cathodal stimulation would also be marked. Next, stimulation would be applied over the median nerve at the elbow. The CMAP evoked from the APB appears later on the display screen, and the longer latency is recorded as L_2 (Fig. 10.9B). The CMAP waveform should be similar in configuration to that elicited by stimulation at the wrist. This similarity in configuration ensures that the same group of nerve fibers is being activated by the stimulus and hence that conduction over the same group of fibers is being assessed. If the configurations of the CMAPs are very different between two points of stimulation, it may mean that the examiner is stimulating different populations of motor nerve fibers and in so doing is leading to an inaccurate result in the calculation of motor-nerve conduction velocity (MNCV).

The point of cathodal stimulation at the elbow is also marked. The last measurement to be made in order to calculate motor conduction velocity is the distance (measured in millimeters) between the two cathodal stimulation points. This is performed by laying a metal tape measure on the skin in a manner that approximates the anatomic course of the nerve. This value represents the distance over which motor-nerve action potentials propagated on the elbow to wrist segment. Since velocity of conduction is expressed in units of length divided by units of time (e.g., meters per second) the conduction velocity of the forearm segment of the median nerve is calculated by dividing the length of the forearm segment by the difference in conduction time between the elbow and thumb and wrist to the thumb; that is,

$$\text{Conduction velocity} = \frac{\text{distance}}{\text{time}} = \frac{\text{forearm segment length (mm)}}{L_2 - L_1 \text{ (msec)}}$$

If the length of the forearm segment of the median nerve is measured as 235 mm, the distal latency (L_1) from wrist to APB is 3.5 msec, and the proximal latency from elbow to APB is 7.8 msec, the conduction velocity (CV) of the forearm median nerve segment becomes 54.7 msec:

$$CV = \frac{235 \text{ mm}}{(7.8 \text{ msec} - 3.5 \text{ msec})}$$

$$= \frac{235 \text{ mm}}{4.3 \text{ msec}} \times \frac{1000 \text{ msec}}{1 \text{ sec}} \times \frac{1 \text{ m}}{1000 \text{ mm}} = 54.7 \text{ m/sec}$$

Note that a conversion is shown to obtain the CV value in meters per second. In fact, if distances and latencies are measured in millimeters and milliseconds, respectively, the numerical value of the division of millimeters by milliseconds is the same as that obtained by performing the measurement unit conversions.

The technique described here can be applied at recommended points to determine CV values for segments of a variety of other peripheral nerves. Figure 10.10 shows examples of stimulation and recording sites for obtaining motor conduction velocities from several other commonly tested peripheral nerves. Table 10.1 shows ranges of normal values for distal latencies and motor conduction velocities of various segments of several major peripheral nerves. Each electrophysiologic testing clinic lab should develop its own set of normative values since variations in the application of the MNCV procedure may alter the results.

Motor-nerve conduction velocities decrease with peripheral nerve pathology. Nerve compression, demyelination, and axonal degeneration as well as other disorders slow axonal conduction. Some environmental factors such as cold may dramatically reduce CV and the examiner must be careful to control for such transient influences. No pathology that we are aware of increases CV in peripheral fibers. Only vigorous heating of peripheral nerve may result in CV increases and this effect is not as dramatic as the influence of cold.

It is important to point out that motor conduction velocities cannot be accurately determined for the segments of peripheral nerves lying between the most distal stimulation points and the innervated muscle for two reasons. First, the length of the most distal segment is difficult to accurately measure. Second, the distal latency measure includes not only actual conduction time along the distal segment, but also synaptic delay time at the NMJ and conduction time along the innervated muscle fibers. For these reasons, the distal latency values for peripheral nerves are simply compared to the normal range of established values to gain a sense as to whether conduction in the distal segment

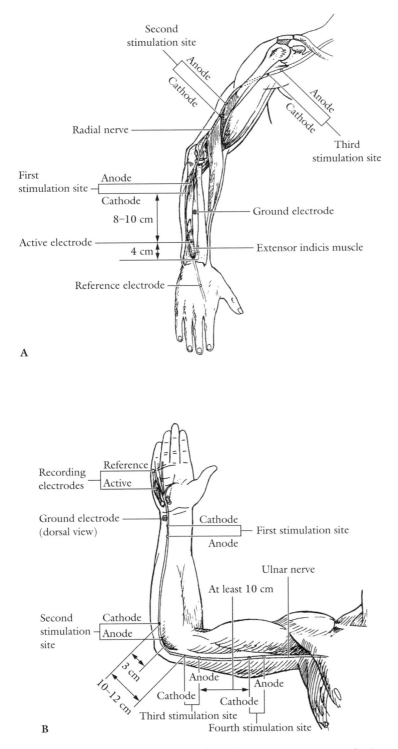

Figure 10.10 Stimulation and recording sites for performing motor nerve conduction studies in **A**, the *radial nerve*, **B**, the *ulnar nerve*.

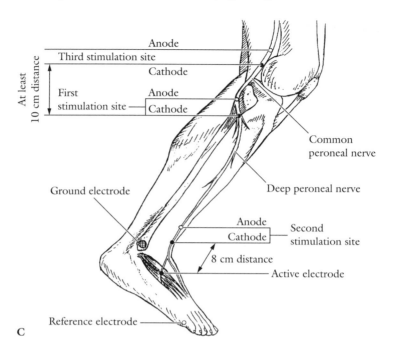

C

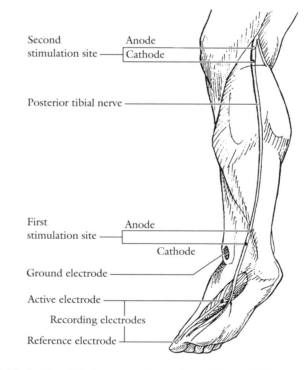

D

Figure 10.10 *Continued* **C**, the *common peroneal nerve* **D**, the *tibial nerve.*

Table 10.1
Normal Values for Motor Conduction Studies of Commonly Tested
Peripheral Nerves

Nerve/ segment	Distal Latency (msec)	Conduction Velocity (m/sec)	CMAP amplitude (mV)
Median			
Wrist-APB	3.5 ± 0.5		
Elbow-wrist		55.0 ± 5.0	7.0 ± 3.0
Axilla-elbow		64.0 ± 5.0	''
Erb's-axilla		64.0 ± 5.0	''
Ulnar			
Wrist-ADM	2.7 ± 0.5		
Elbow-wrist		57.0 ± 5.0	5.5 ± 2.0
Axilla-elbow		63.0 ± 2.0	''
Erb's-axilla		62.0 ± 5.0	''
Deep peroneal			
Ankle-EDB	5.9 ± 1.3		5.6 ± 2.0
Knee-ankle		45.0 ± 5.0	''
Posterior tibial			
Ankle-AH	4.5 ± 1.0		
Knee-ankle		43.3 ± 3.5	5.0 ± 2.0

APB, abductor pollicis brevis; *ADM,* abductor digiti minimi; *EDB,* extensor digitorum brevis; *AH,* abductor hallucis.

is normal or slowed. A longer than "normal" distal latency may reflect a slowing of distal segment conduction, an NMJ delay, or a muscle action potential (MAP) conduction slowing. Table 10.1 presents normal motor distal latency values for the distal segments of several major peripheral nerves.

Several other measurements are made to characterize the evoked CMAP (Fig. 10.11). They include: peak amplitude of each phase (in microvolts), duration of each phase (in milliseconds), and configuration (biphasic, triphasic, etc).

The **amplitude** of the CMAP is measured in microvolts (μV) or millivolts (mV) from the baseline (isoelectric, or zero-voltage, line) to the negative peak of the CMAP and is a function of the total number of muscle fibers innervated by the nerve being stimulated. The CMAP amplitude should not change with stimulation distance. Occasionally, the CMAP amplitude may be slightly attenuated at proximal stimulation sites due to differing rates of conduction within the motor axon dispersing their arrival times. Nerve pathology that reduces the number of motor axons activated would result in an attenuation of the amplitude of the response.

The **duration** of the CMAP is measured from the initial negative deflection to the point where the CMAP again crosses the baseline. The duration is a function of the different speeds of the axons and the nearly synchronous activation of their muscle fibers. Any pathological process that slows the conduction velocity of a portion of the axons ($<$ 100%) may cause the CMAP duration to increase.

The normal **configuration** of the CMAP is biphasic when the recording electrode has been properly placed over the motor point or motor endplate region and triphasic when not optimally placed. Triphasic CMAPs may also

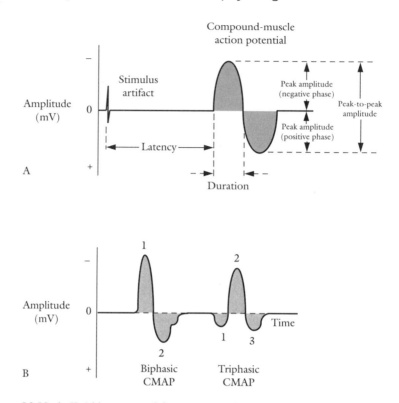

Figure 10.11 A, Variables measured from compound-muscle action potentials. **B,** Common configurations of compound muscle action potentials.

represent a disease state or may result from volume conduction from a nearby muscle. When this occurs the three measurements above are less reliable since one can not always reproduce the CMAP. Stimulating at proximal sites should not change the configuration of the CMAP.

Sensory nerve conduction studies

To examine the conduction properties in peripheral sensory nerve fibers, a slightly different stimulation and recording procedure is used.

In performing sensory-nerve conduction studies, peripheral sensory fibers are electrically stimulated with single short-duration shocks at levels sufficient to produce action potentials simultaneously in all sensory axons beneath the stimulator. The evoked sensory-nerve action potentials are propagated along the sensory fibers and the summated effect of these sensory action potentials propagated along the nerve may be monitored using surface or needle electrodes as the **compound sensory-nerve action potential (CSNAP).**

The CSNAP is either biphasic or triphasic in shape. When the CSNAP is triphasic, an initial positive deflection is followed by a much larger negative deflection. The positive deflection may represent local current flow before depolarization occurs under the recording electrodes. The amplitude of the negative phase is proportional to the number of sensory nerve fibers activated in response to the applied stimulus. The sensory response (compound sensory-nerve action potential or CSNAP) is smaller in amplitude than the CMAP and may require the use of signal averaging in order to obtain a clear display of this response. Stimulating with small needle electrodes placed near the nerve or recording with needle electrodes placed near the nerve may enhance the sensory response considerably and mitigate the need for signal averaging.

When a peripheral nerve is electrically stimulated, nerve action potentials are propagated in both directions (proximally and distally) from the point of stimulation. The action potentials propagated in the same direction as normal physiologic propagation are referred to as **orthodromically** transmitted. Those action potentials traveling opposite to the normal physiologic direction are said to be propagated **antidromically**. Orthodromic conduction in sensory fibers is from the periphery toward the CNS; antidromic conduction in sensory fibers is toward the periphery.

Sensory-nerve conduction may be studied by stimulating the nerve at some point and monitoring either the orthodromically propagated potentials or antidromically transmitted potentials (Fig. 10.12). Sensory compound action potentials are usually larger in amplitude when monitoring the antidromically propagated potentials as opposed to monitoring orthodromically conducted

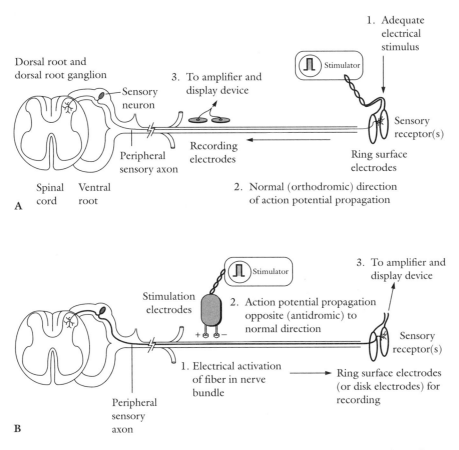

Figure 10.12 Diagrammatic representation of the stimulation and recording sites for performing sensory-nerve conduction studies using the orthodromic sensory conduction technique (**A**) and the antidromic sensory conduction technique (**B**).

signals. This difference in amplitude is generally the result of the recording electrodes being closer to the signal source (sensory axons) in the antidromic procedure. On the other hand, the chances of producing interference from the activation of muscle fibers are reduced by using the orthodromic technique. Actual clinical measurements of sensory conduction are performed using several approaches including: *(a)* stimulating and recording from a purely sensory cutaneous nerve, *(b)* recording from a mixed nerve (containing both sensory and motor components) while stimulating a sensory cutaneous nerve (orthodromic technique), or *(c)* recording from a sensory cutaneous nerve while stimulating a mixed nerve (antidromic technique).

Example application: Ulnar nerve sensory conduction

Figure 10.13 shows the stimulation and recording electrode arrangement for performing the orthodromic sensory-nerve conduction study in the median nerve (3). In this procedure ring electrodes are connected to a stimulator and placed over a finger innervated exclusively on the volar surface by digital sensory branches of the ulnar nerve. For consistency in implementation of the technique, the distal ring electrode is placed over the distal interphalangeal crease and the more proximal ring electrode is place over the midpoint of proximal phalanx. The more proximal electrode is connected to the cathodal output of the stimulator. Surface recording electrodes may be positioned directly over the ulnar nerve at points where the nerve is close the skin surface.

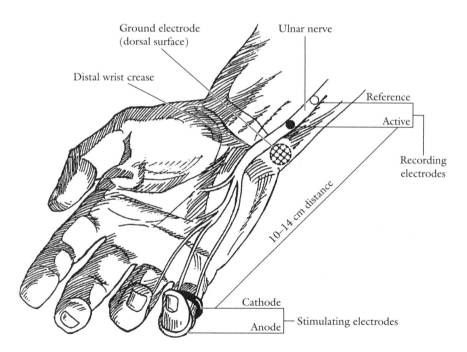

Figure 10.13 Stimulation and recording sites for performing orthodromic sensory-nerve conduction studies in the ulnar nerve.

A ground electrode is placed on the dorsum of the hand or over the styloid process of the ulna.

Stimulation is applied at a frequency of 1 pps, a pulse duration of 0.1 msec, and an intensity that is gradually increased to supramaximal levels. Action potentials are elicited only in sensory fibers because no motor fibers are present in the fingers at the site of stimulation. The sensory potentials are transmitted toward the CNS and the summated effect of these potentials passing along the nerve may be monitored as the CSNAP using recording electrodes placed at recommended positions along the nerve as shown.

The time between the application of the stimulus and the peak of the negative phase of the CSNAP is called the **sensory latency** and is measured in milliseconds. The sensory latency is measured to this negative peak as opposed to the onset of the potential because very often one cannot accurately determine the precise point at which the potential begins (leaves the isoelectric baseline). The latency value represents the time of conduction of sensory potential between the cathode and the most distal of the recording electrodes. The longer the distance between the stimulating cathode and the distal recording electrode, the greater will be the latency value measured. Each position of the distal recording electrode should be clearly marked so that the distance between the stimulating and recording electrode (conduction distance) may be measured using a metal tape.

Knowledge of the conduction distances and conduction latencies for various segments of the nerve allows one to calculate conduction velocities. For example, if the distance between the stimulating cathode on the finger and the distal recording electrode at the wrist is 120 mm, and the latency of the CSNAP is 2.8 msec, the conduction velocity along this segment would be calculated to be:

$$CV_{\text{digit to wrist}} = \text{distance}/\text{time} = 120 \text{ mm}/2.8 \text{ msec} = 42.9 \text{ m}/\text{sec}$$

If the conduction distance between the digit and a recording electrode placed over the ulnar nerve at the elbow is 375 mm and the sensory conduction latency is measured to be 9.3 msec, the conduction along this entire segment would be calculated to be:

$$CV_{\text{digit to elbow}} = \text{distance}/\text{time} = 375 \text{ mm}/9.3 \text{ msec} = 40.3 \text{ m}/\text{sec}$$

Even though these calculations may be a good representation of the conduction over these segments, some argue that the exact site of stimulation is not necessarily directly beneath the stimulating cathode but rather may lie more proximal to the cathode or in fact between the cathode and anode. This concern can be allayed for ulnar nerve segments from wrist to below the elbow, across the elbow, and above the elbow to the axilla by using a calculation similar to that used on motor-nerve conduction studies. For example, if we were interested in learning the sensory conduction velocity for only the segment from the wrist to below the elbow, we would first measure the distance between the distal recording electrodes at the wrist and elbow (255 mm). Next, the conduction latency from digit to wrist would be subtracted from the conduction latency from digit to elbow (9.3 msec − 2.8 msec). The sensory conduction velocity of the forearm segment of the median nerve would then be calculated as follows:

$$CV_{\text{wrist to elbow}} = 255 \text{ mm}/6.5 \text{ msec} = 39.2 \text{ m/sec}$$

When performing sensory conduction studies using stimulation at several locations along a nerve, the evoked compound-nerve action potentials may not be identical in amplitude or duration when comparing evoked sensory potentials. The longer the distance of the stimulating electrode from the recording electrode, the smaller (lower amplitude) and more dispersed (longer duration) will be the response. The attenuation and dispersion is due to the difference in the conduction velocity of the axons and the difference in the time taken to reach the recording electrodes. Because of the dispersion of the CSNAP, durations are not generally reported. Amplitudes of CSNAPs are measured from the peak of the largest negative deflection to the peak of the largest positive deflection. Reductions in the amplitude of elicited CSNAPs along a nerve segment as compared to other regions in the nerve may represent conduction block in sensory fibers.

Table 10.2 shows the range of normal sensory-nerve conduction velocities and CSNAP amplitudes for a variety of upper and lower extremity peripheral nerves. Since sensory nerves contain some of the largest axons and are most vulnerable in neuropathies, the earliest indication of pathology may be abnormalities in these nerves. Electrode placements for determining sensory conduction velocities in other commonly tested nerves are shown in Figure 10.14.

Table 10.2
Normal Values for Sensory Conduction Studies of Commonly Tested Peripheral Nerves

Nerve/ segment	Distal latency (msec)	Conduction velocity (m/sec)	CSNAP amplitude (µV)
Median			
Wrist-digit	3.5 ± 0.4		
Elbow-wrist		50.0 ± 5.0	35.0 ± 15.0
Ulnar			
Wrist-digit	2.7 ± 0.5		
Elbow-wrist		50.0 ± 5.0	32.5 ± 15.0
Radial			
Wrist-thumb	2.5 ± 0.3		
Elbow-wrist		55.7 ± 4.0	5.5 ± 3.0
Axilla-elbow		71.0 ± 5.0	4.0 ± 1.4
Med. antebrachial			
cutaneous	2.7 ± 0.2	63.0 ± 5.0	12.0 ± 5.0
Lat. plantar (tibial)	4.0 ± 0.2	43.4 ± 3.8	3.0 ± 0.6
Superfical peroneal	2.3 ± 0.5	47.3 ± 3.5	14.0 ± 4.0
Sural			
Calf-malleolus	2.3 ± 0.4	46 ± 4.0	16.4 ± 5.5

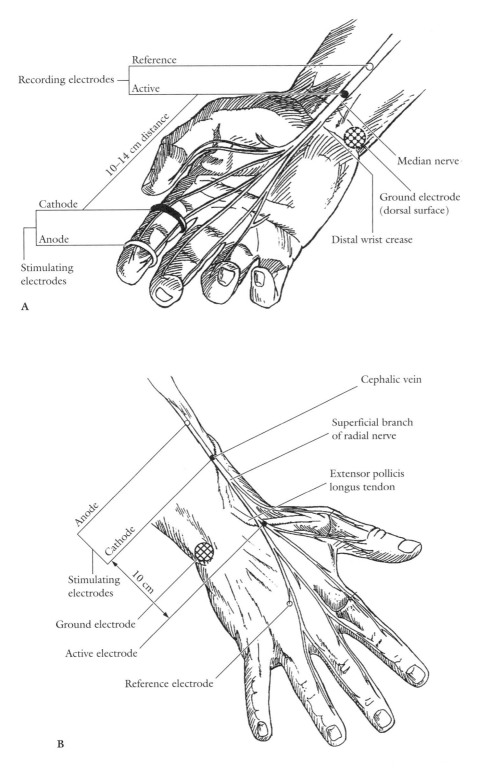

Figure 10.14 Stimulation and recording sites for performing sensory-nerve conduction studies in **A**, the *median nerve* (orthodromic technique), **B**, the *superficial radial nerve*.

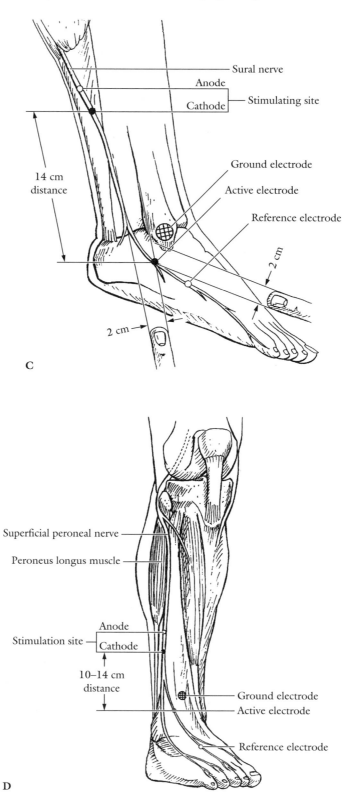

Figure 10.14 *Continued* **C**, the *sural nerve*, **D**, the *superficial peroneal nerve*.

General principles of nerve conduction testing

Whenever NCV testing is selected as an evaluative procedure, several important testing principles should be observed. These include the following:

1. Testing should examine both motor and sensory conduction when possible.

2. Testing should be performed over several segments of the nerve(s) suspected to be involved.

3. Testing should be performed on nerves contralateral to those suspected of being involved.

4. Examination of nerves in both upper and lower limbs may be appropriate.

5. Testing should be performed at the appropriate time in the context of the suspected disorder.

Types of nerve lesions and interpretation of the results of motor- and sensory-nerve conduction studies

Nerve conduction studies are used to examine the ability of peripheral nerves to propagate action potentials. In disease or injury, this primary function of nerve fibers may be decreased or abolished depending upon the extent of involvement of the tissues. Clinically, three levels of nerve injury have been defined (4). The most mild disturbance of peripheral nerve function is called **neuropraxia** (type I lesion, transient neuropraxia, delayed reversible neuropraxia). Neuropraxia is characterized by localized conduction block (or slowing) without interruption of axon continuity and without axonal degeneration. Neuropraxias are often caused by excessive nerve compression, stretching, or inflammation, all of which are thought to produce changes in axon membrane function due to localized demyelination, compression of membrane channel proteins, or closing of nodes of Ranvier due to invagination of the axonal membrane. In neuropraxic disorders, peripheral nerve axons and the connective tissues associated with the nerve remain intact. Regardless of the underlying nature of the disturbance, the primary effect is reduced nerve conduction as reflected by increased distal latencies or reduced nerve conduction velocities. If conduction is actually blocked in a significant number of sensory or motor axons, the amplitudes of evoked compound action potentials may also be reduced. Because the nature of the lesion is mild and often temporary, over time, speed of conduction and compound action potential amplitudes will frequently return to normal values.

The second type of nerve injury is called **axonotmesis** (type II lesion). Axonotmetic lesions are those caused by more severe injury or disease and are characterized by a disruption in axon continuity without significant damage to the connective tissue layers in peripheral nerves. In axonotmesis, the endoneurial tubes remain intact. The axonal disruption associated with this type of lesion subsequently results in degeneration of the distal axonal segments and diminished sensation if sensory axons are involved, or muscle weakness if motor axons are involved. The sensory receptors or muscle fibers normally innervated by the disrupted axons are referred to as denervated. If significant numbers of nerve axons are disrupted while remaining axons are intact (partial denervation), the amplitudes of evoked compound action potentials (with stimulation above the level of the lesion) may be reduced once degeneration of the distal segments has occurred. Incomplete disruption of all axons in a peripheral nerve may not produce detectable changes in the conduction velocity of nerves distal to the site of pathology because conduction remains normal in uninvolved fibers. Complete disruption of all axons in an axonotmetic injury followed by degeneration of distal axon segments (complete denervation) will be clinically apparent by the absence of compound action potentials in the distal distribution of the nerve. If NCV testing is performed before degeneration has taken place in distal nerve segments (within 3–5 days following injury), nerve conduction velocities and amplitudes of evoked motor and sensory responses may be normal with stimulation below the level of the lesion because distal segments may retain conduction capacity until the axon membranes undergo significant degeneration. However, stimulation proximal to the lesion will show complete loss of evoked sensory and/or motor responses.

The third and most severe type of nerve injury is called **neurotmesis** (type III, IV, or V lesion). This type of nerve lesion is characterized by axonal disruption accompanied by damage to one or more of the connective tissue layers within the peripheral nerve. The effects of neurotmetic lesions on the conduction characteristics of peripheral nerves are the same as those noted above for axonotmetic lesions.

From the previous discussion, it should be apparent that interpretation of NCV data depends upon the nature of the nerve lesion (neuropraxia vs. axonotmesis or neurotmesis), the extent of the nerve damage (partial vs. complete involvement), and the time at which testing is performed following the injury. To illustrate the significance of these factors on NCV data interpretation, Table 10.3 shows the changes that may be noted in response to different types of lesions of the median nerve at the midforearm level.

Factors affecting motor and sensory conduction

Besides injury, ischemia or disease, many different factors may influence the results of nerve conduction studies (5). Body **temperature** has long been known to alter nerve conduction. Increases in temperature increase conduction velocity by about 5% per degree and reduce distal latencies. Conversely, cooling of peripheral nerves reduces conduction velocity and increases distal latencies. For these reasons, it is important for the practitioner applying a nerve conduction test to perform these procedures in a temperature-controlled environment (21–23°C) with skin temperatures in the range of 28–30°C.

Upper-extremity nerves generally have conduction velocities that range from 7–10 m/sec faster than lower-extremity nerves. More proximal segments of peripheral nerves are usually faster-conducting than more distal segments, but are generally within 5–10 m/sec of the most distal segment.

Age is another factor that affects nerve conduction. Infants and children less than 3–5 years of age have conduction velocities as low as 50% of normal adult levels. Gradual slowing of conduction may appear after age 40. Conduction velocities in persons in their sixth or seventh decade of life are approximately 10 m/sec less than those of healthy middle-aged individuals.

F-Wave testing

As stated above, when a nerve is stimulated, action potentials are generated in nerve fibers in both directions from the site of stimulation. When record-

Table 10.3
NCV Results for Median Nerve with Different Types of Lesions at the Mid-Forearm Level

		Extent of lesion	Nature of lesion	Day 1	Day 7
Stimulation at wrist	*Amplitude*	Partial <	Neuropraxia	Normal	Normal
			Axonotmesis	Normal	Decreased
		Complete <	Neuropraxia	Normal	Normal
			Axonotmesis	Normal	Absent
	Latency	Partial <	Neuropraxia	Normal	Normal
			Axonotmesis	Normal	Normal
		Complete <	Neuropraxia	Normal	Normal
			Axonotmesis	Normal	Absent
Stimulation at elbow	*Amplitude*	Partial <	Neuropraxia	Decreased	Decreased
			Axonotmesis	Decreased	Decreased
		Complete <	Neuropraxia	Absent	Absent
			Axonotmesis	Absent	Absent
	Latency	Partial <	Neuropraxia	Delayed	Delayed
			Axonotmesis	Slight delay	Slight delay
		Complete <	Neuropraxia	Absent	Absent
			Axonotmesis	Absent	Absent
	Velocity	Partial <	Neuropraxia	Slowed	Slowed
			Axonotmesis	Slightly slow	Slightly slow
		Complete <	Neuropraxia	Absent	Absent
			Axonotmesis	Absent	Absent
Stimulation at upper arm	*Amplitude*	Partial <	Neuropraxia	Decreased	Decreased
			Axonotmesis	Decreased	Decreased
		Complete <	Neuropraxia	Absent	Absent
			Axonotmesis	Absent	Absent
	Latency	Partial <	Neuropraxia	Delayed	Delayed
			Axonotmesis	Slight delay	Slight delay
		Complete <	Neuropraxia	Absent	Absent
			Axonotmesis	Absent	Absent
	Velocity	Partial <	Neuropraxia	Slowed	Slowed
			Axonotmesis	Slightly slow	Slightly slow
		Complete <	Neuropraxia	Absent	Absent
			Axonotmesis	Absent	Absent

ing the motor response at the muscle, the CMAP (also called the M-wave) is seen in addition to a second response that often occurs much later. (Fig. 10.15*B*). This second potential, called the **F-wave** is recorded 20–55 msec after stimulation of the nerve. Whereas the M-wave results from the orthodromic propagation of action potentials in the motoneuron axons, the F-wave

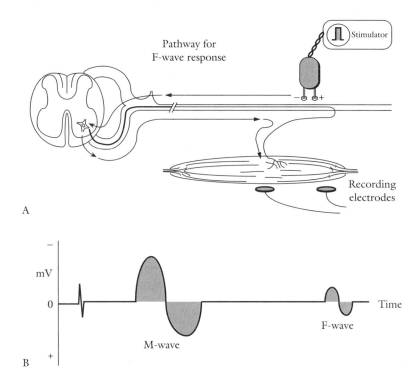

Figure 10.15 Diagrammatic representation of the stimulation and recording sites for performing F-wave testing (**A**); diagram of the M-wave and F-wave evoked in F-wave tests (**B**).

is thought to be produced as the result of antidromically propagated action potentials transmitted to the anterior horn cells (α motoneuron) or initial segment (axon hillocks) of the α motoneuron axons. This antidromic invasion is thought in turn to produce action potentials at the axon hillock which are then propagated orthodromically back along the motor fibers to the muscle being examined (Fig. 10.15A). This second set of action potentials reactivates the muscle, initiating a second CMAP, the F-wave.

F-waves recorded during motor-nerve conduction studies have variable latencies and are smaller in amplitude than the M-waves produced by the direct activation of the motor fibers at the point of stimulation. Since F-waves are smaller in amplitude (usually <500 μV), this signal reflects the activation of a smaller population (1 to 5%) of the motor units in the muscle. F-wave latencies are dependent on the distance from the point of stimulation to the spinal cord and range from 20–32 msec in the upper extremities and 42–58 msec in the lower extremities. The longer the limb in which testing is performed, the longer the latency of the F-wave response.

Although the F-wave response may be produced during a routine motor-nerve conduction study, this procedure may be modified by reversing the stimulation electrodes (placing the cathode proximal to the anode). Recording electrodes are placed in a fashion identical to that for performing motor conduction studies. Interestingly, the F-wave is not evoked in response to

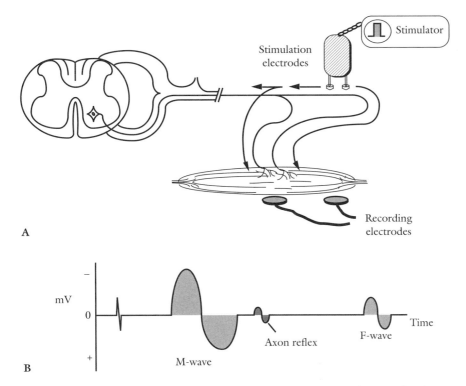

Figure 10.16 Diagrammatic representation of the stimulation and recording sites for eliciting the axon reflex (**A**). Diagram of the M-wave, axon reflex potential, and F-wave evoked during motor nerve testing (**B**).

each and every stimulus. In addition, the latency of the F-wave is variable and the F-wave shape does not have a consistent configuration. In general, only the range of F-wave latencies is reported.

The purpose of performing F-wave testing is to examine conduction in the proximal nerve segments. F-wave testing is primarily used when disease or injury is thought to involve proximal segment of nerves where conventional conduction studies are difficult or impossible to perform. A primary example would be in the case of the nerve compression associated with thoracic outlet syndrome.

Reflex testing

Axon Reflex

Small, time-locked potentials are often seen in peripheral nerve diseases in which there has been some proximal-nerve collateral branching. This response is poorly named in that the pathway for the response does not involve the spinal cord or reflex mechanisms. Rather, stimulation evokes antidromically propagated action potentials in motor axons that make a U-turn at the point where the axons branch (Fig. 10.16A). The action potentials then travel orthodromically down the axon branches and reactivate the innervated muscle fibers. The evoked CMAP that appears immediately following the M-wave is called the axon reflex (Fig. 10.16B).

The waveform latency falls between the M-wave and F-wave latencies. With each submaximal stimulus, the amplitude and latency of the axon reflex waveform remain stable. The latency of the axon reflex response depends on the distance between the stimulating and recording electrodes. These small, time-locked potentials occur infrequently but may be seen in peripheral nerve disease or injury where proximal nerve branching has occurred. The axon reflex is of little value in electrophyiologic assessment and simply reflects the reinnervation of previously denervated muscle fibers.

H-Reflex Tests

Another electrophysiologic test can be performed to quantitatively examine the spinal cord reflex pathway associated with the phasic stretch reflex (myotatic reflex, deep tendon reflex, monosynaptic reflex). In routine neurological testing, the phasic stretch reflex is elicited by providing a quick stretch to a muscle. This is often done by tapping a muscle tendon or muscle belly using a reflex hammer (e.g., patellar tendon reflex, Achilles tendon reflex). The quick stretch activates stretch receptors in the muscle called muscle spindles. One particular class of muscle spindle afferent (sensory) fibers, the Ia afferents that terminate in the primary endings in spindles, produce a burst of action potentials that are propagated toward the CNS. The central processes of Ia afferent axons synapse directly on the α motoneurons that innervate the muscle stretched. In normal individuals, the quick stretch produces enough Ia afferent activation to excite some of the α motoneurons to the stretched muscle. Action potentials produced in the α motoneurons (MN) then pass along the α MN axons to the muscle, resulting in a brief twitchlike contraction, the characteristic reflex response.

Activation of the monosynaptic reflex pathway can be elicited by not only providing quick muscle stretches, but also by applying electrical stimulation to the mixed nerves supplying skeletal muscle. For example, if recording electrodes are positioned over the soleus muscle and the posterior tibial nerve is electrically stimulated at the popliteal fossa with progressively larger-amplitude stimuli, the first observed response may be a CMAP that appears at a latency considerably longer than that for the M-wave produced by the direct activation of motoneuron axons at the site of stimulation. This long-latency CMAP is called an **H-wave, Hoffman response**, or **H-reflex** and results from stimulation of Ia muscle spindle afferents which reflexly activate soleus MNs at the spinal cord (Fig. 10.17). Such reflex activation of soleus MNs send action potentials to the soleus resulting in the excitation of soleus muscle fibers.

As the amplitude of stimulation is increased, the M-wave response appears simultaneously on the display preceding the H-wave at the time when the slightly smaller-diameter α motoneuron axons are activated directly at the site of stimulation. As the amplitude of stimulation is increased even more, the amplitudes of both the M-wave and H-wave increase. The M-wave size increases as more motor units to the muscle are recruited directly at the stimulation site. The H-wave increases in size as greater numbers of soleus muscle spindle afferents are activated, resulting in greater reflex recruitment of soleus MNs. During this procedure, the direct activation of muscle producing the

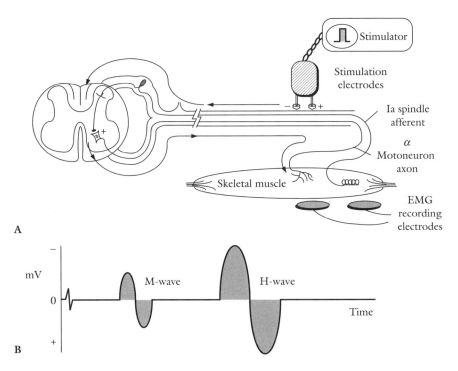

Figure 10.17 Diagrammatic representation of the stimulation and recording sites for eliciting the H-reflex (**A**). Diagram of the M-wave and H-wave evoked during reflex testing (**B**).

M-wave tends to activate axons of larger diameter first and hence the largest motor units (e.g., type FF). The reflex activation of the soleus motoneurons that produces the H-wave, in contrast, tends to activate the smallest of the motoneurons first and hence the smallest motor units first (e.g., type S).

As the amplitude of stimulation is increased to even higher levels, the amplitude of the H-wave will begin to diminish in amplitude. This reduction in H-wave size is thought to result when the reflexly activated soleus motoneuron (MN) action potentials traveling orthodromically collide with antidromically activated action potentials in the same axons initiated at the site of stimulation. Reflexly produced action potentials elicited by spindle afferent feedback that may have initially contributed to the H-wave are blocked and are no longer allowed to contribute to the H-wave response. If stimulation amplitudes are very high, the reflex activation of the soleus may be completely blocked and the H-wave may disappear.

Because the H-reflex test employs the same neural pathways as the stretch reflex, this procedure is also referred to as the **electrically elicited monosynaptic reflex**. H-reflex testing allows one to gain quantitative information regarding the integrity of the stretch reflex pathway. In contrast, the evaluation of the standard deep tendon reflex using a reflex hammer is highly qualitative in nature.

The amplitude and latency of the H-wave is dependent upon the integrity of the sensory afferent neurons, the synapses between Ia afferents and MNs,

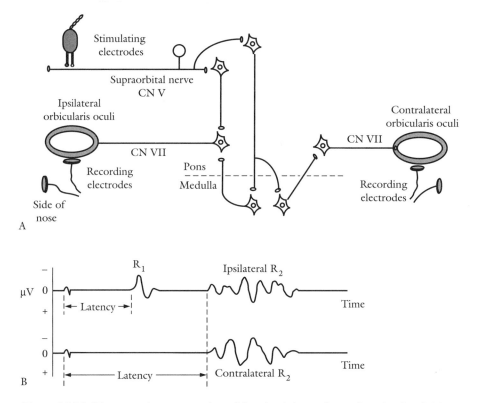

Figure 10.18 Diagrammatic representation of the stimulation and recording sites for eliciting the blink reflex (**A**). Diagram of the R_1 and R_2 responses evoked during blink reflex testing (**B**).

the α MN cell bodies and axons, the NMJs at the muscle, and the muscle fibers themselves. The H-wave may be tested in both lower and upper limbs. The latency of the soleus response normally ranges from 24–32 msec and is dependent on leg length and age. The soleus reflex response is useful for determining pathology of the S_1 nerve root and identifying peripheral neuropathies. This reflex tests large-diameter myelinated motor and sensory nerve fibers, which are usually the first involved in peripheral neuropathies.

Blink Reflex Tests

The **blink reflex test** is used to assess the functional integrity of both the trigeminal nerve (CN V) and the facial nerve (CN VII) (6). The afferent limb of the reflex is the trigeminal nerve whereas the efferent limb is the facial nerve (Fig. 10.18A). The central relay nuclei of the reflex are located in the sensory trigeminal nucleus and the motor nucleus of the facial nerve.

To perform this test, recording electrodes are placed bilaterally over the orbicularis oculi muscles and stimulating electrodes are positioned over the supraorbital branch of the trigeminal nerve. Electrical stimulation to the trigeminal nerve evokes two separate CMAP responses of the ipsilateral orbicularis oculi (Fig. 10.18B). The earliest CMAP response is called R_1 while the later response is called R_2. On the contralateral side only R_2 will be recorded. The latency of R_1 is a reflection of the conduction over the reflex pathway which includes delays associated with brainstem nuclei of cranial nerves V (trigeminal) and VII (facial). R_2 responses represent the ascending trigeminal nerve, pontine relay, and descending facial nerve. The latencies of the responses range from 9–12 msec for R_1 and 26–35 msec for R_2. By determining the respective response latencies, a lesion of either the facial or trigeminal nerve in the reflex loop can be detected. The function of the distal segment of the facial nerve can be assessed by direct stimulation distal to the stylomastoid foramen and measuring the latency of the M-wave response. In most cases of facial neuropathy (e.g., Bell's palsy), the site of pathology is proximal to the stimulation site. With the blink reflex, facial neuropathy will result in diminished ipsilateral R_1 and R_2 whereas the contralateral R_2 will be unaffected. A lesion of the trigeminal nerve will affect both the ipsilateral R_1 and R_2 and the contralateral R_2 component. This test is useful in identifying pathologies affecting the cranial nerves: Bell's palsy, cerebellar-pontine angle tumors, Guillain–Barré syndrome, and central demyelinating processes.

Central evoked-potential testing

Evoked potentials are voltage changes monitored from the electrically excitable tissue of the cerebral cortex, brainstem, and spinal cord in response to various applied sensory stimuli. The function of three different CNS sensory areas can be evaluated using electrophysiologic tests: the somatosensory cortex, the visual cortex, and the auditory region of the brainstem. In general, to test these areas, appropriate sensory stimuli are applied consistent with the sensory modality examined. Under normal circumstances, the sensory stimuli activate the respective sensory receptors and action potentials are initiated and propagated along peripheral and/or central nervous system pathways and subsequently alter the electrical activity of the cerebral cortex cells that are associated with processing the incoming sensory information. The change in electrical activity of the cortical area is monitored by the use of surface recording electrodes placed over the appropriate regions of the cortex or brainstem.

Somatosensory evoked-potential testing

The integrity of these ascending somatosensory pathways may be assessed using electrophysiological procedures by activating sensory fibers in the skin and monitoring the propagation of evoked compound sensory action potentials at convenient sites along these pathways. Stimulation of peripheral sensory fibers in mixed nerves such as the median, ulnar, peroneal, or tibial is applied at amplitudes sufficient to activate sensory responses and results in a simultaneous twitchlike contraction of muscles innervated by the motor fibers of these nerves. Frequencies of stimulation are normally in the 1–4 pps range and monophasic pulsed currents of 0.1–0.2 msec duration are used. In general, such stimuli are applied with surface electrodes although the use of needle electrodes to stimulate near nerve trunks is not uncommon. Stimulation is provided first unilaterally near a particular nerve and later bilaterally. The sensory action potentials elicited by such stimulation are monitored by placing active surface recording electrodes over Erb's point (lower inner angle of the supraclavicular fossa), the spinal cord (C5 for upper extremity, T12 for lower extremity), and/or cortex. A number of different recording site combinations are used in evoked-potential testing and the reader is referred to more detailed texts on this topic for further information on recording electrode placement. The evoked sensory potentials, called **somatosensory evoked potentials (SSEPs)**, are so small that in some cases up to 2000 responses must be averaged in order to clearly visualize the response on a display device.

Once SSEPs are clearly visualized from a particular recording site, measurements of latencies to the onset of the potential, to the peaks of the potential, and between peaks are made, in addition to peak-to-peak amplitudes. The latency values measured using cortical electrodes reflect both the peripheral and central conduction times for sensory action potential transmission. To gain insight into the central pathway conduction, the peripheral conduction latency measured with C5 or T12 electrodes is subtracted from the total conduction latency measured from cortical electrode recordings. This test is useful in the identification of demyelinating diseases and spinal cord, cortical, and brainstem dysfunction. The waveforms of SSEPs are often complex and characterized by combinations of positive and negative peaks. The latency to each positive and negative peak is measured to the nearest millisecond. Each peak is then labeled according to the polarity of the deflection (P for positive, N for negative) and the measured latency. For example, a peak labeled P13 would apply to a positive deflection peak occurring at a latency of 13 msec. The SSEP waveform recorded in response to peripheral stimulation depends upon the relative positions of the active and reference recording electrodes. Analysis and interpretation of the various SSEP waveforms is complex and beyond the scope of this introductory text. The reader is referred to other texts for more detailed discussion (7).

Brainstem auditory evoked-potential testing

Brainstem auditory evoked potentials (BAEPs) are electric waveforms recorded from the brain elicited in response to sound. To produce BAEPs, a clicklike sound is applied through one ear. Surface scalp electrodes are applied over the vertex on the same side as stimulation. Individual evoked potentials are very small in amplitude; up to 2000 responses to stimuli may need to be averaged in order to visualize the evoked waveform. Subjects must remain very still during the test because acts such as neck movement or swallowing may contaminate the recording. A detailed presentation of the interpretation of BAEPs is beyond the scope of this text. The test is applied to assess the integrity of the auditory pathway from the cochlea to the auditory cortex in cases where demyelinization, neuromas, tumors, or other disorders of auditory pathways are suspected. Brainstem evoked-potential testing is generally considered a primary assessment tool of the clinical audiologist.

Visual evoked-potential testing

Visual evoked potentials are bioelectric potentials recorded during electrical activity in the occipital lobe cortex in response to light stimuli. Compared to the other cortical evoked potentials, visual evoked potentials are technically easier to record. Surface electrodes are placed on the scalp overlying the occipital cortex. A controlled light pattern generator is used to stimulate either one eye or both eyes simultaneously. The stimuli are presented from 100 to 200 times with signal averaging processing successive cortical responses. The test is clinically useful in determining the integrity and conduction characteristics of the visual pathways. Changes in the waveform or latency provide information concerning disruption of the visual system.

Repetitive nerve stimulation testing

The **repetitive nerve stimulation test (RNS)**, or Jolly test, has been used for more than 90 years to assess the function of the neuromuscular junction. In this procedure, supramaximal stimuli are applied to peripheral nerves and the CMAPs evoked are recorded from innervated muscles. The stimulation frequency range is 2–3 pps. The evoked CMAP's peak-to-peak amplitude or peak negative amplitude is observed while stimulation is applied. In subjects with normal neuromuscular junctions, repeated stimulation (2–3 pps) will slightly deplete the junction of acetylcholine (ACH) resulting in a small drop in muscle fiber recruitment and a small drop (5 to 8%) in the amplitude of the recorded CMAPs. In diseases such as myasthenia gravis where neuromuscular junction transmission is impaired, the drop in CMAP amplitude is more marked (>10% drop in amplitude between the 1st CMAP and the 4th–6th CMAP). Figure 10.19 illustrates the responses of both normal and abnormal neuromuscular junction function on the amplitude of CMAPs during repetitive stimulation. The test is generally repeated several times and on several muscles in order to establish the consistency of the response and improve its sensitivity.

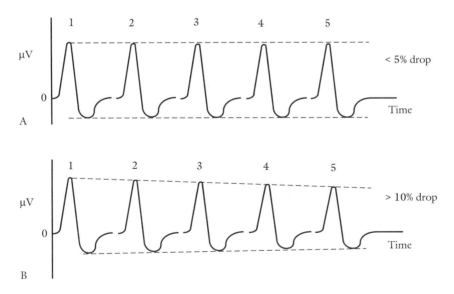

Figure 10.19 A series of muscle action potentials evoked during repetitive nerve stimulation testing in normal muscle (**A**). A series of muscle action potentials evoked during repetitive nerve stimulation testing in muscle with NMJ disorder (**B**).

Like many of the more sophisticated electrophysiologic testing procedures, repetitive nerve stimulation testing procedures are technically difficult to perform. For example, false positive results may be obtained when electrode and limb stabilization are inadequate. For further details on RNS testing, the reader is referred to Oh's extensive dissertation on the topic (8).

Electromyography

All of the procedures described above have activated excitable tissues by electrical stimulation of peripheral nerve axons. These techniques produce electrical activity in nerve and muscle via artificial stimulation. The procedure of electromyography (EMG) compares the electrical activity of skeletal muscle fibers at rest and during voluntary activation of muscle. Electromyography is used clinically to answer the following types of questions:

1. Is the muscle normally innervated, partially innervated, or denervated?
2. Does evidence of reinnervation exist?
3. Are EMG findings consistent with neuropathic or myopathic disease?
4. Is the pattern of EMG abnormality consistent with root, plexus, anterior horn cell, or peripheral nerve distribution?
5. In neuropathy, what is the specific location of the nerve lesion?
6. Does the problem involve a muscle in the extremities (anterior primary rami innervation), paraspinal musculature (posterior primary rami innervation), or head/neck musculature (cranial nerve innervation)?

Electromyographic recording techniques employ three electrodes connected to the recording instruments. These electrodes are connected to three inputs of the recording instrument labeled Active (negative), Reference (positive), and Ground. The three most common types of EMG needle electrodes are defined by the number of conductors contained within the needle. Generally, the more electrodes, the larger the gauge. Bipolar needle electrodes contain electrodes that can be connected to the active, reference, and ground inputs of the recording system. Concentric needle electrodes contain wires for positive and negative input and require a surface ground electrode. Monopolar electrodes contain one conductor connected to the active input and require not only a second electrode connected to external ground, but also another surface or needle electrode connected to the reference input, placed within 2 cm of the active electrode. Reusable, sterilizable concentric electrodes have been the most widely used until recently because of their clean signal, relatively large gauge (small diameter), and economy. Currently, disposable monopolar and concentric electrodes have become more commonly used because of concerns about the transmission of infectious diseases and their smaller size (larger gauge).

The EMG examination is divided into four segments per area of muscle studied: *(a)* **insertional activity,** *(b)* **activity at rest,** *(c)* **activity upon minimal activation** and **recruitment patterns**, and *(d)* **activity during maximal activation.**

Normal and abnormal electrical
activity in skeletal muscle at rest

As the EMG needle electrode is gradually inserted into muscle, a brief electrical discharge can be seen on a visual display and heard on an audioamplifier. The discharge is characterized as a high-frequency burst of positive and negative spikes associated with a "crisp static sound" from the audioamplifiers (9). This normal insertional activity can be reproduced by tapping the muscle and is seen any time the needle electrode is repositioned in the muscle. This electrical activity is thought to be associated with mechanical stimulation of muscle fiber membranes which may or may not cause membrane damage. Insertional activity lasts about 300 msec and slightly outlasts the movement of the electrode (9).

A reduction in normal insertional activity is commonly associated with a loss of normal muscle fibers and is often seen in fibrotic or severely atrophied muscles. A prolongation of insertional activity is indicative of muscle membrane hyperexcitability associated with acute denervation and inflammatory muscle disorders such as myositis. Within 2 to 3 weeks following denervation of muscle, insertional activity may be followed by a series of positive spikelike waveforms that persist for up to several minutes after needle movement in the muscle is stopped. These **insertional positive waves** (Fig. 10.20) discharge at rates ranging from 3 to 30 spikes per second (10). Such prolonged discharges may also be found in cases of acute polymyositis or in muscles that have been denervated for extended periods of time.

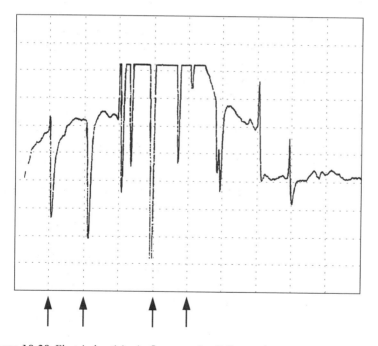

Figure 10.20 Electrical activity in flexor carpi radialis muscle upon insertion of needle electrodes; insertional positive waves (*arrows*) cease when the needle is no longer moving (10 msec/div. sweep speed; 100 μV/div. gain).

Normal muscle at rest is electrically silent when needle electrodes are not moving. If the recording electrodes are positioned at the motor endplate (neuromuscular junction) however, very small, spontaneous electrical potentials may be observed (Fig.10.21*A*). These very low amplitude (10–50 μV) potentials with initial negative deflection are called **miniature end-plate potentials (MEPPs)** and are thought to be produced by localized transient depolarization of the muscle membrane near the NMJ (9). MEPPs are a normal electrophysiologic finding. Passing such signals through an audioamplifier produces a sound described as that heard when a sea shell is held to the ear (9). Often a subject reports a dull pain at the electrode site when MEPPs are recorded; this pain usually disappears if the electrode is withdrawn slightly. Another normal electrical potential that may be occasionally noted when the electrode is near the motor endplate is the **end-plate spike** (Fig.10.21*B*). These initially negative waveforms have amplitudes ranging from 100–200 μV, 3–4 msec durations, and discharge at 5–50 impulses per second. The end-plate spikes are thought to occur when the electrodes activate single muscle fibers. Like MEPPs, end-plate spikes disappear when the electrode is moved away from the NMJ.

Other electrical activity in muscle at rest is indicative of neuropathy or myopathy (10). One such type of electrical potential seen or heard at rest in patients with suspected neuropathy or myopathy is called **fibrillation potentials** (Fig. 10.22*A*). Fibrillation potentials have amplitudes usually ranging from 20 to 200 μV and durations in the 1 to 5 msec range. These biphasic potentials (with an initial positive deflection) generally discharge irregularly at rates between 1 and 30 per second. Fibrillation activity is a manifestation of motor nerve disruption and usually occurs 14–21 days after denervation. Fibrillation potentials result from the generalized spread of acetylcholine receptors over the muscle membrane, making the denervated muscle fibers hypersensitive to acetylcholine. Acetylcholine present near the muscle binds to these receptors and produces depolarizations of single muscle fiber membranes which result in the generation of this type of short-duration, low-amplitude potential observed in the absence of volitional effort.

Positive sharp waves are another spontaneously occurring electromyographic finding associated with denervation or myopathy (10). These potentials are characterized as sawtooth in appearance with an initial marked positive deflection followed by a low-amplitude, long-duration negative phase (Fig. 10.22*B*). The amplitude of positive sharp waves usually increases and decreases within the range of 10 μV to 1 mV with discharge rates in the 50–100 per second range.

At times during electrophysiologic assessment, spontaneous repetitive twitchlike contractions of the examined muscle at rest called fasciculations may be observed. Such contractions reflect the discharge of either the fibers of a single motor unit or fibers within a muscle fascicle. When these spontaneous twitches arise, they are accompanied by the production of electrical potentials called **fasciculation potentials** (10). Characterization of fasciculation potentials is difficult due to the broad variation in potential parameters. In general, however, fasciculation potentials have characteristics of normal motor unit potentials.

Fasciculation potentials are commonly associated with disease of the α motoneuron but may also be encountered in other disorders such as

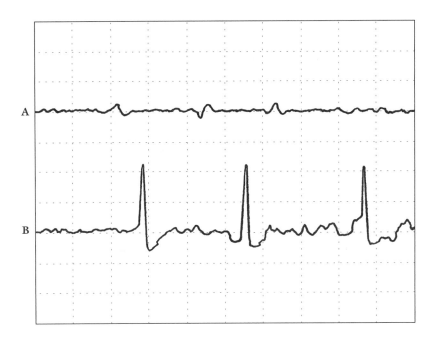

Figure 10.21 Normal electrical activity in resting muscles recorded with needle electrodes in a stable position. Miniature end-plate potentials (**A**); end-plate spikes (10 msec/div. sweep speed; 100 μV/div. gain) (**B**).

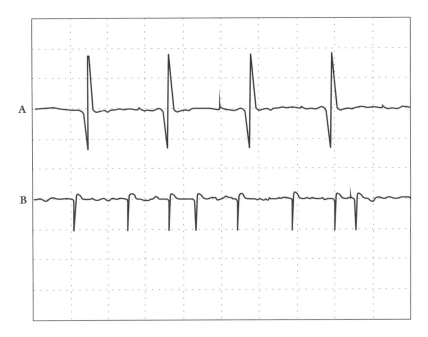

Figure 10.22 Abnormal electrical activity in resting muscle recorded with needle electrodes in a stabilized position. Fibrillation potentials (2 msec/div. sweep speed; 100 μV/div. gain) (**A**); positive sharp waves (5 msec/div. sweep speed; 100 μV/div. gain) (**B**).

radiculopathies and entrapment neuropathies. Fasciculation potentials may be present for a few days following denervation of a muscle but after 3 or 4 days these potentials are no longer recorded in a totally denervated muscle.

Myotonic discharges are rhythmic electrical discharges often initiated by muscle tapping or needle electrode insertion that continue in spite of no further needle movement (10). They arise from spontaneous, repeated activation of muscle fibers. Discharge frequency may initially be as high as 150 pps and may drop to 20 to 30 pps; often discharge frequency increases and decreases spontaneously. In cases where the frequency drops rapidly, the sound from an audioamplifier has been described as a WWII dive-bomber, hence the description of these discharges as "dive-bomber potentials." Individual discharges may have waveform characteristics similar to fibrillation potentials and positive sharp waves.

In diseases such as polymyositis or anterior horn cell disease, needle EMG examination may reveal the presence of polyphasic waveforms of fixed amplitude and relatively high, stable discharge rates (5–100 pps) (Fig. 10.23*A*). Such findings are referred to as **complex repetitive discharges** or **bizarre, high-frequency discharges.** Like myotonic discharges, these signals are often initiated by EMG needle electrode movement but usually both start and stop abruptly without marked change in discharge frequency. Passing such signals through an audioamplifier often results in a sound described as like "machine-gun fire."

Myokymia is a muscle disorder characterized by wormlike or undulating spontaneous contractions of long strip segments of muscle. Such muscular contractions are associated with electrical activity in the muscle called **myokymic discharges** (Fig 10.23*B*). These bursts of electrical activity are like those occurring when several motor units (2–10) discharge at nearly the same moment in time at rates of 30–40 impulses per burst. Volitional activation does not alter the discharge pattern which persists even during sleep. Unlike myotonic discharge, myokymic discharge may be reduced or stopped by chemical block of the innervating motor nerve fibers. Myokymic discharges are commonly found in facial musculature in patients with disorders such as multiple sclerosis.

Normal and abnormal electrical activity in skeletal muscle during volitional activation

When a needle recording electrode is placed within a muscle and a normal person voluntarily activates the muscle at progressively higher levels, a series of electrical waveforms can be recorded on a display device. In normal, innervated muscle, these electrical representations of the activation of the various types of individual motor units during contraction are called **motor-unit potentials**. One can see and hear this electrical activity in muscle on the oscilloscope and loudspeaker, respectively, as described previously. Certain characteristics of the motor units (Fig. 10.24*A*) are measured in clinical electrophysiologic assessment in order to gain insight into muscle function. These motor unit characteristics include: the number of phases in the motor-unit potentials, the amplitude and duration of the motor-unit potentials, and the motor units' firing rates or discharge frequencies. These characteristics vary between the different types of motor units. The variations in motor-unit po-

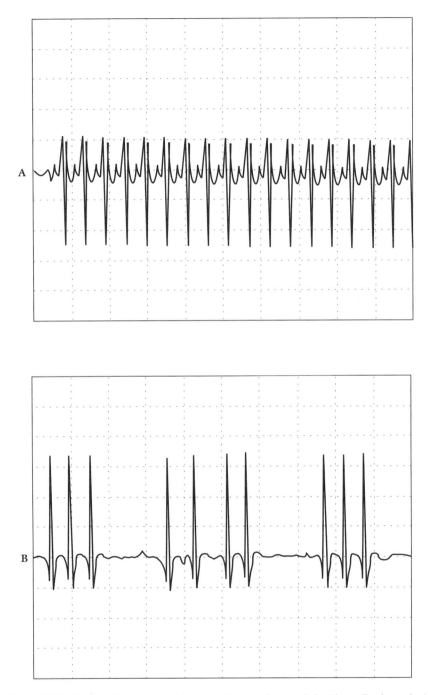

Figure 10.23 Abnormal electrical activity in resting muscle recorded with needle electrodes in a stabilized position. **A**, complex repetitive discharges (20 msec/div. sweep speed; 200 μV/div. gain); **B**, myokymic discharges (200 msec/div. sweep speed; 1 mV/div. gain).

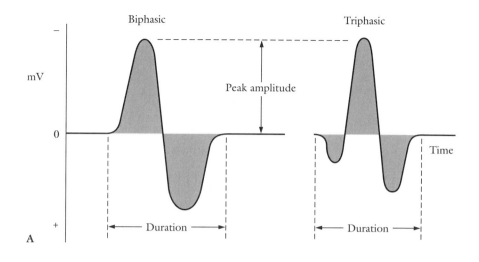

Normal motor unit parameters

Amplitude: 200 mV – 3 mV

Duration: 5 – 15 msec

Rise time:*   100 – 200 msec (<500 msec)

Frequency: 5 – 15 per sec (< 60/sec)

*Measured from initial positive peak to negative peak for triphasic potentials

Figure 10.24 Normal motor-unit potentials that arise during volitional muscle contractions. **A,** Motor unit potentials parameters; **B,** MUP characteristics.

tential characteristics are due to differences in muscle fiber size, the number of fibers in each motor unit, and the spatial distribution of the motor unit fibers within the muscle. For example, the MUP duration is thought to be representative of the spatial distribution of muscle fibers of the motor unit within the muscle with larger units having longer durations than smaller units. As the needle recording electrode is passed through the muscle, these MU characteristics will change. Each muscle has a different mix of the types of motor units, so each has unique characteristics.

In a volitional muscle contraction, the first motor units activated by the CNS are those with the lowest threshold. These are generally thought to be type S motor units consisting of a relatively small number of slowly contracting, very fatigue-resistant muscle fibers (type I or type SO) innervated by relatively small α motoneurons. When a higher level of volitional contraction is required, the type S motor units increase their discharge frequency (rate coding) and the central motor pathways simultaneously recruit (activate) more motor units (recruitment). In general, as moderate levels of contraction force are required, the somewhat larger type FR motor units (fast-twitch, fatigue-resistant) consisting of type IIA muscle fibers and slightly larger α motoneurons are the next to be activated. When very strong voluntary contractions are needed, type S and FR units may discharge at even higher rates and finally the third motor unit class, type FF (fast-twitch, readily fatigable) consisting of type IIb fibers and the largest α motoneurons are recruited.

In electrophysiologic assessment, after observation of a muscle's electrical activity during needle insertion and at rest, electrical activity is observed during mild to moderate volitional contraction of the muscle. Normally, motor-unit potentials are observed with smaller-amplitude potentials recruited first followed by progressively larger-amplitude potentials as the force of contraction is increased. Figure 10.25 illustrates normal potentials, which are triphasic or biphasic and have amplitude, duration, and frequency characteristics as described in Figure 10.24 (9). Motor-unit potentials with four or more phases are thought to reflect a desynchronization of discharge in the muscle fibers making up the motor unit. Peak-to-peak amplitudes of motor unit potentials

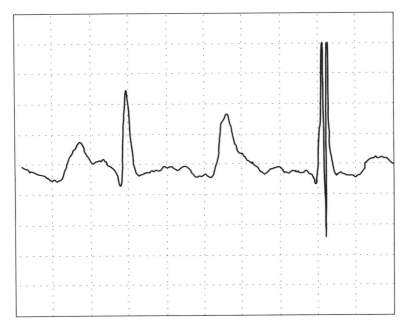

Figure 10.25 Normal motor-unit potentials recorded from the brachioradialis during modest volitional effort. A normal biphasic motor unit is on the far left, a more distal motor unit is in the center, and a triphasic motor unit is on the far right (5 msec/div. sweep speed, 100 μV/div. gain).

range from 200 μV to 5 mV. The **rise time** of the MUP is measured from the first positive peak to the top of the first negative peak. Normal rise times should not exceed 500 μsec and most should fall in the 100 to 200 μsec range. **MUP duration** is measured from the beginning of the initial positive deflection from the baseline to the point at which the potential returns to baseline after all phases and normally ranges from 5 to 15 msec. Motor units normally do not discharge at frequencies in excess of 30 per second, with most discharging at much lower rates. The observation that larger-amplitude potentials are recruited before some smaller-amplitude potentials is not uncommon and may merely represent relative differences in the distance of the recording electrode(s) from activated motor unit fibers rather than a true reversal in the recruitment order of motor units.

Small-amplitude motor-unit potentials of short duration are often noted in myopathic disease. Smaller than normal MUPs with longer than normal durations are thought to reflect early reinnervation of muscle following peripheral nerve lesions. In contrast, motor-unit potentials that are very large in amplitude are thought to be indicative of later stages of reinnervation where regenerating axons reinnervate either a larger than normal number of muscle fibers or a group of muscle fibers that are closer together than muscle fibers in a normal motor unit (10).

With vigorous strong contraction of normal muscle, so many motor units are recruited that an **interference pattern** is observed on the display device (Fig. 10.26A) (9). The discharge of the number of motor units required to produce strong contraction obliterates the baseline and does not allow the examiner to accurately distinguish individual MU characteristics. Normal muscle shows a stepwise increase in interference patterns as the magnitude of voluntary contraction is gradually increased. Interference patterns are often reduced in cases of peripheral nerve injury in spite of maximum volitional contraction efforts by the subject. In contrast, full interference patterns are generated with only minimal to moderate effort in subjects with myopathic disease.

A variety of other electrical potentials may be observed during mild to moderate volitional contraction of muscle and are reflective of dysfunction of the neuromuscular system. Although a few normal motor unit potentials may be polyphasic (having four or more phases in the waveform), the regular occurrence of **polyphasic potentials** (Fig. 10.26B) during muscular contraction is an abnormal finding. Polyphasic potentials are believed to represent the presence of either deteriorating or regenerating motor units where motor axon degeneration or regeneration has occurred and/or primary myopathy exists. The polyphasic appearance is believed to result from desynchronization of distal conduction of the terminal branches of the axon that innervate the muscle fibers of the motor unit (10). Such desynchronization disrupts the normally synchronous activation of the motor unit's muscle fibers. Large-amplitude polyphasics are frequently seen in individuals with chronic neuropathies. Small-amplitude, long-duration polyphasic potentials have been associated with early reinnervation; small-amplitude, short-duration polyphasics are believed to reflect myopathic disease.

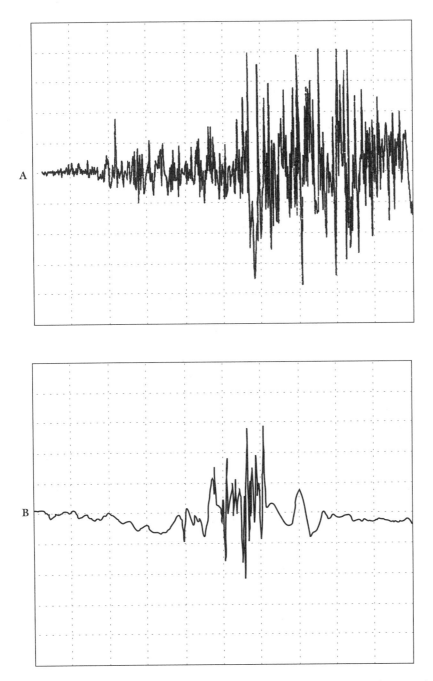

Figure 10.26 Normal interference pattern recorded from extensor carpi radialis (**A**). Note the early recruitment of small-amplitude motor-unit potentials progressing to larger-amplitude potentials as the force of contraction increases (0.5 sec/div. sweep speed, 500 μV/div. gain). Polyphasic motor-unit potential (center) recorded from tibialis anterior muscle undergoing reinnervation following peroneal nerve palsy (**B**) (5 msec/div. sweep speed, 50 μV/div. gain).

General principles of electromyographic testing

When electromyography is selected as an appropriate testing procedure, several general principles apply to the performance of the procedure. These include:

1. Examination of a number of muscles both above and below the suspected site of the pathology.
2. Examination of muscles innervated by other nerves in the same limb.
3. Sampling of EMG activity of the full cross-section of each muscle tested.
4. Examination of muscles in contralateral limbs or both upper and lower limbs may be appropriate.
5. Examination should be performed at the appropriate time in the context of the suspected disorder(s).

Types of nerve lesions and interpretation of electromyographic findings

Abnormal electromyographic findings may be indicative of either myopathic or neuropathic, or combined muscle and nerve disorders. EMG findings alone are difficult to interpret, however, unless they are examined in conjunction with the results of nerve conduction studies.

The presence of abnormal EMG potentials in the absence of abnormal NCV findings suggests that the lesion is confined to the skeletal muscle or, in other words, is a myopathic disorder. However, neuropathic disorders that result in partial axonal degeneration may also give rise to similar findings (abnormal EMG and normal NCV). Numerous muscle diseases and disorders (e.g., muscular dystrophy, polymyositis, myotubular myopathy) can give rise to spontaneous activity at rest such as fibrillation potentials, positive sharp waves, and complex repetitive discharges. Myopathic disorders are also associated with an increase in the number of polyphasic motor unit potentials recorded during voluntary muscle contraction.

Mild to moderate neuropathic problems such as nerve compression and associated partial or complete neuropraxia will often give rise to changes in evoked compound action potential latencies, amplitudes, or durations without any apparent abnormal EMG findings. These changes in NCV parameters may frequently involve sensory axons and evoked CSNAPs before any changes are noted in motor fibers or evoked CMAPs. In more severe neuropathic lesions where axons degenerate distal to the site of the problem, abnormal NCV findings are generally accompanied by abnormal EMG findings. Whether abnormal EMG potentials are noted upon examination following peripheral nerve damage is dependent, though, on the time at which the EMG examination is performed. If, for instance, electromyography is performed within a day or two after a lesion (e.g., partial or complete axonotmesis) to the innervating nerve, fibrillations and positive sharp waves suggestive of denervation will not generally be detected and insertional activity will appear normal. In contrast, EMG testing of the same muscles 14–21 days after partial or complete axonotmesis will usually reveal the presence of those abnormal potentials indicative of muscle fiber denervation as well as the increased insertional activity associated with the hyperexcitable membranes of denervated muscle fibers.

Observation of motor-unit potentials during volitional activation of muscle may allow the examiner to determine whether nerve lesions to muscles involve all of the motor axons or only some of the motor axons. In partial neu-ropraxic and partial axonotmetic lesions, motor-unit potentials recorded from associated muscles will appear normal but interference patterns with strong volitional effort will be decreased. Partial axonotmetic lesions may also be accompanied by the presence of polyphasic motor units. These results occur because some motor units remain normally innervated in partial nerve le-sions and still produce normal-appearing action potentials. Interference pat-terns are reduced in partial nerve lesions because complete conduction block (neuropraxic lesion) or breaks in axon continuity (axonotmetic lesion) do not allow action potentials produced at the α motoneuron cell bodies to reach the muscle through the damaged or degenerating axons distal to the lesion. As such, some of the muscle's motor units cannot be volitionally activated. Complete neuropraxia or complete axonotmesis is present if volitional effort does not produce any motor-unit activity. In these instances, motor-unit po-tentials and interference patterns are absent because the pathways (axons) for activating muscle fibers are completely blocked or broken and therefore no signals are transferred to muscle to initiate excitation.

Table 10.4 summarizes the EMG findings discussed above that allow the differentiation between neuropraxic and axonotmetic lesions and whether these nerve lesions involve all of the motor nerve fibers (complete lesions) or only a portion of the motor nerve fibers (partial lesions). Findings from electromyography may suggest the presence of motor nerve lesions but give no indication of the integrity or level of involvement of sensory or autonomic fibers contained within the peripheral nerve.

Table 10.4
EMG Findings for Various Lesions

	Extent of lesion		Nature of lesion	Day 1	Day 21
Insertional activity	Partial	<	Neuropraxia	Normal	Normal
			Axonotmesis	Normal	Increased
	Complete	<	Neuropraxia	Normal	Normal
			Axonotmesis	Normal	Increased
Spontaneous activity	Partial	<	Neuropraxia	None	None
			Axonotmesis	None	Fibs., +sharps
	Complete	<	Neuropraxia	None	None
			Axonotmesis	None	Fibs., +sharps
Volitional activity	Partial	<	Neuropraxia	Normal	Normal
			Axonotmesis	Normal	Normal
	Complete	<	Neuropraxia	None	None
			Axonotmesis	None	None
Interference patterns	Partial	<	Neuropraxia	Decreased	Decreased
			Axonotmesis	Decreased	Decreased
	Complete	<	Neuropraxia	Absent	Absent
			Axonotmesis	Absent	Absent

Classical Electrophysiologic Tests

Strength–duration test

When the electrically excitable tissues, nerve and muscle, are activated by pulsatile electrical currents, the type of fibers stimulated depends in part on the amplitude ("strength") and duration characteristics of the applied waveforms. For a particular monophasic pulse of a given amplitude and duration applied near normal excitable tissues, the degree of activation is fairly consistent from one stimulus to the next. For example, if a few α motoneuron axons are activated in response to a single stimulus, the twitch contraction (response) elicited will remain constant in amplitude. By varying stimulus amplitudes and durations in a systematic manner, one may determine the set of amplitude and duration combinations necessary to evoke the same magnitude twitch contraction. Plotting the stimulus amplitude and duration combinations just sufficient to evoke a twitch generates a strength–duration (S–D) curve as shown in Figure 10.27. The curve represents all the stimulus amplitude and duration combinations that will evoke the same response, in this case a small twitch contraction. The data reveal that to activate a particular set of excitable fibers one may use stimuli that range from those with high amplitudes with short durations to those with low amplitudes and long durations in order to activate the same subset of fibers. The product of the amplitude and duration (the phase charge) of all applied pulses that fall on the curve are equal. Any stimulus that falls below the curve will not be sufficient in phase charge to activate the same subset of motoneuron axons and hence the size of the evoked contraction will decrease or disappear. Any stimulus that lies

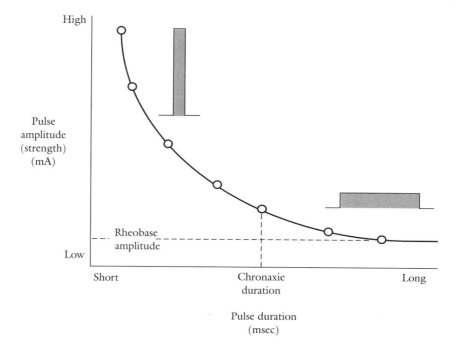

Figure 10.27 Strength–duration curve for a particular muscle showing the rheobase amplitude and chronaxie duration.

above the curve will activate more α motoneuron axons (recruitment) and generally results in larger contractions.

In traditional strength–duration testing procedures, the cathode of a pulsed monophasic stimulator is applied to the motor point of the muscle while the anode of the circuit is placed over the muscle a few centimeters away from the motor point (11). Recall that the motor point is the location over muscle where the nerve innervating the muscle is most easily excited. Stimuli with initial pulse durations of greater than 100 msec and amplitude sufficient to produce a just detectable (e.g., palpable) twitch contraction are delivered. The durations of the stimuli are then systematically reduced in a stepwise manner and the amplitudes required to produce the same response are noted. The combinations of stimulus amplitude and duration sufficient to produce the just palpable response are then plotted to produce the strength–duration curve. The minimum amplitude required to produce the response at the very long pulse durations (100 to 300 msec) is called the **rheobase** strength stimulus (Fig. 10.27). The minimum duration required to produce the twitch when the amplitude is set at twice the rheobase level is called the **chronaxie**. Normal chronaxie durations depend in part upon the type of stimulator used in performing the assessment. For constant-voltage stimulators, Ritchie (12) found that 90% of normal chronaxie values fell between 0.03 and 0.08 msec. For constant-current stimulators, normal chronaxie values ranged from 0.15 to 0.80 msec for 90% of the muscles examined. In contrast, completely denervated muscle may exhibit chronaxie values of more than 30 msec.

Figure 10.28 shows the strength–duration curve for a normally innervated muscle, a denervated muscle, and a partially reinnervated (or partially denervated) muscle. The S–D curve for the denervated muscle is shifted markedly up and to the right because denervated muscle fibers are inherently less excitable than motor-nerve fibers and hence require higher levels of stimulation to reach threshold excitation. In denervated muscle, no contraction will be elicited at very short pulse duration even if near-maximal amplitudes of pulses are applied and a steep rise at the left of S–D curve often appears between 1 to 10 msec pulse duration. If reinnervation begins to take place, the S–D curve for the partially reinnervated muscle shifts down and to the left and is frequently characterized by discontinuities or kinks in the curve. These kinks most commonly occur in the 3–10 msec pulse-duration range.

The purpose of strength–duration testing is to determine the level of innervation of skeletal muscle by α motoneuron fibers following nerve injury or disease. That is, the test results are thought to reflect the ratio of innervated to denervated muscle fibers. The test may reveal normal innervation, partial innervation, or complete denervation. S–D testing is of primary value in assessing progress or deterioration of motor innervation; hence serial testing is recommended. Testing is generally not performed more often than every 2 to 3 weeks in order to allow time for significant reinnervation (or further denervation) to manifest itself. The test will reveal which nerves are involved but does not reveal the precise location of the nerve lesion (11).

A number of factors may influence the results of S–D tests. These include temperature, blood supply, edema, and electrode position. Excitation thresholds are elevated with decreased temperature, decreased blood supply, increased edema, or poor localization of the motor point.

Strength–duration testing was used in patient populations as early as 1916 to assess the extent of nerve injury (13). Widespread clinical use of S–D

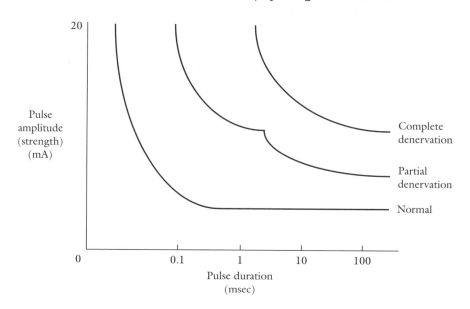

Figure 10.28 Strength–duration curves for normally innervated muscle, partially denervated muscle, and totally denervated muscle.

testing continued through the 1960s. With the development of nerve conduction study and electromyographic techniques, the use of this testing procedure today is rare.

Reaction of degeneration test

The reaction of degeneration (R.D.) test is used to assess the level of innervation of skeletal muscles. Like S–D testing, R.D. tests provide information on the integrity of α motoneurons innervating skeletal muscle.

To perform this test, either monophasic or biphasic pulsed currents have been used to electrically stimulate over the normal motor point of a muscle suspected of having impaired innervation. The individual current waveforms applied have a pulse duration of 1 msec and are delivered initially at frequencies in excess of 20 pps. For monophasic or asymmetric biphasic pulsed currents, the cathode is placed over the region of the normal motor point of the muscle. In normal muscle, such a pattern of stimulation will, with sufficient amplitude of stimulation, evoke a smooth tetanic contraction. In a totally denervated muscle, this form of stimulation will not evoke a muscle contraction and this result is reported as a **reaction of degeneration**. The presence of the R.D. simply reflects the loss of innervation to the muscle. The current described above does not activate the denervated muscle directly because the charge of applied pulses is not sufficient to bring muscle fiber membranes to a threshold level of depolarization.

In selected clinical cases, some but not all of the motor nerve fibers innervating skeletal muscle may be lost. In such situations, the stimulation used in R.D. testing may elicit a contraction but the magnitude of the response (force of contraction) may be dramatically reduced. This type of result in R.D. testing is called a **partial reaction of degeneration**. The percentage of remaining motor axons innervating the muscle when a partial R.D. is noted cannot be determined from the results of this test.

One of the more common uses of R.D. tests was in the evaluation of individuals with Bell's palsy, the unilateral degeneration of facial nerve fibers and associated atrophy of the muscles of facial expression. Oester and Licht have written that motor degeneration is nearly "complete two weeks from the onset" of the nerve lesion and that "no R.D. after the fourteenth day, even in the presence of frank clinical palsy" represents a favorable prognosis for full recovery of facial muscle function (14). Like classical S–D testing, R.D. tests are no longer commonly used in electrophysiologic assessment.

Galvanic twitch–tetanus ratio test

Another classical electrophysiologic test used to examine whether skeletal muscle is normally innervated is the galvanic twitch–tetanus ratio test. In this procedure, direct current is intermittently applied over the motor point of a muscle suspected to be denervated using the cathode of the stimulator circuit. The duration of DC stimulation is controlled manually by use of a handheld electrode with a manual switch. The amplitude of stimulation is gradually increased until a twitch contraction is produced and the rheobase level of stimulation is recorded. Next, the current amplitude is increased to the point where a sustained tetaniclike contraction is evoked and this amplitude is recorded.

In normally innervated muscle, the ratio of the amplitude that elicited the twitch to the amplitude that produced the tetaniclike response ranges from 1 : 3.5 to 1 : 6.5 (14). In a completely denervated muscle, this ratio ranges from 1 : 1 to 1 : 1.5 (14).

Since this procedure can be extremely uncomfortable to the subject and modern assessment procedures are safer and more reliable, the galvanic twitch–tetanus ratio test is, like many other classical tests, rarely if ever used.

Contraindications/Precautions in Electrophysiologic Testing

As with any clinical procedure, a variety of conditions or situations may limit or preclude the use of clinical electrophysiologic assessment procedures. Other conditions may simply call for additional caution in application. These factors include (9):

1. Abnormal blood clotting factors/anticoagulant therapy
2. Extreme swelling
3. Dermatitis
4. Uncooperative patient
5. Recent myocardial infarction
6. Blood-transmittable disease (Jakob–Creutzfeldt, HIV, hepatitis, etc.)
7. Immune-suppressed condition
8. Central-going lines
9. Pacemakers
10. Hypersensitivity to stimulation (open wounds, burns).

In performing needle electromyography, latex gloves and eye protection should be worn in order to protect the examiner from possible blood-borne transmission of infectious agents. In cases where transmittable diseases are known to exist, additional measures such as gowns may be required for all ENMG procedures.

Additional Clinical Tests of Neuromuscular Function

Cutaneous somatosensory tests

The procedures of electrophysiologic assessment are of greatest value in the establishment of a differential diagnosis when findings are integrated with those of other clinical tests and the clinical signs and symptoms presented by an individual. The clinician involved in the use of these procedures should be aware of a variety of other tests of neuromuscular function in order to best select the ENMG procedures most appropriate for the patient.

In ENMG, only the sensory nerve conduction studies provide insight into whether or not the sensory systems are involved. In the routine neurologic exam, qualitative and rather crude tests are often used to assess sensory function. These tests include the use of pinwheels, brushes, or pins to examine tactile acuity. Today, more sensitive and reliable clinical tests are available to evaluate touch–pressure sensibility. For example, Semmes–Weinstein monofilaments (the modern version of the Von Frey hair) allow the clinicians to accurately determine cutaneous touch–pressure psychophysical thresholds in injured or diseased areas and compare these thresholds with those in uninvolved areas or with the established normal values (15). Similarly, two-point discrimination testing, which assesses one's ability to differentiate between two points applied at various distances apart, provides another index of sensory function. The repeated use of such tests of sensory function will assist the clinician in determining whether sensory function is improving, deteriorating, or remaining unchanged. In some cases, these tests may provide the earliest indication of dysfunction in the somatosensory system. A detailed discussion of the correct application of these procedures is beyond the scope of this text and the reader is referred to other references for more specific information on the clinical application of these tests (15, 16).

Reports of the results of Semmes–Weinstein monofilament testing and two-point discrimination testing are best reported on diagrammatic representations of the regions tested. Such cutaneous sensibility diagrams are often color-coded to give a clear picture of the nature and extent of sensory nerve involvement. Changes in the color patterns as sensory function improves or deteriorates provide a clear picture of the progress or decline in sensory function and allow the clinician to make adjustments in the treatment program or in functional activities to better accommodate to the sensory loss.

Cutaneous tests of autonomic function

Electrophysiologic tests are not commonly used to assess the function of autonomic nerve fibers in peripheral nerves. Several simple clinical tests are available to provide some insight into the involvement of autonomic fibers in injury or disease (17). These tests all employ staining techniques to examine sweating, an activity controlled by the sympathetic division of the autonomic nervous system. One of the more common of these tests is the **ninhydrin sweat test**. To perform the test, sweating is induced by heating the area, performing exercise, or some other technique. The area sweating is then carefully placed on filter paper. Acidified 1% ninhydrin in acetone is sprayed on the paper, which is then dried and heated. If the record is to be preserved for long periods of time, the tester may choose to fix the test with acidified copper nitrate in acetone. The test will reveal those areas in which sweat production remains and those areas where sweating is absent. The test is very valuable in following the recovery of a nerve after laceration and to monitor the extent of reinnervation of sweat glands.

Muscle tests

Although electromyography provides information regarding motor nerve fiber and muscle fiber function, the results of EMG testing should be integrated with a variety of other clinical tests of muscle function. Specific manual muscle tests should be performed by a trained practitioner for each muscle or muscle group suspected to be involved (18). Clinical **dynamometry** (hand dynamometers, pinch meters, isometric and isokinetic dynamometers, etc.) may also be useful in identification of deficits in muscle force production in limb and axial musculature.

Provocative tests

Provocative tests consist of specific procedures performed by an examiner or a patient in order to reproduce or "provoke" the symptoms of a nerve compression syndrome. These tests are useful in initial evaluation to assist in localization of peripheral nerve lesions and help guide the planning of an electrophysiologic examination. Table 10.5 lists some examples of the more commonly applied provocative tests used in initial evaluations and the disorders with which they are linked.

Observational and functional assessment

The experienced clinician will attest to the fact that no other clinical testing procedure may be more valuable than observation of patient's posture and movement during the functional activities of daily living. Observation begins as soon as a client enters the clinic during the initial visit. Functional assessment may proceed with a variety of specific tests that are becoming more commonplace in clinical practice and are detailed in other references. The observation of function may well include a visit to the home or workplace as soon as possible to determine which postures and movements are prerequisite to the individual's occupation and to identify those factors that may predispose the individual to additional or repeated injury.

Although detailed descriptions of the supplemental tests of neuromuscular function are beyond the scope of this text, one must realize that these tests must be performed along with electrophysiologic testing to obtain a complete picture of the effects of injury and disease. Reference has been made to these additional procedures to emphasize their importance in the establishment of a differential diagnosis and the development of a comprehensive plan of care.

Problem Solving and the ENMG Examination

The previous sections have detailed the various procedures used to examine motor and somatosensory function electrophysiologically. How does one decide which tests are appropriate to perform in order to assist in the establishment of a diagnosis? What tests of which nerves and muscles must be

Table 10.5
Common Provocative Tests

Name	Test for	Maneuver	Positive response
Phalen's test	Median nerve compression at carpal tunnel	Full wrist flexion held for 60 sec	Tingling sensation (paresthesia) in the median distribution
2-min compression test	Nerve entrapment at compression site	Digital pressure over peripheral nerve	Paresthesia in compressed nerve distribution
Tinel's or percussion	Nerve regeneration or compression	Tapping along course of nerve	Tingling sensation at location of nerve regeneration/ compression
EAST (elevated arm stress test)	Nerve compression in thoracic outlet	Arms abducted to 90° Elbows flexed to 90° Open/close hands	Symptoms increased before 3 min
Adson's test	Nerve compression in thoracic outlet	Rotate neck to affected side Brace shoulder posteriorly Deep inspiration	Radial pulse reduced Reproduces symptoms
Wright's test	Nerve compression in thoracic outlet	Arm abducted 90° Full external rotation of shoulder	Radial pulse reduced Reproduces symptoms
Costoclavicular test	Nerve compression in thoracic outlet	Shoulder retraction and depression	Radial pulse reduced Reproduces symptoms

performed to help determine the site, extent, and severity of the problem? What clues from the clinical evaluation are used to electrophysiologically evaluate the patient's problem? How do ENMG results correlate with other tests and how can ENMG data be used in planning comprehensive care programs?

The next section of this chapter deals with the problem solving that an examiner performs in the ENMG examination. Knowledge of the anatomy of the peripheral nerves, plexuses, nerve roots, and skeletal muscle actions and the dermatomal, sclerotomal, and myotomal innervations is extremely important. Before proceeding, a review of the neuroanatomy of the peripheral nervous system may be appropriate for some readers. Those who would like a brief overview of the sensory and motor innervation of the upper and lower extremities are referred to Appendix A at the back of the text. Appendix A provides several figures illustrating the muscles innervated by major peripheral nerves in the extremities as well as common sites where these nerves may be injured or compressed. In addition, peripheral-nerve sensory dermatome charts and spinal nerve root dermatome charts are also provided.

The **patient interview** is an essential component of the comprehensive evaluation process and should always precede any form of electrophysiologic testing. Questions typically asked during patient interview include:

1. What are the individual's complaints? (Pain, tingling sensation, functional deficit, anesthesia, fatigue, weakness, etc.)
2. Are the symptoms/signs diffuse or are they localized to one area or extremity?
3. Is an entire extremity involved or only one area?
4. Does the problem occur continuously or at certain times of the day or night?
5. Do any activities change the symptoms?
6. How long has the problem been going on?
7. Did symptoms occur suddenly or gradually?
8. Did an accident or injury precipitate the problem(s)?
9. Has the problem(s) worsened or improved?
10. Do any family members have similar problems?

The **clinical neuromuscular evaluation** is also important in clinical problem solving and routinely precedes ENMG procedures. The examiner commonly looks for signs of neuromuscular dysfunction such as:

1. Atrophy (localized vs. diffuse)
2. Trophic or temperature changes
3. Differences in appearance of the extremities (e.g., skin color, skin texture, girth)
4. Joint deformities or limitations in joint movement
5. Functional abnormalities or limitations (e.g., gait deviations, balance disorders, asymmetry of UE/LE movement)
6. Muscle weakness (unilateral vs. bilateral; upper vs. lower limbs vs. trunk)
7. Reflex changes
8. Sensory disturbances (nerve-root dermatome vs. peripheral-nerve dermatome distribution)

Answers to interview questions and findings from the neuromuscular examination dictate the types of ENMG testing procedures to be employed and the specific nerves and muscles to be tested.

Most patients referred for ENMG have some type of peripheral nerve pathology. Primary muscle diseases, neuromuscular junction pathology, and anterior-horn-cell diseases do not occur as frequently. Case studies of patients are presented below to illustrate how the patient interview and the clinical neuromuscular evaluation guide the choice of ENMG testing and show how the results of ENMG testing assist in identification of the problem(s).

In performing the interpretation of ENMG findings, one important question to be addressed is, What do the interview, clinical neuromuscular and ENMG findings have in common? In attempting to answer this question, charts like those developed by Kendall (Fig. 10.29, Fig. 10.30) may be of value especially to those who have less experience in formulating impressions of the findings (18). The use of such charts is described in other texts.

Spinal Nerve and Muscle Chart
trunk and lower extremity

Name _____

KEY

D — Dorsal Prim. Ramus
V — Vent. Prim. Ramus
A — Anterior Division
P — Posterior Division

Peripheral nerve columns (left to right):

- PN1 = (D) T1-12, L1-5, S1-3
- PN2 = (V) T1,2,3,4
- PN3 = (V) T5,6
- PN4 = (V) T7,8
- PN5 = (V) T9,10,11,12
- PN6 = (V) Iliohypogastric T12 L1
- PN7 = (V) Ilioingual T(12) L1
- PN8 = (V) Lumb. Plex. T(12) L1,2,3,4
- PN9 = (P) Femoral L(1) 2,3,4
- PN10 = (A) Obturator L(1) 2,3,4
- PN11 = (P) Sup. Glut. L4,5,S1
- PN12 = (P) Inf. Glut. L5,S1,2
- PN13 = (V) Sac. Plex. L4,5,S1,2,3
- PN14 = (P) Sciatic L4,5,S1,2
- PN15 = (A) Sciatic L4,5,S1,2,3
- PN16 = (P) C. Peroneal L4,5,S1,2
- PN17 = (A) Tibial L4,5,S1,2,3

Spinal segment columns: L1 L2 L3 L4 L5 S1 S2 S3

Group	Muscle	PN	Spinal Segment (L1 L2 L3 L4 L5 S1 S2 S3)
	ERECTOR SPINAE	PN1 •	1 2 3 4 5 · 1 2 3
Thoracic Nerves	SERRATUS POST SUP	PN2 •	
Thoracic Nerves	TRANS THORACIS	PN2 •, PN3 •	
Thoracic Nerves	INT INTERCOSTALS	PN2,3,4,5 •	
Thoracic Nerves	EXT INTERCOSTALS	PN2,3,4,5 •	
Thoracic Nerves	SUBCOSTALES	PN2,3,4,5 •	
Thoracic Nerves	LEVATOR COSTARUM	PN2,3,4,5 •	
Thoracic Nerves	OBLIQUUS EXT ABD	PN3 (•), PN4,5 •	
Thoracic Nerves	RECTUS ABDOMINIS	PN3,4,5 •	
Thoracic Nerves	OBLIQUUS INT ABD	PN4,5,6 •, PN7 (•)	L1 = 1
Thoracic Nerves	TRANSVERSUS ABD	PN4,5,6 •, PN7 (•)	L1 = 1
Thoracic Nerves	SERRATUS POST INF	PN5 •	
Lumbar Plexus	QUAD LUMBORUM	PN8 •	1 2 3
Lumbar Plexus	PSOAS MINOR	PN8 •	1 2
Lumbar Plexus	PSOAS MAJOR	PN8 •	1 2 3 4
Femoral	ILIACUS	PN9 •	(1) 2 3 4
Femoral	PECTINEUS	PN9 •, PN10 (•)	2 3 4
Femoral	SARTORIUS	PN9 •	2 3 (4)
Femoral	QUADRICEPS	PN9 •	2 3 4
Obturator (Ant)	ADDUCTOR BREVIS	PN10 •	2 3 4
Obturator (Ant)	ADDUCTOR LONGUS	PN10 •	2 3 4
Obturator (Ant)	GRACILIS	PN10 •	2 3 4
Obturator (Post)	OBTURATOR EXT	PN10 •	3 4
Obturator (Post)	ADDUCTOR MAGNUS	PN10 •, PN15 •	2 3 4 · · 1
Gluteal (Sup)	GLUTEUS MEDIUS	PN11 •	4 5 1
Gluteal (Sup)	GLUTEUS MINIMUS	PN11 •	4 5 1
Gluteal (Sup)	TENSOR FAS LAT	PN11 •	4 5 1
Gluteal (In)	GLUTEUS MAXIMUS	PN12 •	5 1 2
Sacral Plexus	PIRIFORMIS	PN13 •	(5) 1 2
Sacral Plexus	GEMELLUS SUP	PN13 •	5 1 2
Sacral Plexus	OBTURATOR INT	PN13 •	5 1 2
Sacral Plexus	GEMELLUS INF	PN13 •	4 5 1 (2)
Sacral Plexus	QUADRATUS FEM	PN13 •	4 5 1 (2)
Sciatic (P)	BICEPS (SHORT H)	PN14 •	5 1 2
Sciatic (Tibial)	BICEPS (LONG H)	PN15 •	5 1 2 3
Sciatic (Tibial)	SEMITENDINOSUS	PN15 •	4 5 1 2
Sciatic (Tibial)	SUMIMEMBRANOSUS	PN15 •	4 5 1 2
Common Peroneal (Deep)	TIBIALIS ANTERIOR	PN16 •	4 5 1
Common Peroneal (Deep)	EXT HALL LONG	PN16 •	4 5 1
Common Peroneal (Deep)	EXT DIGIT LONG	PN16 •	4 5 1
Common Peroneal (Deep)	PERONEUS TERTIUS	PN16 •	4 5 1
Common Peroneal (Deep)	EXT DIGIT BREVIS	PN16 •	4 5 1
Common Peroneal (Sup)	PERONEUS LONGUS	PN16 •	4 5 1
Common Peroneal (Sup)	PRONEUS BREVIS	PN16 •	4 5 1
Tibial (Tibial)	PLANTARIS	PN17 •	4 5 1 (2)
Tibial (Tibial)	GASTROCNEMIUS	PN17 •	· · 1 2
Tibial (Tibial)	POPLITEUS	PN17 •	4 5 1
Tibial (Tibial)	SOLEUS	PN17 •	5 1 2
Tibial (Tibial)	TIBIALIS POSTERIOR	PN17 •	(4) 5 1
Tibial (Tibial)	FLEX DIGIT LONG	PN17 •	5 1 (2)
Tibial (Tibial)	FLEX HALL LONG	PN17 •	5 1 2
Tibial (Med Pl)	FLEX DIGIT BREVIS	PN17 •	4 5 1
Tibial (Med Pl)	ABDUCTOR HALL	PN17 •	4 5 1
Tibial (Med Pl)	FLEX HALL BREVIS	PN17 •	4 5 1
Tibial (Med Pl)	LUMBRICALIS I	PN17 •	4 5 1
Tibial (Lat Plant)	ADB DIGITI MIN	PN17 •	· · 1 2
Tibial (Lat Plant)	QUAD PLANTAE	PN17 •	· · 1 2
Tibial (Lat Plant)	FLEX DIGITI MIN	PN17 •	· · 1 2
Tibial (Lat Plant)	OPP. DIGITI MIN	PN17 •	· · 1 2
Tibial (Lat Plant)	ADDUCTORS HALL	PN17 •	· · 1 2
Tibial (Lat Plant)	PLANT INTEROSSEI	PN17 •	· · 1 2
Tibial (Lat Plant)	DORSAL INTEROSSEI	PN17 •	· · 1 2
Tibial (Lat Plant)	LUMB II, III, IV	PN17 •	(4) (5) 1 2

Figure 10.30 Lower-limb spinal nerve and peripheral nerve muscle innervation chart (Kendall, FP, McCreary EK, Prouance PG. Muscles testing and function, 4th edition. Williams & Wilkins, 1993 p 393).

Spinal Nerve and Muscle Chart
trunk and lower extremity

Name

KEY

D	Dorsal Prim. Ramus
V	Vent. Prim. Ramus
A	Anterior Division
P	Posterior Division

Peripheral nerve column headers (type / nerve / spinal segment):

- PN1 — D — (Erector spinae) T1-12, L1.5, S1-3
- PN2 — V — T1,2,3,4
- PN3 — V — T5,6
- PN4 — V — T7,8
- PN5 — V — T9,10,11,12
- PN6 — V — Iliohypogastric T12 L1
- PN7 — V — Ilioingual T(12) L1
- PN8 — V — Lumb. Plex. T(12) L1,2,3,4
- PN9 — P — Femoral L(1) 2,3,4
- PN10 — A — Obturator L(1) 2,3,4
- PN11 — P — Sup.Glut. L4,5,S1
- PN12 — P — Inf. Glut. L5,S1,2
- PN13 — V — Sac. Plex. L4,5,S1,2,3
- PN14 — P — Sciatic L4,5,S1,2
- PN15 — A — Sciatic L4,5,S1,2,3
- PN16 — P — C. Peroneal L4,5,S1,2
- PN17 — A — Tibial L4,5,S1,2,3

Group	MUSCLE	PN1	PN2	PN3	PN4	PN5	PN6	PN7	PN8	PN9	PN10	PN11	PN12	PN13	PN14	PN15	PN16	PN17	L1	L2	L3	L4	L5	S1	S2	S3
	ERECTOR SPINAE	•																	1	2	3	4	5	1	2	3
Thoracic Nerves	SERRATUS POST SUP		•																							
	TRANS THORACIS		•	•	•																					
	INT INTERCOSTALS		•	•	•	•																				
	EXT INTERCOSTALS		•	•	•	•																				
	SUBCOSTALES		•	•	•	•																				
	LEVATOR COSTARUM		•	•	•																					
	OBLIQUUS EXT ABD			(•)	•	•																				
	RECTUS ABDOMINIS			•	•	•																				
	OBLIQUUS INT ABD				•	•	•	(•)											1							
	TRANSVERSUS ABD				•	•	•	(•)											1							
	SERRATUS POST INF					•																				
Lumbar Plexus	QUAD LUMBORUM								•										1	2	3					
	PSOAS MINOR								•										1	2						
	PSOAS MAJOR								•										1	2	3	4				
Femoral	ILIACUS								•										(1)	2	3	4				
	PECTINEUS								•	(•)										2	3	4				
	SARTORIUS								•											2	3	(4)				
	QUADRICEPS								•											2	3	4				
Obturator Ant.	ADDUCTOR BREVIS									•										2	3	4				
	ADDUCTOR LONGUS									•										2	3	4				
	GRACILIS									•										2	3	4				
Obturator Post.	OBTURATOR EXT									•											3	4				
	ADDUCTOR MAGNUS									•						•				2	3	4	5	1		
Gluteal Sup.	GLUTEUS MEDIUS										•											4	5	1		
	GLUTEUS MINIMUS										•											4	5	1		
	TENSOR FAS LAT										•											4	5	1		
Gluteal In.	GLUTEUS MAXIMUS											•											5	1	2	
Sacral Plexus	PIRIFORMIS												•										(5)	1	2	
	GEMELLUS SUP												•										5	1	2	
	OBTURATOR INT												•										5	1	2	
	GEMELLUS INF												•									4	5	1	(2)	
	QUADRATUS FEM												•									4	5	1	(2)	
Sciatic P.	BICEPS (SHORT H)														•								5	1	2	
Sciatic Tibial	BICEPS (LONG H)															•							5	1	2	3
	SEMITENDINOSUS															•						4	5	1	2	
	SUMIMEMBRANOSUS															•						4	5	1	2	
Common Peroneal Deep	TIBIALIS ANTERIOR																•					4	5	1		
	EXT HALL LONG																•					4	5	1		
	EXT DIGIT LONG																•					4	5	1		
	PERONEUS TETIUS																•					4	5	1		
	EXT DIGIT BREVIS																•					4	5	1		
Common Peroneal Sup.	PERONEUS LONGUS																•					4	5	1		
	PRONEUS BREVIS																•					4	5	1		
Tibial	PLANTARIS																	•				4	5	1	(2)	
	GASTROCNEMIUS																	•						1	2	
	POPLITEUS																	•				4	5	1		
	SOLEUS																	•					5	1	2	
	TIBIALIS POSTERIOR																	•				(4)	5	1		
	FLEX DIGIT LONG																	•					5	1	(2)	
	FLEX HALL LONG																	•					5	1	2	
Tibial Med Pl	FLEX DIGIT BREVIS																	•				4	5	1		
	ABDUCTOR HALL																	•				4	5	1		
	FLEX HALL BREVIS																	•				4	5	1		
	LUMBRICALIS I																	•				4	5	1		
Lat Plant	ADB DIGITI MIN																	•						1	2	
	QUAD PLANTAE																	•						1	2	
	FLEX DIGITI MIN																	•						1	2	
	OPP. DIGITI MIN																	•						1	2	
	ADDUCTORS HALL																	•						1	2	
	PLANT INTEROSSEI																	•						1	2	
	DORSAL INTEROSSEI																	•						1	2	
	LUMB II, III, IV																	•				(4)	(5)	1	2	

Figure 10.30 Lower-limb spinal nerve and peripheral nerve muscle innervation chart (Kendall, FP, McCreary EK, Prouance PG. Muscles testing and function, 4th edition. Williams & Wilkins, 1993 p 393).

Case Studies

Case 1

The patient is a 42-year-old female with a 1-year history of progressively worsening right hand paresthesias. She reports no history of trauma. Her symptoms are worse at night and with prolonged driving. She states that it feels like "my hand goes to sleep" and that symptoms ease with changing her hand position or shaking her hand. She reports a prior episode of similar, but less severe, symptoms 12 years ago while pregnant. She has not noticed any change in her daily functional abilities. She has recently had a regular physical exam and has no history of chronic orthopedic, neurological, or metabolic disease. She is taking no medication at this time.

Clinical Exam: The patient ambulates independently with a normal gait. Her posture is good, without visible abnormality. She has full cervical spine range of motion (ROM) without pain or reproduction of symptoms and full upper-extremity ROM without pain. Manual muscle testing reveals that strength is normal in both upper extremities. Deep-tendon reflexes (DTRs) are intact and equal bilaterally in upper and lower extremities. Cutaneous sensation to light touch and two-point discrimination appear normal in both upper extremities. Tinel's sign is negative over median and ulnar nerves bilaterally. Phalen's sign is positive on right after 2 minutes.

Summary of ENMG Studies: Table 10.6 summarizes the results of the NCV studies, which include prolonged distal sensory latencies of the right median

Table 10.6
Results of Nerve Conduction Velocity Studies Performed

Nerve	Latency (msec)		Distance (mm)		Velocity (m/sec)		Amplitude (μV)	
	Right	Left	Left	Right	Right	Left	Right	Left
Median motor								
Wrist–ABP	3.8	3.8	80	80			7000	9000
Elbow	8.0	7.8	220	220	52	55	7000	9000
Upper arm	10.2	9.8	120	120	54	60	6800	9000
F-wave	27.8	28.0						
Median sensory								
Wrist–index	3.6[a]	3.2	140	140			30	42
Wrist–long	3.6[a]	3.2	140	140			28	40
Transcarpal	2.4[a]	2.0	80	80			48	90
Ulnar motor								
Wrist–ADM	3.0		80				6000	
Below elbow	6.8		210		55		6000	
Above elbow	8.6		100		55		5800	
Upper arm	10.6		125		56		5600	
F-wave	27.6							
Ulnar sensory								
Wrist–little	2.8		120				30	
Radial sensory								
Forearm–web	2.4		120				25	

[a]Indicates abnormal value

Table 10.7
Results of EMG Studies Performed

Muscle	Insertional activity	Spontaneous activity	Motor units	Interference patterns
R. abd. pol. brev.	Normal	None	Normal	Full
R. opp. pol.	Normal	None	Normal	Full
R. flex. pol. long.	Normal	None	Normal	Full
R. pron. teres	Normal	None	Normal	Full
R. 1st dor. inter.	Normal	None	Normal	Full
R. abd. dig. min.	Normal	None	Normal	Full
R. ext. carpi rad.	Normal	None	Normal	Full

nerve to the index and long fingers, with normal sensory response amplitudes. Also the right median transcarpal sensory response latency is prolonged, but of normal amplitude. All other motor and sensory conductions were normal. EMG study of the right upper extremity (Table 10.7) showed no signs of denervation. Motor units were of normal amplitude, duration, and form, with complete interference patterns.

Impression of ENMG Findings: ENMG/NCV findings as listed above are consistent with mild involvement of the median nerve at or about the right wrist, primarily neuropraxic in nature, without signs of axonal degeneration.

Discussion: This is actually a fairly typical electrophysiological study showing a mild or early carpal tunnel syndrome. In referring to prior discussions of peripheral nerve injury, note that the only abnormalities associated with this case are the prolonged distal sensory latencies for the right median nerve. All calculated conduction velocities were within normal limits, as were the evoked motor and sensory response amplitudes. These normal amplitudes and the absence of spontaneous activity noted during the EMG needle exam are all consistent with a demyelinating process. Also note how the summary describes the neurological lesion in anatomical perspective, and avoids the diagnostic label of carpal tunnel syndrome.

Case 2

The patient is a 39-year-old white male referred for EMG/NCV study of the right upper extremity to rule out carpal tunnel syndrome. The patient reports a several-month history of paresthesias, cramping, and pain in his right hand. Patient is a surgeon and notes symptoms to be worse during long operative procedures. He denies any history of trauma, but does complain of episodic neck and back pain without frank radicular symptoms. He indicates that his general state of health is good and that he takes no medication at this time.

Clinical Exam: The patient is a healthy-appearing male, ambulating independently, in no apparent distress. His posture shows a mild forward head but otherwise is unremarkable. Neck ROM is mildly restricted in right rotation but pain free. Spurling's test was negative. The upper-extremity ROM is normal and without pain. Strength in the upper extremities was also found to be normal. DTRs were absent in the upper and lower extremities, even with facilitory techniques. Cutaneous sensation to light touch was grossly intact. Tinel's over median and ulnar nerves was negative whereas Phalen's sign was positive.

Summary of ENMG findings: Nerve conduction studies of both upper and lower extremities show diffuse widespread abnormalities of conduction and amplitude (Table 10.8). H-reflexes were absent bilaterally. The changes in

Table 10.8
Results of Nerve Conduction Velocity Studies Performed

Nerve	Latency (msec) Right	Left	Distance (mm) Left	Right	Velocity (m/sec) Right	Left	Amplitude (µV) Right	Left
Median Motor								
Wrist–ABP	5.6ᵃ	5.2ᵃ	80	80			4000ᵃ	5000
Elbow	11.6ᵃ	11.0ᵃ	250	250	41ᵃ	43ᵃ	3000ᵃ	4600ᵃ
Upper arm	14.1ᵃ	13.5ᵃ	120	120	48ᵃ	48ᵃ	2600ᵃ	4400ᵃ
F-wave	34.0ᵃ	32.2ᵃ						
Median sensory								
Wrist–index	5.0ᵃ	4.6ᵃ	140	140			10ᵃ	14ᵃ
Ulnar motor								
Wrist–ADM	5.0ᵃ	4.8ᵃ	80	80			3000ᵃ	4000ᵃ
Below elbow	10.4ᵃ	10.0ᵃ	220	220	40ᵃ	42ᵃ	3000ᵃ	3800ᵃ
Above elbow	13.0ᵃ	12.2ᵃ	120	110	38ᵃ	50	2200ᵃ	3400ᵃ
F-wave	27.6ᵃ	32.0ᵃ						
Ulnar sensory								
Wrist–little	4.2ᵃ	3.8ᵃ	120	120			8ᵃ	12ᵃ
Radial sensory								
Forearm–web	3.6ᵃ	3.4ᵃ	120	120			10ᵃ	12ᵃ
Tibial motor								
Ankle–ABD hall	8.2ᵃ	8.0ᵃ	100	100			800ᵃ	1200ᵃ
Popliteal	22.0ᵃ	21.8ᵃ	400	410	28ᵃ	29ᵃ	200ᵃ	400ᵃ
H-reflex	Absent	Absent						
Sural-sensory								
Calf-foot	Absent	Absent						

ᵃIndicates abnormal value

amplitude and conduction values indicated a mixed axonal and demyelinating process. EMG study shows chronic changes more pronounced in a distal distribution, with the lower extremities worse than the upper extremities (Table 10.9).

Impression of ENMG Findings: EMG/NCV study as listed above is consistent with a peripheral polyneuropathy involving both the motor and sensory nerve fibers. These findings are more severe in a distal distribution, involving the lower more than the upper extremities. The chronic changes noted during the needle EMG study would indicate that this process has been present for a long period of time.

Discussion: Case 2 is a good example of a patient presenting with what appears to be a simple peripheral nerve compression syndrome, when in fact the underlying pathology is anything but simple. This patient was ultimately diagnosed as having a hereditary motor sensory neuropathy. Of particular interest is the manner in which he had adapted to his level of decreased capabilities. On clinical exam, he presented as essentially normal except for the absent DTRs. This was in part a result of decreased thoroughness during the clinical exam, in that we focused upon his complaints and the most likely probable causes. Because we did not examine his lower extremities, we missed the exceptionally high arched feet and obvious loss of lower leg muscle mass. During the electrophysiologic exam it became readily obvious that there was a diffuse process occurring. The widespread abnormal findings in the upper extremities dictated that the lower

Table 10.9
Results of EMG Studies Performed

Muscle	Insertional activity	Spontaneous activity	Motor units	Interference patterns
R. abd. pol. brev.	Slight decr	Occ. sharps	Large amp.	Decr
L. abd. pol. brev.	Slight decr	Occ. sharps	Large amp.	Decr
R. 1st dorsal int.	Decr	Occ. sharps	Large amp.	Decr
L. 1st dorsal int.	Decr	Occ. sharps	Large amp.	Decr
R. pronator teres	Normal	None	Occ. poly	Decr
L. pronator teres	Normal	None	Occ. poly	Decr
R. ext. carpi rad.	Normal	None	Normal	Decr
R. deltoids	Normal	None	Normal	Slight decr
R. vast. medialis	Decr	None	Large amp. poly.	Decr
R. gastroc.	Decr	Occ. sharps	Large amp. poly.	Decr
R. abd. hall.	Absent	Occ. sharps	Increased duration	Severely decr
L. abd. hall.	Absent	Occ. sharps	Increased duration	Severely decr
L. ant. tib.	Increased	Sharps, fibs.	Large Amp. poly.	Decr

extremities be examined, subsequently leading to the impression of a peripheral polyneuropathic process.

Case 3

A 19-year-old male is referred with a 3-day history of weakness in the right ankle and numbness over the dorsal right foot. He does not complain of pain. The patient reports that he awoke with the above symptoms. He reports no history of trauma, but indicates that 4 days ago he went on a long hike in rough terrain and sustained multiple falls. His general state of health is good and he has no prior history of similar symptoms.

Clinical Exam: Patient ambulates independently with a high steppage gait on right with obvious inability to dorsiflex his right ankle. He has full active range of motion of the trunk, hips, knees, and left ankle. Right ankle shows full passive range of motion but the patient is unable to actively dorsiflex or evert the ankle, or extend the toes. Right manual muscle testing shows 5/5 hip groups, 5/5 quads/hamstrings, 5/5 gastroc/soleus, 5/5 posterior tibialis, 5/5 toe flexors, 1/5 peroneus longus, 0/5 anterior tibialis, 0/5 extensor digitorum longus and brevis. Left lower extremity was 5/5 throughout. Sensation was severely diminished over the dorsal right foot and between the great and long toes. DTRs were intact and equal bilaterally for quadriceps and triceps surae. No specific point tenderness was noted, and no Tinel s about the ankle or around the knee could be elicited.

Summary of ENMG Findings: Table 10.10 summarizes the NCV results, which show that motor-nerve conduction of the deep portion of the right peroneal nerve is absent. A response from the right superficial peroneal sensory nerve on the right is not recorded, but left peroneal and right tibial nerve motor conduction studies are normal. H-reflex is present with normal latency. Right sural sensory response is intact, with borderline-normal amplitude.

EMG study of the right lower extremity shows complete absence of motor-unit recruitment in the deep portion of the peroneal nerve, with severely reduced interference in the superficial motor branch of the peroneal nerve distal

Table 10.10
Results of Nerve Conduction Velocity Studies for Case #3

Nerve	Latency (msec)		Distance (mm)		Velocity (m/sec)		Amplitude (μV)	
	Right	Left	Left	Right	Right	Left	Right	Left
Peroneal motor								
Ankle–EDB	Absent[a]	4.0	80	80			None	5200
Fib. neck	Absent[a]	10.2	340	340	None	54	None	5000
Popliteal	Absent[a]	12.4	120	120	None	54	None	5000
Peroneal superficial sensory								
Lat. fibula	Absent[a]	3.6	120	120			None	22
Tibial motor								
Ankle–ABD hal.	4.6		100				10600	
Popliteal	13.8		480			52	9400	
H-reflex	28.4							
Sural sensory								
Calf–lat. foot	3.8		120				10	

[a]Indicates abnormal value

to the knee (Table 10.11). The short head of the biceps femoris muscle was normal, with full recruitment/interference patterns.

Impression of ENMG Findings: EMG/NCV findings as listed above are consistent with a complete lesion of the deep branch of the peroneal nerve and severe involvement of the superficial peroneal nerve at or about the knee. Repeat study in 3 weeks to assess for denervation changes in the peroneal innervated muscles is recommended.

Discussion: The above study is a clear example of a nearly complete peripheral nerve lesion. Sensory and motor loss patterns along with the significant gait pattern are textbook versions of a peroneal nerve lesion about the knee. It is not unusual to have some sparing of specific branches, such as the few fibers that were functioning in the peroneus longus muscle. The key differential muscle to study in this case was the short head of the biceps femoris. The muscle receives its innervation prior to the point where the peroneal nerve enters the popliteal fossa. With the complete loss of distal motor and sensory response to electrical stimulation, repeat study in 3 weeks on the peroneal musculature would most

Table 10.11
Results of EMG Studies

Muscle	Insertional activity	Spontaneous activity	Motor units	Interference patterns
R. ext. dig. brev.	Normal	None	None	None
R. ext. dig. long.	Normal	None	None	None
R. ant. tib.	Normal	None	None	None
R. peroneus long.	Normal	None	Normal	Single unit
R. short bic. fem.	Normal	None	Normal	Full
R. abd. hal.	Normal	None	Normal	Full
R. post. tib.	Normal	None	Normal	Full
R. gastroc.	Normal	None	Normal	Full
R. quadriceps	Normal	None	Normal	Full

likely show spontaneous activity at rest. These spontaneous discharges (positive sharp waves and fibrillation potentials) become evident after the nerve has undergone complete Wallerian degeneration, causing the individual muscle fibers to become hyperexcitable to stimulation.

Case 4

Patient is a 42-year-old female with a long history of episodic low back and right leg pain that has usually resolved with rest, medication, and home exercises. Her current bout of pain started 3 months ago while lifting at work. She noted immediate minor low back pain that did not limit her mobility. Upon awakening the next morning she was in severe pain and unable to move because of low back and right leg pain. The pain in her leg is worse with flexion motions, and extends to her ankle and foot. She notes a sense of numbness along the right lateral foot border. She reports that the leg feels weak, but in no particular pattern. She denies any loss of bowel or bladder control. Pain arises with coughing or sneezing and shoots into her leg.

Clinical Exam: The patient ambulates slowly, with a slight limp on the right. Posture examination reveals trunk held rigidly and slightly shifted to left. The trunk range of motion is limited in all planes, but with pain greatly increased with flexion and right rotation. Strength is difficult to assess secondary to pain with testing, but is grossly intact and at least 4/5 throughout the lower extremities. DTRs are intact at quadriceps and left triceps surae; right triceps surae response is absent. Sensation is decreased over the right lateral aspect of the foot. Sitting and supine straight-leg raise tests were positive.

Summary of ENMG Findings: Nerve conduction studies of right peroneal and tibial motor nerves and left tibial motor nerves were normal (Table 10.12). Normal sensory responses of the right superficial peroneal and sural nerves were recorded. H-reflex was absent on the right; left H-reflex had normal latency.

Table 10.12
Results of Nerve Conduction Velocity Studies

Nerve	Latency (msec)		Distance (mm)		Velocity (m/sec)		Amplitude (µv)	
	Right	Left	Left	Right	Right	Left	Right	Left
Peroneal motor								
Ankle–EDB	4.2		80				5600	
Fib. neck	10.2		300		50		5400	
Popliteal	12.0		100		55		5200	
Peroneal superficial sensory								
Lat. fibula	3.6		120				20	
Tibial motor								
Ankle–ABD hal.	4.4	4.6	100	100			6400	12000
Popliteal	12.2	12.2	400	410	51	53	5800	11000
H-reflex	Absent[a]	27.4						
Sural sensory								
Calf–lat. foot	3.8		14				16	

[a]Indicates abnormal value

Table 10.13
Results of EMG Studies

Muscle	Insertional activity	Spontaneous activity	Motor units	Interference patterns
R. vastus medialis	Normal	None	Normal	Full
R. ant. tib.	Normal	None	Normal	Full
R. ext. hall.	Normal	None	Normal	Full
R. ext. dig. brev.	Normal	None	Normal	Full
R. abd. hall.	Increased	+Sharps fibs.	Frequent Polys.	Decreased
R. gastroc.	Increased	+Sharps	Frequent Polys.	Decreased
R. glut. max.	Increased	+Sharps	Frequent Polys.	Decreased
L. gastroc.	Normal	None	Normal	Full
L. glut. max.	Normal	None	Normal	Full
R. paraspinals				
L3–L4	Normal	None	Normal	
L4–L5	Normal	None	Normal	
L5–S1	Increased	+Sharps	Occ. polys.	
S1–S2	Increased	+Sharps	Freq. polys., fibs.	
L. paraspinals				
S1–S2	Normal	None	Normal	

EMG study showed increased spontaneous activity in the right S1 myotomal distribution, including the paraspinal musculature, along with an increased percentage of polyphasic motor units in the same distribution (Table 10.13). Muscles outside of the S1 distribution were normal.

Impression of ENMG Findings: ENMG/NCV study as listed above is consistent with a lesion at or about the right S1 nerve root level, with changes consistent with axonal degeneration.

Discussion: The above study illustrates findings consistent with a nerve root lesion. In the case of single nerve root lesions, nerve conduction velocities will be normal with normal to borderline-normal evoked motor response amplitudes. Sensory responses will usually be normal, even in the clinical absence of sensation. This is a result of the lesion being proximal to the dorsal root ganglion. The presence of positive sharp waves and fibrillation potentials indicates that the nerve fibers have undergone Wallerian degeneration. The presence of polyphasic motor units indicates that the surviving nerve fibers are reinnervating the denervated muscle fibers via collateral sprouting. The unilateral loss of the H-reflex fits with the clinical picture of an absent ankle deep-tendon reflex.

Summary

This chapter has addressed the ENMG examination and its uses in the establishment of a differential diagnosis of neuromuscular disorder. Both modern procedures in ENMG and classical approaches to electrophysiologic assessment have been described. The goal has been to provide the reader with an understanding of the procedures as well as an appreciation for the meaning of the results of ENMG studies. Findings from ENMG may be used in conjunction with a spectrum of other clinical tests in order to assist in the establishment of a differential diagnosis and subsequent comprehensive plan of care including medical, surgical, and rehabilitative procedures. The chapter has emphasized that the results of an ENMG examination should never be used alone to define the site, extent, or nature of pathology. That is, the results of electrophysiologic tests are not diagnostic when viewed in isolation. Education and training in musculoskeletal and neuromuscular evaluation is prerequisite for the performance and interpretation of the ENMG examination. Clinical competence in the actual performance of ENMG may be attained by further study and extensive clinical practice, preferably under the guidance of an experienced practitioner in the field. For those interested in developing competence in ENMG, guidelines have been established by several organizations including the Section on Clinical Electrophysiology of the American Physical Therapy Association (Alexandria, VA) and the American Association of Electrodiagnostic Medicine (Rochester, MN).

Study Questions

1. Compare and contrast the intracellular and extracellular recording techniques used to monitor activity in excitable tissues.

2. Describe the placement of stimulation and recording electrodes for a median motor-nerve conduction study.

3. Compare and contrast the orthodromic and antidromic sensory nerve conduction techniques for the median nerve.

4. List three factors (not technique related) that may influence conduction velocity in a peripheral nerve.

5. Define *neuropraxia, axonotmesis,* and *neurotmesis.*

6. What types of EMG potentials are considered "abnormal" findings during an examination?

7. What is the name of the test that is analogous to the Achilles-tendon reflex test?

8. Draw a strength–duration curve for a normally innervated muscle. What does the strength–duration curve for normally innervated muscle represent?

9. Name two clinical tests for cutaneous sensory function.

10. The blink reflex test examines the integrity of which two cranial nerves?

11. Compare and contrast EMG and NCV findings for a neuropraxia and an axonotmesis injury. How would these findings change with time (1 day, 5 days, and 21 days postinjury)?

12. List the contraindications/precautions to performing ENMG examination.

13. What is an F-wave and what is thought to occur to give rise to this finding?

14. What is the purpose of somatosensory evoked-potential testing?

15. What electrical potentials are noted in normal muscle during voluntary contraction? What are the characteristics of these potentials?

References

1. Kimura J. Principles of nerve conduction studies. In: Electrodiagnosis of diseases of nerve and muscle. 2nd ed. Philadelphia: F. A. Davis Co. 1989:79-80.

2. Performing motor and sensory neuronal conduction studies in adult humans: A NIOSH technical manual. 1990:(NIOSH) Publication No. 90-113. U.S. Dept. of Health and Human Services, p 8.

3. Performing motor and sensory neuronal conduction studies in adult humans: A NIOSH technical manual. (NIOSH) Publication No. 90-113. Dept. of Health and Human Services, 12-15.

4. Seddon HJ: Three types of nerve injury. Brain 1943;66:237.

5. Kimura J. Principles of nerve conduction studies. In: Electrodiagnosis of diseases of nerve and muscle. 2nd ed. Philadelphia: F. A. Davis Co., 1989;94-97.

6. Kimura J. Principles of nerve conduction studies. In: Electrodiagnosis of diseases of nerve and muscle. 2nd ed. Philadelphia: F. A. Davis Co., 1989;315, 307-331.

7. Kimura J. Principles of nerve conduction studies. In: Electrodiagnosis of diseases of nerve and muscle. 2nd ed. Philadelphia: F.A.Davis Co., 1989;375-426.

8. Oh SJ. Electromyography—neuromuscular transmission studies. Baltimore: Williams & Wilkins, 1988:1-180.

9. Kimura J. Techniques and normal findings. In: Electrodiagnosis in diseases of nerve and muscle: principles and practice. 2nd ed. Philadelphia: F. A. Davis Co., 1989:227-248.

10. Kimura J. Techniques and normal findings. In: Electrodiagnosis in diseases of nerve and muscle: principles and practice. 2nd ed. Philadelphia: F. A. Davis Co., 1989:249-274.

11. Parry CBW. Strength–duration curves. In: Licht S, ed. Electrodiagnosis and electromyography. Baltimore: Waverly Press, 1971:255.

12. Ritchie AE. Peripheral nerve injuries. London: H.M. Stationery Office, 1954.

13. Adrian ED. The electrical reactions of muscle before and after nerve injury. Brain 1916;39:1-33.

14. Oester YT, Licht S. Routine Electrodiagnosis. In: Electrodiagnosis and electromyography. Licht S, ed. Baltimore:Waverly Press, 1971.

15. Bell-Krotoski J, Weinstein S, Weinstein C. Testing sensibility, including touch–pressure, two-point discrimination point localization and vibration. J Hand Ther 1993;6:114-123.

16. Dellon AL. Evaluation of sensibility and re-education of sensation in the hand. Baltimore: Williams & Wilkins, 1981.

17. Lister G. The hand: diagnosis and indications. Edinburgh: Churchill Livingstone, 1977:73.

18. Kendall FP, McCreary EK, Provance PG. Muscle testing and function, 4th edition. Baltimore: Williams & Wilkins, 1993.

Bibliography

Aminoff MJ. Electrodiagnosis in clinical neurology. New York: Churchill Livingstone, 1980.

Brooke MH. A clinician's view of neuromuscular diseases. Baltimore: Williams & Wilkins,1986.

Brown WM. The physiological and technical basis of electromyography. Stoneham, MA: Butterworth, 1984.

Chusid J, McDonald JJ. Correlative neuroanatomy and functional neurology. Los Altos, CA: Lange Medical Publications, 1967.

Dawson DM, Hallett M, Millender LH. Entrapment neuropathies. Boston: Little, Brown & Co., 1983.

Oh SJ. Electromyography–neuromuscular transmission studies. Baltimore: Williams & Wilkins, 1988.

Hoppenfeld S. Orthopaedic neurology. Philadelphia: JB Lippincott, 1977.

Kimura J. Electrodiagnosis in diseases of nerve and muscle: principles and practice. Philadelphia: FA Davis, 1989.

Leffert RD. Brachial plexus injuries. New York: Churchill Livingstone, 1985.

Oh SJ. Clinical electromyography: nerve conduction studies. Baltimore: University Park Press, 1984.

Schaumburg HH, Spencer PS, Thomas PK. Disorders of peripheral nerves. Philadelphia: FA Davis, 1983.

Smorto MP, Basmajian, JV. Electrodiagnosis. New York: Harper & Row, 1977.

Task Force on SCE Guidelines for EMG. Guidelines for clinical electromyography, section on clinical electrophysiology. American Physical Therapy Association, 1988.

Chapter 11

Electromyographic Biofeedback to Improve Voluntary Motor Control

Stuart A. Binder-Macleod

$\Lambda\Lambda\Lambda$ **B**iofeedback is the use of electronic instrumentation to provide objective information (feedback) to an individual about a physiological function or response so that the individual becomes aware of his or her response. The individual then attempts to alter the feedback signal in order to modify the physiological response (1). Though the clinical application of biofeedback includes the use of the electromyograph (EMG), the electroencephalograph (EEG), blood pressure, heart rate, and visceral and vasomotor responses, the present chapter will address only the application most widely used in physical rehabilitation—EMG biofeedback. EMG biofeedback is the use of electronic instrumentation to detect and feed back the myoelectric signals from skeletal muscle in order to allow the patient to gain better volitional control over the muscle. EMG biofeedback is used to train patients to relax hyperactive muscles or to increase the discharge rate and number of motor units activated to increase the strength of contraction.

This chapter discusses the advantages of using biofeedback, technical considerations regarding the application of EMG biofeedback, selection of appropriate patients for the application of EMG biofeedback, and development of training strategies during the use of EMG biofeedback.

Advantages of Using EMG Biofeedback

EMG biofeedback is not a treatment. Rather, it is a **tool** that clinicians can use to help their patients learn new tasks or modify existing motor patterns by providing useful information both to the clinician and patient (see Fig. 11.1). The actual treatment is the activities or exercises that patients perform. To illustrate this point, parallels may be drawn between the use of a mirror during posture training and the use of biofeedback. One would never say that the mirror is being used to treat a patient. Rather, the mirror is merely a tool that is used to provide feedback to the patient. Similarly, in EMG biofeedback, the electromyographic signal is a tool that clinicians and patients use to provide information about the electrical activity of specific muscles.

One advantage of EMG biofeedback is the **speed** and **continuity** with which the information is provided to the clinician and patient. Without biofeedback clinicians must rely on palpation or visual inspection to determine if the appropriate muscles are being recruited or relaxed during an exercise. At best, the detection, processing, and formulation of a response by the health care worker takes several hundred milliseconds. Given the ephemeral nature of most motor responses, by the time the patient receives and processes the verbal or manual feedback that the clinician provides, the patient may be performing at a very different level than was originally perceived by the clinician. For feedback to be effective it must be coincidental with the task that is to be modified. EMG biofeedback can be nearly instantaneous, thus reflecting the existing state of muscle contraction. Related to the speed of processing information is the ability of the EMG biofeedback to provide continuous feedback. If verbal feedback requires several hundred milliseconds to be processed and presented to the patient, the fastest rate that verbal feedback could be updated and presented to the patient would be one to two times per second. In contrast, most biofeedback machines can inform patients of their responses in a nearly continuous manner.

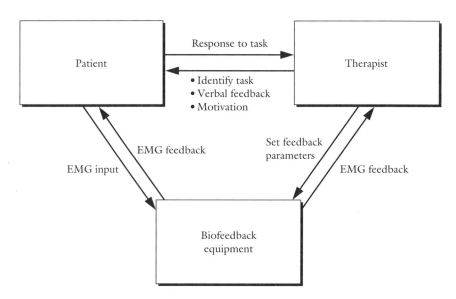

Figure 11.1 Schematic representation of flow of information between patient, therapist, and biofeedback equipment.

The **sensitivity, objectivity, accuracy,** and **quantitative nature** of the feedback signal are also major advantages of EMG biofeedback. Only with biofeedback can relatively subtle changes in the recruitment of muscles be detected. Small changes in motor unit recruitment are particularly difficult to detect with palpation or visual inspection when patients are contracting at either high or low force levels. Knowledge of these subtle changes, however, may be necessary to allow patients to make appropriate changes in recruitment. For example, if a muscle contains relatively few active motor units, due to either a peripheral or central nervous system problem, the recruitment may not be sufficient to produce any joint displacement. The activity in the muscle would thus be very difficult to detect or quantify by the clinician without EMG biofeedback. The use of manual detection and verbal feedback to train motor unit recruitment may not reflect the true changes in the recruitment. Even if small increases in motor unit recruitment are produced, clinician sensitivity may not detect the change, and appropriate positive feedback may not be provided. On the other hand, positive verbal reinforcement in the absence of additional recruitment during maximal effort also is not effective in recruitment training. In contrast, EMG biofeedback is sensitive enough to detect small changes in the level of recruitment and accurately reflects the actual level of recruitment. EMG biofeedback is also not biased by the subject's effort. The quantitative nature of EMG feedback clearly shows which efforts serve to increase recruitment and which efforts show less recruitment. The clinician can objectively observe which techniques or activities really help recruitment and when the patient is beginning to fatigue.

Modern feedback devices can provide a variety of **novel feedback signals** that can serve to motivate the client. These signals range from the "raw" or unprocessed visual and auditory EMG signal, to tones whose frequencies increase or decrease in proportion to the level of EMG activity, to computer-controlled images on a video display terminal. In addition, biofeedback devices can be used to turn on or off other electronic devices, such as radios or tape recorders, which can be used as positive reinforcements for young children.

Technical Considerations Regarding the Application of EMG Biofeedback

This section reviews the factors that determine the amplitude of the EMG, outlines the rationale for specific electrode selection and placement, discusses the method and purpose of each step in the processing of the EMG feedback signal, and presents the various methods of displaying the biofeedback signal and the advantages/indications for each.

The EMG is the recording of the electrical activity of the muscle membrane in response to the physiological activation of skeletal muscles. The amplitude of the EMG reflects the size and number of active motor units as well as the distance of the active muscle fibers from the recording electrodes. Although no direct information is contained within the EMG regarding the force or torque that a muscle produces, a nearly linear relationship does exist between the EMG and the force that a muscle produces under carefully controlled isometric conditions (2, 3). The clinician should be aware, however, that this linear relationship no longer holds when contractions change from isometric to nonisometric or as the muscle fatigues. Similarly, because the EMG only records from a limited area of a muscle, the EMG cannot be used to compare the strength of contraction across muscle groups or even within the same muscle if different electrode placement or types of electrodes are used (see below). To illustrate this concept, recording electrodes may be applied over the abductor digiti minimi muscle of one person and over the quadriceps femoris muscle of another. Depending on the electrode size and spacing, the electrical activity from the abductor of one subject's little finger may approximate the activity from the other subject's knee extensor during volitional activation despite the marked differences in force output between the two muscles.

In addition to physiological factors, such as the size and number of active motor units, the size of the recording area and the interelectrode distance of the recording electrodes also affect the amplitude of the EMG. The larger the recording area, the greater the volume of muscle that is monitored and hence the greater the EMG recorded. Similarly, the larger the interelectrode distance, the larger the volume of muscle that is monitored and the larger the EMG. Thus, to increase the specificity of the EMG recording electrodes, small recording areas and close interelectrode spacing could be used. The use of close spacing thus minimizes the recording of electrical activity from muscles other than the targeted muscle. This may be particularly helpful if the EMG from the targeted muscle is being contaminated by input from a muscle that is an antagonist of the targeted muscle. This phenomenon is termed *cross talk*. Subcutaneous recording electrodes, such as fine wire electrodes, are examples of small, closely spaced electrodes that allow precise localization from within the muscle. Subcutaneous electrodes also offer the advantage of being able to record from deep muscles without interference from more superficial muscles and show greater sensitivity than surface electrodes due to their proximity to the active muscle fibers. Skin, subcutaneous fat, and fascia all serve to attenuate the EMG recorded by surface electrodes. Nevertheless, inserted electrodes are rarely used with EMG biofeedback (4). Surface electrodes are much more convenient for the clinician, more acceptable to the patient, and produce much less movement artifact than subcutaneous electrodes. Movement artifact is the high-voltage, nonphysiological contamination of the EMG due to the physical perturbation of the electrodes, input cables, and wires. To minimize the recording from unwanted muscle groups, the spacing between the recording electrodes should be as small as is practicably possible. Interelectrode spacing of one to two centimeters is generally adequate.

Essentially, five steps are involved in the processing of the EMG feedback signal: amplification, filtering, rectification, integration, and level detection. The processes of amplification, filtering, and integration are discussed in Chapter 10. A schematic representation of the changes in the EMG biofeedback signal is shown in Figure 11.2. Most feedback devices allow the clinician to modify most of these processes. The **amplification, gain,** or **sensitivity** are all terms used to describe the relationship between the input and output voltages of the amplifier. The greater the amplification, the more sensitive the device. That is, with a high amplification, even very small EMG signals produce discernible changes in the output displayed to the patient. In general, the greatest sensitivity that does not saturate the output signal is used. When training a patient to increase recruitment, and given a choice of sensitivities from an output meter of 10, 100, or 1000 μV to produce full-scale deflection, if the patient has a maximum recruitment of 80 μV, then the best choice would be the 100-μV sensitivity.

Differential amplification is used by virtually all biofeedback devices. As described in Chapter 10, differential amplification requires the use of two

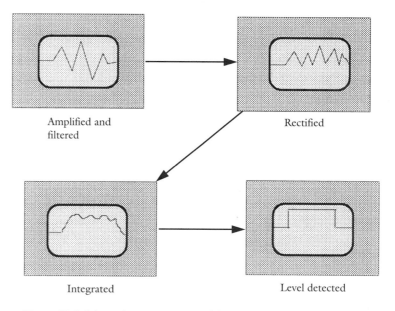

Amplified and
filtered

Rectified

Integrated

Level detected

Figure 11.2 Schematic representation of the processing of the EMG signal.

recording electrodes and a reference electrode. Also, the use of close spacing between the recording electrodes serves to help minimize the noise recorded and give the cleanest signal possible. Thus, particularly when attempting to record low levels of electrical activity, close spacing is used to minimize the noise that would be amplified.

The **filtering** characteristics of most feedback devices can be modified. By restricting the frequency range of the signal that is passed through the amplifier, we can attempt to reduce noise and make the recording more selective. Movement artifact tends to be low frequency (<100 Hz), and much of the electronic noise is high frequency (>1000 Hz). Because most of the EMG signal falls within the 100 to 1000 Hz range, this is the range most often used. However, if high-frequency noise is a problem, narrower frequency range may be required to eliminate more of the high-frequency signal (e.g., only pass signal between 100 to 500 Hz). Because the EMG actually includes a fairly wide frequency range, the disadvantage of using a narrower frequency band is that some of the EMG is lost when a narrower range is used. There are times, however, when some of the EMG is purposely eliminated. Because muscle attenuates high-frequency signals more than lower frequencies, the EMG signals from distant motor units are lower frequency than nearby motor units (5). Eliminating more of the low-frequency signal allows the amplifier to reduce the contribution made by distant motor units to the EMG. Surface electrodes therefore become more selective (record from a narrower area) if the lower limit of the frequency band passed is raised.

If the filtered output from a differential amplifier is fed into an audio speaker or oscilloscope, a **raw** EMG is displayed. This is the signal that is used to see the actual EMG or to listen for 60-Hz interference. Having access to the raw EMG is particularly helpful if there is a question of whether the processed feedback signal is of physiological origin or not. With modern amplifiers, even with surface electrodes, single motor unit potentials can easily be identified.

The next steps in the processing of the EMG are the **rectification** and **integration** of the signal. The signal needs to be full-wave rectified to be integrated (see Chapter 10 for an explanation). The integration of the signal involves the summing of the signal over some period of time. If a leaky capacitor is used to accomplish this task, what is seen is a smoothing of the signal, as shown in Figure 11.2. Other integrators can be made to sum over a period of time or until some preset maximum voltage is reached before the integrator is reset to zero. The rate at which the EMG sums and declines is a function of the time constant of the integrator. A short time constant will allow the integrated EMG to closely follow the peaks and valleys of the rectified signal. A longer time constant will produce much greater smoothing of the signal and require a longer time for the integrated signal to reach its peak and a longer time to relax back to baseline. An integrated signal is required to display anything other than the raw EMG.

Setting an appropriate time constant is important in producing an appropriate feedback signal. If the time constant is too short, the display (e.g., a digital or analog voltage meter) will fluctuate too rapidly (display jitter); little sense can be made from such an output. In contrast, a time constant that is too long will cause the display to lag behind the actual activity of the muscle. As an example, even if the subject relaxes, it may take several seconds for the display to return to zero. Neither of these situations is acceptable. An appropriate time constant will help to accurately reflect the overall state of activation of a muscle but will not show the wide and rapid fluctuations seen within the raw or rectified EMG. For most muscle training applications a time constant of approximately one-third of a second works well. Longer time constants are often used for general relaxation training in which the activity of a specific muscle (e.g., frontalis muscle) is being used to reflect the overall state of relaxation of the patient.

The last step in signal processing is the use of a **threshold detector** to determine if a preset level of integrated EMG activity has been met. The output of a threshold detector is a binary function, that is, "on" or "off." The logic of the output can be set to current the output, whatever it may be, to be on or off when the threshold is exceeded. For example, when training a young child with cerebral palsy to relax his plantarflexor muscle while standing at a table, the feedback can be set to allow an electric train to run as long as the EMG is below a preset threshold. Whenever the EMG exceeds this threshold the train can be made to stop. The logic would thus have been set to give an on signal whenever the EMG voltage was below threshold and an off signal whenever the voltage exceeded the threshold.

Many devices allow a combination of feedback signals. For instance, the output of the integrator may simultaneously be sent to a light meter display (i.e., a series of lights is turned on) and a threshold detector. The meter can provide continuous visual feedback and the output of the threshold detector can be used to trigger an audio signal. Thus, the audio signal can be turned on when the threshold is exceeded.

As already noted, the **feedback signals** can be raw or processed, auditory or visual, continuous or threshold-triggered. Within the limits of the available equipment, the clinician and patient must decide on the most appropriate signal. The raw output (i.e., amplified and filtered only) can give the experienced clinician considerable information regarding the source of the signal. That is, is the source truly physiological, or is it primarily noise that is being recorded? Other than identifying the peak voltages from an oscilloscope screen, the raw signal cannot be quantified. This is a limitation when attempting to objectively document progress or to identify targeted levels of recruitment for the client. When deciding to use an auditory (e.g., raw EMG, tone, or beep) or a visual display (e.g., digital meter or light bar) patient preference and other practical factors need to be considered. If a lower extremity muscle is being monitored in preparation for ambulation training, auditory feedback may be preferred because visual feedback is not practical during ambulation (i.e., the patient needs to watch where he or she is going). Similarly, during relaxation training most patients prefer auditory feedback because they may want to close their eyes to help them relax.

The use of a threshold is necessary whenever EMG levels are used to turn on or off another device, such as a radio or tape player. The use of an audio threshold during targeted or general relaxation is also generally preferred. Most patients find the audio signal annoying and unnecessary if they are able to relax below the target. Only when the activity exceeds that target does the patient need to be alerted.

As previously noted, more than one feedback signal can be used simultaneously, especially if more than one muscle group is monitored. When two muscle groups are monitored simultaneously (dual-channel monitoring), generally continous feedback is provided from one channel, whereas the other channel uses a threshold detector to "sound an alarm" only if the second muscle exceeds the threshold. This technique is commonly used when training for recruitment of one muscle and relaxation of its antagonist. As an example, to train for increased active finger extension from a patient who shows spasticity as the result of cerebral vascular accident (CVA), the finger flexor and extensor muscles may be simultaneously monitored. Continuous auditory and visual feedback to train for recruitment of the extensors could be provided while using a threshold detector to provide a separate auditory signal from the flexors. Only when the flexor activity exceeds a level that is believed to be interfering with finger extension would feedback from the flexors be provided to the patient.

Selection of Appropriate Patients for the Application of EMG Biofeedback

EMG biofeedback is one of the best-researched tools that is presently used in rehabilitation. Publications began appearing in the early 1960s supporting the use of EMG biofeedback in physical rehabilitation. The rate of publication reached its peak in the late 1970s and began to decline by the mid-1980s (6–8). Although the most thoroughly investigated application of EMG biofeedback involves the treatment of patients following CVAs (8,9), numerous reports exist for the treatment of a plethora of conditions, including spinal cord injury, cerebral palsy, spastic torticollis, peripheral nerve injuries, low back pain, and ligament injuries. A review of the clinical efficacy of each of these applications is beyond the scope of this chapter. Instead, the interested reader is referred to a number of related textbooks or review articles (1, 8).

The selection of appropriate patients for the application of EMG biofeedback basically involves answering the following three simple questions.

1. Does the patient demonstrate a motor impairment that would suggest that the information provided by the feedback would be of benefit?
2. Does the patient demonstrate the ability for voluntary control?
3. Is the patient sufficiently motivated and cognitively aware to utilize the feedback information?

One common concern of clinicians is the amount of time required to prepare the patient (i.e., prepare the skin and apply the electrodes) and administer the biofeedback "treatment." Thus, many clinicians who would agree that their patients would benefit from the information provided by EMG biofeedback are reluctant to use the modality. Given the present quality of the amplifiers and filters used in most biofeedback devices and the availability of disposable self-adhering electrodes, the time required for skin preparation and electrode application is minimum. In fact, recording an EMG from a subject may begin in as little as one minute after the individual is seated at a table. Furthermore, consistent with the perspective that biofeedback should be thought of as a tool and not an isolated treatment, specific training objectives can generally be reached faster with biofeedback than without. Admittedly, some additional, initial patient training is required to explain the purpose of the equipment. However, the information provided by most feedback signals is so intuitive to most patients that long or wordy explanations are usually not necessary. A simple demonstration using an uninvolved muscle of the patient is usually sufficient.

The advantages of this "faster learning" with the use of EMG biofeedback is most easily demonstrated in patients who have intact nervous systems yet, due to a prior injury or trauma, are having a difficult time either recruiting or relaxing a specific muscle. One common clinical problem for which EMG biofeedback has been suggested is the inability of patients to recruit their vastus medialis muscles following knee surgery (10, 11). Biofeedback can be very helpful in quickly training a patient to perform a "quad set" (i.e., isometric contraction of the quadriceps femoris muscle with the knee in full extension) as well as training for greater quadriceps femoris muscle recruitment during dynamic exercises.

In contrast, the use of EMG biofeedback with patients with impaired motor control due to central nervous system pathology is much more difficult to demonstrate. Although numerous clinical reports and studies have supported the use of EMG biofeedback as a helpful tool to assist in the rehabilitation of patients with motor impairment due to CNS pathology (8, 9, 12–17), EMG biofeedback is not a treatment that can cure patients with CNS pathology. EMG biofeedback can help patients reach their true potential, but there are physiological limitations that both patients and clinicians must be aware of.

For biofeedback training to be appropriate, the patient must have the potential to control the targeted muscle. The inappropriateness of the use of biofeedback training with patients with complete spinal cord injuries or complete peripheral nerve impairment prior to reinnervation by the peripheral nerve is obvious. Other conditions may not make the selection of appropriate patients so apparent. Wolf and Binder-Macleod (16) demonstrated that in a group of patients who had sustained CVAs at least one year prior to treatment, only patients who demonstrated voluntary finger extension prior to the initiation of therapy were able to show any improvement in hand function as a result of 60 sessions using EMG biofeedback. That is, none of the patients who were unable to perform active finger extension prior to commencement of training demonstrated any improvement as a result of treatment. This suggests that at least a minimum amount of voluntary control must be present for patients to be able to use biofeedback to improve their function. However, several of these patients who lacked even minimum finger extension did show improvement in shoulder, elbow, and wrist function.

In addition to having the ability to volitionally control a muscle, the patient must be motivated and have sufficient cognitive ability to learn to use the feedback signal. Training with the use of feedback is generally not a passive process; it requires the active participation of the patient. One exception is when the clinician uses the EMG for his or her own feedback to determine the effectiveness of a particular intervention. For instance, a clinician may use a Swiss ball to help reduce the tone in a young child with cerebral palsy. EMG biofeedback delivered to the therapist could be used to provide quantitative information to the clinician if the specific techniques being used are actually producing the desired responses.

Thus far only the appropriateness of EMG biofeedback has been discussed. Recently, other forms of feedback relevant to physical rehabilitation have been developed; these include the use of position and force feedback. In general, EMG biofeedback should be used when information regarding the activity of a specific muscle or muscle group is desired. As an example, if patients are very weak and little force or joint displacement is produced, position or force feedback would not be sufficiently sensitive to provide any meaningful information for these patients. EMG biofeedback is also generally most appropriate in situations where training specific muscles to relax while patients perform a particular task is desired. In contrast, the training of a specific muscle or muscle group may not be appropriate when the patient is trying to perform a task that requires the coordination of multiple muscle groups. For instance, when training a child with cerebral palsy to maintain proper head position, head position feedback would be much more helpful than EMG feedback from any specific muscle group. Similarly, in the training of patients to shift their weight either onto or off of an involved lower extremity, force feedback, providing the exact amount of weight bearing by the involved extremity, has been found to be most appropriate.

Development of Training Strategies during the Use of EMG Biofeedback

Although the information provided through the use of EMG biofeedback generally serves to motivate patients, because of its objective nature this information can also serve as a source of frustration. Clinicians are, therefore, encouraged to **consider all factors related to learning theory** when developing their specific training strategies. Positive is better than negative reinforcement when training patients. Obtainable short- and long-term goals must be clearly communicated to patients. Clinicians should listen to each patient to be certain that the established goals are important to him or her. Experience has shown that if patients are told to simply try their best, no matter how well they perform, they are always disappointed that they did not do better. In contrast, if specific tasks or goals are identified within and across sessions, then a real sense of accomplishment can be achieved. Tasks that demonstrate achievement of each goal must be specific enough so that the patient knows all of the relevant conditions, and the criteria must be specific enough so that the patient knows when the task is accomplished. One of the skills that the clinician must acquire is the setting of goals that are sufficiently difficult to truly challenge and motivate patients but are attainable within a specified time frame.

Several considerations need to be made regarding the **sequencing** or **progression** of any treatment program. As an example, when treating a patient who has sustained a CVA, clinicians must decide if it is better to train for relaxation of spastic muscles prior to training recruitment of a weak antagonist or if training should begin directly with weak or poorly recruited muscles.

Without the use of EMG biofeedback, exercises to train spastic muscles to relax are difficult to design and evaluate. As previously noted, the addition of EMG biofeedback makes the monitoring and training of spastic muscles much more objective and straightforward. For this reason, when using EMG biofeedback to train patients with disturbances in muscle tone, treatments have traditionally begun with training for relaxation of spastic muscles before working on recruitment of weak antagonist muscles (12–15). Recently, however, the need for targeted relaxation training has been questioned (18, 19). Similar decisions regarding the progression of training need to be made concerning the choices to train *(a)* proximal muscles first and to then progress distally or to begin distally and progress proximally, *(b)* stability first and then progress to mobility training or reverse this order, or *(c)* component movements first and then integrate the components into a functional movement pattern or to commence training with functional movement patterns. These, as well as other choices, need to be made by clinicians based on their own treatment philosophy and as objective research findings support various approaches to treatment.

The use of biofeedback requires several additional considerations regarding the progression of training. Should only one muscle group be monitored or should a dual-channel system be used? When should the patient be weaned from using the feedback signal? After all, the goal of training is the performance of functional tasks without the use of biofeedback. Thus, the benefits of training with feedback need to be weighed against the long-term need to perform without feedback. One option would be to begin with a continuous feedback signal and progress to the use of some form of threshold feedback in an attempt to wean the patient from the need for any feedback.

What level of success should the patient demonstrate before increasing the level of difficulty? That is, does a patient have to reach a targeted level of recruitment 100% or 50% of the time before we raise the targeted microvolt level that the patient is to achieve? These questions must be decided by the clinician during each training session. Unfortunately, little objective information is presently available to help clinicians answer these and other relevant questions.

The final strategy that will be considered is the **selection of appropriate sites for electrode placement.** To record an EMG, the recording electrodes must be placed over or near the belly of the relevant muscle. In contrast, the placement of the reference electrode is not so critical. Some workers in this field have suggested that the reference electrode be placed equidistant from the two recording electrodes; however, the exact placement is not critical as long as good contact between the skin and electrode is maintained (4). Nevertheless, a number of factors, including goals of training, level of control, available muscle mass, subcutaneous fat, movement artifact, and cross talk, must be considered when selecting the electrode sites. All of these factors interact, so it is impossible to determine the optimal electrode sites without considering all of them. As previously noted, the smaller the distance between the recording electrodes, the greater the specificity of the recording and the less noise that is recorded. Also, the greater the distance of the recording electrodes to the active muscle, the greater the attenuation of the EMG. Thus, when

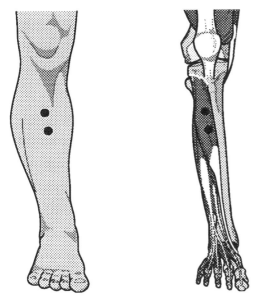

Figure 11.3 Recommended placement for recording from the anterior tibialis muscle.

placing electrodes over a muscle, the following requirements must be met:

1. Areas that have a thickened layer of adipose tissue must be avoided.
2. The distance between the recording electrodes and any muscles that are producing unwanted electrical activity (i.e., cross talk) must be maximized.
3. The smallest interelectrode distance that is practical must be used.

As an example, if the clinician wants to record from the anterior tibialis muscle, the best placement may be to have the recording electrodes less than one centimeter apart and over the most medial aspect of the muscle (see Fig. 11.3). This placement puts the recording electrodes over the targeted muscle, while still being as far away as possible from other active muscles that may contaminate the intended feedback signal.

Also, the electrodes should be placed over the muscle when the limb is in the position that it will assume when the patient is performing the exercise. If a patient supinates his or her forearm while electrodes are placed over the forearm flexor muscle mass but then pronates his or her forearm during training, the electrodes may no longer be lying over the flexors; rather, the electrodes may now be over the brachioradialis muscle. In addition, electrodes and unshielded lead wires should be placed in a position so that they will not be jostled during training. This prevents movement artifact from contaminating the feedback signal.

Within limits, if a patient has poor control over a muscle, the interelectrode distance can be used to advantage by sampling a larger or smaller area of the muscle. If a patient has difficulty recruiting from any of the heads of his or her quadriceps femoris muscle, training may begin using a relatively wide spacing to sample from a large portion of the muscle. However, care should be taken that the spacing is not so wide that activity from the hip adductor or hamstring muscles is erroneously fed back to the patient. As the patient's control increases, closer spacing may be used to monitor individual heads of the quadriceps femoris muscle. In contrast, if a patient displays spasticity and the goal is to train for relaxation of his or her finger and wrist flexors during passive stretch of the muscle to maintain range of motion, training should begin with electrodes that are relatively closely spaced, so as to limit the recording area. As the patient gains better control, a slightly wider spacing could be employed to sample more of the forearm flexor muscle mass.

Case Studies

Case 1

The patient is a 35-year-old male from India who contracted poliomyelitis at age 7. He has never received physical therapy. He now has severe foot-drop on the right side and wears a short leg brace (SLB). He is highly motivated and would like to strengthen his ankle dorsiflexors to shed his brace. Electrodiagnostic testing reveals several small motor units present in his anterior tibialis and extensor digitorum longus muscles. No visible contraction of any of his ankle dorsiflexors can be observed.

Assessment: Although the patient appears to be a good candidate for using EMG biofeedback to help increase motor unit recruitment, the probability for success is limited.

Plan:

1. Initial treatment in the PT clinic using EMG biofeedback to work on increased motor unit recruitment

2. Assess progress and evaluate patient for use of portable EMG biofeedback for independent home training

3. After visible contractions can be produced, begin on resistive strengthening program

Detailed Treatment Plan:

Mode of feedback: Initially provide both auditory and visual EMG biofeedback. Make the transition to auditory feedback prior to gait training with the feedback.

Short-term goal of training: To increase EMG from targeted muscles. Will set specific targeted levels of recruitment to encourage an increase in the discharge rate of already active motor units and to attempt the recruitment of additional motor units.

Electrode placement: Over the ankle dorsiflexor muscles. Moderately wide interelectrode distance. Could monitor plantarflexors to determine if cross talk is a problem.

Duration of treatment: To patient tolerance. When EMG recruitment levels begin to decline, the patient is fatiguing. Allow patient to rest. When recovery from fatigue is incomplete, terminate treatment.

Case 2

The client is a 56-year-old female who has been referred to PT for ROM exercises for her right upper extremity. The client sustained a Colles's fracture of her right wrist approximately 6 months ago and maintained her entire right arm nearly totally immobilized for the first 3 months while in her cast. Following removal of her cast, the client presented with marked limitation in all active movements of her right shoulder, elbow, forearm, and wrist. The patient was then briefly instructed in an exercise program and followed by her surgeon. Due to a lack of progress in ROM, the client underwent a closed manipulation under anesthesia 2 weeks ago. The surgeon's report indicates that the client was able to achieve nearly full passive ROM in all joints.

The client is presently alert, pleasant, and cooperative, though obviously quite apprehensive. As you begin your evaluation, you note marked splinting at all joints during all active or passive movements.

Assessment: Patient appears to be good candidate for EMG biofeedback for targeted muscle training.

Plan:

1. EMG biofeedback to promote relaxation and decrease splinting during passive ROM to all affected joints

2. Compare EMG from involved and uninvolved upper extremities during active ROM exercises to assess recruitment pattern

3. Train appropriate muscle groups to produce more "normal" recruitment levels during active ROM

4. Progress to functional training using EMG biofeedback to help normalize movements

Detailed Treatment Plan:

Mode of feedback: Initially provide both auditory and visual EMG biofeedback. Then have patient use signal that is most effective.

Short-term goal of training: To have patient display similar recruitment patterns from comparable involved and uninvolved muscle groups during passive and active movements. Will set specific targeted levels of recruitment to train for either relaxation or recruitment.

Electrode placement: Over the targeted muscles. Narrow interelectrode distance to increase recording specificity.

Duration of treatment: Use EMG to indicate patient tolerance. When patient begins to show decreasing ability to recruit or relax muscles, terminate treatment to prevent patient from becoming frustrated.

Summary

The EMG can be used by both the clinician and patient to help provide information regarding the activation state of a muscle. EMG biofeedback is a valuable tool whose use should be considered by clinicians whenever a patient displays poor volitional motor control. Although a number of technical and practical considerations need to be taken into account when using EMG biofeedback, this tool can easily be incorporated into most traditional treatment approaches.

Study Questions

1. Define biofeedback and discuss its advantages over simple verbal feedback.

2. Identify the specific characteristics of the client that would suggest an appropriate use of EMG biofeedback.

3. Review the physiological factors that determine the amplitude of the raw EMG.

4. Identify the relationship between both the size and interelectrode distance of the recording electrodes to the amplitude and specificity of the EMG.

5. List each step in the processing of the EMG feedback signal.

6. Discuss the advantages and limitations of using the raw EMG during biofeedback training.

7. Identify the factors that need to be considered in selection of the appropriate sites for electrode placement.

References

1. Basmajian JV. Introduction. Principles and background. In: Basmajian JV, ed. Biofeedback: principles and practices for clinicians. 3rd ed. Baltimore: Williams & Wilkins, 1989:1-4.

2. Lippold OCJ. The relationship between integrated action potential in a human muscle and its isometric tension. J Physiol 1952;117:492-499.

3. Bigland B, Lippold OCJ. The relation between force, velocity and integrated electrical activity in human muscles. J Physiol 1954;123:214-224.

4. Basmajian JV, Blumenstein R. Electrode placement in electromyographic biofeedback. In: Basmajian JV, ed. Biofeedback: principles and practices for clinicians. 3rd ed. Baltimore: Williams & Wilkins, 1989:369-382.

5. Clamann HP, Lamb RL. A simple circuit for filtering single motor unit action potentials for electrograms. Physiol Behav 1976;17:149-151.

6. Hatch JP, Saito I. Growth and development of biofeedback: a bibliographic update. Biofeedback Self Regul 1990;15:37-46.

7. Hatch JP, Saito I. Declining rates of publications within the field of biofeedback continue: 1988-1991. Biofeedback Self Regul 1993;18:174.

8. Wolf SL. Electromyographic biofeedback applications to stroke patients: a critical review. Phys Ther 1983;63:1448-1459.

9. Basmajian JV. Research foundations of EMG biofeedback in rehabilitation. Biofeedback Self Regul 1988;13:275-298.

10. Draper V. Electromyographic biofeedback and recovery of quadriceps femoris muscle function following anterior cruciate ligament reconstruction. Phys Ther 1990;70:11-17.

11. Krebs DE. Clinical electromyographic feedback following meniscectomy. A multiple regression experimental analysis. Phys Ther 1981;61:1017-1021.

12. Brudny J, Korein J, Grynbaum BB, Sachs-Frankel G. Sensory feedback therapy in patients with brain insult. Scand J Rehabil Med 1977; 9:155-163.

13. Kelly JL, Baker MP, Wolf SL. Procedures for EMG biofeedback training in involved upper extremities of hemiplegic patients. Phys Ther 1979;59:1501-1507.

14. Baker MP, Regenos E, Wolf SL. Developing strategies for biofeedback: applications in neurologically handicapped patients. Phys Ther 1977;57:402-408.

15. Binder SA, Moll CB, Wolf SL. Evaluation of EMG biofeedback as an adjunct to therapeutic exercise in treating the lower extremities of hemiplegic patients. Phys Ther 1981;61:886-893.

16. Wolf SL, Binder-Macleod SA. Electromyographic biofeedback applications to the hemiplegic patient: changes in upper extremity neuromuscular and functional status. Phys Ther 1983;63:1393-1403.

17. Wolf SL, Binder-Macleod SA. Electromyographic biofeedback applications to the hemiplegic patient: changes in lower extremity neuromuscular and functional status. Phys Ther 1983;63:1404-1413.

18. Wolf SL, Catlin PA, Blanton S, Edelman J, Lehrer N, Schroeder D. Overcoming limitations in elbow movement in the presence of antagonist hyperactivity. Phys Ther 1994;74;826-835.

19. Gowland C. de Bruin H, Basmajian JV, Plews N, Burcea I. Agonist and antagonist activity during voluntary upper-limb movement in patients with stroke. Phys Ther 1992;72:624-633.

Appendix A

Peripheral Neuroanatomy of the Upper and Lower Extremities

Andrew J. Robinson

Innervation of the Upper Extremity

$\bigvee\!\!\bigvee\!\!\bigvee$ The sensory and motor innervation of the upper extremities and shoulder girdle in humans typically arises from the anterior primary rami of four cervical spine levels and one thorasic spine level (C5–T1). As these spinal nerves exit the intervertebral foramina, nerve axons unite, separate, and recombine to form the complex meshwork of nervous tissue called the **brachial plexus** (Fig A.1). The brachial plexus is divided into trunks, divisions, and cords as described below. The peripheral nerves that innervate the upper extremity and shoulder girdle branch from either the spinal nerve roots or some part of the brachial plexus. The C5–6 nerve roots unite to form the **upper trunk,** the C7 nerve root forms the **middle trunk,** and the C8-T1 nerve roots unite to form the **lower trunk.** At times, the C4 nerve root may contribute to the upper trunk and the T2 nerve root may contribute to the lower trunk. Each trunk of the brachial plexus subsequently divides into **anterior** and **posterior divisions,** which then go on to form the three cords of the brachial plexus.

The anterior divisions of the upper and middle trunk form the **lateral cord,** and the anterior division of the lower trunk forms the **medial cord.** The posterior divisions of the upper, middle, and lower trunks contribute to the formation of the **posterior cord.** The medial cord does not usually receive contributions from the C7 nerve root or middle trunk.

Although most peripheral nerves are branches from the cords of the brachial plexus, several peripheral nerves branch directly from nerve roots or trunks before the nerve fibers form the anterior and posterior divisions. The **long thoracic nerve,** which innervates the serratus anterior muscle, is formed by the C5–C7 nerve roots just after they exit the intervertebral foramen. The **dorsal scapular nerve,** which innervates the rhomboid musculature, originates form the C5 root. The **suprascapular nerve** originates form either the C5 root or the upper trunk and innervates the supraspinatus and infraspinatus muscles. The **subclavian nerve** branches from the upper trunk and innervates the subclavius.

The posterior cord is the origin of several peripheral nerves, including the **upper** and **lower subscapular,** the **thoracodorsal,** the **axillary,** and the **radial nerve.** The lower subscapular innervates the subscapularis and teres major. The thoracodorsal nerve innervates the latissimus dorsi. The axillary nerve branches, supplying the deltoid and teres minor muscles. The terminal continuation of the posterior cord is the radial nerve.

The first nerve branching from the lateral cord of the brachial plexus is the **lateral pectoral nerve.** Emanating from the lateral cord more distally are the **musculocutaneous nerve** (Fig. A.4) and a branch of the cord that contributes fibers to the formation of the median nerve. After innervating the brachialis and biceps brachii muscles, the musculocutaneous nerve contin-

---

Adapted from Eddy JG, Snyder-Mackler L. Clinical electrophysiologic testing. In: Snyder-Mackler L and Robinson AJ, eds, *Clinical electrophysiology: electrotherapy and electrophysiologic testing.* Baltimore: Williams & Wilkins, 1989:261-292.

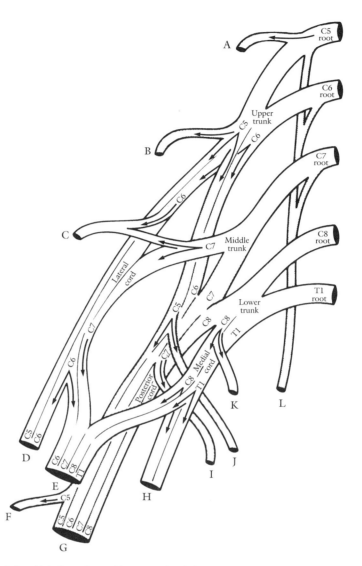

Figure A.1 Brachial plexus formed by cervical and thoracic nerve roots and divided into trunks, divisions, and cords. Peripheral nerves that innervate the shoulder girdle branch from every level of the brachial plexus. *A,* dorsal scapular; *B,* suprascapular; *C,* lateral pectoral; *D,* musculocutaneous; *E,* median; *F,* axillary; *G,* radial; *H,* ulnar; *I,* thoracodorsal; *J,* subscapular; *K,* medial pectoral; *L,* long thoracic. (Reprinted with permission from Kimura J, Electrodiagnosis of diseases of nerve and muscle. Philadelphia: F.A. Davis, 1989.)

ues as the **lateral antebrachial cutaneous nerve,** innervating the skin on the ventral forearm to the level of the hand. The lateral cord component of the median nerve contributes fibers originating from C6–7 nerve root levels.

The **medial pectoral nerve** arises from the medial cord and supplies the pectoralis major and minor muscles. Also arising from the medial cord are the **medial brachial** and **antebrachial nerves,** which supply sensation to the medial arm and forearm to the level of the wrist. Another major branch of the medial cord forms the ulnar nerve, composed of fibers originating from the C8–T1 nerve roots. The most distal branch from the medial cord joins with fibers from the lateral cord to form the complete median nerve.

All of the muscles innervated by the **median nerve** are below the elbow (Fig. A.2). In the forearm after innervating and passing through the pronator teres muscle, the median nerve sends off a motor branch called the **anterior interosseous nerve,** which innervates the flexor digitorum profundus I and II, flexor pollicis longus, and pronator quadratus muscles. The main trunk supplies innervation to the superficial musculature (flexor digitorum superficialis, flexor carpi radialis, palmaris longus) and continues into the hand. Pronators of the arm are innervated by the lateral portion of the median nerve form the C6–C7 nerve roots to the pronator teres, and the medial portion of the median nerve through the C8 nerve root to the pronator quadratus. In the hand, the median nerve innervates the muscles of the thenar eminence (abductor pollicis brevis, flexor pollicis brevis, and opponens pollicis) as well as the first and second lumbricals.

The median nerve sends sensory branches only to the hand. The median nerve passing through the carpal tunnel has already divided into cutaneous and muscular branches to supply the abductor pollicis brevis, opponens pollicis, superficial head of flexor pollicis brevis, and the first two lumbricales and supply sensation to the first through third digits and the dorsal distal interphalangeal joint. Sensation to the lateral palm is via the median palmar cutaneous nerve, which has its origins above the wrist and is external to the carpal tunnel.

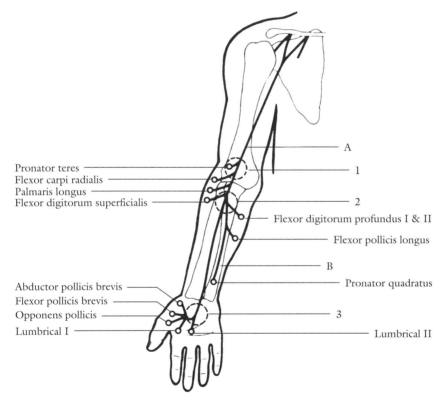

Figure A.2 Muscular innervation by the median nerve *(A)* and branching anterior interosseous *(B)* along with common sites of nerve compression *(dashed circles)* including *(1)* within the pronator teres, *(2)* at the branch point for anterior interosseous nerve, and *(3)* at the carpal tunnel. (Reprinted with permission from Kimura J, Electrodiagnosis of diseases of nerve and muscle. Philadelphia: F.A. Davis, 1989.)

The **radial nerve** supplies both sensation and muscle innervation to the posterior arm and forearm (Fig. A.3), with the exception of the lateral dorsal aspect of the upper arm, which is supplied by the sensory branch of the axillary nerve. The posterior brachial cutaneous nerve arises in the axilla along with muscular branches to the medial and long heads of the triceps. At the level of the spiral groove, muscular innervation is given to the medial and lateral heads of the triceps with a small branch continuing below the elbow to innervate the anconeous muscle. In addition, the posterior antebrachial cutaneous nerve arises, supplying sensation to the dorsal aspect of the forearm to the wrist. Before the radial nerve crosses the axis of the elbow joint it gives off branches to the brachioradialis, extensor carpi radialis, and brachialis muscles. Remember that the brachialis is innervated mostly by the musculocutaneous nerve. In the forearm, the radial nerve divides into a superficial branch (which supplies skin over the dorsal lateral aspect of the lower forearm and wrist) and a muscular branch, the **posterior interosseus nerve,** supplying the extensor muscles of the forearm.

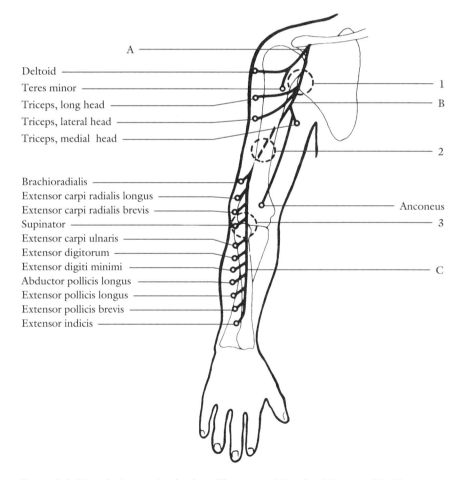

Figure A.3 Muscular innervation by the axillary nerve *(A)* and radial nerve *(B)* with common sites of injury *(dashed circles)* at *(1)* spiral groove and *(2)* elbow. (Reprinted with permission from Kimura J, Electrodiagnosis of diseases of nerve and muscle. Philadelphia: F.A. Davis, 1989.)

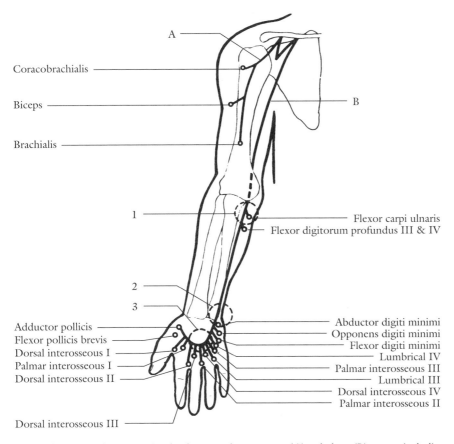

Figure A.4 Muscular innervation by the musculocutaneous *(A)* and ulnar *(B)* nerves, including common sites of compression or injury at *(1)* ulnar groove, *(2)* Guyon's canal, and *(3)* palm. (Reprinted with permission from Kimura J, Electrodiagnosis of diseases of nerve and muscle. Philadelphia: F.A. Davis, 1989.)

The **ulnar nerve** is formed by the roots of C8–T1. Occasionally there may be a small branch from C7 that contributes via the medial cord to supply the flexor carpi ulnaris muscle. The ulnar nerve alone supplies sensation to the ventral and dorsal medial hand.

The classic division of ulnar-median cutaneous innervation is that the ulnar supplies sensory innervation to the skin of the fifth and the medial aspect of the fourth digits on both volar and dorsal surfaces, and the median nerve supplies sensory branches to the lateral palmar aspect of the fourth digit and palmar surface of the other digits. The radial nerve supplies cutaneous sensation on the dorsal aspect of digits 1–3.

As the ulnar nerve (Fig. A.4) approaches the elbow, a motor branch is provided to the flexor carpi ulnaris and then passes through the ulnar groove to dive between the heads of the flexor carpi ulnaris, a common point of entrapment. Above the wrist, the **ulnar dorsal cutaneous nerve** arises. It is external to the tunnel of Guyon and innervates the dorsal medial aspect of the hand. The **ulnar palmar cutaneous nerve** also arises within the forearm

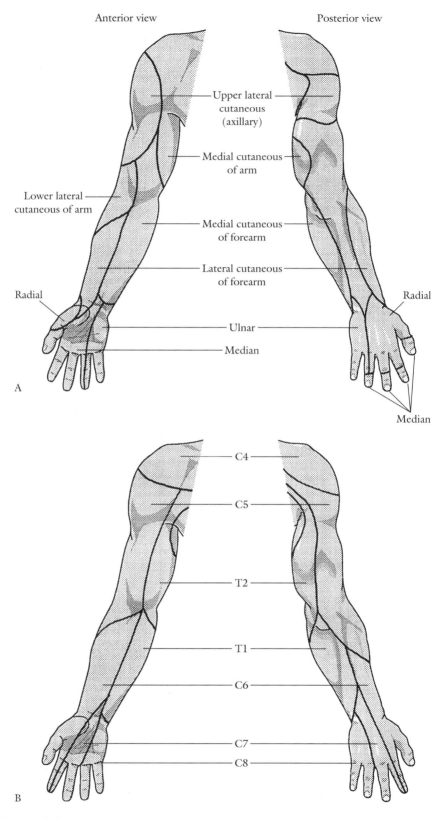

Figure A.5 Sensory innervation of the upper extremity. **A,** Peripheral nerve dermatomes. **B,** Cervical nerve root dermatomes.

to supply the volar aspect of the palm and is external to the tunnel of Guyon. As the ulnar nerve passes through the tunnel it divides into a superficial and a deep branch. The superficial branch supplies the hypothenar eminence, and the deep branch supplies the interossei and adductor pollicis brevis muscles.

Fig. A.5 shows the sensory innervation of the upper extremities by individual peripheral nerves and by spinal nerve root level.

Innervation of the Lower Extremity

The innervation of the lower extremities is provided by nerve fibers derived from the second lumbar to second sacral levels of the spinal cord. The spinal cord nerve roots form these levels combine to form the lumbar (Fig. A.6) and sarcal plexuses (Fig. A.7), which is composed primarily of the **anterior rami divisions of L2–S2** nerve roots. These spinal nerve roots divide into anterior and posterior components. The anterior components of L_2–L_4 unite to form the **obturator nerve,** and the posterior components of L_2–L_4 form the **femoral nerve** (Fig. A.8). Fibers from the posterior component of L2–L3 also form the **lateral femoral cutaneous nerve** to supply sensation to the skin on the anterolateral thigh from the hip to the knee. The L_4–S_2 roots unite to form the **sciatic nerve** (Fig. A.9), composed of the anterior components (**posterior tibial division**) and the posterior components (**peroneal division**). Also, from the posterior components of L_4–S_1 the **superior gluteal nerve** arises, and from L_5–S_2, the **inferior gluteal nerve.** From the anterior and posterior components of S_1–S_3, the **posterior femoral cutaneous nerve** arises, innervating the skin on the posterior thigh to the level of the knee.

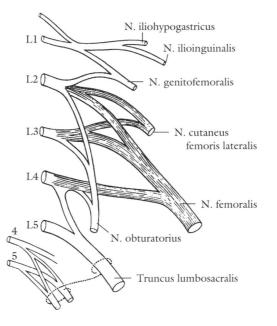

Figure A.6 Components of the lumbar plexus with the major terminal branch the femoral nerve. (Reprinted with permission from Kimura J, Electrodiagnosis of diseases of nerve and muscle. Philadelphia: F.A. Davis, 1989.)

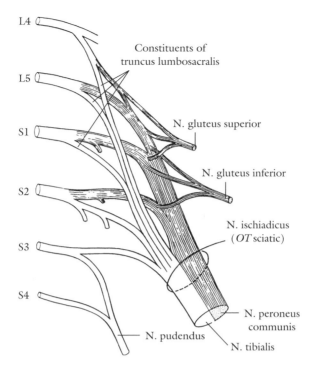

Constituents of
truncus lumbosacralis

N. gluteus superior

N. gluteus inferior

N. ischiadicus
(*OT* sciatic)

N. peroneus
communis

N. pudendus

N. tibialis

Figure A.7 Components of the sacral plexus with the major terminal branch the sciatic nerve. (Reprinted with permission from Kimura J, Electrodiagnosis of diseases of nerve and muscle. Philadelphia: F.A. Davis, 1989.)

The **obturator nerve** supplies the adductor muscles of the thigh after exiting from the pelvis. If present, an accessory obturator nerve supplies the pectineous and the obturator supplies the other thigh adductors. The femoral nerve, after branching to the iliopsoas muscles, enters the thigh under the inguinal ligament. It innervates the pectineous and sartorius muscles and divides into sensory and other motor branches. The nerve innervates the quadriceps musculature and supplies the skin via the anterior femoral cutaneous nerve on the anteromedial thigh to just below the knee. The saphenous branch of the femoral nerve, arising as it exits from the inguinal ligament, lies deep in the thigh and supplies the skin on the medial aspect of the lower leg from the knee to the ankle.

The **sciatic nerve** separates into two divisions and has no cutaneous supply above the knee. The posterior tibial division innervates all of the hamstrings except the short head of the biceps femoris, which is innervated by the peroneal division. At the knee, the **posterior tibial nerve** supplies branches to the medial and lateral gastrocnemius, soleus, and other smaller muscle. As the nerve continues into the lower leg, it branches to the posterior tibialis, flexor digitorum longus, and flexor hallicus longus muscles. It gives an additional branch to the soleus in the lower leg. A branch unites with a branch from the peroneal division to form the **sural nerve,** innervating the skin on the lateral foot.

The continuation of the posterior tibial nerve, after giving off muscular branches in the lower leg, winds around under the medial malleolus and passes through the tarsal tunnel. The **medial plantar nerve** supplies the abductor hallicus brevis, the flexor hallicus brevis, and first lumbrical muscles. It then supplies the skin on the medial plantar aspect of the foot and first three digits.

The **lateral plantar nerve** branches to the quadratus plantae and abductor digiti minimi quinti muscles and then divides into a superficial and a deep component. The superficial component supplies the lateral interossei and flexor digiti quinti muscles and the skin on the lateral plantar aspect of the foot. The deep branch supplies all other deep muscles of the foot.

The peroneal division of the sciatic nerve, after branching to the short head of the biceps femoris muscle, supplies the skin over the upper anterolateral leg and then divides into superficial and deep branches after passing around the neck of the fibula. The **superficial peroneal nerve** innervates the peroneus longus and brevis and then continues as the medial dorsal and intermediate

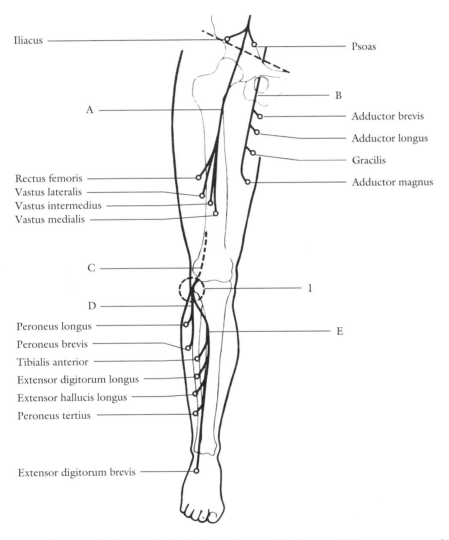

Figure A.8 Muscular innervation by the femoral nerve *(A)*, obturator *(B)*, common peroneal *(C)*, superficial peroneal *(D)*, and deep peroneal *(E)* nerves, with common site of compression *(dashed circles)* at the head of the fibula *(1)*. (Reprinted with permission from Kimura J, Electrodiagnosis of diseases of nerve and muscle. Philadelphia: F.A. Davis, 1989.)

dorsal cutaneous nerves. These supply the lower lateral leg and dorsal surface of the foot with the exceptions of the lateral fifth digit (sural) and the web space between the first and second digit (deep peroneal).

The **deep peroneal nerve** gives branches to the anterior tibialis, extensor digitorum longus, extensor hallicus longus, and possibility peroneus tertius muscles. In the foot, it supplies the extensor digitorum brevis muscle and then terminates in the first dorsal interosseous muscle and supplies the skin over the web space of the foot. The more proximal nerves of the hip are the superior gluteal nerves supplying the gluteus minimus and gluteus medius muscles and terminating in the tensor fascia lata muscle. The inferior gluteal nerve supplies the gluteus maximus muscle.

Figure A.10 shows the sensory innervation of the lower extremities by individual perpheral nerves and by spinal nerve root level.

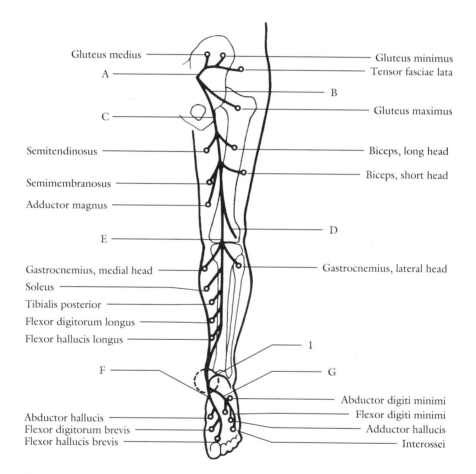

Figure A.9 Muscular innervation by the superior gluteal *(A)*, inferior gluteal *(B)*, sciatic nerve *(C)*, common peroneal *(D)*, tibial *(E)*, medial plantar *(F)*, and lateral plantar *(G)*, nerves. Also shown is the common site of compression *(1)* for the tibial nerve at the tarsal tunnel. (Reprinted with permission from Kimura J, Electrodiagnosis of diseases of nerve and muscle. Philadelphia: F.A. Davis, 1989.)

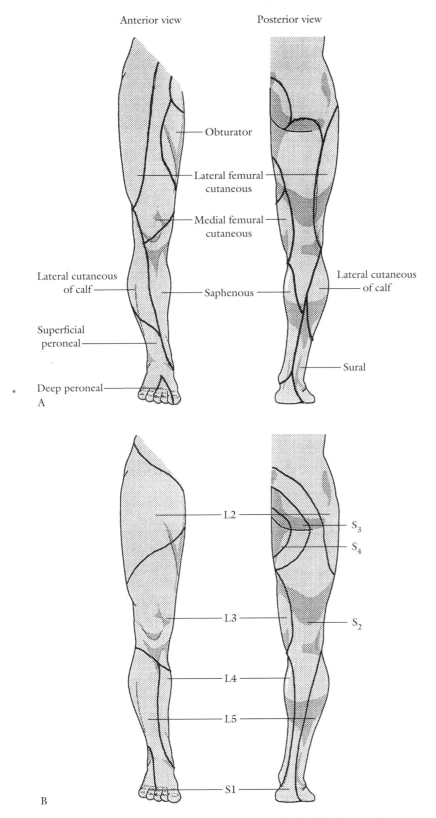

Anterior view

Posterior view

Obturator

Lateral femural
cutaneous

Medial femural
cutaneous

Lateral cutaneous
of calf

Lateral cutaneous
of calf

Saphenous

Superficial
peroneal

Sural

Deep peroneal
A

L2

S_3

S_4

L3

S_2

L4

L5

S1

B

Figure A.10 Sensory innervation of the lower extremity. **A,** Peripheral nerve dermatomes, **B,** Cervical nerve root dermatomes.

Appendix B

Answers to Study Questions

Chapter 1

1. (a) voltage
   (b) electromotive force
   (c) electrical potential difference

2. current

3. resistance

4. impedance

5. (a) cathode
   (b) anode

6. (a) cations
   (b) anions

7. anode cathode

8. (a) current
   (b) voltage
   (c) resistance

9. current

10. increase

11. (a) direct current (DC)
    (b) alternating current (AC)
    (c) pulsed current (PC)

12. waveform

13. a. ampere (amp)
    b. volt
    c. ohm
    d. farad
    e. ohm
    f. mho
    g. pulses per second (pps)
    h. hertz (Hz), or cycles per second
    i. ampere (mA or μA)
    j. millisecond (msec) or microsecond (μsec)
    k. coulomb
    l. seconds

14. a. rectangular, symmetric alternating current
    b. rectangular, unbalanced, asymmetric, biphasic pulsed current
    c. sinusoidal, unbalanced, asymmetric, biphasic pulsed current
    d. twin-spike, monophasic pulsed current

15.

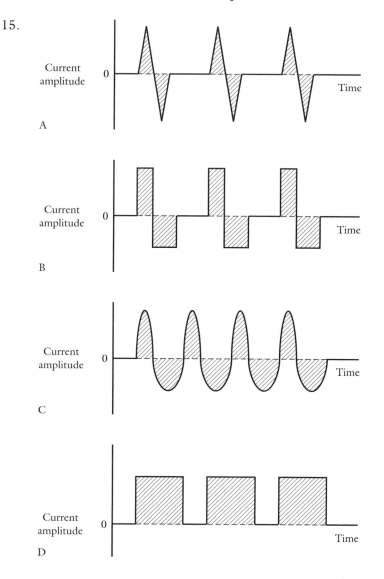

16. a. phase duration
    b. pulse duration
    c. interpulse interval
    d. rise time
    e. decay time
    f. pulse charge
    g. phase charge

17. 25%

Chapter 2

1. The power supply of the stimulator is either a direct current source or an alternating current source. DC power supplies provide low voltages

(1.5 V–9.0 V), most commonly from batteries that chemically store electrical charge. Conventional AC power supplies provide 115 V, 60 Hz, sinusoidal, symmetric, alternating current or "line current" available from outlets in most home and commercial wiring systems.

2. amplitude control(s)

3. waveform generators

4.   a. allows selection of the waveform shape (rectangular, sinusoidal, etc.)

   b. controls the number of seconds over which the amplitude of stimulation is automatically increased (or decreased) to a preset amplitude.

   c. allows selection of the number of seconds a pattern of stimulation is passed to a patient and the number of seconds during which stimulation ceases.

   d. allows selection of the time period (msec or μsec) from the onset of a single pulse to the end of that pulse.

   e. allows selection of the number of pulses per second to be delivered in continuous or interrupted trains of pulses.

   f. allows selection of the number of bursts (brief trains of pulses) per second to be delivered.

   g. allows selection and adjustment of the peak (or peak-to-peak) magnitude of stimulus waveforms.

   h. allows selection of relative timing of stimulation on two or more stimulation channels; usually allows user to set two channels of stimulation on and off simultaneously or alternately on and off.

5.   a. pulse amplitude controls for each of the two separate channels available for stimulation; regulates amplitude of pulses produced

   b. on time/off time controls; allows user to adjust duration of stimulation and duration of no stimulation as stimulator automatically cycles on and off

   c. ramp time control to adjust the number of seconds over which the amplitude rises to a preset level

   d. frequency control; adjusts pulse frequency for both channels

   e. pulse duration control; adjusts the time from the onset to the end of each pulse delivered

   f. waveform selector; allows selection of either rectangular monophasic pulsed current, rectangular symmetric biphasic pulsed current, or rectangular unbalanced, asymmetric biphasic pulsed current

   g. Channel output controls; for selection of both channels producing continuous train of pulses without interruption (continuous), both channels cycling on and off simultaneously (synchronous), or output channels cycling on and off alternately (reciprocal).

6. Surface electrodes serve as the interface between the electrical stimulator and the subject so that driving forces produced by the stimulator may produce movement of charged particles in the tissues. Electrodes differ in construction, size, shape, impedance, and durability.

7. Two

8. Monopolar electrode orientations: one electrode of a single stimulating circuit placed in the region where a therapeutic effect is desired, with the second electrode of the circuit positioned away from the target region.

Bipolar orientation: both surface electrodes from a stimulator channel are placed over the target area.

Quadripolar electrode orientation: both electrodes from two separate stimulating circuits are positioned in the primary target area for stimulation.

Refer to Figures 2.13–2.15 for diagrams of these electrode configurations.

9.

Characteristics	TENS device	NMES device
no. of output channels	two	two
no. of amplitude controls	two	two
on time/off time controls	no	yes
frequency controls	yes	yes
pulse duration controls	yes	sometimes
synchronous/reciprocal modes	rarely	yes
ramp modulation control	automatic	yes
frequency modulation	usually	no
pulse duration modulation	occasionally	no

10. sinusoidal, symmetric alternating current

11. Iontophoresis devices produce direct current and usually have only amplitude controls and a treatment timer. In addition, a meter is generally included to monitor stimulator output.

12. The two-prong supply plug does not connect the electrical device chassis directly to ground as is the case with the three-prong plug, which should be used with all electrical instruments.

13. A ground fault is the passage of current to ground by a pathway other than through power supply lines. A GFI monitors the current carried to an electrical device and the current returning from the device. If the supply and return currents differ by more than 3–5 mA, the GFI trips and stops power supply to the device.

14. a. all switches, dials, push buttons, pressure-sensitive switches, and other types of stimulator controls

   b. meters, auditory and visual signals

   c. connections at stimulator or electrode interface

   d. wires, broken-lead insulation, power plugs, power outlets

   e. electrodes

Every stimulator applied clinically should receive regular inspections by a qualified biomedical technician.

Chapter 3

1. a. sodium and potassium

   b. sodium is higher in concentration in the extracellular fluid; potassium is higher in concentration in the intracellular fluid

2. voltage-gated sodium and potassium channel proteins

3. The sodium–potassium pump consists of special membrane-bound proteins that use energy in the form of ATP to actively transport sodium ions out of the cell while simultaneously moving potassium ions into the cell in a 3:2 ratio.

4. See Figure 3.3*A*.

5. sensory, motor, and autonomic

6. Orthodromic propagation of action potentials occurs in the same direction as occurs with physiologic activation, that is, toward the CNS in sensory fibers and away from the CNS in motor fibers.
Antidromic propagation of action potentials occurs opposite to that normally occurring in peripheral axons, that is, toward the CNS in motor fibers and away from the CNS in sensory fibers.

7. epineurium: protects nerve fibers from compressive forces
perineurium: provides much of elasticity and tensile strength to peripheral nerve
endoneurium: contains Schwann cells and endoneural fluid

8. a. muscle fiber

   b. myofibrils

   c. actin

   d. myosin

   e. troponin

   f. tropomyosin

9. a. rate coding: the frequency of activation of muscle fibers

   b. recruitment: the number of muscle fibers or motor units activated

10. a. sensory-level stimulation: electrical stimulation at levels sufficient to activate only superficial cutaneous sensory nerve fibers and generally giving rise to a tingling, or "pins and needles," sensation

    b. motor-level stimulation: electrical stimulation at levels sufficient to activate alpha motoneuron axons and/or muscle fibers directly resulting in some form of muscle contraction

    c. noxious-level stimulation: electrical stimulation at levels sufficient to activate peripheral pain fibers (A_δ or C) and give rise to the perception of pain

11. a. frequency

    b. amplitude or pulse/phase duration

12. The sarcoplasm reticulum serves as a storage site for intracellular calcium ions. The calcium ions are released into the intracellular fluid when action potentials invade the t-tubules.

Chapter 4

1. Stimulation Characteristics:
amplitude: to at least 60% of the MVC of the muscle being stimulated
pulse/phase duration: 200–300 μsec
frequency of stimulation: 65–80 pps

on/off time of stimulation: 10 sec on and 50–120 sec off
NMES training program characteristics:
number of contractions per training session: 10
number of training sessions per week: 3

2. a. increased muscle strength, perhaps hypertrophy

b. increased fatigue resistance, although this has not been demonstrated in studies of human subjects using intermittent stimulation

3. Carrier frequency: 2500-Hz sine wave
burst frequency: 50 bps; 50% duty cycle
on/off times: 15 sec on (includes 5-sec ramp)/50 sec off
15 contractions daily at supramaximal contraction intensity

4. 30–100 pps or bursts per second

5. fatigue

6. Superimposing a supramaximal stimulation burst on a voluntary contraction allows the clinician to determine whether the patient is truly maximally activating the tested muscle.

7. Pulse duration: at least 200 μsec
frequency: variable 30–100 pulses or bursts per second
amplitude: capable of producing a 100% MVC contraction in the target muscle; sufficient reserve amplitude to present a continually increasing training dose

8. They are directly correlated.

9. They are not correlated.

10. a. NMES over the carotid sinus

b. NMES in patients who are unable to provide clear feedback

c. NMES over the abdomen of a pregnant woman

Chapter 5

1. a. an enhanced capability to oxidatively metabolize muscle fats, carbohydrates, and protein, the fuels that produce ATP due to increases in the activity of oxidative metabolic enzymes

b. an increase in the number of mitochondria where enzymes are stored

c. an increase in the content of the oxygen transport protein, myoglobin

d. a rise in the number of capillaries that bring oxygen to the muscle fibers

2. an increase in the content of muscle's contractile proteins, actin and myosin

3. No single current is best for all clinical applications

4. Pulse amplitude: 0–100 mA RMS for AC; 0–40 μC (microcoulombs) for PC

Phase and/or pulse duration: 10 μsec–500 μsec (may require longer durations for EMS)

Frequency controls: 0–100 pps for PC

On time/off time controls: 0–60 sec for both on time and off time
Ramp modulation controls: 0–5 sec
Treatment duration timer: 0–60 min
Number of output channels: two

5. a. over the antagonist(s) to the spastic muscle
   b. over the spastic muscle itself
   c. dermatome associated with the same level as the motor nerves to the spastic muscle
   d. nerve trunk innervating antagonist of spastic
   e. paraspinal stimulation
   f. over dorsal columns
   g. rectally

6. a. pendulum test
   b. Ashworth scale
   c. Wartenburg spasticity test
   d. H-reflex
   e. Achilles tendon reflex (ATR)
   f. passive resistance to stretch and clonus

7. Functional electrical stimulation is the use of NMES as an orthotic substitute (to take the place of a brace or support), to maintain posture or to produce limb movements important for activities of daily living.

8. a. ankle-foot orthosis (AFO, dorsiflexion assist brace)
   b. either directly over the peroneal nerve near the head of the fibula or over the motor points of the tibialis anterior and peroneal muscles.

9. a. The patient's ability to cooperate and communicate
   b. Limited spasticity of the ankle plantar flexors
   c. NMES-induced improvement of gait pattern significantly greater than the conventional AFO approach
   d. No significant limitation in ankle passive ROM
   e. No significant knee or hip volitional movement limitations
   f. No hypersensitivity to NMES
   g. A high level of subject motivation

10. a. electrodes placed over the midaxillary line on the convex side of the curve
    b. frequency of pulsed current at >25 pps sufficient to produce tetanic contraction
    c. on time/off time of stimulation: 6 sec on/6 sec off
    d. stimulation amplitude: gradually increased until muscular contraction is strong enough to produce spinal movement in the "straightening" direction
    e. duration of stimulation: increased until the patient can tolerate 8 continuous hours of stimulation
    f. patient uses this stimulation each night until skeletal maturity is reached

11. a. supraspinatus
    b. posterior deltoid

12. a. pressure sores
    b. circulatory disorders
    c. diminished cardiovascular capacity
    d. muscle atrophy/fibrosis

13. a. reduction/prevention contracture; maintenance of ROM in LE; reduction in spasticity
    b. prevention of osteoporosis
    c. improvement in bowel, renal, and bladder function
    d. stimulation of circulation
    e. reduction in seating pressure and hence pressure sore development
    f. increased ability to functionally reach
    g. psychological benefit

14. a. Stimulate many of the lower extremity muscles alternately to produce the desired movements.
    b. Direct stimulation of certain muscles (e.g., hip and knee extensors during stance phase) and reflex activation of other muscles (hip flexors, knee flexors, and dorsiflexors during swing phase) by activation of flexion reflexes.
    c. Maintain fixed levels of stimulation to antigravity muscles and employ swing-to or swing-through gait with braces and forms of external support.

15. a. use of supramaximal stimulus intensities; all of the denervated muscle fibers must be activated
    b. isometric contraction in response to stimulation during training
    c. initiation of regular sessions of stimulation as soon as possible after denervation
    d. use of bipolar electrode placement

Chapter 7

1. Sensory-level stimulation is best achieved through short phase durations

2. Anywhere from 2 to 50 microseconds

3. Sensory-level stimulation uses short phase durations, higher frequencies, lower amplitudes, and shorter treatment times as compared to motor-level stimulation. Sensory-level stimulation proposes that either the direct peripheral block of pain transmission or a central inhibition of pain transmission by large-diameter fiber stimulation is responsible for immediate but short-lived pain relief. Motor-level stimulation proposes that due to strong, rhythmic muscle contractions the endogenous opiate mechanisms of analgesia are activated.

4. Sensory-level stimulation provides immediate pain relief and is typically used in the acute pain situation. However, there is little carryover of the

analgesic effect after treatment. Noxious-level stimulation is typically used in chronic pain patients or in patients who no longer respond to sensory-level stimulation. It provides a more prolonged pain relief which begins well after administering the treatment. Noxious stimulation can be performed at the site of pain or in an unrelated area; in either location a systemic release of endogenous opiates is believed to increase a patient's threshold for pain.

5. 1–5 pps

6. The placement of electrodes in sensory-level stimulation is typically over the site/source of pain. In this case, placement would be over the site of pain in the low back region. In motor-level stimulation, electrode placement is usually away from the area of pain, but it is recommended that electrodes be placed in anatomically or physiologically related areas. In this case, placement along the contralateral dermatome or myotome may be indicated.

7. a. when pain is serving a protective or useful function

   b. over the abdomen of a pregnant female

   c. over the thoracic area in the presence of a pacemaker

   d. skin irritation in the location of electrode placement

Chapter 8

1. Two theories are thought to govern edema control. The first may help alleviate edema that is already present through muscular contractions. Set the stimulation parameters to achieve a rhythmic muscle contraction, i.e., a phase duration >100 μsec and an amplitude $>$motor threshold. The second option involves SLS and is thought to prevent the onset or formation of edema in an acute situation. Cathodal stimulation at a frequency of approximately 100 pps, a phase duration <100 μsec, and an amplitude just below the motor threshold has been shown to prevent edema.

2. Anodal DC stimulation has been demonstrated clinically to enhance wound healing. Amplitudes <1 mA for treatment durations of 1–2 hr BID. Cathodal stimulation at >100 pps and >150 V for 1–3 hr of treatment has been shown to reduce the rate of growth of more common gram-negative and gram-positive microorganisms in an open wound.

3. *Large vessel:* Requires that the stimulation parameters be set to elicit a rhythmical muscular contraction (pulsed or burst AC mode, phase duration >100 μsec, amplitude $>10\%$ mvc for >10 min) that will allow for increases in blood flow on the cessation of the contraction. *Microcirculation:* The majority of findings suggest that pulsed or burst-mode AC with a >2-μsec phase duration SLS provides a decrease in the sympathetic activity, which allows for changes in microvascular perfusion.

4. Current evidence exists to support initiating cathodal stimulation to the wound site at high frequencies (>100 pps), at intensities ranging from 200 μA to 800 μA and for 4–6 total treatment hours per day to facilitate reduction of the infection. Further evidence supports the use of anodal stimulation after several treatment sessions with the cathodal stimulation (on resolution of the infection) for continued enhancement of healing (amplitude <1 mA).

5. a. By inducing a rhythmical muscle contraction, electrical stimulation aids in lymphatic and venous drainage.

    b. The utilization of HVPC cathodal stimulation at approximately 120 pps with an intensity just below the motor threshold is thought to reduce edema formation in an acute situation.

6. The use of SLS over the spinal cord to induce a sympatholytic effect would be the treatment of choice for a patient with RSD.

7. The "current of injury" theory proposes that wounds are initially positive in polarity, which triggers the repair process, and that by maintaining this positive polarity via anodal stimulation the healing process will be expedited.

Chapter 9

1. repulsion; anode; cathode

2. direct; 1.0; 4.0

3. amplitude, duration; mA·min

4. 40 mA·min; 80 mA·min

5. 10 additional minutes

6. cathode: NaOH, alkaline

7. convenience; ease of application; possibly safer for the patient because of better skin conformance and chemical buffering.

8. 0.5; 1.0

9. False; increasing the concentration beyond the recommended level may not increase the amount of ions administered and may actually retard ion transfer.

10. negative; positive

11. decrease; less

12. damaged or broken skin, decreased sensation, implanted electrical devices

13. antiinflammatory; catabolic

14. local anesthetics, salicylates, opioids

15. acetic acid; cathode

Chapter 10

1. The **intracellular recording technique** is used primarily in the research laboratory experiments on very large axons or cell bodies by actually penetrating the excitable cell membrane with a glass microelectrode and placing a second recording electrode outside the cell. Hence, one electrode is intracellular and a second electrode is extracellular. This system records changes in transmembrane (across the membrane) potentials.

In the **extracellular recording technique,** one or two small conductors (called recording electrodes) are placed near but outside nerve or muscle cells. These electrodes monitor the changes in ionic concentration outside of electrically excitable cells as action potentials propagate along membranes. The extracellular electrodes are used to monitor the electrical activity in many nerve or muscle fibers; hence, the electrical signals recorded are referred to as compound nerve or muscle action potentials.

Refer to Figure 10.1 for graphs of voltage changes recorded using each technique.

2. Active recording electrodes placed over a muscle innervated by a terminal branch of the median nerve (e.g., abductor pollicis brevis, APB). Reference recording electrode placed distal to active over muscle tendinous insertion or over distal thumb joint. Ground electrode placed on the dorsal surface of the hand over ulnar styloid.

Stimulation electrodes may be placed over median nerve at the wrist no less than 8 cm proximal to the active recording electrode, at the elbow before the median nerve penetrates the pronator teres, or in the axilla immediately lateral and anterior to the brachial artery.

3. Orthodromic conduction technique in the median nerve: distal ring electrode is placed over the distal interphalangeal crease and the more proximal ring electrode is placed over the midpoint of proximal phalanx on the second or third digit. The more proximal electrode is connected to the cathodal output of the stimulator, and the more distal electrode is connected to the anode. Surface recording electrodes may be positioned directly over the median nerve at points where the nerve is close to the skin surface (just above the wrist, at the elbow, over the median nerve in the upper arm or axilla). A ground electrode is placed on the dorsum of the hand or over the styloid process of the ulna.

Antidromic conduction technique in the median nerve: for the antidromic recording technique, the ring electrodes on the second or third digit are used as recording electrodes. Stimulation is provided at sites as described in the answer to question 2 for median nerve motor conduction studies.

4. a. temperature
   b. age
   c. injury, ischemia, disease

5. Neuropraxia is characterized by localized conduction block (or slowing) without interruption of axon continuity and without axonal degeneration.

Axonotmesis is characterized by disruption in axon continuity without significant damage to the connective tissue layers in peripheral nerve.

Neurotmesis is characterized by axonal disruption accompanied by damage to one or more of the connective tissue layers within the peripheral nerve.

6. prolonged insertional activity
   insertional positive waves
   fibrillation potentials
   positive sharp waves
   fasciculation potentials
   myotonic discharges
   "complex repetitive discharges"
   myokymic discharges
   polyphasic potentials

7. H-reflex test

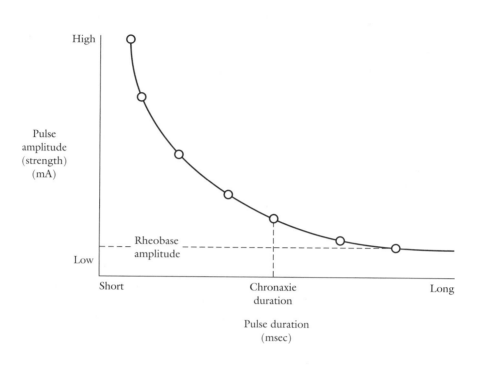

8. The strength–duration curve represents the family of stimulus amplitude/stimulus duration combinations that are sufficient to activate the particular type of excitable tissue.

9. Semmes–Weinstein monofilament testing
   Two-point discrimination testing

10. Cranial nerves V and VII

11. See Table 10.3.

12. a. abnormal blood clotting factors/anticoagulant therapy
    b. extreme swelling
    c. dermatitis
    d. uncooperative patient
    e. recent myocardial infarction
    f. blood-transmittable disease (Jakob–Creutzfeldt, HIV, hepatitis, etc.)
    g. immune-suppressed condition
    h. central-going lines
    i. pacemakers
    j. hypersensitivity to stimulation (open wounds, burns).

13. An F-wave is a compound muscle action potential recorded in response to stimulation of motor nerve fibers, which is thought to result from the antidromic propagation of action potentials in motor axons. The antidromically propagated potentials are thought to "reactivate" a small

number of motor neurons, which then send these action potentials back to the muscle, causing the muscle contraction, which is monitored as the F-wave.

14. Somatosensory-evoked potential testing allows the examiner to test the integrity of CNS transmission pathways for sensory signals along the spinal cord, brain stem, and brain.

15. Motor unit potentials. Peak-to-peak **amplitudes of motor unit potentials** range from 200 μV to 5 mV. The **rise time** of the MUP is measured from the first positive peak to the top of the first negative peak. Normal rise times should not exceed 500 μsec and most should fall in the 100 to 200 μsec range. **MUP duration** is measured from the beginning of the initial positive deflection from the baseline to the point at which the potential returns to baseline after all phases and normally ranges from 5 to 15 msec. Motor units normally do not discharge at frequencies in excess of 30 per second, with most discharging at much lower rates.

Chapter 11

1. Biofeedback is the use of electronic instrumentation to provide objective information (feedback) to an individual about a physiological function or response so that the individual becomes aware of his or her response. The individual then attempts to alter the feedback signal in order to modify the physiological response.

Advantages of EMG biofeedback are the speed and continuity with which the information is provided to the clinician and patient. The sensitivity, objectivity, accuracy, and quantitative nature of the feedback signal are also major advantages of EMG biofeedback. In addition, modern feedback devices can provide a variety of novel feedback signals that can serve to motivate the client.

2. The selection of appropriate patients for the application of EMG biofeedback basically involves answering three simple questions.

   a. Does the patient demonstrate a motor impairment that would suggest that the information provided by the feedback would be of benefit?
   b. Does the patient demonstrate the ability for voluntary control?
   c. Is the patient sufficiently motivated and cognitively aware to utilize the feedback information?

3. The amplitude of the EMG reflects the size and number of active motor units as well as the distance of the active muscle fibers from the recording electrodes.

4. The larger the recording area the greater the volume of muscle that is monitored and hence the greater the EMG recorded. Similarly, the larger the interelectrode distance, the larger the volume of muscle that is monitored and the larger the EMG. Thus, to increase the specificity of the EMG recording electrodes, small recording areas and close interelectrode spacing could be used.

5. amplification
   filtering
   rectification

integration
level/threshold detection

6. The raw output (i.e., amplified and filtered only) can give the experienced clinician considerable information regarding the source of the signal. That is, is it truly physiological or is it primarily noise that is being recorded? It is difficult to use the raw signal to objectively document progress or to identify targeted levels of recruitment for the client.

7. goals of training
level of control
available muscle mass
subcutaneous fat
movement artifact
cross talk

Table and Figure Credits

Table 5.1   Reprinted with the permission of *Paraplegia*. Adapted from Halstead LS, Seager SWJ. The effects of rectal probe electrostimulation on spinal cord injury spasticity. Paraplegia 1991;29:43-47.

Figure 1.6   Redrawn with permission from Urone PP. Physics with health science applications. New York: Harper & Row, 1986.

Figure 3.4   Redrawn with permission from Kuffler SW, Nicholls JG. From neuron to brain: A cellular approach to the function of the nervous system. Sunderland, MA: Sinauer Associates, Inc., 1992.

Figure 3.6   Reprinted with the permission of W.B. Saunders Co. From Lundburg G, Dahlin L. The pathophysiology of nerve compressions. Hand Clinics 1992;8(2):219.

Figure 3.7   Reprinted with permission from Bloom W, Fawcett DW. A textbook of histology. Philadelphia: W.B. Saunders, 1975.

Figure 3.8   Redrawn with permission from Cormack DH. Ham's histology. Philadelphia: J.B. Lippincott, 1987.

Figures 3.9 and 3.10   Redrawn with permission from Vander et al. Human physiology: The mechanisms of body function. New York: McGraw-Hill, 1990.

Figure 3.11   Redrawn with permission from Cormack DH. Ham's histology. Philadelphia: J.B. Lippincott, 1987.

Figure 3.12   Redrawn with permission from Vander et al. Human physiology: The mechanisms of body function. New York: McGraw-Hill, 1990.

Figure 3.23   Redrawn with permission from Benton LA, Baker LL, et al. Functional electric stimulation—A practical clinical guide. Downey, CA: The Professional Staff Association of the Rancho Los Amigos Hospital, 1981.

Figure 4.2   Reprinted with permission from Snyder-Mackler L, Ladin Z, Schepsis A, et al. Electrical stimulation of the thigh muscles after reconstruction of the anterior cruciate ligament. Effects of electricity elicited contractions of the quadriceps femoris and hamstring muscles on gait and strength of the thigh muscles. J Bone Joint Surg (Am) 1991;73:1025-1036.

Figure 5.7   Redrawn with permission from *Physical Therapy*. American Physical Therapy Assoc., 1984, Vol 64, p 486.

Figures 10.29 and 10.30   From Kendall FP, McCreary EK, Prouance PG. Muscle testing and function. Baltimore: Williams & Wilkins, 1993, p 389.

Figures A.1–A.4 and A.6–A.9   Reprinted with permission from Kimura J. Electrodiagnosis of diseases of nerve and muscle. Philadelphia: F.A. Davis, 1989.

Index

Page numbers in *italics* denote figures; those followed by "t" denote tables.